The Mycota

Edited by
K. Esser and P.A. Lemke

Springer
Berlin
Heidelberg
New York
Barcelona
Budapest
Hong Kong
London
Milan
Paris
Santa Clara
Singapore
Tokyo

The Mycota

I *Growth, Differentiation and Sexuality*
Ed. by J.G.H. Wessels and F. Meinhardt

II *Genetics and Biotechnology*
Ed. by U. Kück

III *Biochemistry and Molecular Biology*
Ed. by R. Brambl and G. Marzluf

IV *Environmental and Microbial Relationships*
Ed. by D. Wicklow and B. Söderström

V *Plant Relationships*
Ed. by G. Carroll and P. Tudzynski

VI *Human and Animal Relationships*
Ed. by D.H. Howard and J.D. Miller

VII *Systematics and Evolution*
Ed. by D.J. McLaughlin, E.G. McLaughlin, and P.A. Lemke

VIII *Cell Structure and Function*
Ed. by C.E. Bracker and L. Dunkle

IX *Fungal Associations*
Ed. by B. Hock

The Mycota

A Comprehensive Treatise on Fungi as Experimental Systems for Basic and Applied Research

Edited by K. Esser and P.A. Lemke†

V *Plant Relationships Part A*

Volume Editors:
G.C. Carroll and P. Tudzynski

With 74 Figures and 18 Tables

Springer

Series Editors

Professor Dr. Dr. h.c. mult. KARL ESSER
Allgemeine Botanik
Ruhr-Universität
D-44780 Bochum
Germany

Professor Dr. PAUL A. LEMKE†, Auburn, USA

Volume Editors

Professor Dr. George C. Carroll
Department of Biology
University of Oregon
Eugene, OR 97403
USA

Professor Dr. Paul Tudzynski
Institut für Botanik
Westfälische Wilhelms-Universität
Schloßgarten 3
D-48149 Münster
Germany

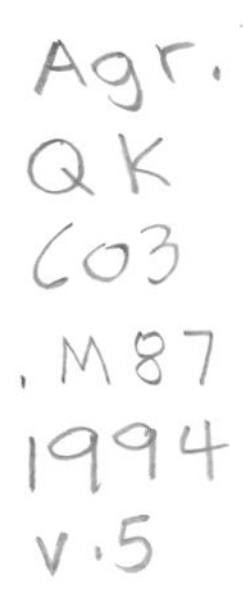

ISBN 3-540-58006-9 Springer-Verlag Berlin Heidelberg New York

Library of Congress Cataloging-in-Publication Data.

The Mycota. Includes bibliographical references and index. Contents: 1. Growth, differentiation, and sexuality/editors, J.G.H. Wessels and F. Meinhardt – 2. Genetics and biotechnology. 1. Mycology. 2. Fungi. 3. Mycology – Research. 4. Research. I. Esser, Karl, 1924– . II. Lemke, Paul A., 1937– . QK603.M87 1994 589.2 ISBN 3-540-57781-5 (v. 1: Berlin: alk. paper) ISBN 0-387-57781-5 (v. 1: New York: alk. paper) ISBN 3-540-58003-4 (v. 2: Berlin) ISBN 0-387-58003-4 (v. 2: New York)

Printed in Germany

Production Editor:

Cover design: Springer-Verlag, E. Kirchner

Typesetting by Best-set Typesetter Ltd., Hong Kong

SPIN: 10043369 31/3137 – 5 4 3 2 1 0 – Printed on acid-free paper

Series Preface

Mycology, the study of fungi, originated as a subdiscipline of botany and was a descriptive discipline, largely neglected as an experimental science until the early years of this century. A seminal paper by Blakeslee in 1904 provided evidence for self-incompatibility, termed "heterothallism", and stimulated interest in studies related to the control of sexual reproduction in fungi by mating-type specificities. Soon to follow was the demonstration that sexually reproducing fungi exhibit Mendelian inheritance and that it was possible to conduct formal genetic analysis with fungi. The names Burgeff, Kniep and Lindegren are all associated with this early period of fungal genetics research.

These studies and the discovery of penicillin by Fleming, who shared a Nobel Prize in 1945, provided further impetus for experimental research with fungi. Thus began a period of interest in mutation induction and analysis of mutants for biochemical traits. Such fundamental research, conducted largely with *Neurospora crassa*, led to the one gene: one enzyme hypothesis and to a second Nobel Prize for fungal research awarded to Beadle and Tatum in 1958. Fundamental research in biochemical genetics was extended to other fungi, especially to *Saccharomyces cerevisiae*, and by the mid-1960s fungal systems were much favored for studies in eukaryotic molecular biology and were soon able to compete with bacterial systems in the molecular arena.

The experimental achievements in research on the genetics and molecular biology of fungi have benefited more generally studies in the related fields of fungal biochemistry, plant pathology, medical mycology, and systematics. Today, there is much interest in the genetic manipulation of fungi for applied research. This current interest in biotechnical genetics has been augmented by the development of DNA-mediated transformation systems in fungi and by an understanding of gene expression and regulation at the molecular level. Applied research initiatives involving fungi extend broadly to areas of interest not only to industry but to agricultural and environmental sciences as well.

It is this burgeoning interest in fungi as experimental systems for applied as well as basic research that has prompted publication of this series of books under the title *The Mycota*. This title knowingly relegates fungi into a separate realm, distinct from that of either plants, animals, or protozoa. For consistency throughout this Series of Volumes the names adopted for major groups of fungi (representative genera in parentheses) are as follows:

Pseudomycota

Division: Oomycota (*Achlya, Phytophthora, Pythium*)
Division: Hyphochytriomycota

Eumycota

Division: Chytridiomycota (*Allomyces*)
Division: Zygomycota (*Mucor, Phycomyces, Blakeslea*)

Division:	Dikaryomycota
Subdivision:	Ascomycotina
Class:	Saccharomycetes (*Saccharomyces, Schizosaccharomyces*)
Class:	Ascomycetes (*Neurospora, Podospora, Aspergillus*)
Subdivision:	Basidiomycotina
Class:	Heterobasidiomycetes (*Ustilago, Tremella*)
Class:	Homobasidiomycetes (*Schizophyllum, Coprinus*)

We have made the decision to exclude from *The Mycota* the slime molds which, although they have traditional and strong ties to mycology, truly represent nonfungal forms insofar as they ingest nutrients by phagocytosis, lack a cell wall during the assimilative phase, and clearly show affinities with certain protozoan taxa.

The Series throughout will address three basic questions: what are the fungi, what do they do, and what is their relevance to human affairs? Such a focused and comprehensive treatment of the fungi is long overdue in the opinion of the editors.

A volume devoted to systematics would ordinarily have been the first to appear in this Series. However, the scope of such a volume, coupled with the need to give serious and sustained consideration to any reclassification of major fungal groups, has delayed early publication. We wish, however, to provide a preamble on the nature of fungi, to acquaint readers who are unfamiliar with fungi with certain characteristics that are representative of these organisms and which make them attractive subjects for experimentation.

The fungi represent a heterogeneous assemblage of eukaryotic microorganisms. Fungal metabolism is characteristically heterotrophic or assimilative for organic carbon and some nonelemental source of nitrogen. Fungal cells characteristically imbibe or absorb, rather than ingest, nutrients and they have rigid cell walls. The vast majority of fungi are haploid organisms reproducing either sexually or asexually through spores. The spore forms and details on their method of production have been used to delineate most fungal taxa. Although there is a multitude of spore forms, fungal spores are basically only of two types: (i) asexual spores are formed following mitosis (mitospores) and culminate vegetative growth, and (ii) sexual spores are formed following meiosis (meiospores) and are borne in or upon specialized generative structures, the latter frequently clustered in a fruit body. The vegetative forms of fungi are either unicellular, yeasts are an example, or hyphal; the latter may be branched to form an extensive mycelium.

Regardless of these details, it is the accessibility of spores, especially the direct recovery of meiospores coupled with extended vegetative haploidy, that have made fungi especially attractive as objects for experimental research.

The ability of fungi, especially the saprobic fungi, to absorb and grow on rather simple and defined substrates and to convert these substances, not only into essential metabolites but into important secondary metabolites, is also noteworthy. The metabolic capacities of fungi have attracted much interest in natural products chemistry and in the production of antibiotics and other bioactive compounds. Fungi, especially yeasts, are important in fermentation processes. Other fungi are important in the production of enzymes, citric acid and other organic compounds as well as in the fermentation of foods.

Fungi have invaded every conceivable ecological niche. Saprobic forms abound, especially in the decay of organic debris. Pathogenic forms exist with both plant and animal hosts. Fungi even grow on other fungi. They are found in aquatic as well as soil environments, and their spores may pollute the air. Some are edible; others are poisonous. Many are variously associated with plants as copartners in the formation of lichens and mycorrhizae, as symbiotic endophytes or as overt pathogens. Association with animal systems varies; examples include the predaceous fungi that trap nematodes, the microfungi that grow in the anaerobic environment of the rumen, the many insect-associated fungi and the medically important pathogens afflicting humans. Yes, fungi are ubiquitous and important.

There are many fungi, conservative estimates are in the order of 100000 species, and there are many ways to study them, from descriptive accounts of organisms found in nature to laboratory experimentation at the cellular and molecular level. All such studies expand our knowledge of fungi and of fungal processes and improve our ability to utilize and to control fungi for the benefit of humankind.

We have invited leading research specialists in the field of mycology to contribute to this Series. We are especially indebted and grateful for the initiative and leadership shown by the Volume Editors in selecting topics and assembling the experts. We have all been a bit ambitious in producing these Volumes on a timely basis and therein lies the possibility of mistakes and oversights in this first edition. We encourage the readership to draw our attention to any error, omission or inconsistency in this Series in order that improvements can be made in any subsequent edition.

Finally, we wish to acknowledge the willingness of Springer-Verlag to host this project, which is envisioned to require more than 5 years of effort and the publication of at least nine Volumes.

Bochum, Germany
Auburn, AL, USA
April 1994

Karl Esser
Paul A. Lemke
Series Editors

Addendum to the Series Preface

In early 1989, encouraged by Dieter Czeschlik, Springer-Verlag, Paul A. Lemke and I began to plan *The Mycota*. The first volume was released in 1994, three other volumes followed in the years 1995 and 1996. Also on behalf of Paul A. Lemke, I would like to take this opportunity to thank Dieter Czeschlik, his colleague Andrea Schlitzberger, and Springer-Verlag for their help in realizing the enterprise and for their excellent cooperation for many years.

Unfortunately, after a long and serious illness, *Paul A. Lemke* died in November 1995. Without his expertise, his talent for organization and his capability to grasp the essentials, we would not have been able to work out a concept for the volumes of the series and to acquire the current team of competent volume editors. He also knew how to cope with unexpected problems which occurred after the completion of the manuscripts. His particular concern was directed at Volume VII; in this volume, a posthumous publication of his will be included.

Paul A. Lemke was an outstanding scientist interested in many fields. He was extremely wise, dedicated to his profession and a preeminent teacher and researcher. Together with the volume editors, authors, and Springer-Verlag, I mourn the loss of a very good and reliable friend and colleague.

Bochum, Germany
January 1997

KARL ESSER

Volume Preface

The number of fungal species has been loosely estimated to be on the order of 1 million, while the number of vascular plants is known with considerably greater certainty to lie between 300000 and 350000. Clearly, any volume which purports to deal with interactions between these two vast assemblages of organisms must do so concisely and selectively. In the chapters to follow, we have made no attempt to be all-inclusive, but rather have chosen examples from which general conclusions about fungus/plant interactions might be drawn. The materials presented here come from the core literature on plant pathology and from research on fungal mutualisms and on evolutionary biology. A variety of approaches are evident: biochemistry, molecular biology, cellular fine structure, genetics, epidemiology, population biology, ecology, and computer modeling. The frequent overlap of such approaches within single reviews has resulted in a rich array of insights into the factors which regulate fungus/plant interactions. In these chapters, such interactions have also been considered on a variety of scales, both geographic and temporal, from single plant cells to ecosystems, from interactions which occur within minutes of contact to mechanisms which have presumably evolved during the course of several hundred million years.

Volume V consists of two parts: Volume V, Part A, and Volume V, Part B. While section headings provide signposts, we wish to make the rationale for the organization of these volumes absolutely clear. Part A begins with a brief introduction to both volumes. A series of reviews follows (Chaps. 1–6) which deal with the temporal sequence of events from the time fungal spores make contact with a host plant until the point where fungal hyphae are either firmly ensconced within a host or the attempted infections have been repulsed. Chapters 7–12 deal with metabolic interactions between host and fungus within the host plant after infection and particularly with the roles played by low molecular weight fungal metabolites such as toxins and phytohormones in pathogenic as well as mutualistic associations.

Chapters 1–8 of Part B are grouped in a section labeled, "Profiles in Pathogenesis and Mutualism"; here, interactions between fungi and host plants are explored in a variety of important model systems. These reviews focus less on processes per se and more on the specific fungi or groups of fungi as examples of pathogens or mutualists on plants. Chapters 9–12 of Part B move from discussions of physiological interactions between individuals to considerations of interactions at an expanded geographic scale, within populations of plants. Here, Chapter 9 provides a treatment of classical plant epidemiology, while Chapter 11 provides the same focus for mutualistic mycorrhizal associations. Chapter 10 covers the fuzzy area between population biology and microevolution in a genus of ubiquitous and pleurivorous pathogens; Chapter 12 offers much the same approach for mutualistic endophytes of grasses.

Chapters 13–16 of Part B offer a view of an expanded temporal scale and consider the evolution of plant/fungus interactions. Chapter 13 considers the flexibility of the fungal genome, the ultimate substrate on which evolutionary forces must act. Chapter 14 discusses the evolutionary relationships between pathogenic and mutualistic fungi in one situation which has been particularly well worked out, the clavicipitaceous endophytes of grasses. Chapter 15 considers the evolutionary interplay between fungi and plants as illuminated through the use of mathematical and computer-driven models. The final chapter in the volume (Chap. 16) deals with overall evolution of fungal parasitism and plant resistance and provides an appropriate coda for this series of essays.

Who is the audience for these volumes? Who might and will read them with profit? Basic literacy in mycology, in particular, and in modern biology, in general, has been assumed as a background for these chapters, and they clearly are not intended for the biological novice. However, we do expect that these volumes will be appreciated by a wide variety of professional biologists including, for example: teachers of upper division courses in general mycology engaged in the valiant (but often futile) attempt to keep their lectures up-to-date; graduate students contemplating literature reviews in connection with a thesis project; nonmycologists who wish to know what the fungi might have to offer in the way of model systems for the study of some fundamental aspect of host/parasite interactions; evolutionary biologists who have just become aware that fungi offer advantages in studying the evolutionary consequences of asexual reproduction. These, and many others, will read these chapters with pleasure. On the whole we are very pleased with the contributions presented here and believe they will prove informative and useful as entrées into the literature on fungus/plant interactions for some years to come.

Eugene, Oregon, USA
Münster, Germany
March 1997

George Carroll
Paul Tudzynski
Volume Editors

Contents Part A

Contents Part B

List of Contributors

BELLINCAMPI, D., Dipartimento di Biologia Vegetale, Università di Roma "La Sapienza", Piazzale A. Moro, 5 – I-00185 Roma, Italy

BUSH, L.P., Department of Agronomy, University of Kentucky, Lexington, Kentucky 40546, USA

CASTORIA, R., Dipartimento di Scienze Animali, Vegetali e dell'Ambiente, Università degli Studi del Molise, Viale Manzoni, I-86100 Campobasso, Italy

CERVONE, F., Dipartimento di Biologia Vegetale, Università di Roma "La Sapienza", Piazzale A. Moro, 5 – I-00185 Roma, Italy

DE LORENZO, G., Dipartimento di Biologia Vegetale, Università di Roma "La Sapienza", Piazzale A. Moro, 5 – I-00185 Roma, Italy

EBEL, J., Botanisches Institut, Universität München, Menzinger Straße 67, D-80638 München, Germany

EPSTEIN, L., Department of Plant Pathology, University of California, Davis, California 94720-3112, USA. Current address: Department of Plant Pathology, University of California, Davis, California 95616-8680, USA

HOCH, H.C., Department of Plant Pathology, New York State Agricultural Experiment Station, Cornell University, Geneva, New York 14456, USA

HOHN, T.M., Mycotoxin Research Unit, USDA/ARS, National Center for Agricultural Utilization Research, 1815 N. University St., Peoria, Illinois 61604, USA

HONEGGER, R., Institute of Plant Biology, University of Zürich, Zollikerstrasse 107, CH-8008 Zürich, Switzerland

HOWARD, R.J., Chemical and Biological Sciences, Central Research and Development, The DuPont Company, P.O. Box 80402, Wilmington, Delaware 19880-0402, USA

KOMBRINK, E., Max-Planck-Institut für Züchtungsforschung, Abteilung Biochemie, Carl-von-Linné-Weg 10, D-50829 Köln, Germany

LAPEYRIE, F., Laboratoire de Microbiologie Forestière, Centre INRA de Nancy, F-54280, Champenoux, France

Macko, V., Boyce Thompson Institute, Cornell University, Ithaca, New York 14853, USA

Martin, F., Laboratoire de Microbiologie Forestière, Centre INRA de Nancy, F-54280 Champenoux, France

Nicholson, R., Department of Botany and Plant Pathology, Purdue University, West Lafayette, Indiana 47907, USA

Scheel, D., Institut für Pflanzenbiochemie, Abteilung Stress- und Entwicklungsbiologie, Weinberg 3, D-06120 Halle (Saale), Germany

Siegel, M.R., Department of Plant Pathology, University of Kentuky, Lexington, Kentucky 40546, USA

Somssich, I.E., Max-Planck-Institut für Züchtungsforschung, Abteilung Biochemie, Carl-von-Linné-Weg 10, D-50829 Köln, Germany

Staples, R.C., Department of Plant Pathology, New York State Agricultural Experiment Station, Cornell University, Geneva, New York 14456, USA

Tagu, D., Laboratoire de Microbiologie Forestière, Centre INRA de Nancy, F-54280, Champenoux, France

Tudzynski, B., Institut für Botanik, Westfälische Wilhelms-Universität Münster, Schlossgarten 3, D-48149 Münster, Germany

Turgeon, B.G., Department of Plant Pathology, 334 Plant Science Building, Cornell University, Ithaca, New York 14853, USA

Wolpert, T.J., Department of Botany and Plant Pathology, Oregon State University, Corvallis, Oregon 97331, USA

Yoder, O.C., Department of Plant Pathology, 334 Plant Science Building, Cornell University, Ithaca, New York 14853, USA

Introduction to Part A and Part B: Fungus/Plant Interactions – An Overview

George C. Carroll

I. Introduction

As a primary source of fixed carbon in the biosphere, living plants provide a large arena for interaction with the fungi, particularly in terrestrial biomes, where plants and fungi may comprise the dominant groups of autotrophs and heterotrophs, respectively. Both fossil (Sherwood-Pike and Gray 1985) and molecular (Berbee and Taylor 1992) evidence suggest that such interactions are ancient and have occurred from the time both groups emerged into terrestrial environments. Indeed, primitive plants may have depended on associations with fungi for successful colonization of dry land (Pirozynski and Dalpé 1989). The present volumes (Vol. V, Parts A and B) explore the modes and mechanisms of accommodation which have arisen during the 400–500 million years in which plants and fungi have coexisted. This initial chapter provides both a preview and summary of the themes which emerge in the chapters to follow, focusing on those attributes which are unique in plants and fungi and which have thus molded their interactions during the course of evolution.

II. Interactions at the Levels of Cells and Individuals

A. The Consequences of Rigid Cell Walls

1. The Importance of Polysaccharide-Degrading Enzymes

Both fungi and plants produce cells which are surrounded by rigid walls composed largely of mixtures of intercalated polysaccharides supplemented with proteins and in plant cells with lignin. Polysaccharides in fungal cell walls include glucans, chitin, chitosan (in the Zygomycota), cellulose (in the Oomycota), mannans (in ascomycetous yeasts), and xylans (in Basidiomycota) (Peberdy 1990; Gooday 1995). In plant cells cellulose dominates, together with complex arrays of heteropolysaccharides containing generous moieties of polygalacturonans as well as xyloglucans and mannans (Carpita and Gibeaut 1993). For both groups, the cell wall constitutes a first line of defense which must be breached or destroyed by pathogens and/or herbivores. Thus, it comes as no surprise that polysaccharide-degrading enzymes figure prominently in fungus/plant interactions. For the fungi such enzymes act to facilitate access to living plant cytoplasm or to hasten death and lysis of host cells. In wood-rotting fungi such enzymes further serve to mobilize large amounts of fixed carbon tied up in wood. In plants enzymes active on fungal walls such as chitinase can cause lysis of invading fungal cells. In addition, these, as well as enzymes which depolymerize elements of the walls of plant cells themselves, form an integral part of the signaling system which allows plant cells to recognize that they are under attack (Chaps. 4, 5, this Vol.).

2. Molecular Sieving

Mature cell walls are fine-grained macromolecular matrices; as such, they impose constraints on the sizes of molecules which can be exchanged between plant and fungal cells. As a consequence, fungal toxins and plant phytoalexins have all turned out to be low molecular weight substances which can easily penetrate cell walls and reach the targeted elements of the cells within. The same argument can be made for hormones of both plant and fungal origin. Significantly, thionins, a class of proteins produced by plants in response to microbial attack, are also of low molecular weight. (Chap. 6, this Vol.)

Department of Biology, University of Oregon, Eugene, Oregon 97403, USA

The Mycota V Part A
Plant Relationships
Carroll/Tudzynski (Eds.)

Elicitors comprise an important, if heterogeneous, class of molecules which serve to initiate signal cascades leading to defensive responses by plants as a result of a wide variety of environmental and biotic stresses (Chaps. 4 and 5, this Vol.). No native polysaccharides can serve as elicitors, since they are too large to reach the signal receptors on plant plasma membranes. However, oligosaccharides derived from polysaccharides, either in fungal or plant cells walls through the action of wall-degrading enzymes, can penetrate the plant cell wall matrix and serve as potent elicitors of defense reactions (Chaps. 4 and 5, this Vol.).

Given that enzymes are proteins of relatively high molecular weight, how can wall-degrading enzymes get through the walls of the cells producing them, fungal or plant? Trevithick and collaborators addressed this question some years ago (Trevithick and Metzenberg 1966; Trevithick et al. 1966; Chang and Trevithick 1970, 1972, 1974) and found that such enzymes are subject to "molecular sieving" – that is, enzymes of high molecular weight become trapped in fungal cell walls and hence can operate only on low molecular weight substrates which move in solution to the site of enzyme action. Conversely, enzymes acting on insoluble substrates at some distance from the cells must be of low molecular weight or be composed of subunits which can be assembled after passage through the cell wall. Beyond this, it was found that hyphal tips, where cells walls are thinnest and still incompletely synthesized, proved the sites of most intensive enzyme secretion, while little was secreted from cells with mature walls. Presumably, similar constraints act on enzymes of plant origin which degrade fungal cell walls; passage of enzymes through heavily lignified secondary plant cell walls should prove impossible.

3. Defense Systems and Cell Motility

Both plants and animals have evolved elaborate biochemical defenses against herbivores and microbial parasites which prey upon them. These systems differ fundamentally in their operation, however, and the presence or absence of cell motility may go far to explain why the two systems have evolved in such radically different directions. Animals are characterized by efficient circulatory systems and complex assemblages of motile cells which serve to identify and kill populations of invading microorganisms. Cell motility is an absolute requirement for these systems to work, since cells must be capable of migrating to sites of microbial invasion and making contact with foreign antigens in order for recognition and selective amplification of antibody production to occur (Golub 1989). In such a situation switching of proteinaceous surface antigens constitutes one effective means of evading the immune system. For most fungi, surface antigens are polysaccharides, antigens which are not easily switched. This may explain the relative paucity of systemic fungal diseases in normal vertebrates.

In contrast, plants are constructed of immobile cells encased in rigid walls. As a consequence, migration of specialized defensive cells to sites of fungal invasion is impossible. Instead, plant defense systems have evolved along quite different lines, such that the capability for biochemical response is "hard-wired" into each cell leading to rapid local responses mediated by the induction of a suite of pathogenesis-related proteins (Chap. 6, this Vol.). Fungi respond to these defenses not by biochemical evasion, but by directly counteracting them. Indeed, a certain symmetry can be recognized in the patterns of attack and defense evident in interactions between fungi and their host plants.

B. Symmetries and Asymmetries in Fungus/ Plant Interactions

Typical responses of plant cells to microbial attack include the production of phytoalexins and of wall-degrading enzymes such as glucanases and chitinases (Chap. 6, this Vol.). Similarly, fungi produce low molecular weight toxins (Chaps. 7, 8, this Vol.) and wall-degrading enzymes (Chap. 4, this Vol.) during invasion of plant hosts. Thus, a certain symmetry is seen in the biochemical interplay between fungus and host. Consideration of countermeasures further reinforces this impression. Pectic polymers (polygalacturonans) play a crucial role in the integrity of plant cell walls. Pectin-degrading enzymes are a key to dismantling such walls, and "pectinases" are almost ubiquitous among fungi associated with plants. In one instance and probably quite generally, a fungal endopolygalacturonase is secreted first in a sequence of plant wall-degrading enzymes; plant cell walls prove resistant to the action of enzymes secreted later in the sequence if they are not first treated with the polygalacturonase (Wijesundera et al. 1989). Plants counter the action of fungal pectinases through the synthesis of polygalacturonase-inhibitor proteins in plant cell walls (Chap. 4, this Vol.). In a similar fashion, fungal cell walls may contain proteins which block the action of glucanases and chitinases of plant origin (Albersheim and Valent 1974; Ham et al. 1995). Lichen metabolites can inhibit the activity of wall-degrading enzymes produced by a lichen parasite (Torzilli and Lawrey 1995). Both fungi and plants synthesize toxins during antagonistic interactions. Frequently, fungi produce enzymes which detoxify plant phytoalexins, enzymes such as pisatin demethylase produced by *Fusarium solani* (Van Etten et al. 1991); fungi can also detoxify constitutively produced antifungal compounds known as phytoanticipins (Van Etten et al. 1995). In at least one instance a plant has been found to detoxify a fungal toxin (Meeley et al. 1992), and the phenomenon may prove more widespread than presently realized.

One can carry analogy only so far. Fungi and plants differ in many respects in their mechanisms of antagonistic interaction. Plant phytoalexins are induced by elicitors, while fungal toxins are produced constitutively and may even occur in conidia prior to infection. Nothing comparable to some of the pathogenesis-related proteins (e.g., thionins) have been found in the fungi. Fungi produce cutin esterases to attack the outermost protective layer of plant leaves (Kolattukudy 1985); plants have not been shown to synthesize enzymes which attack the proteinaceous hydrophobins which form a comparable surface layer on fungal hyphae, perhaps because plant defenses are directed at hyphal tips where polymerized hydrophobins have not yet been deposited.

C. Mutualistic Interactions

Fungi form important mutualistic associations with plants, notably in mycorrhizal roots (Chap. 12, this Vol.; Chaps. 6 and 7, Vol. V, Part B), as endophytes in the leaves of certain grasses (Chap. 10, this Vol.; Chaps. 12 and 14, Vol. V, Part B), and as mycobionts in lichens (Chap. 11, this Vol.). In many cases, these associations are near-obligate: in natural environments one partner is seldom found without the other. Untangling the biochemical and genetic basis of such associations has proved extremely difficult, leading to basic questions of cause and effect. Can the higher levels of auxin seen in mycorrhizal roots be attributed to fungal synthesis of the hormone or do fungi preferentially colonize roots that already show elevated auxin levels? Where symbiota produce unique compounds (e.g., lichen acids, Lawrey 1984; loline alkaloids, Chap. 10, this Vol.), it may be unclear whether the substance is produced by the plant, the mycobiont, or both through a synergism of biochemical pathways. Recently, such questions have become more tractable with the application of molecular approaches to defined experimental systems manipulated under laboratory conditions (Chap. 12, this Vol.; Chap. 6, Vol. V, Part B). Indeed, the auxins of importance for mycorrhizal morphogenesis and function appear to derive from the fungus (Chap. 6, Vol. V, Part B). Where uncertain biochemical provenance of compounds is concerned (e.g., lichen acids), mycobionts alone have often proved capable of their synthesis under appropriate conditions (Culberson et al. 1992), and such may turn out to be the case for other substances as well.

A central question in studies of microbial mutualisms has been the degree of altruism, of constructive give and take, operative in these associations. Studies on the fine structure of fungus/host interfaces in lichens (Chap. 11, this Vol.) and mycorrhizae (Massicotte et al. 1987; Gianinazzi-Pearson 1996) show little evidence for the destructive interplay evident at the interfaces between virulent pathogens and their hosts. Nevertheless, there must be costs to both partners in any mutualistic association (Carroll 1991, 1992), and molecular studies are confirming that mutualism is a special case of pathogenesis in which damage by the parasite and host responses to invasion are damped, but nevertheless present. The early rise and subsequent decrease in chitinase production by roots during the early stages of ectomycorrhizal formation can be taken as evidence for this view (Chap. 12, this Vol.)

III. Interactions at the Level of Populations

A. The Consequences of Passive or Vectored Dispersal

The absence of self-motility is an overriding fact of life for both fungi and plants. Both groups are incapable of the kind of purposive rapid migration and dispersal which characterize the animal kingdom. Although the Chytridiomycota and Oomycota do produce motile spores, their range is limited by comparison with the landscape scales over which plant distributions are measured. The same can be said of the foraging patterns evident in some of the rhizomorph-producing Basidiomycota. Both plants and their associated fungi are thus largely dependent on nonmotile seeds or spores for dispersal. Fungi have evolved several strategies to cope with this situation, the frequent production of huge numbers of rain- or wind-dispersed spores being chief among them. While dispersal of such spores may not be purposive, it can prove highly effective; witness the devastation produced on crop plants by rusts with wind-dispersed spores or the nearly ubiquitous occurrence of fungi with rain-dispersed conidia (e.g., *Fusarium* and *Colletotrichum*) on plants. However, under certain circumstances, fungi and their plant associates may become separated from each other. Crop or plantation plants exported to distant continents can and have been grown under pathogen-free conditions for varying periods depending on the efficacy of quarantine measures and the vagaries of long-distance atmospheric spore transport. Patterns of crop plant colonization by spore-dispersed fungal pathogens is a chief concern of plant epidemiology (Chap. 9, Vol. V, Part B).

Fungal mutualists typically produce fewer spores than pathogens, and situations in which asymbiotic plants become isolated from their fungal mutualists are not uncommon, particularly with mycorrhizal fungi. Such situations arise particularly in areas where the natural vegetation has been completely eradicated from an area for a lengthy period as a result of natural (e.g., volcano) or anthropogenic catastrophes (e.g., strip mine, pollution, or tropical deforestation; Chap. 8, Vol. V, Part B). Exotic tree plantations may also suffer from a lack of appropriate mycorrhizal fungi.

Dispersal of pathogenic fungi can be less effective than seed dispersal of its host, particularly if the pathogen is highly specialized and produces spores that must be vectored by insects, for example, an anther smut. Under such circumstances plant populations may be subdivided into metapopulations whose geographical separation insures that some proportion is pathogen-free. Pathogens within a given metapopulation may prove highly virulent and produce local extinction of the host plant, without affecting the success or fitness of the overall population. Such a situation depends, however, on the inefficient dispersal of the pathogen; if for some reason dispersal efficiency were to increase, the pathogen would increasingly select for resistant genotypes in the host population (Chap. 15, Vol. V, Part B).

B. Host Specificity and Foraging Strategies

Fungal pathogens can be highly host-specific, such that their occurrence is restricted not only to single species of plants, but to single cultivars (Chaps. 1 and 3, Vol. V, Part B). Studies of specificity have been confined mostly to parasites of agricultural plants; the situation in any patch of natural vegetation is far less clear. No studies have critically enumerated the total mycoflora in any sample of natural vegetation to allow a realistic mapping of fungal species onto host plants. Although several countries have produced impressive host indices (Farr et al. 1989), any experienced mycologist can culture fungi from common native plants in any flora which prove clearly different from anything listed in a host index.

Records in host indices have figured prominently in extrapolations from the number of described fungal species to estimates of the total number which might exist (Hawksworth 1991). The strategy most commonly invoked involves calculation of the mean number of unique fungal species/host plant in a host index multiplied by the number of species in the world vascular plant flora. This procedure usually generates estimates between 1 and 2 million for the total world mycoflora. In fact, this approach rests on many untested assumptions, among them that plant entries in host indices represent an unbiased sample of the world vascular plant flora. Indeed, cultivated or numerically abundant native plants from the north temperate zone are vastly overrepresented in such indices. Since a majority of plant species occur in the tropics and a majority of these are endemic or rare, the above estimates may be severely compromised. May (1988, 1991, 1994) has argued that rare plants should have few or no host-specific fungal associates, precisely for the reasons of inefficient dispersal discussed above. In agreement with this prediction, a recent study by Lodge et al. (1996) revealed a large assemblage of nonhost-specific fungi as endophytes in leaves of a member of the Sapotaceae in Puerto Rico. No assessment of host specificity has been carried out in an area such as the fynbos in the western Cape region of South Africa, where an astounding diversity of congeneric vascular plants occur in genera such as *Erica* and *Protea* (Cowling et al. 1995); again, the expectation is that a relatively small assemblage of fungi may be associated with a large number of closely related plant species.

Many plant-associated fungi are Ascomycota and asexual stages thereof. Ascomycetous species are based on the sexual stage, and in plant-associated Ascomycetes these stages are typically restricted in host range and are produced seasonally. Many Ascomycetes also form asexual spores, which are produced more copiously and over longer periods than ascospores. Conidia can initiate limited asymptomatic infections on a wide variety of nonhost plants and, in fact, such infections comprise a large proportion of the fungal isolates recorded as leaf and stem endophytes (Carroll 1995; Chap. 8, Vol. V, Part B). These endophytic infections should be considered as foraging states of Ascomycetes which serve to maintain a fungus in a habitat in the absence of the preferred host until it appears again during the course of succession or as a result of human activities. Crop pathogens which form latent infections in roadside weeds should be thought of in this context.

Finally, it should be mentioned that a number of fungi have formed specific associations with insects as vectoring agents. In so doing, they have adopted the foraging patterns of their insect partners, often with catastrophic consequences for the host plant. Dutch elm disease represents merely one example in which fungal pathogenicity has been linked with purposive insect host location to produce a highly effective pathogenic mutualism (Webber and Brasier 1984). These associations belong more in the realm of fungus/animal relationships than fungus/plant relationships, and so are not discussed in these volumes.

IV. Interactions over Evolutionary Time

A. Life Cycles and Evolution

For over 400 million years evolutionary forces have acted continually on fungi and plants to effect a remarkable degree of overall accommodation. Recent breakdowns in balanced coexistence between the groups lie in the realm of agricultural blights, introduced plant diseases, and maltreatment of plant habitats, all artifacts of human activities and of our perception of biological events on a human instead of an evolutionary time scale. What aspects of plant and fungal biology have led to the kinds of interaction between these groups evident in the biosphere now? Life cycles are considered important in structuring the genetic systems on which evolution acts. While the forerunners of fungi and plants may once have exhibited similar life cycles, these now differ in many important respects. Plants are diploid or polyploid, while fungi (Oomycota and some yeasts excepted) exist predominantly in a haploid state. Plants can store genetic variation as multiple copies of the same gene at single loci within the same nucleus, while many fungi exhibit a heterokaryotic state with genetic variation stored in genetically different nuclei within the same cytoplasm. Models of coevolution between plant pathogens and their hosts suggest that such differences may exert a strong effect on the stability of pathosystems (Chap. 15, Vol. V, Part B).

Beyond this, in plants, sexually produced seeds function most importantly in reproduction and dispersal, while many fungi reproduce in large measure asexually. Asexual reproduction allows rapid exploitation of susceptible plant genotypes but not the generation of genetic variation. Asexual fungi may participate in somatic recombination events which result in recombinant genotypes, although at frequencies orders of magnitude lower than possible with sexual reproduction. Further, such events may require fusions between genetically distinct mycelia, events which can be drastically curtailed by somatic incompatibility genes.

If asexuality carries with it the rewards of rapid dissemination and occupation of empty niches, it also incurs costs. Several lines of thought/evidence hold this to be the case. Asexuality allows the propagation of chromosomal abnormalities and aneuploids (Chap. 13, Vol. V, Part B); while some regard such genomic flexibility as an advantage, in the short term, chromosomal imbalances are usually manifest as a loss of vigor or as deleterious mutations. Theoretical arguments suggest that organisms which lack sexual recombination should experience a gradual, eventually fatal, decline because they are unable to purge themselves of deleterious mutations through recombination, a phenomenon termed Muller's ratchet (Gabriel et al. 1994). Computer simulations suggest that, although this process may be slowed in haploid populations because highly deleterious mutations are expressed and eliminated immediately, it nevertheless continues, culminating in "mutational meltdown" (Lynch et al. 1993). Anecdotal evidence from curators of fungal culture collections is certainly in accord with these predictions. The Oomycete *Phytophthora infestans* reproduced in Europe asexually for about 130 years because one of the mating types in this heterothallic species was never introduced to the continent. This strain was replaced almost completely within 5 years with the introduction of the opposite mating type in the early 1980s, suggesting that new strains generated through sexual reproduction were more fit than the older one, which may have suffered a loss of competitive fitness due to the accumulation of deleterious mutations (Chap. 2, Vol. V, Part B). Construction of evolutionary trees from DNA sequence data has revealed that asexual clades are of relatively recent origin (Chapela et al. 1994; Geiser et al. 1996; Lobuglio et al. 1993).

B. The Evolution of Mutualistic Interactions

Mutualistic associations between fungi and plants abound in the biosphere. All available evidence suggests that mutualistic and antagonistic fungi are closely related and that fungal mutualism has arisen many times independently during the course of evolution. Examples of such close relationships include: *Rhizoctonia* species pathogenic on roots (Anderson 1982; Ogoshi 1987) and mycorrhizal fungi on orchids (Harley 1989; Leake 1994), *Rhabdocline* (Sherwood-Pike et al. 1985) and *Lophodermium* (Minter and Millar 1980) with both pathogenic and putatively mutualistic species on conifers, pathogenic *Epichloë* and congeneric asexual isolates mutualistic in grasses (Chap. 14, Vol. V, Part B), and lichen fungi such as *Evernia* which can penetrate and presumably parasitize the trunks of woody plants on which they grow (Orus and Ascaso 1982; Yagué and Estevez 1989). Independent origins are obvious among mycorrhizal fungi, which include both Ascomycetes and Basidiomycetes, among lichens (Gargas et al. 1995), grass endophytes (Chap. 14, Vol. V, Part B), and endophytes of woody plants (Chap. 8, Vol. V, Part B).

It is commonly claimed that mutualism is an inevitable result of evolutionary accommodation between pathogen and host, that an evolutionary arms race must inevitably result in a draw through a process of amelioration of virulence. If this were strictly and always true, present mutualistic associations should be ancient lineages and highly antagonistic ones of recent origin. In fact, this idea contravenes the intuition of many mycologists and plant pathologists, who have long recognized pathogenic lineages as ancient clades, still durably pathogenic (e.g., Luttrell 1974). The notion of inevitable amelioration of virulence has been recently vigorously challenged on theoretical grounds (Ewald 1994). Loss of virulence in pathogenic situations may depend importantly on the frequency of transmission; indeed, in obligate mutualisms between fungi and plants, fungal colonization of new hosts can be an inefficient process (Chaps. 11 and 12, Vol. V, Part B).

A fascinating alternative to the above scenario is that some fungal mutualisms are of comparatively recent origin and that they are evolutionarily transient. This may apply particularly where association with a host has resulted in suppression of sexual reproduction, bringing Muller's ratchet into play. Examples may include lichens which reproduce only by soredia or isidia, mycorrhizal fungi with no known sexual state (e.g., *Cenococcum*), and grass endophytes transmitted vertically from one generation to the next through seed (Chap. 14, Vol. V, Part B). With this last case, the situation has apparently been complicated by instances of presumed somatic hybridization with *Epichloë* strains within grass leaves.

Finally, it should be stressed that the line between pathogen and mutualist can be a fine one and that evolutionary forces could drive a fungal associate in either direction. Plant pathologists should consider examples such as *Fusarium moniliforme* endophytic and occasionally pathogenic in maize (Leslie and Pearson 1990), but mutualistic during the seedling phase (Van Wyk et al. 1988) and *Glomus macrocarpum* pathogenic on tobacco (Modjo and Hendrix 1986; Jones and Hendrix 1987) and ponder whether fungi that are now classified as pathogens might have functioned as mutualists in some progenitor of today's crop plants.

V. Conclusions

Fungi and plants have interacted as mutualists and antagonists at least since they moved into terrestrial habitats. The patterns of these interactions have been constrained and molded by the underlying construction plans of the organisms involved. On the one hand, the fungi have operated as haploid heterotrophs which feed by absorptive nutrition. Their vegetative cells are usually nonmotile and encased in polysaccharide walls. Reproduction occurs both sexually and asexually, the latter usually involving copious production of air- or waterborne spores to colonize an appropriate host. The efficiency of such colonization has been on occasion improved markedly by adaptation to animal vectors. Plants, on the other hand, are diploid or polyploid autotrophs composed of complex organ systems. Plants are also nonmotile and are also composed of cells encased in polysaccharide walls. Reproduction occurs by sexually reproduced seed dispersed by a wide variety of agents, but often stochastically deposited on the landscape. Vegetative reproduction may supplement seed production. These then are the raw materials upon which evolution has acted. Readers are entreated to peruse the chapters which follow and to marvel at the sophistication and sheer baroque creativity of the mechanisms and modes of interaction which have ensued.

References

Albersheim P, Valent BS (1974) Host-pathogen interactions. VII. Plant pathogens secrete proteins which inhibit enzymes of the host capable of attacking the pathogen. Plant Physiol 53:684–687

Anderson NA (1982) The genetics and pathology of *Rhizoctonia solani*. Annu Rev Phytopathol 20:329–347

Berbee ML, Taylor JW (1992) Dating the evolutionary radiations of the true fungi. Can J Bot 71:1114–1127

Carpita NC, Gibeaut DM (1993) Structural modes of primary cell walls in flowering plants: consistency of molecular structure with the properties of the walls during growth. Plant J 3:1–30

Carroll G (1991) Beyond pest-deterrence – alternative strategies and hidden costs of endophytic mutualism in vascular plants. In: Andrews JH, Hirano SS (eds) Microbial ecology of leaves. Springer, Berlin Heidelberg New York, pp 358–375

Carroll GC (1992) Fungal mutualism. In: Carroll GC, Wicklow DT (eds) The fungal community. Its organization and role in the ecosystem. Marcel Dekker, New York, pp 327–354

Carroll G (1995) Forest endophytes: pattern and process. Can J Bot 73 (Suppl 1):S1316–S1324

Chang PLY, Trevithick JR (1970) Biochemical and histochemical localization of invertase in *Neurospora crassa* during conidial germination and hyphal growth. J Bacteriol 102:423–429

Chang PLY, Trevithick JR (1972) Release of wall-bound invertase and trehalase in *Neurospora crassa* by hydrolytic enzymes. J Gen Microbiol 70:13–22

Chang PLY, Trevithick JR (1974) How important is secretion of exoenzymes through apical cell walls of fungi? Arch Mikrobiol 101:281–293

Chapela IH, Rehner SA, Schultz TR (1994) Evolutionary history of the symbiosis between fungus-growing ants and their fungi. Science 266:1691–1696

Cowling R, Richardson D, Paterson-Jones C (1995) Fynbos. South Africa's unique floral kingdom. Institute for Plant Conservation, Fernwood Press, Vlaeberg, RSA

Culberson CF, Culberson WL, Johnson A (1992) Characteristic lichen products in cultures of chemotypes of the *Ramalina siliquosa* complex. Mycologia 84:705–714

Ewald PW (1994) The evolution of infectious disease. Oxford University Press, New York, 289 pp

Farr DF, Bills GF, Chamuris GP, Rossman AY (1989) Fungi on plants and plant products in the United States. APS Press, St Paul, Minnesota

Gabriel W, Lynch M, Burger R (1994) Muller's ratchet and mutational meltdowns. Evolution 47:1744–1757

Gargas A, DePriest PT, Grube M, Tehler A (1995) Multiple origins of lichen symbiosis in fungi suggested by SSU of rDNA phylogeny. Science 268:1492–1495

Geiser DM, Timberlake WE, Arnold ML (1996) Loss of meiosis in *Aspergillus*. Mol Biol Evol 13:809–817

Gianinazzi-Pearson V (1996) Plant cell responses to arbuscular-mycorrhizal fungi: getting to the roots of the symbiosis. Plant Cell 8:1871–1883

Golub, ES (1989) Immunology: a synthesis. Sinauer Associates, Sunderland, 550 pp

Gooday GW (1995) Cell walls. In: Gow NAR, Gadd GM (eds) The growing fungus. Chapman and Hall, London, pp 43–62

Ham KS, Albersheim P, Darvill AG (1995) Generation of beta-glucan elicitors by plant enzymes and inhibition of the enzymes by a fungal protein. Can J Bot 73 (Suppl):S1100–S1103

Harley JL (1989) The significance of mycorrhiza. Mycol Res 93:129–139

Hawksworth DL (1991) The fungal dimension of biodiversity: magnitude, significance, and conservation. Mycol Res 95:641–655

Jones K, Hendrix JW (1987) Inhibition of root extension in tobacco by the mycorrhizal fungus *Glomus macrocarpum* and its prevention by benomyl. Soil Biol Biochem 19:297–299

Kolattukudy PE (1985) Enzymatic penetration of the plant cuticle by fungal pathogens. Annu Rev Phytopathol 23:233–250

Lawrey JD (1984) Biology of lichenized fungi. Praeger, New York

Leake JR (1994) The biology of myco-heterotrophic ("saprotrophic") plants. New Phytol 127:171–216

Leslie J, Pearson C (1990) *Fusarium* spp. from corn, sorghum, and soybean fields in the central and eastern United States. Phytopathology 80:343–350

Lobuglio KF, Pitt JI, Taylor JW (1993) Phylogenetic studies of two ribosomal DNA regions indicate multiple independent losses of *Penicillium* species in subgenus *biverticillium*. Mycologia 85:592–604

Lodge DJ, Fisher PJ, Sutton BC (1996) Endophytic fungi of *Manilkara bidentata* leaves in Puerto Rico. Mycologia 88:733–738

Luttrell ES (1974) Parasitism of fungi on vascular plants. Mycologia 66:1–15

Lynch M, Burger R, Butcher D, Gabriel W (1993) Mutational meltdowns in asexual populations. J Hered 84:339–344

Massicotte HB, Ackerley CA, Peterson RL (1987) The root-fungus interface as an indicator of symbiont interaction in ectomycorrhizae. Can J For Res 17:846–854
May RM (1988) How many species are there on earth? Science 241:1441–1449
May RM (1991) A fondness for fungi. Nature 352:475–476
May RM (1994) Conceptual aspects of the quantification of the extent of biological diversity. In: Hawksworth D (ed) Biodiversity, measurement and estimation. Chapman and Hall, London, pp 13–20
Meeley RB, Johal GS, Briggs SP, Walton JD (1992) A biochemical phenotype for a disease resistance gene of maize. Plant Cell 4:71–77
Minter DW, Millar CS (1980) Ecology and biology of three *Lophodermium* species on secondary needles of *Pinus sylvestris*. Eur J For Pathol 10:169–180
Modjo HS, Hendrix JW (1986) The mycorrhizal fungus *Glomus macrocarpum* as a cause of tobacco stunt disease. Phytopathology 76:988–961
Ogoshi A (1987) The ecology and pathogenicity of anastomosis and intraspecific groups of *Rhizoctonia solani* Kuhn. Annu Rev Phytopathol 25:125–143
Orus MI, Ascaso C (1982) Localization of lichen hyphae in the conducting tissues of *Quercus rotundifolia* Lam. Collect Bot Barc Bot Inst Ed 13:325–338
Peberdy JF (1990) Fungal cell walls – a review. In: Kuhn PJ, Trinci AP, Jung MJ, Goosey MW, Copping LG (eds) Biochemistry of cell walls and membranes in fungi. Springer, Berlin Heidelberg New York, pp 5–30
Pirozynski KA, Dalpé Y (1989) Geological history of the Glomaceae with particular reference to mycorrhizal symbiosis. Symbiosis 7:1–36
Sherwood-Pike MA, Gray J (1985) Silurian fungal remains: probable records of the class Ascomycetes. Lethaia 18:1–20
Sherwood-Pike M, Stone JK, Carroll GC (1985) *Rhabdocline parkeri*, a ubiquitous foliar endophyte of Douglas-fir. Can J Bot 64:1849–1855
Torzilli AP, Lawrey JD (1995) Lichen metabolites inhibit cell wall-degrading enzymes produced by the lichen parasite *Nectria parmeliae*. Mycologia 87:841–845
Trevithick JR, Metzenberg RL (1966) Molecular sieving by *Neurospora* cell walls during secretion of invertase enzymes. J Bacteriol 92:1010–1015
Trevithick JR, Metzenberg RL, Costello DF (1966) Genetic alteration of pore size and other properties of the *Neurospora* cell wall. J Bacteriol 92:1016–1020
Van Etten HD, Matthews DE, Matthews PS (1991) Phytoalexin detoxification: importance for pathogenicity and practical implications. Annu Rev Phytopathol 29:214–223
Van Etten HD, Sandrock RW, Wasmann CC, Soby SD, McCluskey K, Wang P (1995) Detoxification of phytoanticipins and phytoalexins by phytopathogenic fungi. Can J Bot 73 (Suppl 1):S518–S525
Van Wyk PS, Scholtz DJ, Marasas WFO (1988) Protection of maize seedlings by *Fusarium moniliforme* against infection by *Fusarium graminearum* in the soil. Plant Soil 107:251–257
Webber JF, Brasier CM (1984) The transmission of Dutch elm disease: a study of the processes involved. In: Anderson JM, Rayner ADM, Walton DWH (eds) Invertebrate microbial interactions. Cambridge Univ Press, Cambridge, pp 271–306
Wijesundera RLC, Bailey JA, Byrde RJW, Fielding AH (1989) Cell wall degrading enzymes of *Colletotrichum lindemuthianum*: their role in the development of bean anthracnose. Physiol Mol Plant Pathol 34:403–413
Yagüé E, Estevez MP (1989) Beta-1,4-glucanase (cellulase) location in the symbionts of *Evernia prunastri*. Lichenologist 21:147–151

External Interactions

1 Adhesion of Spores and Hyphae to Plant Surfaces

L. EPSTEIN[1] and R.L. NICHOLSON[2]

CONTENTS

I. Introduction

Although fungal adhesion to plants has been recognized for over a century, this aspect of fungal-plant interaction has not been well characterized (Nicholson and Epstein 1991). The importance of adhesion has rarely been critically tested, no fungal adhesive compound that mediates attachment to plants has been fully characterized, and few data are available that describe the molecular bases of fungal-substratum binding. The detailed description of fungal-plant attachment is further complicated by the fact that adhesion occurs at multiple stages of fungal morphogenesis; adhesion can be associated with zoospores and their cysts, conidia, germlings, appressoria, and infection cushions (Nicholson 1984). In some species of rust and anthracnose fungi, several different types of cells (i.e., spores, germlings and appressoria) attach to the host surface before penetration (Chap. 2, this Vol.; Chap. 5, Vol. V, Part B).

Several lines of evidence suggest that adhesion is required for morphogenesis and infection of plants by fungi. As alluded to above, many fungal species produce vegetative and reproductive units which adhere to a variety of plant and synthetic surfaces (Nicholson and Epstein 1991). Electron microscopy often indicates a close physical contact between the fungus and a substratum (Howard et al. 1991a; Braun and Howard 1994; Chap. 3, this Vol.). There is limited experimental evidence that indicates that adhesion is required for infection. Studies with adhesion-reduced mutants indicated that spore adhesion is a virulence factor for the plant pathogen *Nectria haematococca* (Jones and Epstein 1990). Similarly, disease severity caused by *Fusarium solani* f. sp. *phaseoli* (Schuerger and Mitchell 1992) and *Colletotrichum graminicola* (Mercure et al. 1994a) was reduced in environmental conditions in which adhesion was reduced. Experiments using enzymes that selectively degraded the adhesive material on *Uromyces appendiculatus* germlings indicated that adhesion was necessary for oriented growth towards the leaf stoma, the physical recognition of the stoma (the site of fungal penetration), and the induction of the fungal appressorium (the structure from which penetration occurs) (Epstein et al. 1987). Experiments with appresso-

[1] Department of Plant Pathology, University of California, Davis, California 95616-8680, USA
[2] Department of Botany and Plant Pathology, Purdue University, West Lafayette, Indiana 47907, USA

The Mycota V Part A
Plant Relationships
Carroll/Tudzynski (Eds.)

ria of *Magnaporthe grisea* suggest that adhesion is necessary for maintenance of a high internal hydrostatic pressure inside the appressorium and consequently for appressorial function (Howard et al. 1991b; Chap. 3, this Vol; Chap. 3, Vol. V, Part B).

In this chapter, we present an overview of adhesion of fungi to plants and synthetic surfaces, describe the role of adhesion in initiating the infection process, and evaluate the data on the composition of adhesives. The discussion focuses on experimental systems which will allow elucidation of the molecular bases of fungal-plant adhesion. Other related papers concern the adhesion of phytopathogenic, rhizosphere, and mutualistic bacteria to plant surfaces (Smit et al. 1989; Romantschuk 1992) and the adhesion of fungi pathogenic to animals (Calderone and Braun 1991; Barki et al. 1993; Chap. 1, Vol. VI), insects (Boucias and Pendland 1991; Chap. 17, Vol. VI), nematodes (Tunlid et al. 1992), and other fungi (Manocha and Chen 1990; Wisniewski et al. 1991). Readers also are referred to reviews concerning cell-cell adhesion of fungi (Lipke and Kurjan 1992) and cell-substratum adhesion of aquatic plants to rocks (Vreeland and Epstein 1996), aquatic fungi in marine environments (Read et al. 1992; Hyde et al. 1993; Jones 1994) and fungi to bioreactors (Jones and Briedis 1992).

II. Characteristics of Fungal Adhesive Compounds

Although adhesion to plant surfaces is common among fungal species, between fungal species there is variation in the apparent composition of the adhesive materials, the environmental cues which induce the development of adhesiveness, and the stages of fungal development at which adhesion occurs. Differences between species in different taxonomic classes perhaps should be expected; fungi are not a natural taxonomic group. The Oomycota are evolutionarily related to algae rather than to the Ascomyotina and Basidiomycotina (Förster et al. 1990). However, even among Ascomycota, Basidiomycotina, and imperfect Deuteromycota, there is considerable variation in adhesion of different species within and between these classes.

Different environmental stimuli may trigger fungal morphogenesis and the development of adhesion in different fungi. For example, a preformed adhesive is released upon hydration of *M. grisea* conidia. In contrast, *Nectria haematococca* macroconidia are still nonadherent after hydration (Jones and Epstein 1989). Rather, respiration and, apparently, protein synthesis are required before the spores become adhesive. Adhesiveness is a transient event in some fungi. *Phytophthora cinnamomi* and *P. palmivora* encysting zoospores are adhesion-competent for less than 4 min (Bartnicki-Garcia and Sing 1987; Gubler et al. 1989). In contrast, adhesion-competent conidia of *N. haematococca* and *C. musae* can adhere at any time for hours before germ tube emergence (Jones and Epstein 1989; Sela-Buurlage et al. 1991). Many fungi become adhesion-competent very rapidly. Fourier-transform infrared spectroscopic data indicated that a proteinaceous adhesive appears on the surface of the chytridiomycete *Catenaria anguillulae* after only 2.5 min of incubation (Tunlid et al. 1991).

While some extracellular material which is presumably involved in adhesion can be observed with light or fluorescent microscopy (Hamer et al. 1988), some adherent fungi appear to be surrounded only by a relatively thin film of extracellular material visible only with electron microscopy (Mims and Richardson 1989). Although there have been no studies on the relationship between the thickness of the adhesive layer and the adhesive strength, there is no apparent correlation.

The extracellular matrices of fungi are biochemically complex (Ramadoss et al. 1985; Moloshok et al. 1993). Some researchers assume that microscopically detectable material located between the fungus and the substratum is involved in adhesion, but this may not be true. For example, conidia and conidial germlings are often surrounded by an extracellular layer of "mucilage" (Nicholson and Epstein 1991). However, in many cases, this mucilage is water-soluble, whereas fungal adhesives typically are water-insoluble. The mucilage which surrounds *C. graminicola* spores serves multiple functions unrelated to adhesion that ensure conidium survival and assist in the infection process (Leite and Nicholson 1992, 1993). The mucilage contains compounds that protect the conidia from desiccation (Nicholson and Moraes 1980), diverse enzymes which appear to prepare the host surface for penetration (Pascholati et al. 1993), and proteins that protect conidia from toxic phenylpropanoids present in or on the host (Nicholson et al. 1986, 1989). The mucilage also contains mycosporine-alanine, a germination self-inhibitor (Leite and Nicholson 1992).

Only limited information is available on the composition of fungal adhesives. Most fungal adhesives appear to be glycoproteins (McCourtie and Douglas 1985; Bartnicki-Garcia and Sing 1987; Chaubal et al. 1991; Tunlid et al. 1991; Kwon and Epstein 1993; Xiao et al. 1994). Labeling with lectins indicates that fungal adhesives often are glycosylated (Hamer et al. 1988; Durso et al. 1993; Mercure et al. 1995). Polysaccharides are involved in adhesion of the black yeast *Aureobasidium pullulans* (Andrews et al. 1994); galactosaminoglycans appear to mediate adhesion of *Bipolaris sorokiniana* germlings onto glass (Pringle 1981). Additional information on adhesive composition will be discussed in Section IV (this Chap.).

Fungi can adhere either nonspecifically to a variety of substrata or specifically in "adhesin"-receptor or lectin-hapten interactions. Lectins are important in adhesion of mycorrhizal fungi to host roots (Bonfante-Fasolo and Spanu 1992), pathogenic fungi within plant tissue (Hohl and Balsiger 1988), nematophagous fungi to nematodes (Tunlid et al. 1992), and in fungal-fungal interactions such as mycoparasitism (Manocha and Chen 1990; Manocha et al. 1990). Roots are coated with a carbohydrate-rich hydrophilic mucilage and there is evidence that encysting Oomycota bind via lectin-like interactions to fucosyl residues on the root surface (Hinch and Clarke 1980). However, most encysting Oomycota also bind to inert substrata. Thus, encysting Oomycota may adhere to roots by both nonspecific and specific interactions. In contrast, adhesion of *Phialophora radiciola* and *F. moniliforme* to maize root mucilage may depend upon ionic, rather than lectin-hapten, interactions (Northcote and Gould 1989). Similarly, plant lectins apparently are not involved in attachment of *F. solani* f.sp. *phaseoli* conidia to bean roots (Schuerger et al. 1993).

The above-ground portions of terrestrial plants are covered with a hydrophobic cuticle composed of cutin and waxes; aerially dispersed fungi appear to be specifically adapted for adhesion onto hydrophobic substrata. Many fungi, including *M. grisea*, *C. graminicola*, and *U. appendiculatus*, adhere tenaciously to inert hydrophobic substrata such as polystyrene. Indeed, some fungi adhere better to hydrophobic than to hydrophilic surfaces (Sela-Buurlage et al. 1991; Doss et al. 1993; Terhune and Hoch 1993). Thus, adhesion to aerial plant surfaces may be mediated primarily by hydrophobic interactions. Hydrophobins (Templeton et al. 1994), small cysteine-rich proteins on fungal surfaces, may be candidates for fungal adhesive compounds; hydrophobins can engage in hydrophobic interactions and can self-assemble at the interface of a hydrophilic fungal wall and the hydrophobic plant cuticle (Wessels 1994). However, most fungal adhesives appear to be glycosylated proteins and hydrophobins are not markedly glycosylated.

Regardless of whether the fungal adhesives are glycoproteins, proteins, or carbohydrates, we postulate that most adhesives produced by phytopathogenic fungi on leaf surfaces are more analogous to "glues" which adhere nonspecifically to a variety of substrata rather than to "adhesins" or lectins which bind to a specific receptor. If specific receptors for adhesins are present on plant cuticles, they need to be identified. Toti et al. (1992) demonstrated that conidia of the endophyte *Discula umbrinella* attached differentially to host leaf and nonhost surface. However, it was not determined whether differential attachment occurred because of the presence of (1) a specific receptor on host leaves, (2) a host-specific exudate which stimulated the release of a nonspecific glue, or (3) physical differences in the topography of host and nonhost surfaces.

Although many fungi can adhere to hydrophobic inert substrata, spores of some species can modify the hydrophobic, cuticular surfaces of plants. Upon hydration, ungerminated conidia of *Erysiphe graminis* (Nicholson et al. 1988; Pascholati et al. 1992), *C. graminicola* (Pascholati et al. 1993), and *U. viciae-fabae* (Deising et al. 1992) release cutinases and nonspecific esterases. The enzymes degrade the plant cuticle during the same period that adhesion occurs. Deising et al. (1992) suggested that these enzymes are involved in adhesion of ungerminated *U. viciae-fabae* urediospores onto leaves. When dead spores were incubated with a cutinase and two nonspecific esterases isolated from the surfaces of urediospores, adhesion increased. Living spores treated with the serine-esterase inhibitor diisopropyl fluorophosphate were less adherent. Thus, although *U. viciae-fabae* urediospores adhere to inert substrata such as Mylar, adhesion may be stronger on a cutinase-degraded cuticle than a nondegraded substratum. Whether different kinds of bonds between host and pathogen are formed on a cutinase-degraded cuticle is unknown. In addition to cutinases, the spore exudate of *E. graminis* may function as a wetting agent (Nicholson et al. 1993). Liquid exudate released from a conidium makes the hydrophobic leaf surface more hydrophilic, specifically at the site of appressorium induction; importantly, appressorium induction evidently requires a more hydrophilic surface than that of the native barley leaf.

Eagland (1988) reviewed the physical properties of glues and the adhesive process. Briefly, glues typically spread in a liquid state and then polymerize. Since conidia, germ tubes, and appressoria adhere to inert substrata while submerged in water, the fungal glue apparently can displace water and then polymerize into a water-insoluble material. Glues can have multiple components (see Vreeland and Epstein 1996); Knox (1984) suggested that an adhesive produced by plants would contain an adhesive base, plasticizer, thickener, tackifier, and detergent (wetting agent). Fungal adhesives also may have have multiple-components, and the polymerization process might be regulated by Ca^{2+} (Gubler et al. 1989), peroxidases (Leatham et al. 1980) and/or transglutaminases (Bradway and Levine 1993; Ruiz-Herrera et al. 1995).

Ideally, researchers would specify the strength of the attachment of a fungus to a substratum. For example, adhesion could be expressed quantitatively as the force required to dislodge the propagules. In addition, researchers would know the amount of adhesive strength necessary to maintain adhesion on the host surface. In practice, adhesion data are generally presented as the percentage of propagules remaining on a substratum after washing; the force of washing treatments varies widely between laboratories. Most laboratories adjust their washing protocols so that either increases or losses in adhesiveness can be detected. Regardless of the particular washing protocol, the strength of adhesive bonding to a substratum differs between the fungi and between the cell types. For example, *N. haematococca* conidia are dislodged with less force than *C. graminicola* conidia (L. Epstein, unpubl.). *C. graminicola* conidia are dislodged with less force than *C. graminicola* germ tubes or appressoria (E.W. Mercure and R.L. Nicholson, unpubl.). Similarly, *Botrytis cinerea* germlings are more adherent than conidia (Doss et al. 1995).

III. The Role of Fungal Adhesion in Plant Pathogenesis and Fungal Morphogenesis

A. Prevention of Displacement from the Infection Court

With many fungi, spores alight on a plant and germinate when water and nutrients become available. Because fungi can be displaced from the host surface by water or wind, the adhesion of propagules or germlings plays an important role in preventing displacement from the infection court.

B. Appressed Growth of Hyphae

Although there are examples of fungal hyphae that elongate aerially and then fall onto the host surface, more typically, germ tubes elongate in contact with a surface. The morphology of an adherent germ tube differs from a nonattached germ tube; an adherent germ tube is generally flat along the substratum (Epstein et al. 1987), while a germ tube in suspension is rounded. Adherent bean rust germlings have a greater area of surface contact than less adherent germlings (Terhune and Hoch 1993).

C. Thigmotropism and Thigmodifferentiation

Adhesion is important for establishing a physical connection between organisms. In order to successfully penetrate and cause disease, fungi such as rusts and anthracnose-causing fungi must enter their hosts through particular locations (e.g., stomata or junctions between cells) via structures such as appressoria. On the plant surface, growth toward a particular location and the induction of infection cells can be stimulated by physical stimuli (Hoch et al. 1987; Chap. 2, this Vol.). If fungi are not firmly attached to the plant surface, topographical stimuli on the host surface do not induce directed growth (thigmotropism) or morphogenesis (thigmodifferentiation) (Epstein et al.

1987). Terhune and Hoch (1993) demonstrated a correlation between adhesion of bean rust urediospores and germlings, and the number of appressoria that were induced on selected substrata. While adhesion is necessary for the contact-induced responses, it is not known whether adhesives are involved in the reception and transmission of signals from the host surface to the fungal cell.

D. Infection Cell Function

Adhesion is such a common feature of appressoria and other infection structures that often they are defined as "adherent structures." Adherent fungal infection structures include encysting zoospores, appressoria, hyphopodia, and infection cushions. Encysting zoospores are produced by oomycete fungi. Appressoria and hyphopodia are distinguished by whether they are produced by germlings or hyphae (Epstein et al. 1994a), respectively. Infection cushions are defined as multicelled infection structures formed by vegetative hyphae. Adhesion of the infection structure apparently is required to ensure direct penetration of the host surface. The adhesive bond of the infection structure to the plant must exceed the force required to puncture the host; otherwise, the emerging penetration peg would push the appressorium away from the substratum and penetration would not occur (Howard and Ferrari 1989). In *M. grisea* appressoria, the penetration peg is driven through the substratum by water pressure; appressorial melanization is required to maintain a high internal hydrostatic pressure (Howard et al. 1991b). However, at the appressorium-substratum interface, the penetration peg emerges through a central pore, a walless region without melanin. The pore appears to be surrounded by an "O-ring" and adhesive material in *M. grisea* and other fungi such as *C. lindemuthianum* (Bailey et al. 1992). Adhesion of the nonmelanized bottom of the appressorium to the substratum is apparently required to retain the internal water pressure necessary for penetration.

Appressorial adhesion may serve other functions in some fungi. For example, in *C. musae* (Muirhead and Deverall 1981), the appressorium functions as a dormant propagule in which long-term attachment to the host surface is essential for eventual disease development. In fungi in which the appressoria are positioned over guard cells or the junctions between cells, adhesion may serve to position the cell over a required penetration site.

E. Chemical Communication Between Host and Pathogen

The physical contact mediated by adhesion also allows chemical communication between the fungus and the plant (Nichols et al. 1980). Smith and Cruickshank (1987) suggested that firm adhesion of conidia of *Monilinia fructicola* to its host was required for the fungal elicitor to induce synthesis of the phytoalexin pisatin by the pea endocarp. Timely delivery of the elicitor depends on sustained physiological contact between the two organisms, a phenomenon evidently mediated by adhesion. Jones and Epstein (1990) isolated two chemically induced mutants of *N. haematococca* that produced macroconidia with a reduced ability to adhere to the host, i.e., zucchini fruits, and to polystyrene. When inoculated into wounded fruit tissue, the mutants were as virulent as the wild type. However, when inoculated onto unwounded fruit, mutants were less virulent, suggesting that adhesion is a virulence factor. Since macroconidia were not displaced in this experiment, the authors suggested that contact is required for the establishment of a "compatible" or pathogenic interaction. Since *N. haematococca* germlings do not produce appressoria and penetrate the host surface through an unwounded cuticle, adhesion might facilitate both the induction of, and maximize the concentration of, the pathogen's degradative enzymes on the host surface.

IV. Model Systems for Studying Fungal-Plant Substratum Adhesion

A model system should have a sexual cycle to facilitate a genetic analysis of adhesion. Several systems that may be especially useful are described below. Mutational analyses will be more straightforward in the three Ascomycotina, which are culturable and have a haploid vegetative stage. Mutational analyses in *Phytophthora cinnamomi* and *Uromyces appendiculatus* are complicated by a diploid vegetative stage and obligate parasitism, respectively.

A. Oomycotina (see also Chap. 2, Vol. V, Part B)

Phytophthora cinnamomi

Although the Oomycotina are not "true fungi," we will discuss *P. cinnamomi* because it is one of the best-characterized systems and is a root-infecting pathogen (Hardham 1992). Motile zoospores are attracted to plant roots by chemical stimuli. After contact with a substratum (or specific chemical treatments in the laboratory), zoospores are induced to encyst, a process that includes cell wall formation. *P. cinnamomi* cysts adhere well to glass, plastic, and to root tissue. Although cysts do bind nonspecifically onto synthetic substrata, binding onto maize roots occurs at terminal fucosyl (and possibly other sugar monomers) residues in root mucilage (Hinch and Clarke 1980).

If encystment occurs in response to a chemical stimulus, cysts are only transiently adhesive; after 4 min, the free cysts are nonadherent (Gubler et al. 1989). In the laboratory, encystment of *P. cinnamomi* can be induced specifically by the lectin Concanavalin A, which binds to the entire surface of the zoospore (Hardham and Suzaki 1986) and by pectin (Irving and Grant 1984). The process of encystment and becoming adhesion-competent requires Ca^{2+} (Deacon and Donaldson 1993).

Gubler and Hardham (1988) demonstrated that a preformed adhesive is stored in small peripheral vesicles in zoospores. During encystment, the contents are rapidly exocytosed in less than 1 min. At the same time, large peripheral vesicles move toward the center of the cell; these vesicles may contain storage proteins. The adhesive material is a high-molecular-weight protein (ca. 220 kDa).

Adhesion in encysting zoospores was studied in two other Oomycota. In *Saprolegnia ferax*, the adhesive material is apparently released as a fluid and polymerizes extracellularly (Durso et al. 1993). Estrada-Garcia et al. (1990) demonstrated that several high-molecular-weight glycoproteins ($>200 \times 10^3$ M) stored in the large vesicles in *Pythium aphanidermatum* zoospores are incorporated into the adhesive cyst coat.

B. Ascomycotina

We chose three Ascomycotina as model systems, partly because of their genetics. However, the sexual cycle is unimportant in the epidemiology of these pathogens; only the asexually produced conidia are important for disease development. Consequently, the dissemination, adhesion, and development of the conidia have been studied. The three fungi differ in several respects with regards to adhesion. On the plant surface, *C. graminicola* and *M. grisea* produce adherent conidia, germ tubes, and appressoria. *N. haematococca* produces adherent macroconidia and germ tubes but does not produce appressoria. *M. grisea* conidia become adhesion-competent after hydration; *N. haematococca* macroconidia become adhesion-competent only after exposure to certain plant extracts. In all three ascomycetes, the transformation from a nonadherent to an adherent conidium occurs rapidly and hours before germination. These fungi bind nonspecifically to inert substrata such as polystyrene; *M. grisea* and *C. graminicola* bind more strongly to hydrophobic than to hydrophilic substrata.

1. *Magnaporthe grisea* (*Pyricularia grisea*, = *P. oryzae*; see also Chap. 3, Vol. V, Part B)

In *M. grisea* conidia, the adhesive is preformed in the conidial apex in the periplasmic space between the plasma membrane and the cell wall. Upon hydration, the adhesive or "spore tip mucilage" is released and the conidium is anchored at its tip to the substratum (Hamer et al. 1988). Concanavalin A binds to the spore tip mucilage and blocks adhesion, suggesting the adhesive contains α-mannose and/or α-glucose residues. Inoue et al. (1987) suggested that development of adhesive appressoria parallels appressorial melanization and that either melanin catabolism triggered development of adhesiveness or that melanin biosynthesis intermediates are an adhesive component. However, Howard and Ferrari (1989) concluded that melanin is not involved in appressorium adhesion since nonmelanized appressoria, formed by either melanin-less mutants or by the wild-type strain incubated in the melanin biosynthesis inhibitor tricyclazole, were as adherent as melanized appressoria.

2. *Glomerella graminicola* (anamorph = *Colletotrichum graminicola*; see also Chap. 23, this Vol.)

Conidia in acervuli are surrounded by extracellular mucilage which is not involved in adhesion

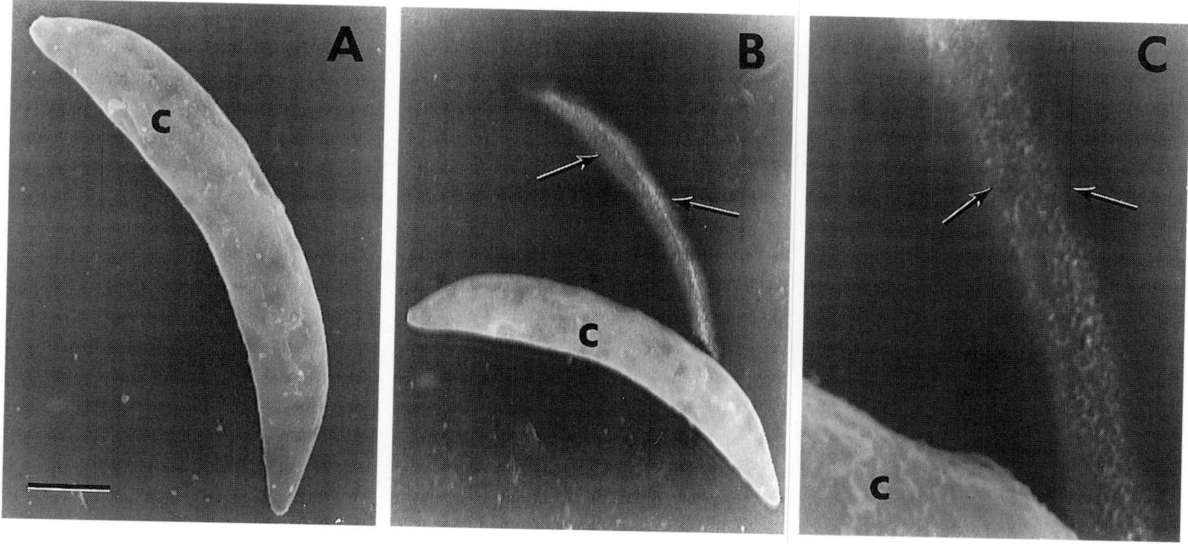

Fig. 1A–C. An ungerminated conidium of *Colletotrichum graminicola* adhered to polystyrene. **A** Thirty minutes after contact with the substratum, the conidium (*c*) was fixed for microscopy. **B** The conidium in **A** was micromanipulated away from the polystyrene. *Arrows* indicate fungal material along the conidium-substratum interface. **C** An enlarged view of the material (*arrows*) released from the conidium (*c*) as shown in **B**. *Bar* 4.5, 6, and 1.3 μm for **A**, **B**, and **C**, respectively. (Photographs H. Kunoh and R.L. Nicholson)

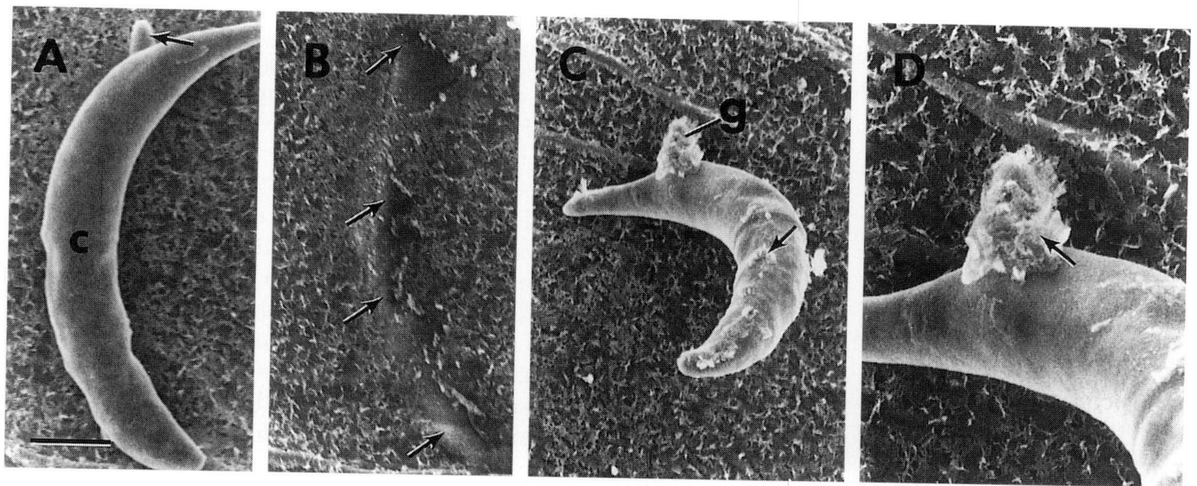

Fig. 2A–D. A germinated conidium of *Colletotrichum graminicola* adhered to a corn leaf. **A** Conidium (*c*) with a germ tube (*arrow*) after a 6-h incubation. **B** The conidium in **A** was micromanipulated off the leaf. Part of the leaf cuticle that was in contact with the conidium was removed (*arrows*) during micromanipulation. The *upper arrow* shows where the germ tube and the cuticle were in contact. **C** The lower surface of the conidium in **A**. Fragments of the leaf cuticle (*arrows*) are adhered to the conidium. The lower surface of the germ tube (*g*) is covered with fibrous material that may include part of the leaf cuticle and/or an adhesive. **D** Close-up of the lower surface of the germ tube (*arrow*) shown in **C**. *Bar* represents 4.5 μm for **A–C**, and 2 μm for **D**. (Photographs H. Kunoh and R.L. Nicholson)

(Nicholson and Moraes 1980; Nicholson and Epstein 1991; Mercure et al. 1994a; Mims et al. 1995). Adhesion of ungerminated conidia of *C. graminicola* is initiated in water. Upon contact with the substratum, conidia release material at the substratum interface (Mercure et al. 1994a). The conidia adhere more strongly to hydrophobic surfaces such as polystyrene (Fig. 1), glass treated with dimethyldichlorosilane, and corn leaves (Fig. 2) than to more hydrophilic substrata (Mercure et al. 1994a,b). Adhesion of conidia was reduced by incubation in Pronase E, Concanavalin A, and inhibitors of protein synthesis (cycloheximide), and glycoprotein transport (Brefeldin A). The transcription inhibitor actinomycin D did not reduce adhesion. The results suggest that a glycoprotein is involved in adhesion. Adhesion in other *Colletotrichum* sp. appears similar to that in *C. graminicola* (Sela-Buurlage et al. 1991). *C. musae* conidia can produce adhesive material for an extended time period; conidia rendered nonadhesive by treatment with an extracellular protease regained adhesiveness after the protease was removed.

3. *Nectria haematococca* Mating Population I (anamorph, *Fusarium solani* f. sp. *cucurbitae* race 1)

Organic compounds are apparently required to induce the development of adhesion in *N. haematococca* macroconidia; within 10 min after addition of some plant extracts, macroconidia become adhesive (Jones and Epstein 1989; Kwon and Epstein 1993). When incubated in zucchini fruit extract or potato dextrose broth, macroconidia rapidly adhere at the conidial apices, produce material at the spore apices which binds with Concanavalin A (Con A; Fig. 3), and produce a 90-kDa glycoprotein which also binds with Con A (Fig. 4; Kwon and Epstein 1993). When incubated in water, V8 broth, or in a defined medium, the macroconidia do not become adherent, and produce neither the tip material nor the 90-kDa glycoprotein (Figs. 3, 4). The 90-kDa glycoprotein is macroconidal-specific; the compound is not present on microconidia, which are relatively nonadherent. Because conidia germinate but do not adhere in V8 broth or in the defined medium, the production of material which is labeled with Con A at the site of initial attachment and the production of the 90-kDa glycoprotein are specifically associated with adhesion and not with the initiation of the germination process. Con A prevents adhesion, but not germination, of macroconidia. Thus, the same compound which labels the 90-kDa glycoprotein and the material at the site of adhesion, also blocks adhesion. Further experiments are required to determine if the 90-kDa glycoprotein is either the adhesive compound or temporally and spatially associated with the development of adhesion competence.

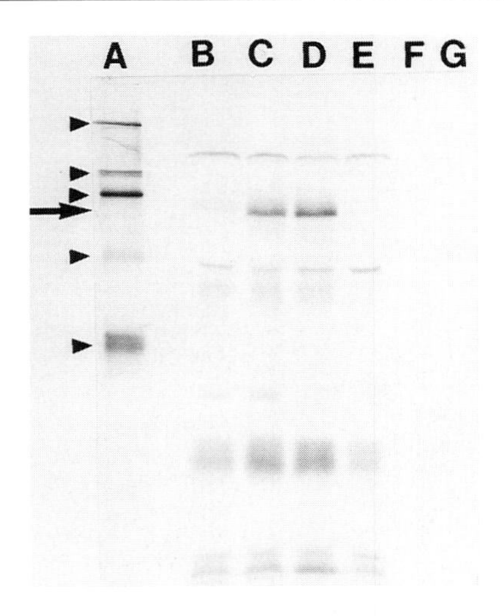

Fig. 4. A cell-surface rich preparation from *Nectria haematococca* macroconidia separated by polyacrylamide gel electrophoresis and then probed on a Western blot with Concanavalin A. *A* Molecular weight markers, labeled with *triangles*, are (from *top to bottom*) 200, 116, 93, 66, and 45 kDa; *B–E* macroconidia incubated in various media; *B* water; *C* zucchini extract; *D* potato dextrose broth; *E* V8 juice broth; *F–G* no-spore controls containing media; *F* zucchini extract; *G* potato dextrose broth. The *arrow* indicates the 90-kDa glycoprotein present in adherent macroconidia, i.e., conidia incubated in zucchini extract (*C*) or potato dextrose broth (*D*) but not present in nonadherent macroconidia, i.e., conidia incubated in water (*B*), V-8 juice broth (*E*), or in control solutions with only media (*F,G*). (After Kwon and Epstein 1993)

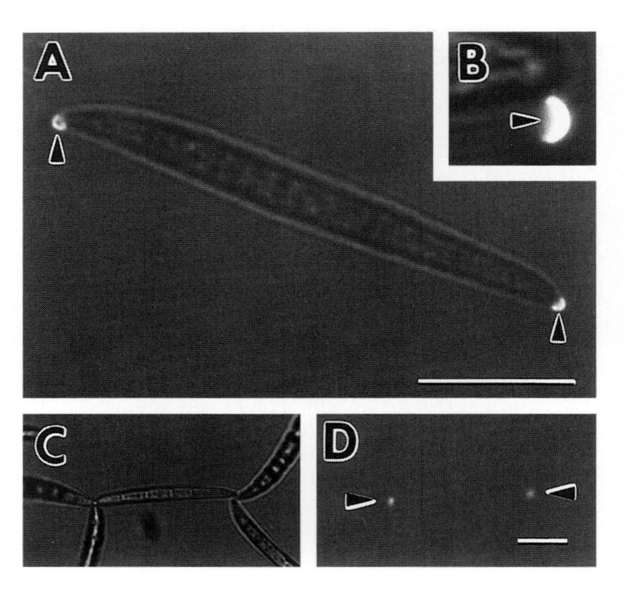

Fig. 3A–D. Adherent *Nectria haematococca* macroconidia stained with Concanavalin A-FITC. **A**, **B**, and **D** Fluorescence microscopy. **C** Light microscopy. **A** ConA-FITC stains the macroconidial tips, the initial points of adhesion. **B** Enlargement of ConA-FITC-stained macroconidial tip. **C–D** (same microscopic field) In suspension, adhesion-competent macroconidia agglutinate at the tips. (Kwon and Epstein 1993)

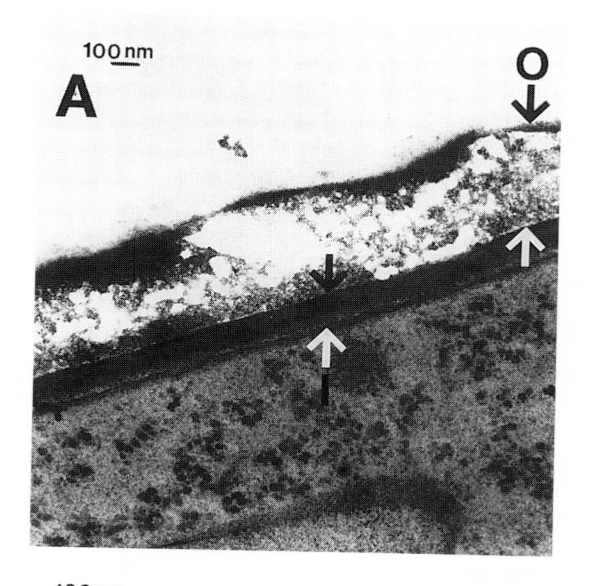

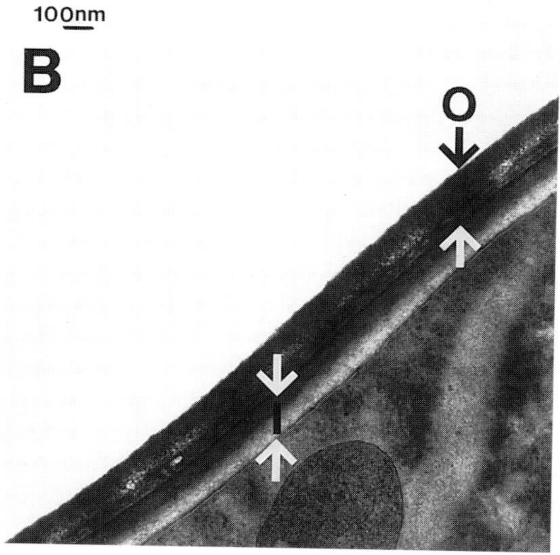

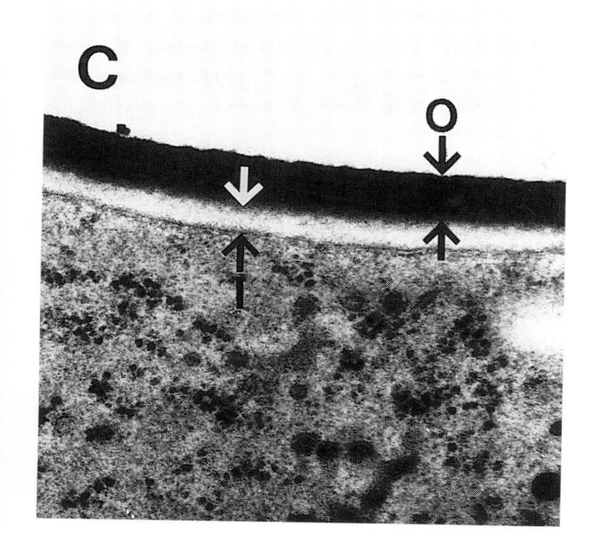

Jones and Epstein (1990) isolated two *N. haematococca* mutants with macroconidia with reduced ability to adhere (Att^-) to zucchini fruits and polystyrene. The adhesion-reduced phenotype in both mutants is temperature-sensitive and dependent upon nutrient concentration. Pathogenicity assays of the mutants and wild type indicated that macroconidial adhesion is a virulence factor. Classical genetic analysis of progeny derived from one mutant (LE1) identified a mutation in a genetic locus, named *Att*1 (Epstein et al. 1994b). Interestingly, the 90-kDa glycoprotein and macroconidial tip mucilage are produced in both Att^+ and Att^- strains. However, after the development of adhesion-competence in the wild type, the ultrastructure of the entire cell surface differs between the wild type and the two mutants (Caesar-TonThat and Epstein 1991). Freshly harvested macroconidia of all strains are nonadhesive and have a similar wall ultrastructure. However, after incubation in zucchini extract for 1, wild-type macroconidia have an outer layer (Fig. 5A) not originally present. After 1 h, the adhesion-reduced mutants also have an outer layer (Fig. 5B, C) not present on freshly harvested conidia. However, the outer layers of the mutants differ from the wild type and each other. In both mutant strains, the outer layer is more electron-dense and more compact. Thus the development of adhesiveness is correlated with the appearance of new material on the conidium surface. In all three strains, additional changes in wall ultrastructure appeared after 3- and 5-h incubations; differences between the adhesive wild type and the adhesion-reduced mutants are still evident at these times. This may be evidence that adhesion occurs in two stages; first at the spore tips and later along the length of the spore.

Fig. 5A–C. Transmission electron micrographs of freeze-substituted *Nectria haematococca* macroconidia after incubation for 1 h. **A** Adhesive wild type. **B** Adhesion-reduced mutant LE1. **C** Adhesion-reduced mutant LE2. *Arrows* indicate the inner (*I*) and outer (*O*) layers of the spore walls. **A** The outer wall in the adhesive wild type is thick ($\bar{x}$ = 400 nm), dispersed, and electron-variable in appearance. **B** Compared to the wild type, the outer wall of the adhesion-reduced mutant LE1 is thinner ($\bar{x}$ = 214 nm), less dispersed, and more electron-dense. **C** Compared to the wild type, the outer wall of the adhesion-reduced mutant LE2 is also thinner, less dispersed, and more electron-dense than the wild type. The outer wall of LE2 is more electron-dense than LE1. All micrographs are the same magnification. (After Caesar-TonThat and Epstein 1991)

Schuerger and colleagues observed adhesive material at the tips of macroconidia and germlings of the related fungus *Fusarium solani* f. sp. *phaseoli* (Fig. 6); this fungus adheres and penetrates through roots (Schuerger and Mitchell 1993).

C. Basidiomycotina

Uromyces appendiculatus (see also Chap. 2, this Vol.; Chap. 5, Vol. V, Part B)

Many species of rust fungi produce adherent spores, germlings, and appressoria (Beckett et al. 1990; Chaubal et al. 1991; Clement et al. 1993); we will limit our few comments here to *U. appendiculatus*, since adhesion of rusts is discussed earlier in this chapter and later in this volume. Urediospores, germlings, and appressoria adhere onto inert, hydrophobic substrata such as polystyrene and plant tissue (Wynn 1981). Adhesion occurs to a greater extent on hydrophobic than on hydrophilic substrata (Terhune and Hoch 1993). Adhesion is mediated by extracellular material. On germlings of *U. appendiculatus*, a protein apparently is involved in adhesion (Epstein et al. 1985). Firm germling adhesion is necessary for directed growth towards stomata and for induction of appressoria (Epstein et al. 1987; Terhune and Hoch 1993).

V. Experimental Approaches to Studying Adhesion

The challenge remains to determine the structure of fungal adhesives and the precise function of adhesion in fungal-plant interactions. While there has been some progress in the biochemical characterization of the fungal wall and extracellular matrices, we know little of the functional significance of most of the components. The problem of characterization of cell surfaces is exacerbated by the fungal lifestyle of active secretion of many compounds into the environment, including toxins and enzymes. Indeed, fungi often solubilize their carbon sources with extracellular hydrolytic enzymes. Thus the demonstration either that a particular compound is released coincidently with fungal adhesion or that a particular compound is present in an adhesive layer is not proof that the compound is the adhesive material.

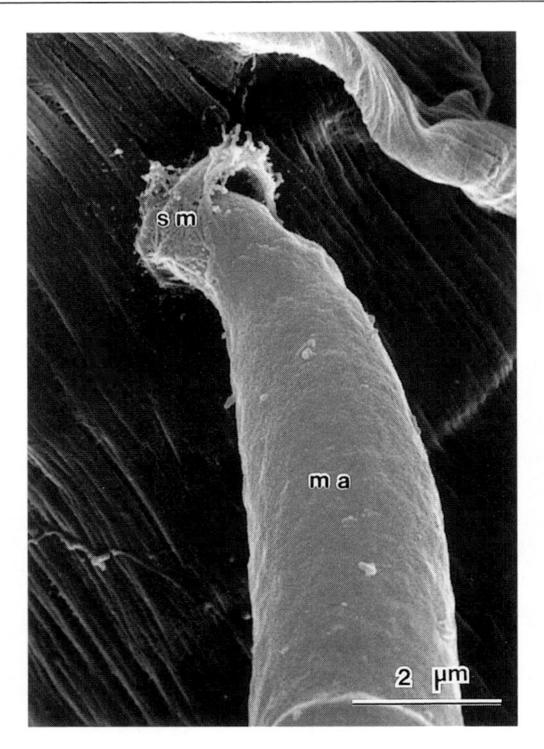

Fig. 6. *A Fusarium solani* f. sp. *phaseoli* macroconidium with a short germ tube. The spore apex and germ tube are adhered to a nitrocellulose membrane. (Schuerger and Mitchell 1993)

A. Correlation of Adhesion with Specific Cellular or Biochemical Events

Although such studies do not conclusively identify the adhesive material, the elucidation of specific subcellular events that are spatially and temporally correlated with adhesion provides valuable information on the adhesive material and particularly on the adhesive process (Kwon and Epstein 1993; Mercure et al. 1994a). In studies with encysting zoospores in the Oomycotina *P. cinnamomi* (Gubler and Hardham 1988, 1990) and *P. aphanidermatum* (Estrada-Garcia et al. 1990), monoclonal antibodies have been useful for identifying putative adhesives. Fixed zoospores, which are wall-less, were used as the antigen. However, since the carbohydrates in the walls of Ascomycotina and Basidiomycotina are highly antigenic, it seems unlikely that fixed conidia or basidiospores would induce antibodies which recognize the fungal adhesives; even with monoclonals, purified adhesives probably will be required to produce anti-adhesive antibodies for the higher fungi.

B. Detection of Adhesive Compound(s) by Disruption of Adhesive Activity

In several studies, compounds involved in adhesion have been deduced by the utilization of enzymes, lectins, or other reagents which prevent binding of normally adherent fungi (Epstein et al. 1985; Gubler et al. 1989; Chaubal et al. 1991; Xiao et al. 1994). However, results of some studies must be considered preliminary because of nonspecific or indirect effects of the treatments. For example, in one well-cited study, an apparently impure "hemicellulase" preparation was used as evidence that a hemicellulose is involved in adhesion (Lapp and Skoropad 1978). In studies using enzymes to identify extracellular compounds, several controls are important. Inactivated enzyme preparations should be used to establish that the protein per se and the buffer do not inhibit adhesion. Enzymes should either be purified to homogeneity or used with specific inhibitors, such as trypsin inhibitor, to establish that the desired enzyme is actually inhibiting adhesion. Experimental evidence should indicate that the enzyme or lectin has only extracellular, and not intracellular, activity. For example, physiological processes such as germination and growth rate could be monitored. Similarly, adhesion-disruption experiments with lectins should include lectin plus hapten and hapten-only controls to demonstrate specificity of the interaction.

C. Nonadhesive Mutants

Suitable adhesion-reduced mutants would provide powerful tools for conclusive identification of adhesive compounds and would allow an experimental assessment of the role of adhesion in fungal development and plant pathogenesis. Adhesion-reduced mutants have been useful in studies on the fungal human pathogen *Candida albicans* (Calderone and Wadsworth 1988; Saxena et al. 1990). Since ascomycete spores are generally haploid, all mutations in genes with single copies are expressed and screening for mutants is straightforward. Hamer et al. (1989) enriched for nonadhesive *M. grisea* and then isolated **s**pore **mor**phology (Smo$^-$) mutants. The same (Smo$^-$) strains were isolated regardless of whether mutants were originally screened for nonadhesion, failure to form appressoria, or cytoskeletal defects; the authors concluded that the SMO locus may be highly mutable. Although all of the *smo* mutations are alleles of a single genetic locus, Smo$^-$ strains have multiple defects; the conidia and appressoria are abnormally shaped, and appressoria are induced on surfaces that are not normally conducive for appressorium formation. Since the Smo$^-$ strains are pleiotropically affected, it is unclear whether they will be useful for studies on adhesion. Using an enrichment screen for nonadhesive *Nectria haematococca* macroconidia, Jones and Epstein (1990) isolated two strains with adhesion-reduced macroconidia (Att$^-$). In contrast to the *M. grisea* mutants, the *N. haematococca* mutants produced normally shaped macroconidia. Identification of the adhesive compound(s) using nonadhesive mutants should be facilitated by the fact that regardless of the nature of the mutation, all Att$^-$ strains should be either deficient in production of an adhesive compound or have at least some biochemical alteration in a cell surface compound. Proteins or carbohydrates on cell surfaces can be labeled by a variety of techniques including biotinylation (Lisanti et al. 1989; Kwon and Epstein 1992) and iodination (Epstein et al. 1987). The labeled macromolecules can be separated biochemically; compounds present in greater quantity in the wild type than in the mutant strains could be tested as potential adhesives. Alternatively, a cell-surface-enriched preparation might be obtained simply by incubating intact spores for several minutes in an extraction buffer (Kwon and Epstein 1993); after this process, spores appeared to be intact. The cell walls may have prevented leakage of intracellular, high-molecular-weight compounds into the extraction buffer.

VI. Potential Applications of Information

Identification of fungal adhesive compounds and adhesins, and determination of the mechanisms of binding may allow the design and development of new classes of chemicals for agricultural use which interfere specifically with adhesion of spores, germlings, or infection structures. Fungal-substratum adhesion seems an attractive point at which to interrupt pathogen development, for the following reasons. Firstly, during the stage of germling-plant adhesion, fungi are exposed to environmental perturbations and are thus vulnerable to control measures. Secondly, since adhesion may not be required for germination, germ tube elongation, or saprophytic growth, adhesion might be inhibited without using the broad-spectrum, eukaryotic metabolic inhibitors characteristic of agricultural fungicides. Thus, one might be able to develop a compound which interfered with adhesion, without injuring the host plant, agricultural workers, or the environment. In addition, synthetic or natural preparations of fungal adhesive compounds may provide useful biocompatible and biodegradable glues for a variety of applications in which adhesion in an aqueous environment is required. Since it is often difficult to obtain good coverage of a crop with either agricultural chemicals or fungi used for biological control, amendment of spray solutions with an adhesive may result in better crop protection with a lower application rate. Finally, we speculate that a greater recognition of the importance of adhesion in the infection court might assist biocontrol specialists in mycoherbicide development and plant pathologists in establishing epiphytotics.

VII. Conclusions

Many species of fungi, including plant and animal pathogens, produce sticky spores and hyphae. Adhesion of phytopathogenic fungi to exterior plant surfaces is often a nonspecific process that involves a glycoprotein glue that spreads, and then forms a water-insoluble, underwater bond. Specific adhesin-receptor interactions also occur among phytopathogenic fungi, but seem less common than nonspecific glues. Adhesion has been studied microscopically in several fungi including *Phytophthora cinnamomi*, *Magnaporthe grisea*, *Glomerella graminicola*, *Nectria haematococca*, and *Uromyces appendiculatus*. In this chapter, we summarized the literature on adhesion of these taxonomically diverse fungi, presented limited experimental evidence on **why** fungi adhere, i.e., the multiple functions of adhesion in pathogenesis and morphogenesis, and reviewed the experimental approaches that can further our present knowledge. Much additional biochemical and genetic work is required to determine **how** fungi adhere: (1) the particular macromolecules that mediate adhesion, i.e., the glues, adhesins, and receptors; and (2) the processes involved in exocellular-substratum attachment, including extracellular modifications such as polymerization that are required for fungal glues "to set." Knowledge gained by such studies may lead to innovative methods of plant disease control.

Acknowledgments. This work was supported in part by grants to L. Epstein from the USDA NRI Competitive Grants Program (#93-00677) and the Office of Naval Research (N00014-95-1-0044), to both authors from the DOE/NSF/USDA multiinstitutional research coordination group (#92-37310-7821) and to R.L. Nicholson from the National Science Foundation (#IBN 9105943). We thank our colleagues T. Caesar-TonThat, H. Kunoh, Y. Kwon, and A. Schuerger for photographs.

References

Andrews JH, Harris RF, Spear RN, Lau GW, Nordheim EV (1994) Morphogenesis and adhesion of *Aureobasidium pullulans*. Can J Microbiol 40:6–17

Bailey JA, O'Connell RJ, Pring RJ, Nash C (1992) Infection strategies of *Colletotrichum* species. In: Bailey JA, Jeger MJ (eds) *Colletotrichum*: biology, pathology and control. CAB International, Wallingford, pp 88–120

Barki M, Koltin Y, Yanko M, Tamarkin A, Rosenberg M (1993) Isolation of a *Candida albicans* DNA sequence conferring adhesion and aggregation on *Saccharomyces cerevisiae*. J Bacteriol 175: 5683–5689

Bartnicki-Garcia S, Sing VO (1987) Adhesion of zoospores of *Phytophthora* to solid surfaces. In: Fuller MS, Jaworski A (eds) Zoosporic fungi in teaching and research. Southeastern Publ Corp, Athens, Georgia, pp 279–283

Bradway SD, Levine MJ (1993) Do proline-rich proteins modulate a transglutaminase catalyzed mechanism of candidal adhesion? Crit Rev Oral Biol Med 4:293–299

Beckett A, Tatnell JA, Taylor N (1990) Adhesion and pre-invasion behaviour of urediniospores of *Uromyces viciae-fabae* during germination on host and synthetic surfaces. Mycol Res 94:865–875

Bonfante-Fasolo P, Spanu P (1992) Pathogenic and endomycorrhizal associations. Methods Microbiol 24:141–168

Boucias DG, Pendland JC (1991) Attachment of mycopathogens to cuticle: the initial event of mycoses in arthropod hosts. In: Cole GT, Hoch HC (eds) The fungal spore and disease initiation in plants and animals. Plenum Press, New York, pp 101–127

Braun EJ, Howard RJ (1994) Adhesion of *Cochliobolus heterostrophus* conidia and germlings to leaves and artificial surfaces. Exp Mycol 18:211–220

Caesar-TonThat T-C, Epstein L (1991) Adhesion-reduced mutants and the wild type *Nectria haematococca*: an

ultrastructural comparison of the macroconidial walls. Exp Mycol 15:193–205
Calderone RA, Braun PC (1991) Adherence and receptor relationships of *Candida albicans*. Microbiol Rev 55:1–20
Calderone RA, Wadsworth E (1988) Characterization of mannoproteins from a virulent *Candida albicans* strain and its derived, avirulent strain. Rev Infect Dis 10: S423–S427
Chaubal R, Wilmot VA, Wynn WK (1991) Visualization, adhesiveness, and cytochemistry of the extracellular matrix produced by urediniospore germ tubes of *Puccinia sorghi*. Can J Bot 69:2044–2054
Clement JA, Martin SG, Porter R, Butt TM, Beckett A (1993) Germination and the role of extracellular matrix in adhesion of urediniospores of *Uromyces viciae-fabae* to synthetic surfaces. Mycol Res 97:585–593
Deacon JW, Donaldson SP (1993) Molecular recognition in the homing responses of zoosporic fungi, with special reference to *Pythium* and *Phytophthora*. Mycol Res 97:1153–1171
Deising H, Nicholson RL, Haug M, Howard RJ, Mendgen K (1992) Adhesion pad formation and the involvement of cutinase and esterases in the attachment of uredospores to the host cuticle. Plant Cell 4:1101–1111
Doss RP, Potter SW, Chastagner GA, Christian JK (1993) Adhesion of nongerminated *Botrytis cinerea* conidia to several substrata. Appl Environ Microbiol 59:1786–1791
Doss RP, Potter SW, Soeldner AH, Christian JK, Fukunaga LE (1995) Adhesion of germlings of *Botrytis cinerea*. Appl Environ Microbiol 61:260–265
Durso L, Lehnen LP Jr, Powell MJ (1993) Characteristics of extracellular adhesions produced during *Saprolegnia ferax* secondary zoospore encystment and cystospore germination. Mycologia 85:744–755
Eagland D (1988) Adhesion and adhesive performance: the scientific background. Endeavour New Ser 12:179–184
Epstein L, Laccetti L, Staples RC, Hoch HC, Hoose WA (1985) Extracellular proteins associated with induction of differentiation in bean rust uredospore germlings. Phytopathology 75:1073–1076
Epstein L, Laccetti L, Staples RC, Hoch HC (1987) Cell-substratum adhesive proteins involved in surface contact responses of the bean rust fungus. Physiol Mol Plant Pathol 30:373–388
Epstein L, Kaur S, Goins T, Kwon YH, Henson JM (1994a) Production of hyphopodia by wild type and three transformants of *Gaeumannomyces graminis* var. *graminis*. Mycologia 86:72–81
Epstein L, Kwon YH, Almond DE, Schached LM, Jones MJ (1994b) Genetic and biochemical characterization of *Nectria haematococca* strains with adhesive and adhesion-reduced macroconidia. Appl Environ Microbiol 60:524–530
Estrada-Garcia MT, Callow JA, Green JR (1990) Monoclonal antibodies to the adhesive cell coat secreted by *Pythium aphanidermatum* zoospores recognize 200 × 10^3 M_r glycoproteins stored within large peripheral vesicles. J Cell Sci 95:199–206
Förster H, Coffey MD, Elwood H, Sogin M (1990) Sequence analysis of the small subunit ribosomal RNAs of three zoosporic fungi and implications for fungal evolution. Mycologia 82:306–312
Gubler F, Hardham AR (1988) Secretion of adhesive material during encystment of *Phytophthora cinnamomi* zoospores, characterized by immunogold labelling with monoclonal antibodies to components of peripheral vesicles. J Cell Sci 90:225–235
Gubler F, Hardham AR (1990) Protein storage in large peripheral vesicles in *Phytophthora* zoospores and its breakdown after cyst germination. Exp Mycol 14:393–404
Gubler F, Hardham AR, Duniec J (1989) Characterizing adhesiveness of *Phytophthora cinnamomi* zoospores during encystment. Protoplasma 149:24–30
Hamer JE, Howard RJ, Chumley FG, Valent B (1988) A mechanism for surface attachment in spores of a pathogenic fungus. Science 239:288–290
Hamer JE, Valent B, Chumley FG (1989) Mutations at the *SMO* genetic locus affect the shape of diverse cell types in the rice blast fungus. Genetics 122:351–361
Hardham AR (1992) Cell biology of pathogenesis. Annu Rev Plant Physiol Plant Mol Biol 43:491–526
Hardham AR, Suzaki E (1986) Encystment of zoospores of the fungus, *Phytophthora cinnamoni*, is induced by specific lectin and monoclonal antibody binding to the cell surface. Protoplasma 133:165–173
Hinch JM, Clarke AE (1980) Adhesion of fungal zoospores to root surfaces is mediated by carbohydrate determinants of the root slime. Physiol Plant Pathol 16: 303–307
Hoch HC, Staples RC, Whitehead B, Comeau J, Wolf ED (1987) Signaling for growth orientation and cell differentiation by surface topography in *Uromyces*. Science 235:1659–1662
Hohl HR, Balsiger S (1988) Surface glycosyl receptors of *Phytophthora megasperma* f.sp. *glycinea* and its soybean host. Bot Helv 98:271–277
Howard RJ, Ferrari MA (1989) Role of melanin in appressorium function. Exp Mycol 13:403–418
Howard RJ, Bourett TM, Ferrari MA (1991a) Infection by *Magnaporthe*: an *in vitro* analysis. In: Mendgen K, Lesemann D-E (eds) Electron microscopy of plant pathogens. Springer, Berlin Heidelberg New York, pp 251–264
Howard RJ, Ferrari MA, Roach DH, Money DP (1991b) Penetration of hard substrates by a fungus employing enormous turgor pressures. Proc Natl Acad Sci USA 88:11281–11284
Hyde KD, Greenwood R, Jones EBG (1993) Spore attachment in marine fungi. Mycol Res 97:9–14
Inoue S, Kato T, Jordan VWL, Brent KJ (1987) Inhibition of appressorial adhesion of *Pyricularia oryzae* to barley leaves by fungicides. Pestic Sci 19:145–152
Irving HR, Grant BR (1984) The effects of pectin and plant root surface carbohydrates on encystment and development of *Phytophthora cinnamomi* zoospores. J Gen Microbiol 130:1015–1018
Jones EBG (1994) Fungal adhesion. Mycol Res 98:961–982
Jones MJ, Epstein L (1989) Adhesion of *Nectria haematococca* macroconidia. Physiol Mol Plant Pathol 35:453–461
Jones MJ, Epstein L (1990) Adhesion of macroconidia to the plant surface and virulence of *Nectria haematococca*. Appl Environ Microbiol 56:3772–3778
Jones SC, Briedis DM (1992) Adhesion and lignin peroxidase production by the white-rot fungus *Phanerochaete chrysosporium* in a rotating biological contractor. J Biotechnol 24:277–290

Knox RB (1984) Pollen-pistil interactions. In: Linskens HF, Heslop-Harrison J (eds) Encyclopedia of plant physiology, new series, vol 17. Cellular interactions. Springer, Berlin Heidelberg New York, pp 508–608

Kwon YH, Epstein L (1992) Biotin-labeling identifies glycoproteins in cell walls of *Nectria haematococca* macroconidia. Phytopathology 82:1127 (Abstr)

Kwon YH, Epstein L (1993) A 90kDa glycoprotein associated with adhesion of *Nectria haematococca* macroconidia to substrata. Mol Plant-Microbe Interact 6:481–487

Lapp MS, Skoropad WP (1978) Nature of adhesive material of *Colletotrichum graminicola* appressoria. Trans Br Mycol Soc 70:221–223

Leatham GF, King V, Stahmann MA (1980) *In vitro* protein polymerization by quinones or free radicals generated by plant or fungal oxidative enzymes. Phytopathology 70:1134–1140

Leite B, Nicholson RN (1992) Mycosporine-alanine: a self-inhibitor from the conidial mucilage of *Colletotrichum graminicola*. Exp Mycol 16:76–86

Leite B, Nicholson RN (1993) A volatile self-inhibitor from *Colletotrichum graminicola*. Mycologia 85:945–951

Lipke PN, Kurjan J (1992) Sexual agglutination in budding yeasts: structure, function, and regulation of adhesion glycoproteins. Microbiol Rev 56:180–194

Lisanti MP, Le Bivic A, Sargiacomo M, Rodriguez-Boulan E (1989) Steady-state distribution and biogenesis of endogenous Madin-Darby canine kidney glycoproteins: evidence for intracellular sorting and polarized cell surface delivery. J Cell Biol 109:2117–2127

Manocha MS, Chen Y (1990) Specificity of attachment of fungal parasites to their hosts. Can J Microbiol 36:69–76

Manocha MS, Chen Y, Rao N (1990) Involvement of cell surface sugars in recognition, attachment, and appressorium formation by a mycoparasite. Can J Microbiol 36:771–778

McCourtie J, Douglas LJ (1985) Extracellular polymer of *Candida albicans*: isolation, analysis and role in adhesion. J Gen Microbiol 131:495–503

Mercure EW, Leite B, Nicholson RL (1994a) Adhesion of ungerminated conidia of *Colletotrichum graminicola* to artificial hydrophobic surfaces. Physiol Mol Plant Pathol 45:421–440

Mercure EW, Kunoh H, Nicholson RL (1994b) Adhesion of *Colletotrichum graminicola* to corn leaves: a requirement for disease development. Physiol Mol Plant Pathol 45:407–420

Mercure EW, Kunoh H, Nicholson RL (1995) Visualization of materials released from adhered, ungerminated conidia of *Colletotrichum graminicola*. Physiol Mol Plant Pathol 46:121–135

Mims CW, Richardson EA (1989) Ultrastructure of appressorium development by basidiospore germlings of the rust fungus *Gymnosporangium juniperi-virginianae*. Protoplasma 148:111–119

Mims CW, Richardson EA, Clay RP, Nicholson RL (1995) Ultrastructure of conidia and the conidium aging process in the plant pathogenic fungus *Colletotrichum graminicola*. Int J Plant Sci 156:9–18

Moloshok TD, Leinhos GME, Staples RC, Hoch HC (1993) The autogenic extracellular environment of *Uromyces appendiculatus* urediospore germlings. Mycologia 85:392–400

Muirhead IF, Deverall BJ (1981) Role of appressoria in latent infection of banana fruits by *Colletotrichum musae*. Physiol Plant Pathol 19:77–84

Nichols J, Beckman JM, Hadwiger LA (1980) Glycosidic-enzyme activity in pea tissue and pea-*Fusarium solani* interactions. Plant Physiol 66:199–204

Nicholson RL (1984) Adhesion of fungi to the plant cuticle. In: Roberts DW, Aist JW (eds) Infection processes of fungi. The Rockefeller Foundation, New York, pp 74–89

Nicholson RL, Epstein L (1991) Adhesion of fungi to the plant surface: Prerequisite for pathogenesis. In: Cole GT, Hoch HC (eds) The fungal spore and disease initiation in plants and animals. Plenum, New York, pp 3–23

Nicholson RL, Moraes WBC (1980) Survival of *Colletotrichum graminicola*: importance of the spore matrix. Phytopathology 70:255–261

Nicholson RL, Butler LG, Asquith TN (1986) Glycoproteins from *Colletotrichum graminicola* that bind phenols: implications for survival and virulence of phytopathogenic fungi. Phytopathology 76:1315–1318

Nicholson RL, Yoshioka H, Yamaoka N, Kunoh H (1988) Preparation of the infection court by *Erysiphe graminis*. II. Release of esterase enzyme from conidia in response to a contact stimulus. Exp Mycol 12:336–349

Nicholson RL, Hipskind J, Hanau RM (1989) Protection against phenol toxicity by the spore mucilage of *Colletotrichum graminicola*, an aid to secondary spread. Physiol Mol Plant Pathol 35:243–252

Nicholson RL, Kunoh H, Shiraishi T, Yamada T (1993) Initiation of the infection process by *Erysiphe graminis*: conversion of the conidial surface from hydrophobicity to hydrophilicity and influence of the conidial exudate on the hydrophobicity of the barley leaf surface. Physiol Mol Plant Pathol 43:307–318

Northcote DH, Gould J (1989) The mucilage secreted by roots and its possible role in cell-cell recognition for the adhesion of fungal pathogens to root surfaces of *Zea mays* L. In: Chantler E, Ratcliffe NA (eds) Mucus and related topics. Society for Experimental Biology, Cambridge, pp 429–447

Pascholati SF, Yoshioka H, Kunoh H, Nicholson RL (1992) Preparation of the infection court by *Erysiphe graminis* f. sp. *hordei*: cutinase is a component of the conidial exudate. Physiol Mol Plant Pathol 41:53–59

Pascholati SF, Deising H, Leite B, Anderson D, Nicholson RL (1993) Cutinase and non-specific esterase activities in the conidial mucilage of *Colletotrichum graminicola*. Physiol Mol Plant Pathol 42:37–51

Pringle RB (1981) Nonspecific adhesion of *Bipolaris sorokiniana* sporelings. Can J Plant Pathol 3:9–11

Ramadoss CS, Uhlig J, Carlson DM, Butler LG, Nicholson RL (1985) Composition of the mucilaginous spore matrix of *Colletotrichum graminicola*, a pathogen of corn, sorghum, and other grasses. J Agric Food Chem 33:728–732

Read SJ, Moss ST, Jones EBG (1992) Germination and development of attachment structures by conidia of aquatic Hyphomycetes: a scanning electron microscope study. Can J Bot 70:838–845

Romantschuk M (1992) Attachment of plant pathogenic bacteria to plant surfaces. Annu Rev Phytopathol 30:225–243

Ruiz-Herrera J, Iranzo M, Elorza MV, Sentandreu R, Mormeneo S (1995) Involvement of transglutaminase in the formation of covalent cross-links in the cell wall of *Candida albicans*. Arch Microbiol 164:186–193

Saxena A, McElhaney-Feser GE, Cihlar RL (1990) Mannan composition of the hyphal form of two relatively avirulent mutants of *Candida albicans*. Infect Immun 58:2061–2066

Schuerger AC, Mitchell DJ (1992) Effects of temperature and hydrogen ion concentration on attachment of macroconidia of *Fusarium solani* f. sp. *phaseoli* to mung bean roots in hydroponic solution. Phytopathology 82: 1311–1319

Schuerger AC, Mitchell DJ (1993) Effects of mucilage secreted by *Fusarium solani* f. sp. *phaseoli* on spore attachment to roots of *Vigna radiata* in hydroponic nutrient solution. Phytopathology 83:1162–1170

Schuerger AC, Mitchell DJ, Kaplan DT (1993) Influence of carbon source on attachment and germination of macroconidia of *Fusarium solani* f. sp. *phaseoli* on roots of *Vigna radiata* grown in hydroponic nutrient solution. Phytopathology 83:1162–1170

Sela-Buurlage MB, Epstein L, Rodriguez RJ (1991) Adhesion of ungerminated *Colletotrichum musae* conidia. Physiol Mol Plant Pathol 39:345–352

Smit G, Logman TJJ, Boerrigter METI, Kijne JW, Lugtenberg BJJ (1989) Purification and partial characterization of the *Rhizobium leguminosarum* biovar. *viciae* Ca^{2+}-dependent adhesin, which mediates the first step in attachment of cells of the family Rhizobiaceae to plant root hair tips. J Bacteriol 171:4054–4062

Smith MM, Cruickshank IAM (1987) Dynamics of conidial adhesion and elicitor uptake in relation to pisatin accumulation in aqueous droplets on the endocarp of pea pods. Physiol Mol Plant Pathol 31:315–324

Templeton MD, Rikkerink EHA, Beever RE (1994) Small, cysteine-rich proteins and recognition in fungal-plant interactions. Mol Plant Microbe Interact 7:320–325

Terhune BT, Hoch HC (1993) Substrate hydrophobicity and adhesion of *Uromyces* urediospores and germlings. Exp Mycol 17:241–252

Toti L, Viret O, Chapela IH, Petrini O (1992) Differential attachment by conidia of the endophyte *Discula umbrinella* (Berk. & Br.) Morelet to host and nonhost surfaces. New Phytol 121:469–475

Tunlid A, Nivens DE, Jansson H-B, White DC (1991) Infrared monitoring of the adhesion of *Catenaria anguillulae* zoospores to solid surfaces. Exp Mycol 15:206–214

Tunlid A, Jansson HB, Nordbring-Hertz B (1992) Fungal attachment to nematodes. Mycol Res 96:401–412

Vreeland V, Epstein L (1996) Analysis of plant-substratum adhesives. In: Jackson JF, Linskens H-F (eds) Modern methods of plant analysis, vol 17. Plant cell-wall analysis. Springer, Berlin Heidelberg New York, pp 95–116

Wessels JGH (1994) Developmental regulation of fungal cell wall formation. Annu Rev Phytopathol 32:413–437

Wisniewski M, Biles C, Droby S, McLaughlin R, Wilson C, Chalutz E (1991) Mode of action of the postharvest biocontrol yeast, *Pichia guilliermondii*. I. Characterization of attachment to *Botrytis cinerea*. Physiol Mol Plant Pathol 39:245–258

Wynn WK (1981) Tropic and taxic responses of pathogens to plants. Annu Rev Phytopathol 19:237–255

Xiao J, Ohshima A, Kamakura T, Ishiyama T, Yamaguchi I (1994) Extracellular glycoprotein(s) associated with cellular differentiation in *Magnaporthe grisea*. Mol Plant Microb Interact 7:639–644

2 Physical and Chemical Cues for Spore Germination and Appressorium Formation by Fungal Pathogens

R.C. STAPLES and H.C. HOCH

CONTENTS

I. Introduction

Fungal plant pathogens penetrate their hosts directly or through wounds, and of those that penetrate directly, a majority use an appressorium to provide a locus of force and enzymatic activity to aid the process. These **infection structures** usually are developed by the spore germling on the tip of the germ tube; however, some fungi also produce **infection cushions** at the tip of mycelia, i.e., *Saprolegnia* spp., where they arise on hyphal apices attracted to the substrata by chemotropism (Willoughby and Hasenjager 1987). The mystery is how fungi decide where to develop infection structures, and how they detect the place for penetration. In fact, selection, often random, may be influenced by the presence of nutrients and a compatible surface. A good review of the process as it was known before 1975 is that by Emmett and Parbery (1975).

Much of the earlier research on the appressoria of facultative parasites was influenced by the extensive work on the rusts, obligate pathogens of higher plants, whose urediospore germlings in some species enter via the stomata; however, recent studies now suggest that each fungal pathogen has probably adapted to the unique defenses of its own host. In this chapter, we will review the preinfection responses that germlings of a broad range of fungi make in response to physical and chemical stimuli. The infection structures that we include develop on the germ tube of spore germlings; we exclude hyphal infection cushions because so little is known regarding them (Armentrout and Downer 1987; Armentrout et al. 1987). As spore germination has a role in preparing the germ tube for appressorium development (Suzuki et al. 1981, 1982), we begin with a discussion of spore germination, in order to incorporate what is known of the physical and chemical cues which guide initiation of the infection structures. Zoospores, some of which are stomatal penetrators, are not included.

II. Spore Germination

A. Regulation of Urediospore Germination

1. Self-Inhibitors

Spores of many fungi germinate poorly in dense suspensions or if crowded on a surface. The most

Department of Plant Pathology, New York State Agricultural Experiment Station, Cornell University, Geneva, New York 14456, USA

The Mycota V Part A
Plant Relationships
Carroll/Tudzynski (Eds.)

detailed studies of this phenomenon of **self-inhibition** before the 1990s were carried out with the rust fungi, beginning with studies reported by Yarwood (Yarwood 1954), who demonstrated that urediospores of *Uromyces appendiculatus* seeded heavily on agar germinated poorly compared with spores seeded at a lower density. The first demonstration that a chemical was involved was made by P.J. Allen (Allen 1955), who showed that water extracts of spores of *Puccinia graminis* f. sp. *tritici* contain an inhibitory principle responsible for the crowding effect. The inhibition of germination was reversible, and he coined the term **self-inhibitor**. The inhibitors apparently serve the role of suppressing germination while the spores are still in the pustule (French 1992). The self-inhibitor of bean, sunflower, corn, snapdragon, peanut, and stripe rust urediospores is methyl *cis*-3,4-dimethoxycinnamate; that for *P. graminis* f. sp. *tritici* is methyl *cis*-ferulate (Macko 1981). Three apparently closely related self-inhibitors were isolated from aeciospores of *Cronartium fusiforme* (Foudin et al. 1977), but they were not identified chemically.

The mode of action of the self-inhibitors has never been resolved satisfactorily; however, Hess et al. (1975) showed that methyl *cis*-ferulate prevents digestion of the pore plug at concentrations similar to those required to inhibit germination (10 to 100 nmol). Removal of the self-inhibitor allows digestion to proceed, a process interrupted if the inhibitor is reintroduced before digestion is completed. The authors suggest that methyl *cis*-ferulate inhibits an enzyme required for digestion of the plug, a process insensitive to inhibitors of RNA and protein synthesis, and that the enzyme is preformed and held in an inactive state by presence of the endogenous self-inhibitor.

2. Germination Stimulators

French and Weintraub discovered germination stimulators (French and Weintraub 1957), and French (1992) has recently reviewed the volatile aroma and flavor compounds associated with spices, condiments, herbs, and perfumes that will stimulate germination of spores from a wide range of fungi. The compounds vary from simple ones like nonanal (Rines et al. 1974) to the more complex 2,3-dimethyl-1-pentene. Small amounts of the volatile compounds placed on filter paper suspended near rusted plants in a dew chamber will cause the spores to germinate in the pustule, overcoming the self-inhibitor. While little is known about their modes of action, stimulants appear to overcome the self-inhibition that prevails in dense populations of spores without reacting with the self-inhibitor molecule (Macko et al. 1976).

Germination stimulators probably are most useful for improving and coordinating the germination performance of a large mass of spores. Thus, if one wants a gram of urediospores of *U. appendiculatus* to germinate and begin to grow in a liter of water in a coordinate manner (as is often required for biochemical work), they can be germinated in water containing a trace of β-ionone (Moloshok et al. 1993).

3. Light

Rust fungi vary considerably in sensitivity to light. Germination of urediospores of *P. graminis* f. sp. *tritici* is inhibited by continuous irradiation (Givan and Bromfield 1964a,b). Light was found to affect germination prior to emergence of the germ tube. Because this stage coincides with pore plug dissolution, an event reversibly inhibited by the native self-inhibitor, methyl *cis*-ferulate, there may be a link between the self-inhibitor and the photocontrol mechanism. Washing, which removes the self-inhibitor, also relieves photoinhibition (Knights and Lucas 1980).

The mechanism by which light affects urediospore germination has not been elucidated, but both blue and far red light have phototrophic effects (Calpouzos and Chang 1971). Germination of urediospores of *P. graminis tritici* is inhibited by white light; however, the inhibition lasts for only about 6 to 8 h. In a study using interference filters with bandwidths of either 4 nm (filters in the 400–699-nm range) or 10 nm (filters in the 720–750-nm range), Calpouzos and Chang (1971) showed that germination was inhibited in two spectral regions: at 420 nm (blue) and 720 nm (far red). The experiments were controlled for spurious effects of heat. Germination of *U. appendiculatus* is also sensitive to intense white light as used in microscopy studies, and, for that reason, filtered light (546 ± 2 nm) is often used (Corrêa and Hoch 1993).

4. Storage Temperature

Probably the most useful response of urediospores to temperature is the phenomenon called cold dormancy (Bromfield 1964). Air-dry urediospores of *P. graminis* f. sp. *tritici* can be stored in liquid nitrogen indefinitely, thus allowing spores to be raised in bulk and used for several years. This is a great aid to programs in plant breeding and molecular genetics alike. While freezing the spores markedly reduces their apparent viability, germination is readily restored by a mild heat shock (i.e., raising the temperature to about 37 °C for 20 min), before the spores are germinated (Bromfield 1964).

5. Ions

Ions in air and water can influence urediospore germination. For example, high concentrations of heavy metal ions, pollutants in air and water, can inhibit the germination of urediospores of *P. striiformis* (Sharp 1967). Germlings are most sensitive during the first hours of germination. Combustion nuclei from automobile exhaust can be an important source of air-borne heavy-metal ions, especially lead. However, ions can be beneficial as well – urediospores of *U. appendiculatus* germinate better in unpurified tap water than in ultrapure laboratory water (Baker et al. 1987). Of the four major cations in the water, calcium, at concentrations between 0.1 and 3 mM, was found to have a stimulatory effect proportional to the amount of cation present. Magnesium had a lesser effect, while the monovalent cations had little stimulatory effect on germination.

B. Germination of Sexual Spores of Rust Fungi

1. Teliospores

Dormancy, an asset that helps carry the fungus over periods unfavorable for growth, is an inherent property of teliospores (i.e., the spore requires an activation process), and is not a consequence of the presence of inhibitory environmental factors (Allen 1965). As reported by Anikster (1986), teliospores remain viable for more than 14 years when stored at 5 °C under dry conditions in partial vacuum. Even so, storage is not a prerequisite for germination; in this sense, teliospore dormancy differs from seed dormancy. Probably the most comprehensive recent review of teliospores is that by Mendgen (1984; see also Chap. 5, Vol. V, Part B). Centered on *P. graminis* f. sp. *tritici*, he includes results from other rusts, and has cited many earlier reviews. Very little is known about the regulation of germination and germling development of the sexual spores of rusts; however, protocols that describe how these spores may be germinated suggest that some controls are involved. We now review the more representative of these by way of illustration.

A general procedure for germinating the teliospores of a wide range of rusts, except *U. appendiculatus*, was developed by Anikster (1986). Spores are floated on water at 16–18 °C for 24–144 h, the teliospores transferred to 2% water agar, and germinated at 16 °C. Teliospores of *P. xanthii*, a microcyclic autoecious rust lacking pycnial, aecial, and uredial stages, germinate within 30 min after hydration, to produce a four-celled metabasidium (Morin et al. 1992). A basidiospore is formed at the tip of a conical sterigma which arises from each cell. A specific procedure for germinating teliospores of *U. appendiculatus* was developed by Groth and Mogen (1978). The procedure differs from that developed by Gold and Mendgen (1983a,b), in which light-to-dark alternation was necessary for teliospore germination, a difference which apparently relates to the diverse races used by the two groups of researchers. The common underlying mechanism of dormancy may be related to water uptake, and an assortment of techniques may be required to weaken the spore coat. Some teliospores, like those of *P. xanthii*, may be fully permeable.

2. Basidiospores

A single, polar germ tube develops from a germinating basidiospore (Gold and Mendgen 1984). The germ tubes tend to grow toward or along the junction lines between host epidermal cells. A mucilaginous exudate is associated with the germ tubes along the lines of contact between host and fungus. While appressoria develop near the junction lines, penetration of the host occurs directly through the epidermis, and thigmotropic signals may not be required for appressorium development.

3. Pycniospores

Pycnia of *U. appendiculatus* begin to form only on the upper surfaces of bean (*Phaseolus vulgaris*) leaves about 10 days after infection by basidiospores (Groth and Mogen 1978). Aecia form directly beneath the pycnia after fertilization if the two opposite mating types of **pycniospores** are present, and infective dikaryotic **aeciospores** arise in the aecium. Thus, the pycniospores are unique in having the ability to recognize the opposite mating type and to initiate sexual fusions which result in dikaryons.

4. Aeciospores

It is well known that aeciospores of some stomatal-penetrating rust species (e.g., *P. graminis* f. sp. *tritici*) will infect only the alternate host, which suggests that they have the ability to discriminate between hosts. In addition, aeciospores of *U. vignae* develop appressoria when germinated on scratched polyethylene or polystyrene membranes with defined ridges of 0.3 μm height (Stark-Urnau and Mendgen 1993). Thus, dikaryotic spores may have thigmotropic receptors which aid in host discrimination.

C. Regulation of Conidial Germination

1. Self-Inhibitors

Numerous reports of the self-inhibition phenomenon have been summarized in reviews published since 1958, in addition to those reviewed above under the rusts (Allen 1965; Cochrane 1958; Staples and Wynn 1965; Sussman and Halvorson 1966). Several factors apparently cause the "crowding effect", but in many cases it has been demonstrated that "self-inhibitors" (Allen 1955) washed from the spores are a primary cause in propagules of *Erysiphe graminis* (Domsch 1954), *Glomerella cingulata* (Lingappa et al. 1973), and *Rhynchosporium secalis* (Ayres and Owen 1970). One self-inhibitor is a compound extracted from *Colletotrichum gloeosporioides*. Given the trivial name **gloeosporone** by Lax et al. (1985), it was synthesized by Seebach et al. (1989), who showed that the (−)-enantiomer inhibited germination of conidia with an ED_{50} of 4 μg/ml. Three additional compounds were synthesized and shown to be physiologically important as germination inhibitors in *C. gloeosporioides* f. sp. *jussiaea* by Tsurushima et al. (1995). One of these [(E)-3-ethylidene-1,3-dihydroindol-2-one (CG-SI1)] reduced germination of conidia with an ED_{50} of 3 μg/ml.

Self-inhibitors from conidia of the anthracnose fungi seem to be located externally to the spore. For example, **mycosporine-alanine**, identified as a self-inhibitor from *C. graminicola*, was shown to be a component of the conidial mucilage; there was no evidence for its presence within the conidia (Leite and Nicholson 1992). Mycosporine-alanine is synthesized during sporogenesis (Leite and Nicholson 1992). If these self-inhibitors do not reside within the conidium, their modes of action may not involve inhibition of hydrolysis of the spore plug; indeed, Leite and Nicholson (1992) suggest that they act by suppression of metabolic activity until conidial dissemination occurs.

Another departure from the model provided by urediospores (Sect. II.A.1 above) is **chloromonilicin**, a growth self-inhibitor isolated from the culture broth of the cherry rot fungus, *Monilinia fructicola* (Horiguchi et al. 1989). The compound, with an unusual β-chloro-α, β, γ, δ-unsaturated ε-lactone ring, has high antifungal activity. That parasitic fungi can secrete such self-toxic compounds may be of great importance in understanding the environment in which wound parasites must grow.

2. Surface Components

The surface wax of white clover (*Trifolium repens*) stimulates conidia of *Cymadothea trifolii* to infect through the stomata (Roderick 1993), and spores fail to germinate and infect the abaxial surface, possibly due to a difference in the structure or composition of the leaf waxes. Among insect pathogens, it has been found that conidia of *Metarhizium anisopliae* isolated from coleopteran hosts are resistant to 2-deoxyglucose and germinate poorly on glucose, while isolates from homopteran and lepidopteran hosts germinate well on glucose but are inhibited by 2-deoxyglucose (St. Leger et al. 1994). Thus, conidia respond to components of various types found on the surfaces of their hosts, and selectable strain variations in response to these may play a role in development of pathogen attack strategies.

3. Physical Cues

Pycnidiospores of *Phyllosticta ampelicida*, the causal agent of black rot of grape, exhibit a unique requirement for germination. They germinate only on substrata on which they are firmly attached. Furthermore, conidia of the fungus attach best to hydrophobic substrata, e.g., grape leaf, polystyrene, Teflon, polycarbonate, collodion, and glass treated with the silanes noctadecyltrichlorosilane, dimethyldichlorosilane, or diphenyldichlorosilane (Kuo and Hoch 1996). When pycnidiospores were deposited on more wettable surfaces, e.g., glass, cellophane, nutrient- and water-agars, polystyrene treated with UV-irradiation or sulfuric acid, and glass silanized with n-2-aminoethyl-3-aminopropyltrimethoxysilane, n-(trimethoxysilylpropyl)ethylenediamine triacetic acid trisodium, or 3-aminopropyltriethoxysilane, they did not attach firmly and did not germinate. The attachment process is passive since pycnidiospores treated with sodium azide, formaldehyde, or boiled in water for 10 min adhered as readily as nontreated conidia. The signaling process responsible for initiation of the germination event is not understood, but possibly involves conformational changes of the wall/plasma membrane at the conidium-substratum interface.

D. Spore Eclosion

Spore eclosion is a term, first proposed by Chapela et al. (1990), to describe a host-specific recognition mechanism which consists of a series of fast movements that precede germ tube formation by ascospores of *Hypoxylon fragiforme*, an endophyte of beech (*Fagus sylvatica*). The fast movements lead to the emergence of a bivalved, flexible structure from an outer rigid shell formed by a differentiated transparent wall layer, and result in the exposure of the cell body. Germ tube growth follows, but eclosion without germ tube outgrowth probably does not require protein synthesis because it can occur in the presence of cycloheximide (Chapela et al. 1990). Eclosion represents the initial steps of infection by the ascospores, and is initiated within minutes of contact with the beech tree.

Eclosion occurs in two phases (Chapela et al. 1993; Chap. 8, Vol. V, Part B). First, an activated spore body is released from a rigid exosporium within milliseconds. This apparently results from a permeability change in the wall or plasmalemma, leading to a rapid net water flow into the cell, and causing it to swell and break out of its exosporial shell. In the second phase with time constants of seconds, the released spore body swells, causing the rim of the germ fissure to gape apart to expose an inner, plastic cell wall layer. The second phase of eclosion is driven principally by osmotic pressure.

The biochemical signature responsible for recognition of the beech tree is provided by extracts of beech bark, and by mixtures of monolignols, including the glycosides Z-isoconiferin and Z-syringin. These compounds are released to the plant surface from small wounds in the bark (Chapela et al. 1991). In nature, E-(*trans*) isomers of the glycosides are fixed as polymerized lignin in the plant cell walls, while their Z-(*cis*) counterparts accumulate unpolymerized in the bark (Lewis et al. 1988).

III. Germling Differentiation

Plant pathogens employ different mechanisms to regulate appressorium development. Emmett and Parbery (1975) seem to have been the first to suggest that few fungi require stimuli for appressorium formation, and they assumed the probability that well-defined environmental conditions are required in most species. They proposed "that appressorium formation is primarily controlled by genotype whose expression may require a specific conducive environment". Suggestions have also been made that signals for appressorium formation include a rigid surface (Dey 1933), ridges between cells (Preece et al. 1967), and patterns of wax on the surfaces of leaves (Ellingboe 1972). It was also suggested that substratum porosity may have a key role in triggering appressorium formation (Hoch and Staples 1991). These ideas require a firmer experimental foundation but remain to be disproved.

A. Physical Cues

A conducive environment for many fungi seems to depend on the characteristics of the surface to which the germ tube is attached. In *B. cinerea* (Hill et al. 1980), as for most fungi, the more hydrophobic the surface, the more readily appressoria develop.

1. Hydrophobicity

Many species of anthracnoses often require a stimulus in order to produce the appressorium; some respond to a purely physical stimulus, while others need a chemical stimulus. For example, *C. graminicola*, which responds only to a purely physical stimulus, develops appressoria on natural barley leaf surfaces, and on plastic replicas of the surfaces fabricated from nail polish or epoxy plastic, relatively hydrophobic surfaces to which the germling readily attaches (Lapp and Skoropad 1978). The distribution of appressoria relative to the groove over the anticlinal walls of the epidermal cells was found to be the same on both the natural and artificial surfaces. The host topography controls appressorium formation, and localized host exudations are not necessarily involved.

Adhesion of urediospores and urediospore germlings of *U. appendiculatus* have been evaluated with regard to surface hydrophobicity in terms of the spread of water or "wettability" (Terhune and Hoch 1993). Urediospores are allowed to adhere to a range of glass or quartz substrates coated with various organosilanes, and evaluated after washing the spore- or germling-laden surfaces. Both urediospores and germlings adhere tenaciously to surfaces with wettability ratings less than 30, such as polystyrene and glass treated with tridecafluoro-1,1,2,2-tetrahydrooctyl)-1-trichlorosilane or diphenyldichlorosilane, and the degree of germling contact to the various surfaces correlates closely with both hydrophobicity and adhesion of germlings. Induction of appressoria on quartz substrates bearing inductive topographies (0.5-μm-deep grooves) is also closely associated with the degree of hydrophobicity.

2. Hardness

That surface hardness might be an important property of a substrate for the induction of appressorium development was originally suggested by Dey (1933), and has been supported by the studies of Dickinson (1971, 1974), Van Burgh (1950), Wynn (1976), Freytag et al. (1988), St. Leger et al. (1989a), and ourselves (Staples et al. 1983). In general, while urediospore germ tubes of some direct-penetrating rust fungi will not differentiate on the surface of either water or soft (2–3%) water-agar, appressoria will form on a surface that is hard, including 7% water-agar.

The confusion between hardness and hydrophobicity is seen most clearly in two recent papers on *M. grisea* (anamorph, *Pyricularia oryzae*). Lee and Dean (1994), concluded from a study using a series of silicon wafers (e.g., very hard surfaces) modified to create various degrees of hydrophobicity, that the hydrophobicity of the contact surface is the primary and sufficient determinant to induce appressorium development. In contrast, Xiao et al. (1994), using a hardness tester to evaluate surface hardness, found that hard hydrophilic and hydrophobic surfaces equally induced appressorium formation by *M. grisea*, and that hardness correlated with development of the appressoria and not hydrophobicity, as measured by the spread of a uniform drop of water.

Hardness has rarely been evaluated objectively, and some usage may actually relate to **porosity** of the surface. Hardness would reduce porosity. For example, it is commonly observed that mycelia will grow through agar-containing growth media, eventually attaching to the bottom and sides of the glass petri dish where appressoria develop. A reduced porosity may limit the loss of ions, sugars, and other nutrients form mycelia, and this restriction in nutrient loss (i.e., reduction in membrane flux) could signal appressorium formation via a type of feedback mechanism.

3. Thigmotropism

Fungal responses to contact stimuli are growth changes that occur when the fungus recognizes the various surface features of the host. The responses fall into two groups: **directional changes**, a reorientation of the direction of germ tube outgrowth as reported by Johnson (1934), and **differentiation**, the formation of specialized infection cells as reported by Dickinson (1949, 1971). These **thigmotropic** responses, which result from contact of the germ tube with changes in thickness of the surface, can be detected by the fungus as signals. Dickinson's conclusions about membrane signals, and signal reception in rust fungi, were finally confirmed and extended by Maheshwari et al. (1967a) after great controversy. While these signals have now been defined for several rust fungi (Allen et al. 1991a), it is not known how they are perceived.

a) Pores

Hyphal development in the dimorphic pathogenic fungus, *Candida albicans*, is thought to facilitate the primary invasion of surface epithelia during superficial infections. When mycelia grown on Nucleopore membrane filters are placed over serum-containing agar, the hyphae grow over the membrane surface and through the pores, crossing to the other side of the membrane (Sherwood et al. 1992). Hyphae that do not contact the lip of a pore do not enter it. Sherwood et al. (1992) concluded that this response was due to contact guidance (thigmotropism), including substrate topography, and tropic movement in relation to changes in contour. Chemotropism towards the nutrients seems not to be involved, since hyphae growing on the underside of the membrane also enter the pores then grow away from the underlying nutrient (serum) agar. The behavior may enable the hyphae of this human pathogen to penetrate epithelia at microscopic wound sites, membrane invaginations, and other foci where integrity of the epithelium is weak. Chemotropism has apparently never been observed in any ascomycetous fungus or imperfect fungus, except in response to mating factors by heterothallic hyphae of sexually active species (Sherwood et al. 1992; Gooday 1975).

b) Stomata

Most fungi penetrate their hosts directly with appressoria, for example *M. grisea* (Dobinson and Hamer 1992); while a few penetrate both directly with appressoria or through wounds without appressoria, as does *Botrytis cinerea* (Hancock and Lorbeer 1963). Penetration of the stomata of a host provides a protected point of entry. Appressoria developed by conidial germlings of *Phyllosticta ampelicida* are shown in Fig. 1.

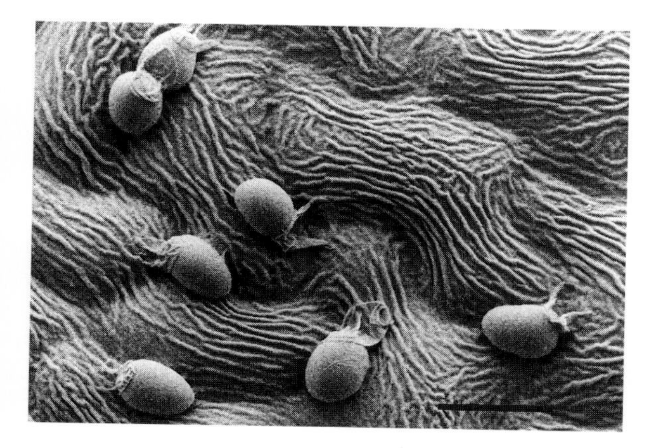

Fig. 1. Appressoria of *Phyllosticta ampelicida*, the causal agent of black rot of grape (*Vitis vinifera*), form on the grape leaf surface predominately over anticlinal walls. The turgid appressoria usually develop immediately upon conidial germination. The conidia in this cryo-SEM preparation are observed as collapsed, cytoplasmically empty structures. *Bar* 15 μm. (Micrograph kindly supplied by Ker-Chung Kuo)

Stomatal penetration by the rust fungi has been known at least since the work of Allen (1923). The demonstration by Wynn in 1975, that appressoria of *U. appendiculatus* germlings form over stomata in response to a surface contact stimulus located in the stomatal guard cell (Wynn 1976), set the stage for our own discovery that the essential signal was a 0.5-μm high ridge (Hoch et al. 1987b).

Urediospore germlings of rust fungi which penetrate their hosts via the stomata (Allen et al. 1991a) develop an **appressorium** over the stomatal opening in response to contact with an inductive ridge (Terhune et al. 1991). The developmental process, **thigmodifferentiation** (Staples et al. 1984a), starts a program of development that leads to the sequential formation of a group of six specialized structures which enable a rust fungus like *U. appendiculatus* to locate a stoma, enter it, colonize its host, and ultimately penetrate the internal leaf mesophyll cells via haustoria for the extraction of nutrients. From our studies, we conclude that the entire process of host infection can be divided into four phases – germination of the spore and attachment of the germ tube to the substrate, mitosis and gene expression within the appressorium, emergence of the penetration peg, and development of the haustorial mother cells. Germ tubes and appressoria of urediospore germlings of *P. graminis* f. sp. *tritici* are shown in Fig. 2.

The parameters for the signal ridge have been measured for a broad range of rust fungi (Allen et al. 1991a). Nine species of rust (*Uromyces appendiculatus*, *U. vignae*, *Melampsora medusae*, *Puccinia antirrhini*, *P. calcitrapae*, *P. carduorum*, *P. melanocephala*, *P. substriata*), and one isolate of *P. recondita* develop appressoria on topographies within an optimal height range of approximately 0.45–0.75 μm (Allen et al. 1991a,b). Appressorium formation is reduced considerably on ridge heights above and below this range. Six species (*P. polysora*, *P. menthae*, *P. hieracii*, *P. sorghi*, *P. arachidis*, and *Physopella zeae*) maintain high levels of appressorium formation on ridges up to 2.25 μm high. Seven rust species (*Coleosporium asterum*, *C. tussilaginis*, *Phragmidium potentillae*, *Puccinia coronata*, *P. graminis* f. sp. *tritici*, *P. graminis* f. sp. *avenae*, and *Tranzschelia discolor*) do not form appressoria on microfabricated topographies of defined heights, but do form very low numbers of appressoria on scratched polystyrene membranes. *Phakopsora pachyrhizi* form

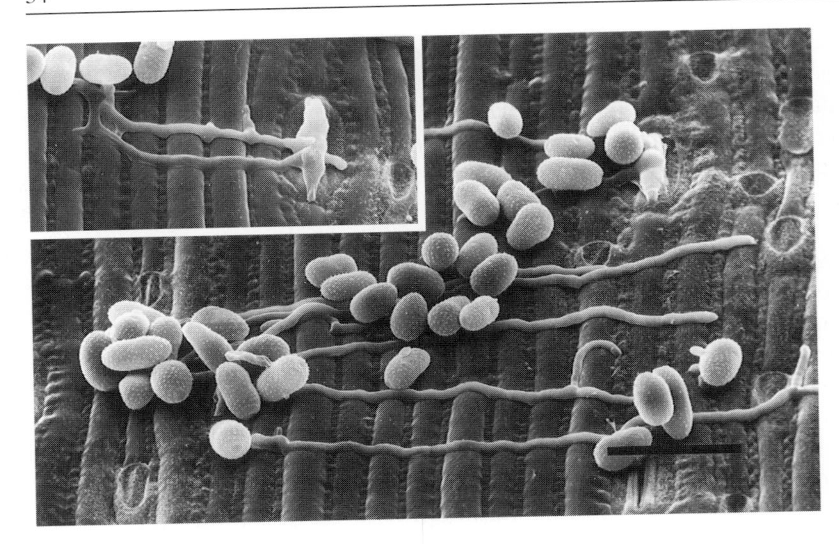

Fig. 2. Urediospore germlings of *Puccinia graminis* f. sp. *tritici*, like most rust urediospore germlings, grow oriented perpendicular to topographical features, the longitudinal ridges and vlleys formed by the epidermal cells of the host leaf. Upon encountering a stoma, the germling forms an appressorium over the stomatal aperture (*inset*). *Bar* 50 μm

appressoria on both smooth membranes and in apparent association with ridges.

Thigmo-sensing of surface topography appears to be unique to rust fungi; furthermore, it is exacting, as the fungal cells must distinguish topographical differences between the 0.5-μm-high signal ridge of the stomatal guard cells and other surface features of the epidermis. An important characteristic of this phase is that a differential expression of genes does not appear to be required to support appressorium development until the end of the first mitosis when DNA replication begins (Kwon and Hoch 1991), and at least four genes (*Inf24*, *Inf56*, *Inf64*, *Inf88*) are newly expressed (Bhairi et al. 1989; Xuei et al. 1992; Bhairi et al. 1990; H.C. Hoch and R.C. Staples, unpubl.). All of the enzymes and substrates needed to construct most of the appressorium appear to be provided during sporogenesis.

Artificial ridges (0.5 μm high and 0.25 μm wide), which mimic cuticular lips of stomatal guard cells of *P. vulgaris*, can be microfabricated of either polystyrene or polycarbonate replicated from silicon plates (Terhune et al. 1993). Such artificial lips induce appressoria, yet become deformed approximately 30 min after initial contact by the germ tube apex, as recorded and observed with time-lapse video light microscopy. The collapsed nature of the ridge suggests that mechanical forces imposed by a combination of cell turgor pressure and adhesion of the appressorium to the substrate are responsible for deformation of the inductive topography. On the stomatal guard cell of the bean leaf, the normally erect stomatal guard cell lips have also been observed to be prostrate at most stages of appressorium development (Mendgen 1973; Terhune et al. 1993).

Infection of white clover (*T. repens*) by conidia of *Cymadothea trifolii* occurs through stomata on the adaxial leaf surface only (Roderick 1993); there was no evidence of cuticular penetration. Spores fail to germinate on the abaxial surface, possibly due to a difference in the composition of leaf waxes. Germ tubes grow preferentially along the epidermal wall junctures, and develop an appressorium upon contact with a stoma. These appressoria then produce a penetration peg which passes through the stoma to form an irregularly shaped vesicle.

c) *Mechanisms of Thigmotropism*

Urediospore germ tubes appear to contain sites that are responsive to topographical induction for appressorium formation. The most responsive region was found to be within 0–10 μm from the cell apex, where more than 90% of germ tubes that were perturbed with a glass micropipette developed appressoria (Corrêa and Hoch 1995). Neither the germ tube surface not in contact with the substratum, nor regions in contact and more than 40 μm from the apex were responsive. Maximum appressorium formation occurred when the perturbing micropipette was left in place for more than 20 min. There also appear to be receptors on the germ tube for peptides having the RGD (Arg-Gly-Asp) sequence, since micropipettes coated with various peptides containing the sequence were no longer inductive (Corrêa et al. 1996).

B. Chemical Cues

Conidia of the rice blast fungus, *M. grisea*, normally differentiate only if germinated on the surface of a rice leaf, (*Oryza sativa*) or a hard hydrophobic surface (Lee and Dean 1993b, 1994; Xiao et al. 1994), and two genes, *nif23* and *nif29*, are expressed in a stage-specific manner as the appressorium develops (Lee and Dean 1993b). Appressoria do not appear on growth medium or other noninductive surfaces. However, addition of cAMP, its analogs (8-bromo-cAMP, N^6-monobutyrlyl-cAMP), or the phosphodiesterase inhibitor, 3-isobutyl-1-methylxanthine, to conidia or to vegetative hyphae induce appressoria to form on noninductive surfaces (Lee and Dean 1993a). The identification of cAMP as a mediator of infection structure formation provides a clue to the regulation of appressorium development.

Calcium ions also have a role in development of appressoria by the alfalfa anthracnose fungus, *C. trifolii*. From a study of the effects of various chemical agents on appressorium formation including calcium ionophores, calcium channel blockers, and calmodulin antagonists, Warwar and Dickman (1996) showed that calcium and calmodulin are essential as signal-transduction components in the overall differentiation process. They have recently cloned the calmodulin gene from *C. trifolii* (V. Warwar and M.B. Dickman, pers. comm.).

Chemicals exuded from ripe pepper fruit, *Capsicum frutescens*, stimulate the differentiation of germlings of *C. piperatum* conidia (Grover 1971). Sucrose, fructose, and thiamine stimulated germling differentiation of appressoria, while amino acids and amides are generally inhibitory, even in the presence of sucrose. Washed spores of *C. piperatum* germinated without forming appressoria in water on glass slides as well as on leaf, stem, and on surfaces of green fruit (i.e., thigmotropism was not involved as a stimulus to appressorium formation). Grover (1971) suggested that appressorium formation and subsequent disease development on red fruits was dependent on the quality and quantity of nutrients and the balance between stimulatory and inhibitory substances present in the infection drop.

The surface wax of avocado (*Persea americana*) fruit stimulates both germination and appressorium formation of conidial germlings of *C. gloeosporioides*, a parasite of the fruit (Podila et al. 1993). Assays of the avocado wax showed that fatty alcohols longer than C23 are the main appressorium-inducing components. Waxes from nonhost plants did not induce appressorium formation, and even inhibited appressorium induction by avocado wax. The authors concluded that a favorable balance between appressorium-inducing very long-chain fatty alcohols and the absence of inhibitors allows the fungus to use the host surface wax to trigger germination and differentiation of infection structures of this pathogen.

C. Nutrient Deprivation

One conducive environment for many fungi seems to be starvation for nutrients. Conidial germlings of *M. grisea*, for example, readily develop appressoria on both hydrophobic surfaces such as Teflon or Mylar, and cellophane which is hydrophilic (Hamer et al. 1988; Bourett et al. 1987; Howard et al. 1991). Yet transcript levels of *MPG1*, a hydrophobin gene whose expression is required for appressorium development, are elevated in RNA from cultures starved for either a carbon or nitrogen source (Talbot et al. 1993). This suggests that one of the main environmental cues for appressorial morphogenesis among facultative parasites may be nutrient starvation rather than the characteristic of the substrate surface.

Nutrient deprivation is also important to several other such fungi. For example, conidia of *C. acutatum* develop appressoria poorly in water droplets on surfaces of glass or polycarbonate (Blakeman and Parbery 1977). However, when the spores are leached to induce nutrient stress, almost all of the germinated conidia produce appressoria, while the presence of sugars together with amino acids suppresses appressorium formation. Conidial differentiation of the insect pathogen, *M. anisopliae*, is stimulated by reduced levels of nutrients such as complex nitrogenous compounds (St. Leger et al. 1987). Analysis of multicomponent media suggests that amino acids and lipid components of the epicuticle act in combination with the hydrophobic surface of the cuticle to stimulate differentiation during pathogenesis. In fact, infection-related morphogenesis occurs only in the presence of low levels of complex nitrogenous nutrients when depletion of endogenous reserves makes it necessary for the pathogen to colonize a host (St. Leger et al. 1989b).

D. Bypass Signals

1. Chemicals

Appressorium formation in *Uromyces* was believed to be initiated only by signal ridges until we demonstrated that cell differentiation could be induced by K^+ and sugars (Staples et al. 1983). Subsequently, reports by our group (Staples et al. 1983, 1984a, 1985a; Laccetti et al. 1987; Stumpf et al. 1991); and others (Kaminskyj and Day 1984a,b), further demonstrated the involvement of K^+, Ca^{2+}, glucose, and sucrose in appressorium formation. We showed that differentiation induced by K^+ occurs only in aerially oriented germlings, apart from the substrate (Hoch et al. 1987a). Potassium ions appear to accelerate protein synthesis and the functioning of a K^+/H^+ antiport (Staples et al. 1985a), a conclusion which is strengthened by the fact that vanadate (K^+/H^+ ATPase pump inhibitor) prevents appressorium formation (H.C. Hoch, unpubl.). These chemical signals seem to bypass the thigmotropic signal because heat- and chemically stimulated appressoria develop without regard to the location of stomata.

Exogenously supplied cAMP or cGMP induce *U. appendiculatus* germlings to undergo one round of mitosis and to form septa, processes normally associated with appressorium formation which in this case do not arise (Hoch and Staples 1984; Hoch et al. 1987a). However, phosphodiesterase inhibitors, which induce cellular cAMP to increase, do induce complete appressoria (Hoch and Staples 1984). Little is known yet about the involvement of cAMP and cGMP in appressorium development, but the urediospore germling contains three cyclic nucleotide-binding peptides, proteins which bind either cAMP or cGMP (Epstein et al. 1989).

Urediospores of the rust fungus, *U. vicea-faba*, exhibit pH-mediated tropism in relation to pH gradients generated as a consequence of electrophysiological activities of the cells of the stomatal complex (Edwards and Bowling 1986). The germ tubes locate the stomata most efficiently when the aperature is closed and the pH gradient largest. Proton gradients also induce germ tube tropisms on inert replicas of the leaf surface.

2. Heat Shock

P. graminis f. sp. *tritici* urediospores germinating at 20 °C for 2 h can be induced to develop infection structures by successive exposures of 1.5 h at 30 °C and 12–16 h at 20 °C (Maheshwari et al. 1967b). Urediospores of the bean rust fungus germinating at 18 °C can be induced to undergo differentiation by shifting them to 28.5 °C and holding them at that temperature for 90 min (Staples et al. 1989). Morphologically, heat shock-induced germlings are indistinguishable from those induced by thigmotropism, but they develop randomly and not over the stomata.

Heat shock induces the synthesis of at least two new polypeptides in germlings of *P. graminis* f. sp. *tritici* during infection structure development (Kim et al. 1982; Wanner et al. 1985). A 2D-PAGE analysis revealed that heat shock induces at least six polypeptides in urediospores of *U. appendiculatus*, none of which corresponds to the thigmotropically induced polypeptides (Staples et al. 1989). Hence, we looked to see if the genes specifically expressed during thigmodifferentiation are also induced by heat shock (Bhairi et al. 1990). We observed that *Inf24*, *Inf56*, and *Inf64* are moderately induced, but *Inf88* is not. *Inf88* appears to be thigmotropically specific, while *Inf24* and *Inf56* are developmentally induced regardless of whether the differentiation is achieved by the touch stimulus or heat shock, i.e., they are developmentally controlled.

IV. Conclusions

Many fungi have developed protective mechanisms to retard spore germination until the conditions for germination are favorable. The oldest and most studied example of this **constitutive dormancy** is the heat-shock requirement for germination of ascospores of *Neurospora tetrasperma* (Goddard and Smith 1938). In conidia, however, chemical inhibitors contribute to spore dormancy. Some of these self-inhibitors are constituents of the spore (Macko 1981), and some are a part of the external mucilage (Leite and Nicholson 1992). While a number of self-inhibitors have been identified, their modes of action are unknown.

The development of most, but not all, infection structures requires some type of signal from the environment. For example, conidial germlings of *Metarhizium anisopliae* develop appressoria when the germling is starved in a nutrient-poor environment (St. Leger et al. 1989a). Other signals include chemicals (Grover 1971) and physical features of the host's epidermis (Hoch et al. 1987b). How fungi recognize these signals is not known for any parasite, but the response mechanisms must be complex because numerous development triggers, such as heat shock (Manners 1981), can bypass the primary signal. Responses of the germling include **germ tube orientation** (Hoch et al. 1993) and **directional peg emergence** (Staples et al. 1985b). They suggest that host-related signals may control germling development at many levels. In those fungi where appressoria appear to develop spontaneously on any surface in response to no obvious external signal, it may merely be that we have not yet recognized the signal.

References

Allen EA, Hazen BE, Hoch HC, Kwon Y, Leinhos GME, Staples RC, Stumpf MA, Terhune BT (1991a) Appressorium formation in response to topographical signals by 27 rust species. Phytopathology 81:323–331

Allen EA, Hoch HC, Stavely JR, Steadman JR (1991b) Uniformity among races of *Uromyces appendiculatus* in response to topographic signaling for appressorium formation. Phytopathology 81:883–887

Allen PJ (1955) The role of a self-inhibitor in the germination of rust urediospores. Phytopathology 45:259–266

Allen PJ (1965) Metabolic aspects of spore germination in fungi. Annu Rev Phytopathol 3:313–342

Allen RF (1923) A cytological study of infection of Baart and Kanred wheats by *Puccinia graminis tritici*. J Agric Res 23:131–152

Anikster Y (1986) Teliospore germination in some rust fungi. Phytopathology 76:1026–1030

Armentrout VN, Downer AJ (1987) Infection cushion development by *Rhizoctonia solani* on cotton. Phytopathology 77:619–623

Armentrout VN, Downer AJ, Grasmick DL, Weinhold AR (1987) Factors affecting infection cushion development by *Rhizoctonia solani* on cotton. Phytopathology 77:623–630

Ayres PG, Owen H (1970) Factors influencing spore germination in *Rhynchosporium secalis*. Trans Br Mycol Soc 54:389–394

Baker CJ, Tomerlin JR, Mock N, Davidson L, Melhuish J (1987) Effects of cations on germination of uredinospores of *Uromyces phaseoli*. Phytopathology 77:1556–1560

Bhairi SM, Staples RC, Freve P, Yoder OC (1989) Characterization of an infection structure-specific gene from the rust fungus, *Uromyces appendiculatus*. Gene 81:237–243

Bhairi SM, Laccetti L, Staples RC (1990) Effect of heat shock on expression of thigmo-specific genes from a rust fungus. Exp Mycol 14:94–98

Blakeman JP, Parbery DG (1977) Stimulation of appressorium formation in *Colletotrichum acutatum* by phylloplane bacteria. Physiol Plant Pathol 11:313–325

Bourett T, Hoch HC, Staples RC (1987) Association of the microtubule cytoskeleton with the thigmotropic signal for appressorium formation in *Uromyces*. Mycologia 79:540–549

Bromfield KR (1964) Cold-induced dormancy and its reversal in uredospores of *Puccinia graminis* var. *tritici*. Phytopathology 54:68–74

Calpouzos L, Chang H-S (1971) Fungus spore germination inhibited by blue and far red radiation. Plant Physiol 47:729–730

Chapela IH, Petrini O, Petrini LE (1990) Unusual ascospore germination in *Hypoxylon fragiforme*: first steps in the establishment of an endophytic symbiosis. Can J Bot 68:2571–2575

Chapela IH, Petrini O, Hagmann L (1991) Monolignol glucosides as specific recognition messengers in fungus-plant symbioses. Physiol Mol Plant Pathol 39:289–298

Chapela IH, Petrini O, Bielser G (1993) The physiology of ascospore eclosion in *Hypoxylon fragiforme*: mechanisms in the early recognition and establishment of an endophytic symbiosis. Mycol Res 7:157–162

Cochrane VW (1958) Physiology of fungi. John Wiley, New York

Corrêa AJ, Hoch HC (1993) Microinjection of urediospore germlings of *Uromyces appendiculatus*. Exp Mycol 17: 253–273

Corrêa AJ, Hoch HC (1995) Identification of thigmoresponsive loci for cell differentiation in *Uromyces* germlings. Protoplasma 186:34–40

Corrêa AJ, Staples RC, Hoch HC (1996) Inhibition of thigmostimulated cell differentiation with RGD-peptides in *Uromyces* germlings. Protoplasma 194:91–102

Dey PK (1933) Studies in the physiology of the appressorium of *Colletotrichum gloesporioides*. Ann Bot (Lond) 47:305–312

Dickinson S (1949) Studies in the physiology of obligate parasitism. II. The behaviour of germ-tubes of certain rusts in contact with various membranes. Ann Bot (Lond) 1:219–236

Dickinson S (1971) Studies in the physiology of obligate paratism. VIII. An analysis of fungal responses to thigmotropic stimuli. Phytopathol Z 70:62–70

Dickinson S (1974) The production of nitrocellulose membranes as used in the study of thigmotropism. Physiol Plant Pathol 4:373–377

Dobinson KE, Hamer JE (1992) *Magnaporthe grisea*. In: Stahl U, Tudzynski P (eds) Molecular biology of filamentous fungi. VCH Verlagsgesellschaft, Weinheim, pp 67–86

Domsch KH (1954) Keimungsphysiologische Untersuchungen mit Sporen von *Erysiphe graminis*. Arch Microbiol 20:163–175

Edwards MC, Bowling DJF (1986) The growth of rust germ tubes towards stomata in relation to pH gradients. Physiol Mol Plant Pathol 29:185–196

Ellingboe AH (1972) Genetics and physiology of primary infection by *Erysiphe graminis*. Phytopathology 62:401–406

Emmett RW, Parbery DG (1975) Appressoria. Annu Rev Phytopathol 13:147–167

Epstein L, Staples RC, Hoch HC (1989) Cyclic AMP cyclic GMP and bean rust uredospore germlings. Exp Mycol 13:100–104

Foudin AS, Bush PB, Macko V, Porter JK, Robbins JD, Wynn WK (1977) Isolation and characterization of

three self-inhibitors of germination from aeciospores of *Cronartium fusiforme*. Exp Mycol 1:128–137
French RC (1992) Volatile chemical germination stimulators of rust and other fungal spores. Mycologia 84:277–288
French RC, Weintraub RL (1957) Pelargonaldehyde as an endogenous germination stimulator of wheat rust spores. Arch Biochem Biophys 72:235–237
Freytag S, Bruscaglioni L, Gold RE, Mendgen K (1988) Basidiospores of rust fungi (*Uromyces species*) differentiate infection structures in vitro. Exp Mycol 12:275–283
Givan CV, Bromfield KR (1964a) Light inhibition of urediospore germination in *Puccinia recondita*. Phytopathology 54:116–117
Givan CV, Bromfield KR (1964b) Light inhibition of urediospore germination in *Puccinia graminis* var. *tritici*. Phytopathology 54:382–384
Goddard DR, Smith PE (1938) The reversible heat activation inducing germination and increased respiration in the ascospores of *Neurospora tetrasperma*. Plant Physiol 13:241–264
Gold RE, Mendgen K (1983a) Activation of teliospore germination in *Uromyces appendiculatus* var. *appendiculatus* II. Light and host volatiles. Phytopathol Z 108:281–293
Gold RE, Mendgen K (1983b) Activation of teliospore germination in *Uromyces appendiculatus* var. *appendiculatus* I. Aging and temperature. Phytopathol Z 108:267–280
Gold RE, Mendgen K (1984) Cytology of basidiospore germination, penetration, and early colonization of *Phaseolus vulgaris* by *Uromyces appendiculatus* var. *appendiculatus*. Can J Bot 62:1989–2002
Gooday GW (1975) Chemotaxis and chemotropism in fungi and algae. In: Carlile MJ (eds) Primitive sensory and communication systems. Academic Press, New York, pp 155–204
Groth JV, Mogen BD (1978) Completing the life cycle of *Uromyces phaseoli* var. *typica* on bean plants. Phytopathology 68:1674–1677
Grover RK (1971) Participation of host exudate chemicals in appressorium formation by *Colletotrichum piperatum*. In: Preece TF, Dickinson CH (eds) Ecology of leaf surface microorganisms. Academic Press, London, pp 509–518
Hamer JE, Howard RJ, Chumley FG, Valent B (1988) A mechanism for surface attachment in spores of a plant pathogenic fungus. Science 239:288–290
Hancock JG, Lorbeer JW (1963) Pathogenesis of *Botrytis cinerea, B. squamosa*, and *B. allii* on onion leaves. Phytopathology 53:669–673
Hess SL, Allen PJ, Nelson D, Lester H (1975) Mode of action of methyl *cis*-ferulate, the self-inhibitor of stem rust urediospore germination. Physiol Plant Pathol 5:107–112
Hill G, Stellwaag-Kittler F, Schlösser E (1980) Substrate surface and appressoria formation by *Botrytis cinerea*. Phytopathol Z 99:186–191
Hoch HC, Staples RC (1984) Evidence that cyclic AMP initiates nuclear division and infection structure formation in the bean rust fungus, *Uromyces phaseoli*. Exp Mycol 8:37–46
Hoch HC, Staples RC (1991) Signaling for infection structure formation in fungi. In: Cole GT, Hoch HC (eds) The fungal spore and disease initiation in plants and animals. Plenum Press, New York, pp 25–46
Hoch HC, Staples RC, Bourett T (1987a) Chemically induced appressoria in *Uromyces appendiculatus* are formed aerially, apart from the substrate. Mycologia 79:418–424
Hoch HC, Staples RC, Whitehead B, Comeau J, Wolf ED (1987b) Signaling for growth orientation and cell differentiation by surface topography in *Uromyces*. Science 235:1659–1662
Hoch HC, Bojko RJ, Comeau GL, Allen EA (1993) Integrating microfabrication and biology. Circuits Devices 9:17–22
Horiguchi K, Suzuki Y, Sassa T (1989) Biosynthetic study of chloromonilicin, a growth self-inhibitor having a novel lactone ring, from *Monilinia fructicola*. Agric Biol Chem 53:2141–2145
Howard RJ, Bourett TM, Ferrari MA (1991) Infection by Magnaporthe: An in vitro analysis. In: Mendgen K, Lesemann DE (eds) Electron microscopy of plant pathogens. Springer, Berlin Heidelberg New York, pp 251–264
Johnson T (1934) A tropic response in germ tubes of urediospores of *Puccinia graminis tritici*. Phytopathology 24:80–82
Kaminskyj SGW, Day AW (1984a) Chemical induction of infection structures in rust fungi I. Sugars and complex media. Exp Mycol 8:63–72
Kaminskyj SGW, Day AW (1984b) Chemical induction of infection structures in rust fungi. II. Inorganic ions. Exp Mycol 8:193–201
Kim WK, Howes NK, Rohringer R (1982) Detergent-soluble polypeptides in germinated uredospores and differentiated uredosporelings of wheat stem rust. Can J Plant Pathol 4:328–333
Knights IK, Lucas JA (1980) Photosensitivity of *Puccinia graminis* f. sp. *tritici* urediniospores in vitro and on the leaf surface. Trans Br Mycol Soc 74:543–549
Kuo KC, Hoch HC (1996) Germination of *Phyllosticta ampelicida* pycnidiospores: prerequisite of adhesion to the substratum and the relationship of substratum wettability. Fungal Genet Biol 20:18–29
Kwon YH, Hoch HC (1991) Temporal and spacial dynamics of appressorium formation in *Uromyces appendiculatus*. Exp Mycol 15:116–131
Laccetti L, Staples RC, Hoch HC (1987) Purification of calmodulin from bean rust uredospores. Exp Mycol 11:231–235
Lapp MS, Skoropad WP (1978) Location of appressoria of *Colletotrichum graminicola* on natural and artificial barley leaf surfaces. Trans Br Mycol Soc 70:225–228
Lax AR, Templeton GE, Meyer WL (1985) Isolation, purification, and biological activity of a self-inhibitor from conidia of *Colletotrichum gloeosporioides*. Phytopathology 75:386–390
Lee Y-H, Dean RA (1993a) cAMP regulates infection structure formation in the plant pathogenic fungus *Magnaporthe grisea*. Plant Cell 5:693–700
Lee Y-H, Dean RA (1993b) Stage-specific gene expression during appressorium formation of *Magnaporthe grisea*. Exp Mycol 17:215–222
Lee Y-H, Dean RA (1994) Hydrophobicity of contact surface induces appressorium formation of *Magnaporthe grisea*. FEMS Microbiol Lett 115:71–76
Leite B, Nicholson RL (1992) Mycosporine-Alanine: A self-inhibitor of germination from the conidial mucilage of *Colletotrichum graminicola*. Exp Mycol 16:76–86
Lewis NG, Inciong MEJ, Ohashi H, Towers GHN, Yamamoto E (1988) Exclusive accumulation of Z-isomers of

monolignols and their glucosides in bark of *Fagus grandifolia*. Phytochemistry 27:2119–2121
Lingappa BT, Lingappa Y, Bell E (1973) A self-inhibitor of protein synthesis in the conidia of *Glomerella cingulata*. Arch Mikrobiol 94:97–107
Macko V (1981) Inhibitors and stimulants of spore germination and infection structure formation in fungi. In: Turian G, Hohl HR (eds) The fungal spore: morphogenetic controls. Academic Press, New York, pp 565–584
Macko V, Staples RC, Yaniv Z, Granados RR (1976) Self-inhibitors of fungal spore germination. In: Weber DJ, Hess WM (eds) The fungal spore. John Wiley, New York, pp 73–100
Maheshwari R, Allen PJ, Hildebrandt AC (1967a) Physical and chemical factors controlling the development of infection structures from urediospore germ tubes of rust fungi. Phytopathology 57:855–862
Maheshwari R, Hildebrandt AC, Allen PJ (1967b) The cytology of infection structure development in uredospore germ tubes of *Uromyces phaseoli* var. *typica* (Pers.) Wint.. Can J Bot 45:447–450
Manners JG (1981) Biology of rusts on leaf surfaces. In: Blakeman JP (eds) Microbial ecology of the phylloplane. Academic Press, New York, pp 103–114
Mendgen K (1973) Feinbau der Infektionsstrukturen von *Uromyces phaseoli*. Phytopathol Z 78:109–120
Mendgen K (1984) Development and physiology of teliospores. In: Bushnell WR, Roelfs AP (eds) The cereal rusts, vol 1. Academic Press, Orlando, pp 375–398
Moloshok TD, Leinhos GME, Staples RC, Hoch HC (1993) The autogenic extracellular environment of *Uromyces appendiculatus* urediospore germlings. Mycologia 85:392–400
Morin L, Brown JF, Auld BA (1992) Teliospore germination, basidiospore formation and the infection process of *Puccinia xanthii* on *Xanthium occidentale*. Mycol Res 96:661–669
Podila GK, Rogers LM, Kolattukudy PE (1993) Chemical signals from Avocado surface wax trigger germination and appressorium formation in *Colletotrichum gloeosporioides*. Plant Physiol 103:267–272
Preece TF, Barnes G, Bayley JM (1967) Junction between epidermal cells as sites of appressorium formation by plant pathogenic fungi. Plant Pathol 16:117–118
Rines HW, French RC, Daasch LW (1974) Nonanal and 6-methyl-5-hepten-2-one: Endogenous germination stimulators of urediospores of *Puccinia graminis* var. *tritici* and other rusts. J Agric Food Chem 22:96–100
Roderick HW (1993) The infection of white clover (*Trifolium repens*) by conidia of *Cymadothea trifolii*. Mycol Res 97:227–232
Seebach D, Adam G, Zibuck R, Simon W, Rouilly M, Meyer WL, Hinton JF, Privett TA, Templeton GF, Heiny DK, Gisi U, Binder H (1989) Gloeosporone – a macrolide fungal germination self-inhibitor: total synthesis and activity. Liebigs Ann Chem 1989:1233–1240
Sharp EL (1967) Atmospheric ions and germination of uredospores of *Puccinia striiformis*. Science 156:1359–1360
Sherwood J, Gow NAR, Gooday GW, Gregory DW, Marshall D (1992) Contact sensing in *Candida albicans* a possible aid to epithelial penetration. J Med Vet Mycol 30:461–469
Staples RC, Wynn WK (1965) The physiology of uredospores of the rust fungi. Bot Rev 31:537–564
Staples RC, Grambow H-J, Hoch HC (1983) Potassium ion induces rust fungi to develop infection structures. Exp Mycol 7:40–46
Staples RC, Hassouna S, Laccetti L, Hoch HC (1984a) Metabolic alterations in bean rust germlings during differentiation induced by the potassium ion. Exp Mycol 8:183–192
Staples RC, Macko V, Wynn WK, Hoch HC (1984b) Terminology to describe the differentiation response by germlings of fungal spores. Phytopathology 74:380
Staples RC, Hassouna S, Hoch HC (1985a) Effect of potassium on sugar uptake and assimilation by bean rust germlings. Mycologia 77:248–252
Staples RC, Hoch HC, Epstein L, Laccetti L, Hassouna S (1985b) Recognition of host morphology by rust fungi: responses and mechanisms. Can J Plant Pathol 7:314–322
Staples RC, Hoch HC, Freve P, Bourett TM (1989) Heat shock induced development of infection structures by bean rust uredospore germlings. Exp Mycol 13:149–157
Stark-Urnau M, Mendgen K (1993) Differentiation of aecidiospore- and uredospore-derived infection structures on cowpea leaves and on artificial surfaces by *Uromyces vignae*. Can J Bot 71:1236–1242
St Leger RJ, Cooper RM, Charnley AK (1987) Production of cuticle-degrading enzymes by the entomopathogen *Metarhizium anisopliae* during infection of cuticles from *Calliphora vomitoria* and *Manduca sexta*. J Gen Microbiol 133:1371–1382
St Leger RJ, Butt T, Goettel MS, Staples RC, Roberts DW (1989a) Production in vitro of appressoria by the entomopathogenic fungus *Metarhizium anisopliae*. Exp Mycol 13:274–288
St Leger RJ, Butt TM, Staples RC, Roberts DW (1989b) Synthesis of proteins including a cuticle-degrading protease during differentiation of the entomopathogenic fungus *Metarhizium anisopliae*. Exp Mycol 13:253–262
St Leger RJ, Bidochka MJ, Roberts DW (1994) Germination triggers of *Metarhizium anisopliae* conidia are related to host species. Microbiology 140:1651–1660
Stumpf MA, Leinhos GME, Staples RC, Hoch HC (1991) The effect of pH and K^+ on appressorium formation by *Uromyces appendiculatus* urediospore germlings. Exp Mycol 15:356–360
Sussman AS, Halvorson HO (1966) Spores: their dormancy and germination. Harper & Row, New York
Suzuki K, Furusawa I, Ishida N, Yamamoto M (1981) Protein synthesis during germination and appressorium formation of *Colletotrichum lagenarium* spores. J Gen Microbiol 124:61–69
Suzuki K, Furusawa I, Ishida N, Yamamoto M (1982) Chemical dissolution of cellulose membranes as a prerequisite for penetration from appressoria of *Colletotrichum lagenarium*. J Gen Microbiol 128:1035–1039
Talbot NJ, Ebbole DJ, Hamer JE (1993) Identification and characterization of *MPG1*, a gene involved in pathogenicity from the rice blast fungus *Magnaporthe grisea*. Plant Cell 5:1575–1590
Terhune BT, Hoch HC (1993) Substrate hydrophobicity and adhesion of *Uromyces* urediospores and germlings. Exp Mycol 17:241–252
Terhune BT, Allen EA, Hoch HC, Wergin WP, Erbe EF (1991) Stomatal ontogeny and morphology in *Phaseolus vulgaris* in relation to infection structure initia-

tion by *Uromyces appendiculatus*. Can J Bot 69:477–484

Terhune BT, Bojko RJ, Hoch HC (1993) Deformation of stomatal guard cell lips and microfabricated artificial topographies during appressorium formation by *Uromyces*. Exp Mycol 17:70–78

Tsurushima T, Ueno T, Fukami H, Irie H, Inoue M (1995) Germination self-inhibitors from *Colletotrichum gloeosporioides* f. sp. *jussiaea*. Mol Plant-Microbe Interact 8:652–657

Van Burgh P (1950) Some factors affecting appressorium formation and penetrability of *Colletotrichum phomoides*. Phytopathology 40:29

Wanner R, Forster H, Mendgen K, Staples RC (1985) Synthesis of differentiation-specific proteins in germlings of the wheat stem rust fungus after heat shock. Exp Mycol 9:279-283

Warwar V, Dickman MB (1996) Effects of calcium and calmodulin on spore germination and appressorium development in *Colletotrichum trifolii*. Appl Environ Microbiol 62:74–79

Willoughby LG, Hasenjager R (1987) Formation and function of appressoria in *Saprolegnia*. Trans Br Mycol Soc 89:373–380

Wynn WK (1976) Appressorium formation over stomates by the bean rust fungus: response to a surface contact stimulus. Phytopathology 66:136–146

Xiao J-Z, Watanabe T, Kamakura T, Ohshima A, Yamaguchi I (1994) Studies on cellular differentiation of *Magnaporthe grisea*. Physicochemical aspects of substratum surfaces in relation to appressorium formation. Physiol Mol Plant Pathol 44:227–236

Xuei X-L, Bhairi S, Staples RC, Yoder OC (1992) Characterization of *INF56*, a gene expressed during infection structure development of *Uromyces appendiculatus*. Gene 110:49–55

Yarwood CE (1954) Mechanism of acquired immunity to a plant rust. Proc Natl Acad Sci USA 40:374–377

Fungal Invasion and Plant Responses

3 Breaching the Outer Barriers – Cuticle and Cell Wall Penetration

R.J. Howard

CONTENTS

I. Introduction

Fungi use many different methods for accessing nutrients within plant hosts. Strictly speaking, each host species may have triggered the evolution of a distinctive penetration strategy for individual fungal taxa. To make sense of the early stages in these fungal-plant interactions, we need to organize our thinking about techniques of fungal penetration and to consider in particular the importance of specific functional elements which facilitate the breaching of surface layers of whole plants, individual plant organs, and cells. This chapter will concentrate on what is known and unknown about how fungi penetrate these outer barriers.

DuPont Company, Central Research & Development, P.O. Box 80402, Wilmington, Delaware 19880-0402, USA

II. Overview of Penetration

A. Indirect and Direct Penetration

Some fungi enter into plant hosts **indirectly,** through openings such as stomata (Fig. 1), hydathodes or wounds, while others enter **directly** into epidermal cells by creating their own avenue of ingress through the plant surface. Those which enter stomata, e.g., some forms of many rust fungi (Chap. 2, this Vol.; Chap. 5, Vol. V, Part B), will subsequently encounter a cell wall of epidermal or mesophyll cells, and perhaps cuticle as well, within the substomatal chamber. These fungi also must then penetrate plant cell surface layers directly before they can establish a typical nutritional relationship with plant cytoplasm (Manners and Gay 1983). They do so, however, without necessarily interacting with epidermal cells, an avoidance which may bypass specific resistance responses (Aist 1981) or a particularly formidable surface barrier such as the thickened cuticle on gymnosperm needles (Millar 1981; Gold and Mendgen 1991). Because the structures of host-pathogen interfaces differ with their location on plant surfaces, various distinct strategies may be required to accomplish penetration at different sites. The complexity of such an interaction is made apparent by the classification of 6 developmental stages and 18 distinct host-pathogen interface types during pathogenesis by a single hypothetical rust pathogen (Bracker and Littlefield 1973). Fungi that penetrate strictly by either direct **or** indirect means are common, but some pathogens can employ both (e.g., Clark and Lorbeer 1976; Myers and Fry 1978), perhaps dependent upon their environment.

B. Penetration Structures

Fungi elaborate a dazzling array of structures for the purpose of breaching plant host surfaces, rang-

The Mycota V Part A
Plant Relationships
Carroll/Tudzynski (Eds.)

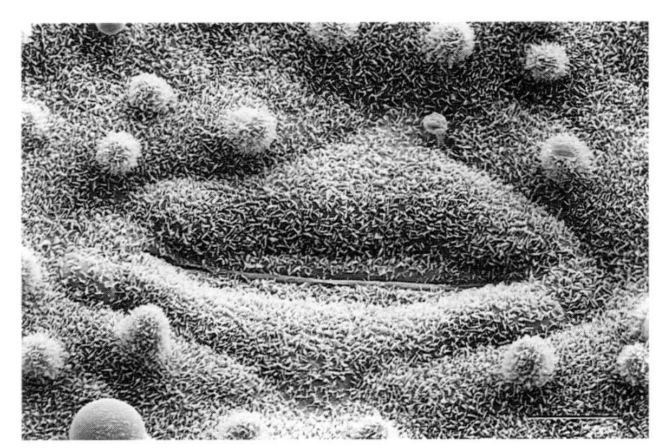

Fig. 1. Microtopography of a rice leaf surface observed with cryoscanning electron microscopy. Junctions between epidermal and guard cells, surrounding a stoma, can be discerned as *depressions*. *Knobs* scattered over the surface are protrusions of epidermal cell wall. A uniform coating of wax crystals covers most structures. *Bar* 5 µm

ing from hyphae or germ tubes, to encysted zoospores (see Hardham 1992) and highly specialized piercing structures (e.g., Aist and Williams 1971). Obviously, all of these forms do not rely upon the same mechanisms for penetration but they do certainly share one important attribute – function. Each provides a means for the fungal pathogen to enter tissues of a plant host.

The morphological diversity of infection structures has been quite useful to taxonomists. Understanding the details of that diversity can also be useful in determining how each type of infection structure accomplishes the formidable task of direct penetration of host. We are just beginning to appreciate the importance of these details, in many cases recognizing them for the first time.

A great variety of morphologies has been described from ascomycetous and basidiomycetous fungi which facilitate direct penetration of substrates. These infection structures, as well as some of those that form subsequently within host cells, have been found to differentiate on both "natural" substrata as well as on artificial surfaces (Freytag et al. 1988; Blaich et al. 1989; Bourett and Howard 1990; Swann and Mims 1991). Studies of in vitro differentiation (Fig. 2) will become increasingly important as an experimental approach to understanding host-pathogen interaction. Such studies could allow for testing the effects of potential signal molecules that might modulate in situ fungal development, or facilitate the detection and characterization of important stimuli (Carver and Ingerson 1987; Jelitto et al. 1994) or differentiation-specific gene expression (Hwang and Kolattukudy 1995; Hwang et al. 1995). It remains an open question whether particular structures might facilitate pathogenesis per se and not saprophytic growth through nonliving cells and tissues. In the case of *Magnaporthe grisea*, it appears that a specific sequence of morphogenesis and gene expression are switched on as a unit, representing all the early events required for infection (Howard et al. 1991a; Chap. 3, Vol. V, Part B).

1. Penetration Pegs

The penetration peg is a thin tip-growing cellular protuberance, generally <1 µm in diameter. These structures have also been termed penetration hyphae, infection hyphae, or infection pegs, although penetration per se does not necessarily result in infection, and thus the use of the word "infection" in this context is inappropriate. What the peg does is penetrate, not infect. The penetration peg is clearly morphologically distinct from the somatic hypha, generally five to ten times narrower in cross section. The peg does extend via tip growth, and in that sense could be considered as a modified hypha. However, because the function of

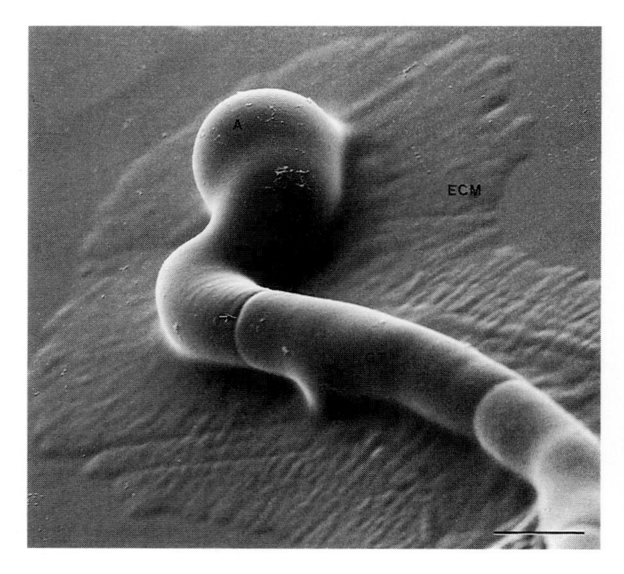

Fig. 2. In vitro development of infection structures by *Cochliobolus heterostrophus* on cellophane membrane. A broad sheath of extracellular matrix material (*ECM*) surrounds the germ tube (*GT*) and appressorium (*A*). Septum (*arrow*). Cryoscanning electron micrograph. *Bar* 5 µm. (Braun and Howard 1994a)

the peg is so specialized, an entirely independent name is preferred. A similar preference is also held for germ tubes. Ergo, just as germ tubes are not germination hyphae, penetration pegs are not penetration hyphae.

Penetration pegs generally arise from structures that are easily recognized and distinguished from conidia and germ tubes, but have been reported to develop directly from germ tubes, spores, and hyphae (e.g., Mckeen and Rimmer 1973; Clark and Lorbeer 1976; Myers and Fry 1978; Hau and Rush 1982; Porto et al. 1988). We must interpret these reports cautiously as we do not have detailed information concerning penetration sites. An ultrastructural analysis might demonstrate the presence of specializations at the subcellular level associated with the generation of penetration pegs. At present, there is no reason to exclude possible variation in structure among penetration pegs of different fungi.

2. Appressoria

Appressorium is a term commonly used to describe fungal structures from which plant surfaces are penetrated with a penetration peg. From the time appressoria were first observed (see Hasselbring 1906) the idea that they function in attachment has been universally held, hence their name; but the term is so broadly applied that it implies very few specific characteristics, the reduced form being no more than a vaguely discernible, semidiscrete, penetration peg-bearing organ.

An appressorium can be a simple or multilobed swelling of a germ tube apex as in basidiospore germlings of rust fungi (Mims and Richardson 1989; Gold and Mendgen 1991), or a discrete swollen region delimited from the germ tube by a septum. Examples of the latter include aeciospore germlings of *Arthuriomyces peckianus* (Swann and Mims 1991), as well as conidial germlings of *Colletotrichum* sp. (Van Dyke and Mims 1991; Bailey et al. 1992) and *M. grisea* (Bourett and Howard 1990). Either type may exhibit additional specializations of cell wall form or composition. An obvious example of cell wall specialization in appressoria is dark pigmentation (see Sect. V), but less visible changes between germ tube and appressorium wall also exist (see Sect. IV.B). The presence of an adhesive layer is one possible modification. As mentioned above, adhesion was one of the original criteria for defining appressoria (Frank 1883) but after more than a century we still have very limited information about the nature of this function.

3. Hyphopodia

In 1975, Emmett and Parbery abandoned the traditional infection structure classification of the time. Relying primarily upon function rather than restricting the term appressoria to those structures arising from germ tubes (Walker 1980), they chose to include the various infection structures that differentiate from somatic hyphae as well. These latter hyphal infection structures have been called hyphopodia, as well as other names, are quite common among leaf fungi in the tropics (particularly among the Meliolaceae), and have served a utilitarian role for taxonomists (e.g., Sutton 1968).

Hyphopodia (Fig. 3) are infection structures arising from epiphytic hyphae. Hyphopodia may be terminal, lateral, or intercalary, simple or

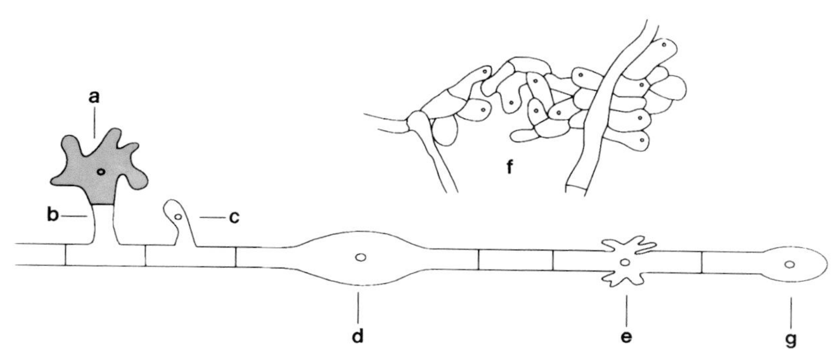

Fig. 3. Various forms of hyphopodia: *a* stigmatocyst and *b* stigmatopodium, together form a stalked, lobed, and pigmented hyphopodium; *c* sessile unlobed hyphopodium; *d* unlobed intercalary hyphopodium; *e* lobed intercalary hyphopodium; *f* infection plate; *g* unlobed terminal hyphopodium. *Open circles* represent penetration pegs. (After Walker 1980 with minor alterations)

lobed, hyaline or pigmented, epiphytic or endotrophic. Hyphopodia and appressoria can be produced by the same organism, but generally only one or the other is found under specific environmental or ecological conditions. Presumably, the role of hyphopodia in pathogenesis during **initial** penetration into the host is relatively minor among pathogens which are disseminated via the production of spores. Still, we do not know how common or important these structures might be under natural conditions. On the other hand, plant pathogens that rely upon epiphytic colonization, generally of plant parts near or below the soil line, utilize hyphopodia more often than not. Root infection arising from soil-borne conidia and involving tropic growth of germ tubes toward roots is also an important means of fungal ingress (see Carlson et al. 1991). In addition, some pathogens, such as those that cause powdery mildew diseases, utilize hyphopodia to spread epiphytically once initial infection has been established (see Bracker 1968; Falloon et al. 1989; Heintz and Blaich 1990). In other fungi, infection structures arising from germ tubes and hyphae can appear identical (e.g., Bourett and Howard 1990; cf. Yaegashi et al. 1987), in which case, as stated above, we must attempt to emphasize differentiation and function under natural conditions. Epstein and coworkers (1994) have studied hyphopodia of *Gaeumannomyces graminis* under laboratory conditions, and never observed them to function in a penetration process. They concluded that hyphopodia, especially those that are melanized, might serve for survival. Similar structures do clearly serve for penetration under some conditions (cf. Huffine 1994). The influence of various environmental conditions on infection structure formation and function remain largely unknown and deserving of further investigation. This line of research may bear fruitful information for the development of novel and safe crop-protectant chemicals.

4. Compound Appressoria

A third category of infection structures, consisting of multicellular aggregates, has been termed the compound appressoria (Emmett and Parbery 1975). These structures take various forms and are referred to by terms including infection cushions, plaques (Fig. 4a), or plates, and can arise from a single hypha or from several. A single compound appressorium can give rise to a multitude of indi-

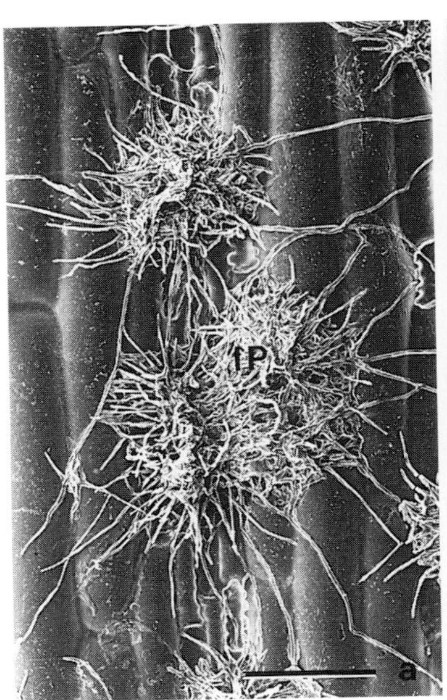

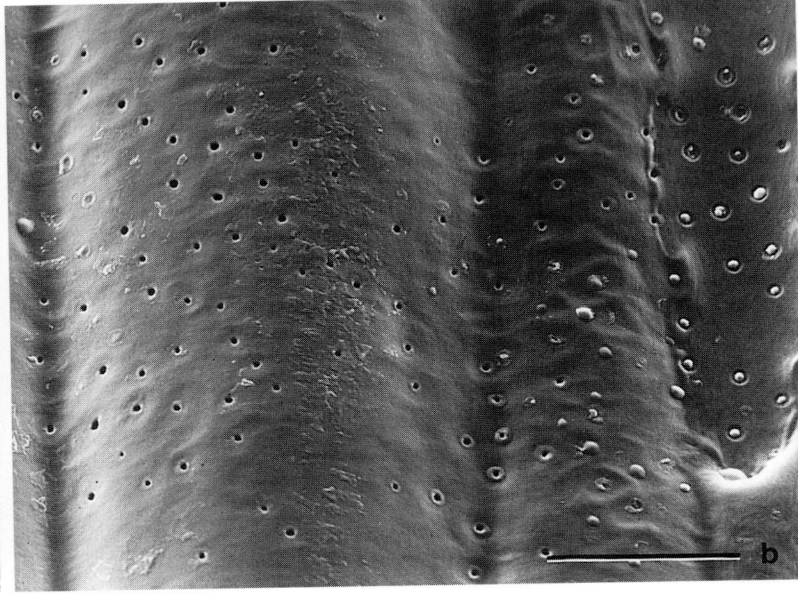

Fig. 4a,b. Cryoscanning electron micrographs depicting stages of infection by *Psuedocercosporella herpotrichoides* on wheat leaf sheaths. **a** Epiphytic runner hyphae differentiate to form discrete clusters of parenchyma-like cells termed infection plaques (*IP*). *Bar* 50 μm. **b** Multiple penetrations occur from a single plaque as seen in this view of a host cell wall after removal of the infection structure. Numerous holes are visible where penetration pegs pierced the host surface. To the *right* are penetration sites where pegs fractured and remained after removal of the infection structure. *Bar* 10 μm. (Daniels et al. 1991)

vidual penetration sites (Fig. 4b), as in *Pseudocercosporella herpotrichoides* (Daniels et al. 1991) and *Rhizoctonia solani* (Armentrout and Downer 1987).

5. Endotrophic Infection Structures

Direct penetration into a host epidermal cell is often followed by formation of a bulbous structure, the **primary** or **intraepidermal vesicle**, from which **primary** and **secondary infection hyphae** subsequently grow intracellularly and colonize the host. In the case of biotrophic parasites, an elaborate structure of determinate growth, a **haustorium**, is formed within the penetrated host cell and is surrounded by an **extrahaustorial matrix** (Littlefield and Bracker 1972; Knauf et al. 1989; Hahn and Mendgen 1992). Some fungi penetrate the cuticle layer and grow intramurally between the cuticle and cell wall, or within epidermal cell walls. When infection hyphae pass from one cell to another, a swelling sometimes forms against the host wall prior to penetration. Of those fungi that gain entry through stomata, an appressorium is sometimes formed over the stomatal opening (see Chap. 2, this Vol.; Chap. 5, Vol. V, Part B), while in other species the germ tube grows into the substomatal chamber without any apparent morphogenetic changes. Once inside, a **substomatal vesicle** forms intercellularly, sometimes followed by formation of what might be considered an endotropic appressorium, the **haustorial mother cell**, from which initial penetration of a host mesophyll cell occurs through production of a penetration peg. The peg continues to grow within the living host cell, forming the **haustorial neck** which connects the haustorial mother cell with the haustorium (see Littlefield and Heath 1979; Manners and Gay 1983).

III. Structure of Host Surface

For purposes of this discussion, there are a number of reasons to consider various characteristics of the host surface. First, it has been established that many fungal propagules and germlings adhere to surfaces on which they germinate and grow (see Braun and Howard 1994b). Fungal adhesion has become a field unto itself (Chap. 1, this Vol.) – after all, the pathogen-substrate interface is, literally, central to our interest of understanding mechanisms of pathogenesis. Second, differentiation of fungal cells, i.e., infection structures, can be influenced by physical and chemical signals provided by the substrate. For example, extracts from host epicuticular waxes have been demonstrated to affect patterns of pathogen morphogenesis (Uchiyama et al. 1979), but it is not known whether these effects can be attributed to physical or chemical changes in substrate surfaces imparted by the extracts (see Xiao et al. 1994). Another example of induced morphogenesis has been reported for the rust *Uromyces* (see Terhune et al. 1991), where host guard cell lip morphology can trigger differentiation when recognized (Chap. 2, this Vol.). This phenomenon may well prove a common aspect of differentiation yet to be described for other pathogens. Third, the biochemical structure and physical character of plant cell walls are important components of host defenses (see Smart 1991). To fathom mechanisms of penetration, we must know what the fungus is up against. At present, plant surfaces are rather poorly characterized.

Roberts (1990) writes: "In order to understand many of the functions, properties and mechanical behaviour of the plant [extracellular matrix], it is desirable to have explanations at the molecular level, and for this we ideally need a model of the cell wall that not only defines what polymers are present, how they are cross linked, with what sort of spacings, in what sort of networks and with what sort of order, but also how, molecule by molecule, that wall changes across its width." From the vantage of a penetration peg (see McCann et al. 1990), one can appreciate the importance of such detailed knowledge. Unfortunately, little to nothing is known about the forces needed to breach **individual** or combined wall layers which protect plant cells. This makes the task of understanding the penetration process very difficult indeed. In addition, too little is known of the chemical structure of walls – which is not to say that little of their structure is known. On the contrary, there exists a huge body of literature pertaining to plant cell wall architecture.

A. Cuticle and Primary Cell Walls

The plant cuticle is a surface layer of variable composition and thickness covering the outer walls of aerial plant parts (Kerstiens 1996). The cuticle is composed primarily of a soluble, complex mixture of long chain aliphatic compounds, or waxes, often visible in the form of prominent platelets on host surfaces (Fig. 1), and an insoluble polymeric fraction known as cutin (Kolattukudy 1980; Holloway 1982; Post-Beittenmiller 1996). A wall layer of similar composition, the suberin-containing casperian strip, is found encircling endodermal cells and encasing the vascular stele in plant roots. The cuticle functions to retard water loss from the plant, but also prevents the necessary exchange of gases between the plant and its environment, thus the need for stomata. It is permeable, but only slightly so (Bukovac and Petracek 1993). Similarly, the casperian strip limits the apoplastic flow of molecules into the vascular cylinder.

Beneath the cuticle, the bulk of primary plant cell wall is constructed of cellulosic microfibrils that are embedded in a matrix of proteins and noncellulosic polysaccharides (see Carpita and Gibeaut 1993). For present purposes, we can disregard constitutive secondary wall structure because epidermal cells rarely develop a secondary wall. Much of what is known about the biochemistry of plant cell walls has been learned from studies of relatively few model systems such as callus, suspension cultures, or regenerating protoplasts of representative taxa. Smart (1991) has made the point that these experimental systems do not provide a true representation of real wall complexity and thus can only offer a gross approximation of the chemical nature of the barriers which stand between fungal pathogens and internal tissues of the host. For not only do wall structures vary with age and location on the plant, but also as a function of host environment and, of course, host taxonomic affiliation. Perhaps a better understanding of wall structure with relevance to mechanisms of fungal penetration will be learned from studies of fungi and the enzymes they secret, rather than of plant hosts (see Bateman and Basham 1976; Anderson 1978; Baker et al. 1980; Cooper 1987; Walton 1994; Deising et al. 1995; Rauscher et al. 1995; Chap. 4, this Vol.).

B. Induced Structural Changes

To make our job even more challenging, a number of localized structural and chemical changes in plant cell walls, or cytoplasm (Snyder et al. 1991), can result during attack by fungi (see Aist 1981; Aist and Bushnell 1991). These changes involve the accretion of various materials of host origin, at pathogen encounter sites, in the form of wall appositions such as **papillae** or **halos**, for example. Accumulated materials are for the most part unknown (see Inoue et al. 1994) but may include autofluorescent phenolic compounds, elevated levels of silicon (Kunoh and Ishizaki 1976; Heath and Stumpf 1986; Carver et al. 1987; Chérif et al. 1992; Winslow 1992; Samuels et al. 1994), various enzymes (e.g., Takahashi et al. 1985; Mauch and Staehelin 1989; Benhamou et al. 1990a,b), and even elemental sulfur (Cooper et al. 1995). Presumably, these host reactions represent defense mechanisms designed to inhibit further ingress by the pathogen, but their actual effects during pathogenesis remain poorly understood. Because these responses are very localized, it has been extremely difficult to characterize the chemical and physical changes they bestow upon the host. Our ignorance is thus compounded.

IV. Design of Infection Structures

The early view of appressoria as adhesive organs supported the idea that nothing much in particular happened inside these cells. However, we are now becoming aware of very sophisticated structural and morphogenetic features which suggest the contrary. Some of these features have not as yet been assigned a function, while others have received some attention and preliminary interpretations.

The importance of design details can be quite subtle, as in the case of many powdery mildew pathogens of a wide variety of hosts (Chap. 4, Vol. V, Part B). A careful and sustained study by numerous individuals has brought to light the very remarkable design features of these infection structures which account for their being described as "exquisitely adapted to develop on and infect epidermal cells of cereals" (Aist and Bushnell 1991). The pathogen structures involved are shown in Fig. 5. Individual conidia of *Erysiphe graminis*, for example, germinate and produce first a single "primary germ tube" which attempts to penetrate the host but is nearly always thwarted by an active plant defense response. A "secondary germ tube" is produced hours later, and penetration from its appressorium may be successful. The fungus enters the host cell cytoplasm and forms the nutrient-gathering haustorium. The attempted

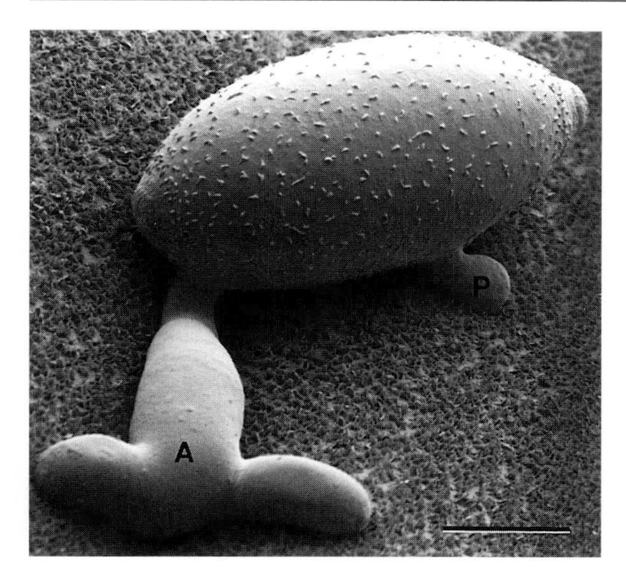

Fig. 5. Infection structures of *Erysiphe graminis tritici* including a primary germ tube (*P*) and lobed appressorium (*A*), formed on a wheat leaf. Cryoscanning electron micrograph. *Bar* 10 μm

penetration from the primary germ tube activates plant defenses. Failure to invade the host from the primary germ tube is good evidence of the potential effectiveness of the plant cell to inhibit invasion by the pathogen. Successful invasion from the secondary germ tube results from a carefully orchestrated sequence of fungal maneuvers which somehow prevail over the plant host defenses (Chap. 4, Vol. V, Part B). Why the second attempted penetration of the same or adjacent host cell is more successful than from the primary germ tube remains unknown, but is quite likely the result of a sophisticated **design** (see Aist and Bushnell 1991).

A. Morphogenesis

Much of what we can discuss concerning infection structure formation comes from one end of a spectrum of structural studies: classification of different forms, observations of changes in cell wall composition during growth, temporally specific expression of genes, documentation of subcellular changes that accompany morphogenesis. Analysis will become quite detailed as we begin to investigate the other end of this spectrum, using the techniques of molecular biology. However, we must not lose sight of the fact that real insight will result only from a holistic bridging of these approaches from the vantage of cell biology. In speaking of microbial morphogenesis, Harold (1990) put it: "Although grounded in the actions of molecules, it must be addressed as a problem in physiology."

Any discussion of morphogenesis among filamentous fungi must, of course, address the hypha as a basis from which the various other forms take shape, literally. A number of recent reviews on tip growth demonstrate all too well that we have yet to understand even this basic growth form. In addition, it has become increasingly apparent that the germ tube, from which appressoria develop, exhibits sufficient morphological, and no doubt biochemical, differences from hyphae to be considered a unique growth form. This leaves us precious little from which to build an understanding of infection structure development. Nevertheless, some obvious changes in germ tubes have been recognized that accompany the development of infection structures.

Both immuno- and cytochemical techniques have become increasingly important in detecting differences in cell wall or surface components of infection structures during morphogenesis. Application of various lectins, probes for specific carbohydrate moieties, to whole infection structures of *Colletotrichum*, for example, has revealed a range of differential labeling patterns (O'Connell et al. 1992). Addition of lectins after enzyme or alkali treatments (Freytag and Mendgen 1991) or as probes of thin-sectioned infection structures (Rohringer et al. 1989; Howard et al. 1991a) has shown additional chemical differences within cell walls of these organs. A similar type of approach with antibodies has recently demonstrated great promise for biochemical characterization of the host-pathogen interface (Pain et al. 1992; Xia et al. 1992). As pointed out by O'Connell and colleagues (1992), antibodies offer a much broader range of binding specificities than lectins, including proteins as well as carbohydrates. The pioneering efforts of Hardham and colleagues (see Hardham et al. 1991) with *Phytophthora* have set a high standard for mycologists interested in these techniques. Combining the use of lectin, antibody, or other probes with cryofixation methods (see Czymmek et al. 1996) offers the prospect of an entirely new class of insights into such questions.

1. Cytoskeleton

An understanding of infection structure morphogenesis requires a basic knowledge of hyphal tip growth as the foundation upon which to build a different fungal cell form. Tip growth, maintenance of cell shape, and morphogenesis are phenomena in which elements of the cytoskeleton are intimately associated. We know from animal systems that cells can reorganize cytoskeletal components rapidly in response to specific stimuli, resulting in changes of cell shape (see Bretscher 1991); we also know that the cytoskeleton is often implicated in cellular events involving cytoplasmic transport of organelles and the exchange of materials with the extracytoplasmic milieu, via both endo- and exocytosis. Light microscope observation of living hyphal tips and the morphogenetic events that lead to infection structure development would convince anyone of the possible involvement of the cytoskeleton, a conclusion strongly supported by recent results from several laboratories.

Actin microfilaments have been studied in hyphal tips using various techniques (Gow 1989; Chap. 3, Vol. I), including the use of fluorescent and ultrastructural immunocytochemical probes (Bourett and Howard 1991; Roberson 1992). At least one form of actin, reacting to a specific monoclonal antibody in thin sectioned embedments, has been localized within filasomes and the central core region of the Spitzenkörper. A role for actin in modulating the exocytosis responsible for tip growth is supported by its presence within the core. Certainly, the type of tip swelling that results from treatment of hyphae with antibiotics which disrupt actin function (Grove and Sweigard 1980), as well as the association of actin with deposition of new cell wall (Kobori et al. 1992), again supports the importance of actin in both tip growth and morphogenesis. During infection structure formation by uredospore germlings of *Uromyces*, dramatic restructuring of both the microtubule and F-actin microfilament cytoskeleton have been documented (Kwon et al. 1991). The reorganizations reflect the altered orientation of cell expansion in close alignment with inductive ridges in the substratum.

Later during the developmental sequence of appressorium formation, actin appears to play other important roles in terms of cell function. In forming appressoria of *Gymnosporangium juniperi-virginianae*, numerous filasomes have been observed near the area of contact with the substratum; and at later stages, when elaborations of the lower appressorium wall begin to form, presumably after the appressorium structure as a whole is nearly complete, filasomes again are abundant (Mims and Richardson 1989). A similar observation has been reported for appressoria of *M. grisea* (Bourett and Howard 1990) and for those from aeciospore germlings of *Arthuriomyces peckianus* (Fig. 6). The similar presence of filasomes in these diverse taxa may reflect the high level of exocytosis, or endocytosis, that probably occurs over this area of contact with the host (see Sect. IV.C).

During penetration peg formation and growth, cytoskeletal components also become prominent. In *M. grisea*, the contents of young penetration pegs are devoid of most organelles, including ribosomes. Peg cytoplasm, as well as the adjacent cytoplasm of the appressorium itself, is filled by a **zone of exclusion** (Fig. 7). Evidence indicates that this zone contains what might be a concentration of actin (Bourett and Howard 1992). As the peg penetrates cellophane and elongates, microtubules can be seen within and aligned parallel to the long axis of the peg (Bourett and Howard 1990). A few vesicles are usually found at the tip of the peg, especially during its earlier growth stages.

B. Adhesion

Adhesion of fungal cells to substrates is a common phenomenon, often beginning early in pathogenesis (Hamer et al. 1988; Deising et al. 1992; Tunlid et al. 1992; Braun and Howard 1994a; Mercure et al. 1995; see also Dijksterhuis et al. 1990), but a phenomenon not widely contemplated. There remains much to be learned about this important aspect of fungal-plant interaction (Chap. 1, this Vol.), especially in terms of function.

The rice blast pathogen *M. grisea* is global in both its distribution and importance as a major problem in the production of lowland rice. As the most important fungal pathogen of what is arguably the world's most important crop, *M. grisea* has been studied extensively since it was first described over 300 years ago (Chap. 3, Vol. V, Part B). Yet, it was not until recently that we learned that *M. grisea* conidia are equipped to glue themselves to the host surface tenaciously, even under water (Hamer et al. 1988). Similar attachment mechanisms have been reported for marine fungi (reviewed by Hyde et al. 1993; Jones 1994) and nematophagous fungi (reviewed by Tunlid et al. 1992), organisms which must overcome more ob-

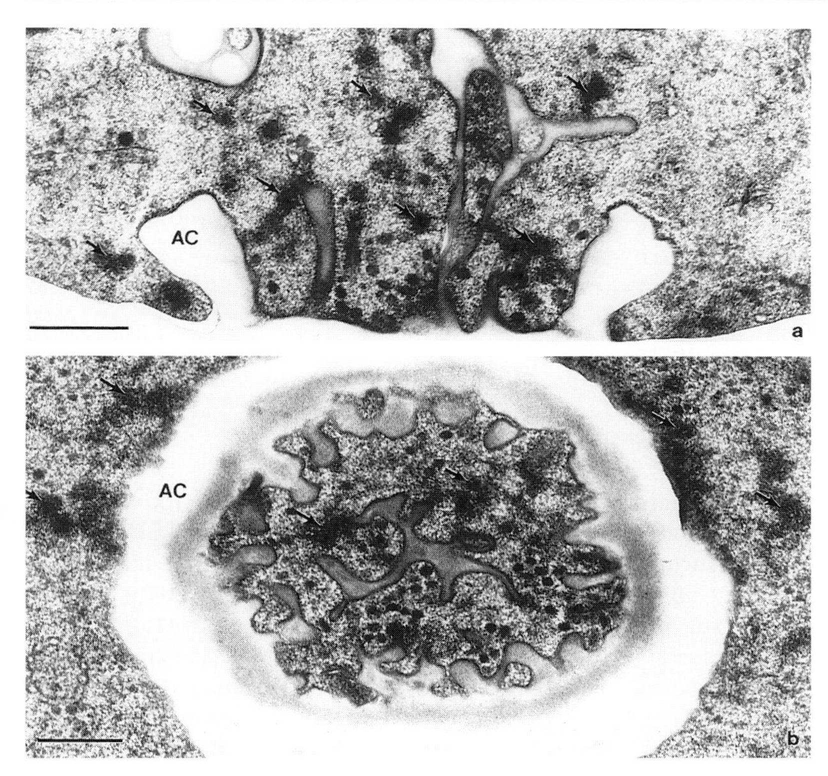

Fig. 6a,b. Region of substratum interface in freeze-substituted appressoria of *Arthuriomyces peckianus*, formed in vitro. A ring of cell wall, the appressorial cone (*AC*), forms in the region where a penetration peg will emerge. The two different vantages are from sections taken perpendicular **a** and parallel **b** to the plane of the substrate. The presence of numerous filasomes (*arrows*) and vesicles, in close association with a highly elaborated plasma membrane, point to the existence of a localized and very active transcellular transport apparatus. *Bars* 0.5 μm. (Swann and Mims 1991)

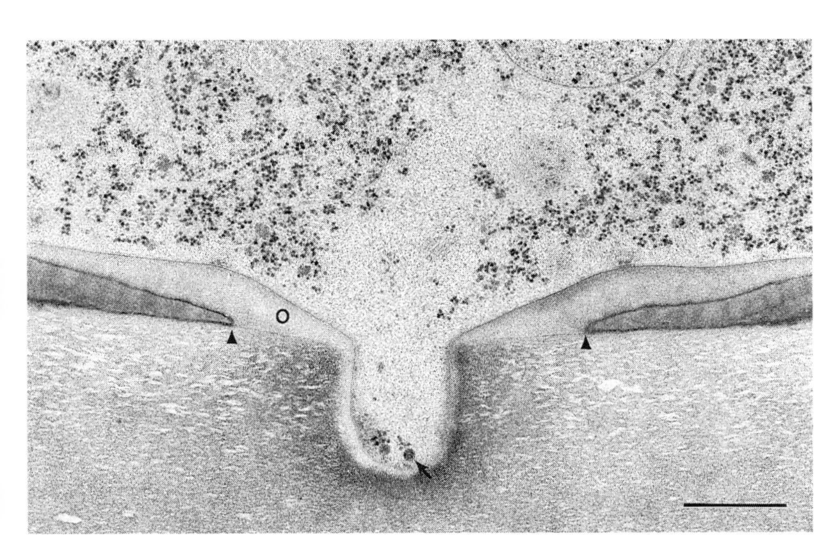

Fig. 7. Freeze-substituted nascent penetration peg of *Magnaporthe grisea* exhibiting a zone of exclusion (area lacking ribosomes) which fills the peg and extends into the adjacent region of appressorium cytoplasm. Perimeter of appressorium pore (*arrowheads*), pore wall overlay (*O*), apical vesicle (*arrow*). *Bar* 0.5 μm. (Bourett and Howard 1992)

vious deterrents to pathogenesis than those typically thought to be important to foliar plant pathogens. Conidia of these fungi exude a droplet of adhesive from the conidial apex, as in the nematode pathogen *Drechmeria coniospora*. An adhesive knob ensures sustained contact with the host surface during the penetration process. These examples involve structures that, even though ephemeral, are quite large, and play an extremely important role during the early, pivotal stages of host-pathogen interaction. The fact that these structures where reported only recently underscores the importance we must place upon a very careful examination, or reexamination, of the details of infection structure design.

Adhesive structures like these might increase the efficacy of inoculum by ensuring that a greater number of propagules remain on a host surface long enough to complete penetration; but adhesive structures might play other roles as well. As discussed in Chapter 2, this Vol. appressorium formation among some rust pathogens is triggered mechanically by specific surface features of the host. In this case, contact, adhesion, and subsequent development must be regarded as a unified process (Beckett et al. 1990). Indeed, adhesion between fungal cells and host walls probably occurs throughout the disease process, not just at the earliest stages of the interaction (Beckett and Porter 1988). In other fungi, the sudden presence of an adhesive layer at the tip of a growing germ tube might have a similar effect: mechanical stress would be placed upon the growing cell's surface. Some evidence for this type of trigger has been documented for infection structure initiation by *M. grisea* (unpubl.). An adhesive extracellular matrix also might play some role in preparing the host surface for penetration, for example by providing a medium within which an enzyme like cutinase would function (Kunoh et al. 1990; McRae and Stevens 1990; Deising et al. 1992).

Early stages of appressorium formation, the cessation of germ tube growth and the swelling of its apex, are accompanied in many species by a change in wall structure. One obvious manifestation of such change is often visible as a new, outer layer of the cell wall or the production of an extracellular layer, mucilaginous substance, or matrix (Mims and Richardson 1989; Bourett and Howard 1990; Kwon et al. 1991; Swann and Mims 1991). This layer may play a significant role in the reception of a signal that is responsible for initiation of infection structure development, for example by ensuring topographical conformity between the substrate and the plasma membrane. In addition, the cluster of apical vesicles normally present near the germ tube apex often persists during appressorium initiation located along the bottom of the cell, presumably where new cell surface material is deposited. Kwon and colleagues (1991) have provided excellent documentation of vesicle dynamics during this particular stage of development. Changes in the cytoskeleton also accompany these morphogenetic events.

C. Fungal-Plant Interface

The term **interface** defines, perhaps too narrowly, the common boundary which lies between the cells of host and pathogen, for that interaction certainly relies upon much more than a two-dimensional surface within a strictly defined area (Bracker and Littlefield 1973); but the whole process of infection structure development aims to provide a highly specialized conformation of cellular organization for the purpose of penetration. For the fungal cell biologist, this interface holds the key to understanding the fundamental processes of penetration. To advance our knowledge in this area we must move beyond the simple observations and classifications of infection structures which have occupied too many mycologists and plant pathologists for too long.

Appressorial structure is most broadly specialized, across taxonomic borders, in the region lying against the substratum. Nusbaum (1938) first noted in appressoria of *Venturia inaequalis* a pore, surrounded by a circular thickening, at the bottom of these infection structures. The well-characterized structure of *M. grisea* appressoria also exhibits a similar feature (Howard and Ferrari 1989; Bourett and Howard 1990; Howard et al. 1991a). In *M. grisea*, the appressorium pore is about 4.5 μm in diameter and for a period of time apparently lacks cell wall altogether (Fig. 8a). A ring of material circumscribing the pore may function to seal appressorial contents, at high pressure (see below), against the underlying substratum. Interestingly, analogous ring structures have also been shown in melanized infection structures from taxonomically and ecologically diverse pathogens (see Figs. 4, 6 of Epstein et al. 1994; Fig. 12 of Khan and Kimbrough 1974). It is from the edge of the pore that the penetration peg eventually

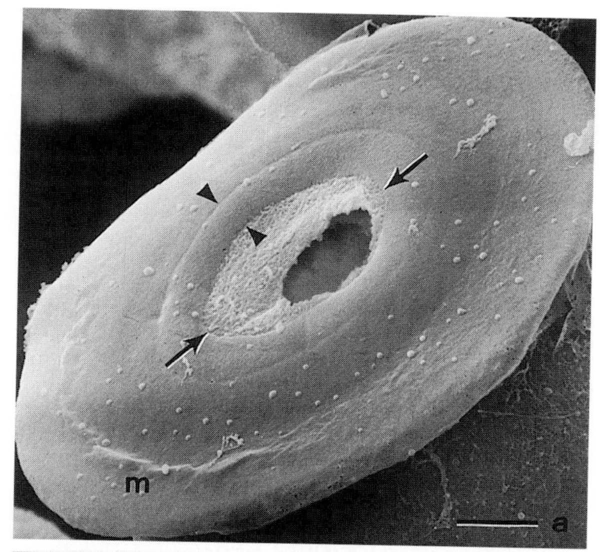

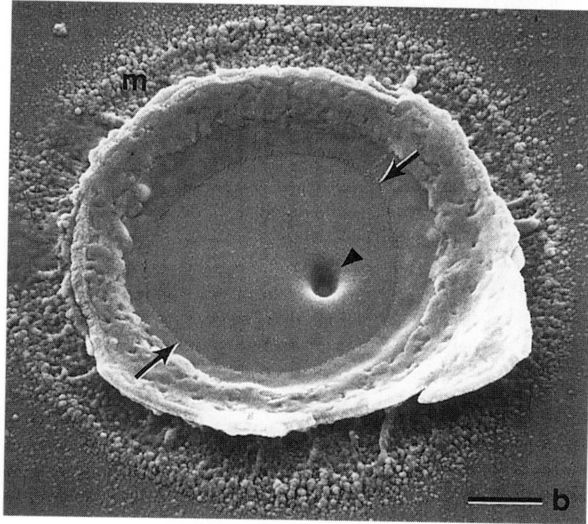

Fig. 8a,b. Appressoria of *Magnaporthe grisea*. **a** Bottom view of a cell detached from the substrate. A hole was torn in the bottom of the appressorium during detachment and marks the location where a peg had emerged and penetrated the substrate. A prominent ring (*between arrowheads*) encircles the appressorium pore (*between arrows*). (M.A. Ferrari and R.J. Howard, unpubl.) **b** Remnants of a sonicated appressorium, still attached to the substratum. The appressorium pore is devoid of cell wall, and includes the site where a peg had penetrated the Mylar substrate (*arrowhead*). Extracellular matrix material (*m*). *Bars* 1.0 μm. (Howard et al. 1991b)

emerges (Fig. 7). With some variation, an area with a similar apparent lack of cell wall has been described in appressoria from many species (Mims and Richardson 1989; Swann and Mims 1991; O'Connell et al. 1992; see discussion in Howard and Ferrari 1989).

So common are these structures among appressoria and hyphopodia (see Fig. 3) that one can be confident in suggesting that the pores must endow infection structures with some special ability. In our consideration of the cytoskeleton, mention was made of the frequent presence of filasomes near these pores. Another common feature are elaborations of cell wall, and plasma membrane, around the periphery of these pores (Fig. 6), also reported from a wide variety of taxa (see Howard and Ferrari 1989; Swann and Mims 1991; O'Connell et al. 1992).

The presence of filasomes and numerous vesicles can be interpreted as indication of localized secretion in the pore regions, as suggested above. The transfer cell-like elaborations of the plasma membrane might also represent yet an additional indication of this process. Rather than invoking a specialized function for growth cones and the like, these wall structures could result from deposition of cell wall at the onset of penetration peg formation. Presumably, hydrolytic enzymes are deposited onto the substratum to aid the penetration process. Cell wall synthesis must also proceed at nearly the same time. It seems possible that any delay in growth of the peg, which might occur during host surface resistance to deformation, could result in the accumulation of peg wall material at the site of peg formation and the deposition of various cell wall elaborations. There is some indication that the chemical composition of these elaborations and of the peg cell wall are similar, but different from the preexisting wall of the appressorium (Bourett and Howard 1990; O'Connell et al. 1992). Growth cones might also be artifacts that arise when localized secretion is blocked by allowing appressoria to form on artificial, nonporous substrates.

A final question concerns possible functions of the pore itself, and, here again, secretion seems to be an obvious candidate. We know that the porosity of appressorial walls can indeed be a factor that limits the exchange of materials with the extracellular environment (see Sect. V). Considering the obvious need to release from appressoria various molecules, for affecting host surface integrity or for synthesizing new fungal wall materials, an efficient means for secretion must exist. A hole through the appressorium wall, in the form of an appressorium pore, is perhaps a solution to this fungal problem: "the proverbial window unto the host" (Howard 1994).

V. Mechanical Forces in Penetration

The synthesis and accumulation by fungi of various solute molecules is of both phylogenetic and physiological interest (Blomberg and Adler 1992; Pfyffer et al. 1990). With regard to infection structures, it has long been speculated that turgor pressures elevated above those of host cells might play a significant role during penetration (see Aist 1976), and since fungi can often accumulate highly elevated levels of, e.g., sugar alcohols (Money 1994), this possibility must be considered seriously. That fungi deform surfaces of host cells or artificial media is well established (e.g., Terhune et al. 1993), but the mechanisms of force generation and application, and the role that force might play during penetration, remain largely unknown. For one example, *M. grisea*, we can state with certainty that mechanical pressures derived from elevated osmotic pressures within specific melanized appressoria, applied through the action of powerful adhesives, **are essential** to the penetration process.

Melanins are dark pigments found in cells of many eukaryotic organisms and exhibit diverse chemical composition, even among fungi (Bell and Wheeler 1986). A melanin derived from pentaketide, 1,8-dihydroxynaphthalene is known to be synthesized by species from diverse fungal genera including *Colletotrichum* and *Magnaporthe*. Appressoria of these fungi become pigmented after they are fully expanded (Bourett and Howard 1990), constituting yet another specialization of appressorial design. The melanin is localized within the appressorial cell wall, in a layer just outside the plasma membrane, except across the appressorial pore.

That the role for this appressorial melanin is important in the early stages of pathogenesis is indicated by the biological activity of compounds that inhibit melanin biosynthesis (Sisler 1986). These compounds, some of which are commercial crop protectants, block pathogenesis by these fungi at the stage of penetration. Using one such inhibitor as well as genetically defined single-gene melanin-deficient mutants of *M. grisea*, Howard and Ferrari (1989) demonstrated that only melanized appressoria develop highly elevated osmotic pressure.

Subsequent experiments measured the porosity of appressorial cell walls with and without melanin, by observing the effect of incubating intact infection structures in solutions containing high concentrations of various solutes (Howard et al. 1991b). If the solute molecules were small enough, they passed through the cell wall, water was drawn out of the cell, and the effect was plasmolysis. However, if the solute molecules were larger than the pores in the cell wall and they did not pass through the cell wall, water was still drawn out of the cell, causing cytorrhysis. This well-established method for measuring wall porosity is known as the solute exclusion technique (Carpita et al. 1979; Money 1990). These experiments indicated that melanin effectively lowered the porosity of appressorial wall to a very small size, <1 nm, thus trapping even fairly small solute molecules and providing for increased turgor. In addition, a modification of these experiments was used to estimate the magnitude of turgor in melanized appressoria of *M. grisea*, an astonishing 80 bar, and to demonstrate penetration of nonbiodegradable surfaces (Fig. 8b).

Clearly, mechanical pressure is an essential force that is generated and applied by the appressoria of this pathogen. Further investigation may show that similar forces are involved during host penetration by other fungi as well, but that is not to say that penetration is accomplished by mechanical forces alone. Despite the enormous turgor pressures which are applied during penetration by *M. grisea*, experimental evidence indicates that some additional factor(s) are necessary for penetration of host surface (Howard et al. 1991b). These additional unknown factors most likely are enzymes which act to modify the host surface.

VI. Enzymes in Penetration

Questions regarding a role for wall-degrading enzymes during penetration are complicated. It is not a question of whether enzymes **can** degrade the wall. "Since all the polymers of a plant cell wall are eventually degraded by microorganisms, for every type of chemical bond in the wall there must be an enzyme that can cleave it" (Walton 1994). Further, it is not just a question of whether a particular strain of a fungal pathogen can produce a certain wall-degrading enzyme, although much has been learned from studies comparing the pathogenicity or virulence of strains and derived mutants which lack the ability to produce an enzyme in question (see Cooper 1987). Complications arise in determining whether a particular fungal enzyme is produced and delivered to the host encounter site and what effects that enzyme or combination of enzymes has upon interaction with the host. This approach, not yet applied in any system, requires the combined use of genetics, cell biology, and physiology, and is very likely the only approach capable of answering our questions definitively. In the meantime, let us briefly consider a few enzymes that could play a critical role during penetration of the host surface – enzymes that represent good subjects for the approach just described.

The outermost layer of aerial host surfaces is the cuticle, which includes a wax layer representing the first plant component encountered by propagules of potential pathogens. Various reports have suggested that some fungi are able to "erode" the cuticle, as evidenced by "erosion tracks", areas devoid of wax crystals, which surround germ tubes or other fungal structures formed on leaf surfaces (see Nicholson and Epstein 1991). In certain cases, it seems likely that alterations in continuity of the wax layer could be caused by the physical displacement of wax crystals during infection structure development (Fig. 9), or during specimen preparation for microscopic examination. Structure of both the pathogen and accompanying extracellular matrices, as well as that of delicate epicuticular wax crystals, are extremely difficult specimens to prepare and examine with a microscope, even when using the most sophisticated techniques available (see Kunoh et al. 1990; Braun and Howard 1994a). At present, we have little evidence to indicate what effects fungi have on cuticular waxes.

Cutinase is one fungal enzyme that has received perhaps the most attention experimentally in terms of determining a possible role during pathogenesis. Yet, controversy surrounds the involvement of cutinase during penetration, even with regard to a single host-pathogen combination (Stahl and Schäfer 1992; Rogers et al. 1994; Chap. 3, Vol. V, Part B). Determining a potential role for any given cutinase is complicated by numerous factors, for example whether or not the particular

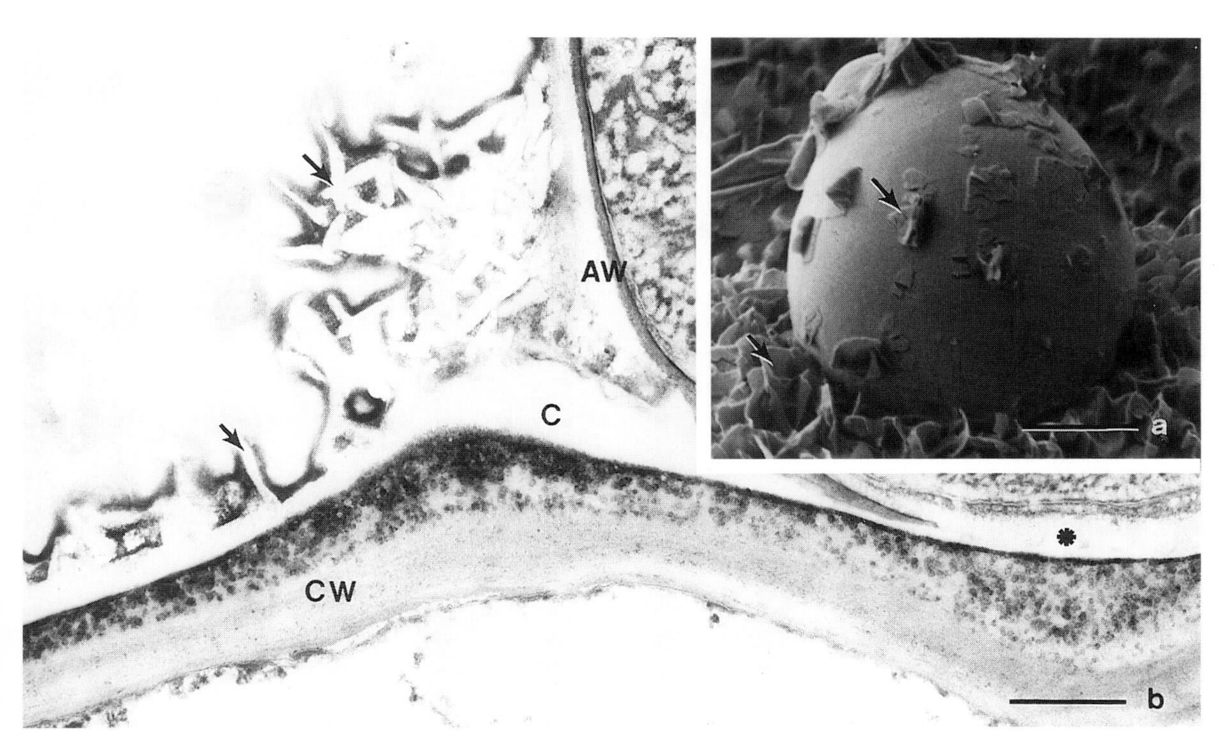

Fig. 9. Movement of epicuticular leaf wax occurs during development of *Magnaporthe grisea* appressoria. **a** Wax crystals (*arrows*) typically decorate the exterior of mature appressoria. Cryoscanning electron micrograph. *Bar* 2.5 μm. **b** Transmission electron micrograph of the region of contact between a freeze-substituted leaf and appressorium. A berm of wax crystals surrounds the appressorium, presumably created during expansion of the infection structure. No wax crystals are evident between the fungus and host surface (*). Appressorium cell wall *AW*; plant cell wall *CW*; cuticle *C*. *Bar* 0.5 μm. (M.A. Ferrari and R.J. Howard, unpubl. micrographs)

gene product is present at the host-pathogen encounter site (e.g., see Sweigard et al. 1992). Nevertheless, persuasive evidence exists in support of a potential role of cutinase during host penetration (Dickman et al. 1989). As mentioned above, the cuticle is responsible for the highly restrictive limits to permeability of both water and gases. Certainly, access to underlying layers at the host surface of any enzyme, which might play a role during penetration, could be restricted by the same permeability limits, considering their large molecular weights. In this regard, it is interesting to note pioneering work by Baker and coworkers (1982), who demonstrated two effects of a cutin esterase on physical characteristics of isolated cutin membranes: permeability to various wall-degrading enzymes was significantly increased, and the strength of the cutin membrane was dramatically decreased. Should either of these effects occur in situ, at the host-pathogen encounter site, one can easily conceive how the progress of pathogenesis would be advanced. Finally, cutinase has also been suggested to play a third possible role, one in fungal adhesion (Deising et al. 1992).

There are many other potential enzymes which might be used by a pathogen to aid in penetration of primary cell walls (Cooper 1987; Walton 1994; Chap. 4, this Vol.). As we have mentioned, there exist enzymes that can effect the integrity of each and every corresponding wall component. As far as is known from biochemical studies of model cell walls, major potential wall-degrading enzymes might include various forms of cellulase, β-1,4-xylanase, arabinosidase, β-1,4-galactanase, β-1,3- and 1,4-glucanases, arabanase, and endopolygalacturonidases, but evidence for their involvement is mostly circumstantial.

VII. Conclusions

An understanding of how an organism develops and responds to its environment at the cellular and molecular levels can be achieved only through a convergence of many disciplines. The questions we must address in our studies of the penetration process are further complicated as they often involve two organisms, the pathogen and the host. As our aim is focused toward broader entities, such as pathogenesis and disease, the challenges we face are formidable.

Several areas of our science are particularly wanting. First, studies of pathogens must include the particular morphogenetic stages of relevance to pathogenesis. From our brief discussions in this chapter it is evident that fungi elaborate extremely specialized structures for the purpose of invading host tissues. Research on penetration mechanisms which utilizes vegetative mycelium for studies of pathogen enzyme production, for example, seems thwarted from the start. Second, there remains much to learn about the host structures which the pathogen must face and breach. Studies of plant cell walls which are based upon **model systems** of cell suspension cultures, for example, have advanced our knowledge considerably. However, the outer barriers of plants vary depending upon host taxonomic affiliation, age, and the particular surface in question, as well as the plant's ability to alter its own structure at the encounter site during fungal attack.

These concerns point to the obvious – we must use the host-pathogen encounter site as the object of our analyses. For example, in-situ immu-

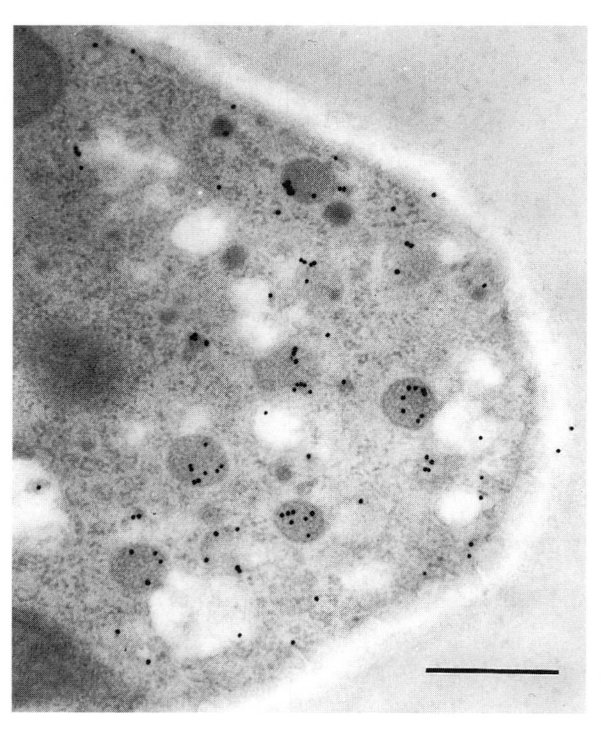

Fig. 10. Protein A-gold, indirect, postembedment immunolocalization of pectinesterase in chemically fixed hyphal tip of *Phytophthora infestans*. Evidence suggests the enzyme is processed through Golgi and delivered to the host-pathogen interface via exocytotic vesicles. *Bar* 0.5 μm. (Förster and Mendgen 1987)

nolocalization of a secreted enzyme within infection structures could offer strong evidence that that enzyme plays some role during pathogenesis (see Fig. 10). Alternatively, host cells could be targeted for analysis or experimentation (e.g., see Kobayashi et al. 1991; Toyoda et al. 1992). If the subjects in this type of anaylsis were genetically defined, one could manipulate the particular gene product in question and elucidate its function.

References

Aist JR (1976) Cytology of penetration and infection – fungi. In: Heitefuss R, Williams PH (eds) Encyclopedia of plant physiology, new series, vol 4. Physiological plant pathology. Springer, Berlin Heidelberg New York, pp 197–221

Aist JR (1981) Development of parasitic conidial fungi in plants. In: Cole GT, Kendrick B (eds) Biology of conidial fungi, vol 2. Academic Press, New York, pp 75–111

Aist JR, Bushnell WR (1991) Invasion of plants by powdery mildew fungi, and cellular mechanisms of resistance. In: Cole GT, Hoch HC (eds) The fungal spore and disease initiation in plants and animals. Plenum Press, New York, pp 321–345

Aist JR, Williams PH (1971) The cytology and kinetics of cabbage root hair penetration by *Plasmodiophora brassicae*. Can J Bot 49:2023–2034

Anderson AJ (1978) Extracellular enzymes produced by *Colletotrichum lindemuthianum* and *Helminthosporium maydis* during growth on isolated bean and corn cell walls. Phytopathology 68:1585–1589

Armentrout VN, Downer AJ (1987) Infection cushion development by *Rhizoctonia solani* on cotton. Phytopathology 77:619–623

Bailey JA, O'Connell RJ, Pring RJ, Nash C (1992) Infection strategies of *Colletotrichum* species. In: Bailey JA, Jeger MJ (eds) *Colletotrichum* biology, pathology and control. CAB International, Wallingford, pp 88–120

Baker CJ, Aist JR, Bateman DF (1980) Ultrastructural and biochemical effects of endopectate lyase on cell walls from cell suspension cultures of bean and rice. Can J Bot 58:867–880

Baker CJ, McCormick SL, Bateman DF (1982) Effects of purified cutin esterase upon the permeability and mechanical strength of cutin membrane. Phytopathology 72:420–423

Bateman DF, Basham HG (1976) Degradation of plant cell walls and membranes by microbial enzymes. In: Heitefuss R, Williams PH (eds) Encyclopedia of plant pathology, new series, vol 4. Physiological plant pathology. Springer, Berlin Heidelberg New York, pp 316–355

Beckett A, Porter R (1988) The use of complementary fractures and low-temperature scanning electron microscopy to study hyphal-host cell surface adhesion between *Uromyces viciae-fabae* and *Vicia faba*. Can J Bot 66:645–652

Beckett A, Tatnell JA, Taylor N (1990) Adhesion and pre-invasion behaviour of urediniospores of *Uromyces viciae-fabae* during germination on host and synthetic surfaces. Mycol Res 94:865–875

Bell AA, Wheeler MH (1986) Biosynthesis and functions of fungal melanins. Annu Rev Phytopathol 24:411–451

Benhamou N, Chamberland H, Pauzé FJ (1990a) Implication of pectic components in cell surface interactions between tomato root cells and *Fusarium oxysporum* f. sp. *radicis-lycopersici*. A cytochemical study by means of a lectin with polygalacturonic acid-binding specificity. Plant Physiol 92:995–1003

Benhamou N, Joosten MHAJ, De Wit PJGM (1990b) Subcellular localization of chitinase and of its potential substrate in tomato root tissues infected by *Fusarium oxysporum* f. sp. *radicis-lycopersici*. Plant Physiol 92: 1108–1120

Blaich R, Heintz C, Wind R (1989) Studies on conidial germination and initial growth of the grapevine powdery mildew *Uncinula necator* on artificial substrates. Appl Microbiol Biotechnol 30:415–421

Blomberg A, Adler L (1992) Physiology of osmotolerance in fungi. Adv Microb Physiol 33:145–212

Bourett TM, Howard RJ (1990) In vitro development of penetration structures in the rice blast fungus *Magnaporthe grisea*. Can J Bot 68:329–342

Bourett TM, Howard RJ (1991) Ultrastructural immunolocalization of actin in a fungus. Protoplasma 163:199–202

Bourett TM, Howard RJ (1992) Actin in penetration pegs of the fungal rice blast pathogen, *Magnaporthe grisea*. Protoplasma 168:20–26

Bracker CE (1968) Ultrastructure of the haustorial apparatus of *Erysiphe graminis* and its relationship to the epidermal cell of barley. Phytopathology 58:12–30

Bracker CE, Littlefield LJ (1973) Structural concepts of host-pathogen interfaces. In: Byrde RJW, Cutting CV (eds) Fungal pathogenicity and the plant's response. Academic Press, London, pp 159–318

Braun EJ, Howard RJ (1994a) Adhesion of *Cochliobolus heterostrophus* conidia and germlings to leaves and artificial surfaces. Exp Mycol 18:211–220

Braun EJ, Howard RJ (1994b) Adhesion of fungal spores and germlings to host surfaces. Protoplasma 181:202–212

Bretscher A (1991) Microfilament structure and function in the cortical cytoskeleton. Annu Rev Cell Biol 7:337–374

Bukovac MJ, Petracek PD (1993) Characterizing pesticide and surfactant penetration with isolated plant cuticles. Pestic Sci 37:179–194

Carlson H, Stenram U, Gustafsson M, Jansson H-B (1991) Electron microscopy of barley root infection by the fungal pathogen *Bipolaris sorokiniana*. Can J Bot 69: 724–2731

Carpita NC, Gibeaut DM (1993) Structural models of primary cell walls in flowering plants: consistency of molecular structure with the physical properties of the walls during growth. Plant J 3:1–30

Carpita N, Sabularse D, Montezinos D, Delmer DP (1979) Determination of the pore size of cell walls of living plant cells. Science 205:1144–1147

Carver TLW, Ingerson SM (1987) Responses of *Erysiphe graminis* germlings to contact with artificial and host surfaces. Physiol Mol Plant Pathol 30:359–372

Carver TLW, Zeyen RJ, Ahlstrand GG (1987) The relationship between insoluble silicon and success or failure of attempted primary penetration by powdery mildew (*Erysiphe graminis*) germilings on barley. Physiol Mol Plant Pathol 31:133–148

Chérif M, Menzies JG, Benhamou N, Bélanger RR (1992) Studies of silicon distribution in wounded and *Pythium ultimum*-infected cucumber plants. Physiol Mol Plant Pathol 41:371–385

Clark CA, Lorbeer JW (1976) Comparative histopathology of *Botrytis squamosa* and *B. cinerea* on onion leaves. Phytopathology 66:1279–1289

Cooper RM (1987) The use of mutants in exploring depolymerases as determinants of pathogenicity. In: Day PR, Jellis GJ (eds) Genetics and plant pathogenesis. Blackwell, Oxford, pp 261–281

Cooper RM, Resende MLV, Flood J, Rowan MG, Beale MH, Potter U (1995) Detection and cellular localization of elemental sulphur in disease-resistant genotypes of *Theobroma cacao*. Nature 379:159–162

Czymmek KJ, Bourett TM, Howard RJ (1996) Immunolocalization of tubulin and actin in thick-sectioned fungal hyphae after freeze substitution fixation and methacrylate de-embedment. J Microsc 181:153–161

Daniels A, Lucus JA, Peberdy JF (1991) Morphology and ultrastructure of W and R pathotypes of *Pseudocercosporella herpotrichoides* on wheat seedlings. Mycol Res 95:385–397

Deising H, Nicholson RL, Haug M, Howard RJ, Mendgen K (1992) Adhesion pad formation and the involvement of cutinase and esterases in the attachment of uredospores to the host cuticle. Plant Cell 4:1101–1111

Deising H, Frittang AK, Kunz S, Mendgen K (1995) Regulation of pectin methlyesterase and polygalacturonate lyase activity during differentiation of infection structures in *Uromyces viciae-fabae*. Microbiology 141:561–571

Dickman MB, Podila GK, Kolattukudy PE (1989) Insertion of cutinase gene into a wound pathogen enables it to infect intact host. Nature 342:446–448

Dijksterhuis J, Veehnuis M, Harder W (1990) Ultrastructural study of adhesion and initial stages of infection of nematodes by conidia of *Drechmeria coniospora*. Mycol Res 94:1–8

Emmett RW, Parbery DG (1975) Appressoria. Annu Rev Phytopathol 13:147–167

Epstein L, Kaur S, Goins T, Kwon YH, Henson JM (1994) Production of hyphopodia by wild-type and three transformants of *Gaeumannomyces graminis* var. *graminis*. Mycologia 86:72–81

Falloon RE, Sutherland PW, Hallett IC (1989) Morphology of *Erysiphe pisi* on leaves of *Pisum sativum*. Can J Bot 67:3410–3416

Förster H, Mendgen K (1987) Immunocytochemical localization of pectinesterases in hyphae of *Phytophthora infestans*. Can J Bot 65:2607–2613

Frank AB (1883) Über einige neue und weniger bekannte Pflanzenkrankheiten. Ber Dtsch Bot Ges 1:I 29–34, II 58–63

Freytag S, Mendgen K (1991) Surface carbohydrates and cell wall structure of in vitro-induced uredospore infection structures of *Uromyces viciae-fabae* before and after treatment with enzymes and alkali. Protoplasma 161:94–103

Freytag S, Bruscaglioni L, Gold RE, Mendgen K (1988) Basidiospores of rust fungi (*Uromyces* species) differentiate infection structures in vitro. Exp Mycol 12:275–283

Gold RE, Mendgen K (1991) Rust basidiospore germlings and disease initiation. In: Cole GT, Hoch HC (eds) The fungal spore and disease initiation in plants and animals. Plenum, New York, pp 67–99

Gow NAR (1989) Control of extension of the hyphal apex. In: McGinnis MR, Borgers M (eds) Current topics in medical mycology, vol 3. Springer, Berlin Heidelberg New York, pp 109–152

Grove SN, Sweigard JA (1980) Cytochalsin A inhibits spore germination and hyphal tip growth in *Gilbertella persicaria*. Exp Mycol 4:239–250

Hahn M, Mendgen K (1992) Isolation by ConA binding of haustoria from different rust fungi and comparison of their surface qualities. Protoplasma 170:95–103

Hamer JE, Howard RJ, Chumley FG, Valent B (1988) A mechanism for surface attachment in spores of a plant pathogenic fungus. Science 239:288–290

Hardham AR (1992) Cell biology of pathogenesis. Annu Rev Plant Physiol Plant Mol Biol 43:491–526

Hardham AR, Gubler F, Duniec J, Elliott J (1991) A review of methods for the production and use of monoclonal antibodies to study zoosporic plant pathogens. J Microsc 162:305–318

Harold FM (1990) To shape a cell: an inquiry into the causes of morphogenesis of microorganisms. Microbiol Rev 54:381–431

Hasselbring H (1906) The appressoria of the anthracnoses. Bot Gaz 42:135–142

Hau FC, Rush MC (1982) Preinfectional interactions between *Helminthosporium oryzae* and resistant and susceptible plants. Phytopathology 72:285–292

Heath MC, Stumpf MA (1986) Ultrastructural observations of penetration sites of the cowpea fungus in untreated and silicon-depleted French bean cells. Physiol Mol Plant Pathol 29:27–39

Heintz C, Blaich R (1990) Ultrastructural and histochemical studies on interactions between *Vitis vinifera* L. and *Uncinula necator* (Schw.) Burr. New Phytol 115:107–117

Holloway PJ (1982) The chemical constitution of plant cuticles. In: Cutler DF, Alvin KL, Price CE (eds) The plant cuticle. Academic Press, New York, pp 45–86

Howard RJ (1994) Cell biology of pathogenesis. In: Zeigler RS, Leong SA, Teng PS (eds) Rice blast disease. CAB International, Wallingford, pp 3–22

Howard RJ, Ferrari MA (1989) Role of melanin in appressorium function. Exp Mycol 13:403–418

Howard RJ, Bourett TM, Ferrari MA (1991a) Infection by *Magnaporthe*: an in vitro analysis. In: Mendgen K, Lesemann D-E (eds) Electron microscopy of plant pathogens. Springer, Berlin Heidelberg New York, pp 251–264

Howard RJ, Ferrari MA, Roach DH, Money NP (1991b) Penetration of hard substrates by a fungus employing enormous turgor pressures. Proc Natl Acad Sci USA 88:11281–11284

Huffine MS (1994) The infection of *Cynodon dactlyon* by *Gaeumannomyces graminis*. MS Thesis, University of Texas, Arlington

Hwang C-S, Kolattukudy PE (1995) Isolation and characterization of genes expressed uniquely during appressorium formation by *Colletotrichum gloeosporioides* conidia induced by the host surface wax. Mol Gen Genet 247:282–294

Hwang C-S, Flaishman MA, Kolattukudy PE (1995) Cloning of a gene expressed during appressorium formation by *Colletotrichum gloeosporioides* and a marked decrease in virulence by disruption of this gene. Plant Cell 7:183–193

Hyde KD, Greenwood R, Jones EBG (1993) Spore attachment in marine fungi. Mycol Res 97:7–14

Inoue S, Aist JR, Macko V (1994) Earlier papilla formation and resistance to barley powdery mildew induced

by a papilla-regulating extract. Physiol Mol Plant Pathol 44:433–440
Jelitto TC, Page HA, Read ND (1994) Role of external signals in regulating the pre-penetration phase of infection by the rice blast fungus, *Magnaporthe grisea*. Planta 194:471–477
Jones EBG (1994) Fungal adhesion. Mycol Res 98:961–981
Khan SR, Kimbrough JW (1974) Morphology and development of *Termitaria snyderi*. Mycologia 66:446–462
Kerstiens G (1996) Plant cuticles. An integrated functional approach. BIOS Scientific Publ, Ltd., Oxford
Knauf GM, Welter K, Müller M, Mendgen K (1989) The haustorial host-parasite interface in rust-infected bean leaves after high-pressure freezing. Physiol Mol Plant Pathol 34:519–530
Kobayashi I, Kobayashi Y, Yamaoka N, Kunoh H (1991) An immunofluorescent cytochemical technique applying micromanipulation to detect microtubules in plant tissues inoculated with fungal spores. Can J Bot 69:2634–2636
Kobori H, Sato M, Osumi M (1992) Relationship of actin organization to growth in the two forms of the dimorphic yeast *Candida tropicalis*. Protoplasma 167:193–204
Kolattukudy PE (1980) Biopolyester membranes of plants: cutin and suberin. Science 208:990–1000
Kunoh H, Ishizaki H (1976) Accumulation of chemical elements around the penetration sites of *Erysiphe graminis hordei* on barley leaf epidermis: (III) micromanipulation and x-ray microanalysis of silicon. Physiol Plant Pathol 8:91–96
Kunoh H, Nicholson RL, Yosioka H, Yamaoka N, Kobayashi I (1990) Preparation of the infection court by *Erysiphe graminis*: degradation of the host cuticle. Physiol Mol Plant Pathol 36:397–407
Kwon YH, Hoch HC, Aist JR (1991) Initiation of appressorium formation in *Uromyces appendiculatus*: organization of the apex, and the responses involving microtubules and apical vesicles. Can J Bot 69:2560–2573
Littlefield LJ, Bracker CE (1972) Ultrastructural specialization at the host-pathogen interface in rust-infected flax. Protoplasma 74:271–305
Littlefield LJ, Heath MC (1979) Ultrastructure of rust fungi. Academic Press, New York
Manners JM, Gay JL (1983) The host-parasite interface and nutrient transfer in biotrophic parasitism. In: Callow JA (ed) Biochemical plant pathology. Wiley, Chichester, pp 163–195
Mauch F, Staehelin LA (1989) Functional implications of the subcellular localization of ethylene-induced chitinase and β-1,3-glucanase in bean leaves. Plant Cell 1:447–457
McCann MC, Wells B, Roberts K (1990) Direct visualization of cross-links in the primary plant cell wall. J Cell Sci 96:323–334
McKeen WE, Rimmer SR (1973) Initial penetration process in powdery mildew infection of susceptible barley leaves. Phytopathology 63:1049–1053
McRae CF, Stevens GR (1990) Role of conidial matrix of *Colletotrichum orbiculare* in pathogenesis of *Xanthium spinosum*. Mycol Res 94:890–896
Mercure EW, Kunoh H, Nicholson RL (1995) Visualization of materials released from adhered, ungerminated conidia of *Colletrotrichum graminicola*. Physiol Mol Plant Pathol 46:121–135
Millar CS (1981) Infection processes in conifer needles. In: Blakeman JP (ed) Microbial ecology of the phylloplane. Academic Press, London, pp 185–209
Mims CW, Richardson EA (1989) Ultrastructure of appressorium development by basidiospore germlings of the rust fungus *Gymnosporangium juniperi-virginianae*. Protoplasma 148:111–119
Money NP (1990) Measurement of hyphal turgor. Exp Mycol 14:234–242
Money NP (1994) Osmotic adjustment and the role of turgor in filamentous fungi. In: Wessels JGH, Meinhardt F (eds) The mycota, vol 2. Growth, differentiation and sexuality. Springer, Berlin Heidelberg New York, pp 67–88
Myers DF, Fry WE (1978) The development of *Gloeocercospora sorghi* in sorghum. Phytopathology 68:1147–1155
Nicholson RL, Epstein L (1991) Adhesion of fungi to the plant surface. Prerequisite for pathogenesis. In: Cole GT, Hoch HC (eds) The fungal spore and disease initiation in plants and animals. Plenum, New York, pp 3–23
Nusbaum CJ (1938) A cytological study of the resistance of apple varieties to *Gymnosporangium juniperi-virginianae*. J Agric Res 51:573–596
O'Connell RJ, Nash C, Bailey JA (1992) Lectin cytochemistry: a new approach to understanding cell differentiation, pathogenesis and taxonomy in *Colletotrichum*. In: Bailey JA, Jeger MJ (eds) *Colletotrichum* biology, pathology and control. CAB International, Wallingford, pp 67–87
Pain NA, O'Connell RJ, Bailey JA, Green JR (1992) Monoclonal antibodies which show restricted binding to four *Colletotrichum* species: *C. lindemuthianum*, *C. malvarum*, *C. orbiculare* and *C. trifolii*. Physiol Mol Plant Pathol 40:111–126
Pfyffer GE, Boraschi-Gaia C, Weber B, Hoesch L, Orpin CG, Rast DM (1990) A further report on the occurrence of acyclic sugar alcohols in fungi. Mycol Res 94:219–222
Porto MDM, Grau CR, de Zoeten GA, Gaard G (1988) Histopathology of *Colletotrichum trifolii* on alfalfa. Phytopathology 78:345–349
Post-Beittenmiller D (1996) Biochemistry and molecular biology of wax production in plants. Annu Rev Plant Physiol Plant Mol Biol 47:405–430
Rauscher M, Mendgen K, Deising H (1995) Extracellular proteases of the rust fungus *Uromyces viciae-fabae*. Exp Mycol 19:26–34
Roberts K (1990) Structures at the plant cell surface. Curr Opin Cell Biol 2:920–928
Roberson RW (1992) The actin cytoskeleton in hyphal cells of *Sclerotium rolfsii*. Mycologia 84:41–51
Rogers LM, Flaishman MA, Kolattukudy PE (1994) Cutinase gene disruption in *Fusarium solani* f. sp. *pisi* decreases its virulence on pea. Plant Cell 6:935–945
Rohringer R, Chong J, Gillespie R, Harder DE (1989) Gold-conjugated arabinogalactan-protein and other lectins as ultrastructural probes for the wheat/stem rust complex. Histochemistry 91:383–393
Samuels AL, Glass ADM, Menzies JG, Ehret DL (1994) Silicon in cell walls and papillae of *Cucumis sativis* during infection by *Sphaerotheca fuliginea*. Physiol Mol Plant Pathol 44:237–242
Sisler HD (1986) Control of fungal diseases by compounds acting as antipenetrants. Crop Prot 5:306–313
Smart MG (1991) The plant cell wall as a barrier to fungal invasion. In: Cole GT, Hoch HC (eds) The fungal spore

and disease initiation in plants and animals. Plenum, New York, pp 47–66

Snyder BA, Leite B, Hipskind J, Butler LG, Nicholson RL (1991) Accumulation of sorhum phytoalexins induced by *Colletrotrichum graminicola* at the infection site. Physiol Mol Plant Pathol 39:463–470

Stahl DJ, Schäfer W (1992) Cutinase is not required for fungal pathogenicity on pea. Plant Cell 4:621–629

Sutton BC (1968) The appressoria of *Colletotrichum graminicola* and *C. falcatum*. Can J Bot 46:873–876

Swann EC, Mims CW (1991) Ultrastructure of freeze-substituted appressoria produced by aeciospore germlings of the rust fungus *Arthuriomyces peckianus*. Can J Bot 69:1655–1665

Sweigard JA, Chumley FG, Valent B (1992) Disruption of a *Magnaporthe grisea* cutinase gene. Mol Gen Genet 232:183–190

Takahashi K, Aist JR, Israel HW (1985) Distribution of hydrolytic enzymes at barley powdery mildew encounter sites: implications for resistance associated with papilla formation in a compatible system. Physiol Plant Pathol 27:167–184

Terhune BT, Allen EA, Hoch HC, Wergin W, Erbe EF (1991) Stomatal ontogeny and morphology in *Phaseolus vulgaris* in relation to infection structure initiation by *Uromyces appendiculatus*. Can J Bot 69:477–484

Terhune BT, Bojko RJ, Hoch HC (1993) Deformation of stomatal guard cell lips and microfabricated artificial topographies during appressorium formation by *Uromyces*. Exp Mycol 17:70–78

Toyoda K, Shiraishi T, Kobayashi I, Yamaoka N, Kunoh H (1992) Treatment of a single plant cell with chemicals. Can J Bot 70:225–227

Tunlid A, Jansson H-B, Nordbring-Hertz B (1992) Fungal attachment to nematodes. Mycol Res 96:401–412

Uchiyama T, Ogasawara N, Nanba Y, Ito H (1979) Conidial germination and appressorial formation of the plant pathogenic fungi on the coverglass or cellophane coated with various lipid components of plant leaf waxes. Agric Biol Chem 43:383–384

Van Dyke CG, Mims CW (1991) Ultrastructure of conidia, conidium germination, and appressorium development in the plant pathogenic fungus *Colletotrichum truncatum*. Can J Bot 69:2455–2467

Walker J (1980) *Gaeumannomyces*, *Linocarpon*, *Ophiobolus* and several other genera of scolecospored ascomycetes and *Phialophora* conidial states, with a note on hyphopodia. Mycotaxon 11:1–129

Walton JD (1994) Deconstructing the cell wall. Plant Physiol 104:1113–1118

Winslow MD (1992) Silicon, disease resistance, and yield of rice genotypes under upland cultural conditions. Crop Sci 32:1208–1213

Xia JQ, Lee FN, Scott HA, Raymond LR (1992) Development of monoclonal antibodies specific for *Pyricularia grisea*, the rice blast pathogen. Mycol Res 96:867–873

Xiao J-Z, Watanabe T, Kamakura T, Oshima A, Yamaguchi I (1994) Studies on cellular differentiation of *Magnaporthe grisea*. Physicochemical aspects of substratum surfaces in relation to appressorium formation. Physiol Mol Plant Pathol 44:227–236

Yaegashi H, Matsuda I, Sato Z (1987) Production of appressoria at the tips of hyphae of *Pyricularia oryzae*. Ann Phytopathol Soc Jpn 53:203–209

4 Fungal Invasion Enzymes and Their Inhibition

G. De Lorenzo, R. Castoria, D. Bellincampi, and F. Cervone

CONTENTS

I. Introduction

The first lines of defense of a plant against phytopathogenic fungi are the external cuticle and the polysaccharide-rich cell wall (Fig. 1). The vast majority of fungi need to breach these barriers to gain access to the plant tissue, and, once inside the tissue, to degrade the cell wall components in order to sustain their growth and to complete the invasion process. It is generally accepted that the enzymatic arsenal of the fungus contributes, together with mechanical forces (Howard et al. 1991; Chap. 3, this Vol.), to the degradation of both cuticle and cell walls.

The interaction between a plant and a fungus does not always result in the invasion of the plant tissues and disease symptoms. Plants are resistant to most of the existing phytopathogenic fungi, and fungal development often arrests soon after penetration. Fungi that have acquired the capability to invade the tissues of a particular plant (the host) and to suppress and/or neutralize its resistance mechanisms have established a so-called basic compatibility. Specific compatibility (race-cultivar specificity) is generally considered as superimposed on basic compatibility (Chap. 16, Vol. V, Part B). In incompatible interactions, blockage of the development of an invading fungus is often caused by active defense responses of the plant. The effectiveness of the plant defense responses against a pathogen lies both in the magnitude and in the rapidity of their onset. The triggering of these events is preceded by the recognition of the microorganism by the plant cell. Fungal molecules involved in the early stages of the plant-pathogen interaction, such as cell wall-degrading enzymes, may therefore play a role in the recognition process and in the trigger or plant defense mechanisms.

Biochemical evidence on the correlation between cell wall-degrading enzymes (CWDE) and disease symptoms has been extensively reviewed (Bateman and Basham 1976; Collmer and Keen 1986; Cooper 1987; Hahn et al. 1989); mechanisms of cutin degradation are discussed elsewhere in Chap. 3, this Vol. In this chapter, we examine several characteristics of CWDE and the interactions of these enzymes with the components of the plant cell wall; we further discuss their possible roles not only in invasion and pathogenesis, but

Dipartimento di Biologia Vegetale, Università di Roma La Sapienza, Piazzale A. Moro, 00185 Rome, Italy

The Mycota V Part A
Plant Relationships
Carroll/Tudzynski (Eds.)

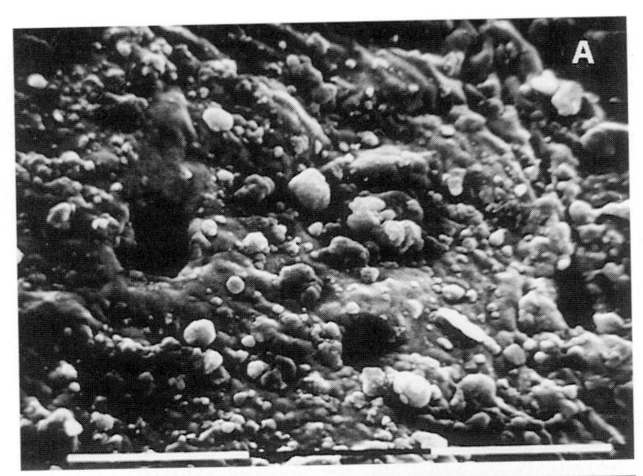

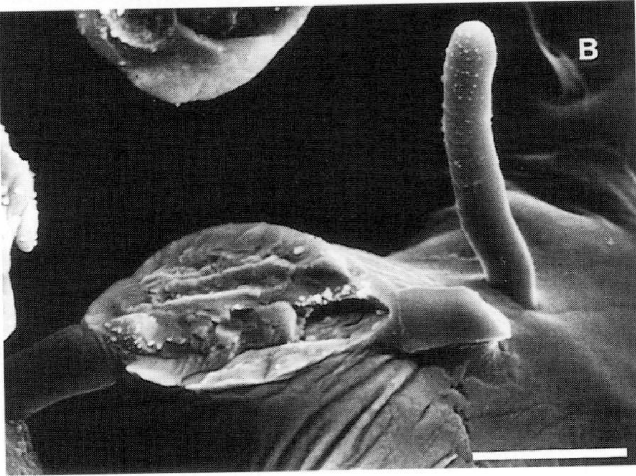

Fig. 1A,B. Fungi penetrating and invading the host plant tissue. **A** Holes caused by *Spilocaea oleagina* in the cuticle of an olive leaf. SEM photograph taken after detachment of three conidiophores. *Bar* 10 μm. (Original photograph Prof. A. Graniti, Dipartimento di Patologia Vegetale, Università di Bari, Bari, Italy). **B** Hyphae of *Botrytis cinerea* which have grown through the walls of mesophyll cells in a rose petal. Low temperature SEM. *Bar* 10 μm. (Original photograph Dr. B. Williamson, Scottish Crop Research Institute, Invergowrie, Scotland)

also in induction of the plant defense mechanisms and determination of specificity.

II. Types of Cell Wall-Degrading Enzymes (CWDE)

Degradation of plant cell walls is caused by many pathogenic and saprophytic fungi that extensively invade the plant tissue either intracellularly or extracellularly. Although minimal, breakdown of cell wall also occurs in diseases associated with nondestructive fungal invasion of the plant tissue. Cell wall degradation may be important to fungi not only for penetration and ramification inside the plant tissue but also for releasing, from the wall polysaccharides, nutrients necessary for growth. Most fungi produce a wide array of enzymes capable of depolymerizing the polysaccharides of the plant cell wall. Many of these enzymes are extracellularly targeted glycoproteins which are inducible upon exposure of the fungus to plant cell walls. The large number of CWDE reflects the complexity of the cell wall polysaccharide structures: activities that have been identified include pectic enzymes, cellulase, arabinase, xylanase, and galactanase (Table 1). Progress on the identification of CWDE with new specificities is restrained by our limited knowledge of the plant cell wall structure. Fungal CWDE that attack uncommon linkages in the wall polymers can be studied only if their substrates (often complex and rare oligosaccharides) are purified from cell walls in large quantities and characterized. Also, since most CWDE are substrate-induced and catabolite-repressed, an additional problem with the identification of novel CWDE is the limited knowledge of the conditions which are optimal for the induction of these enzymes either in vitro and in planta.

Generally, CWDE are considered important for the basic pathogenicity of a fungus; so far there is no evidence that they may be determinant of race-cultivar specificity. The lack of specificity of the action as well as of the mechanisms of the induction of CWDE is suggested by the observed ability of fungi to grow in vitro on cell walls from either host and nonhost plants and from either mono- and dicotyledonous plants (Collmer and Keen 1986).

A. Pectic Enzymes

Since pectin is a major component of the cell wall of dicotyledonous plants, pectin-degrading enzymes, often termed pectic enzymes, have been the group of CWDE most extensively studied. Nevetheless, their role in invasion and pathogenesis is not yet well defined.

1. Pectic Enzymes as General Pathogenesis Factors

When fungi are grown on plant cell wall material in vitro, pectic enzymes are invariably the first CWDE to be secreted, followed by hemicellulases

Table 1. Types of cell wall-degrading enzymes produced by fungi

Enzyme	Reference[a]
Pectic enzymes	
Hydrolases	
*endo*polygalacturonase	Stratilová et al. (1993)
*exo*polygalacturonase	Valsangiacomo and Gessler (1992)
rhamnogalacturonase	Schols et al. (1990)
Lyases	
*endo*pectate lyase	Robertsen (1989)
*exo*pectate lyase	Linhardt et al. (1986)
*endo*pectin lyase	Linhardt et al. (1986)
*exo*pectin lyase	Linhardt et al. (1986)
Esterases	
pectinmethylesterase	Schejter and Marcus (1988)
rhamnogalacturonan acetylesterase	Searle-van Leeuwen et al. (1992)
Cellulase	
endo-β-1,4-glucanase (carboxymethylcellulase)	Vincken et al. (1994)
exo-β-1,4-glucanase	Collings et al. (1988)
endo-β-1,4-glucan cellobiohydrolase	Blanchette et al. (1989)
β-glucosidase	Collings et al. (1988)
Depolymerases for neutral matrix polisaccharides	
endo-α-1,5-arabinase	Flipphi et al. (1993)
α-L-arabinofuranosidase	Cooper (1987)
endo-β-1,4-xylanase	Apel et al. (1993)
β-xylosidase	Tan et al. (1987)
endo-β-1,4-glucanase	Gilkes et al. (1991)
exo-β-1,4-glucanase	Collings et al. (1988)
β-glucosidase	Rapp (1989)
exo-β-1,3 glucanase (laminarinase)	Stahmann et al. (1993)
exo-β-1,3 1,6 glucanase (cinereanase/laminarinase)	Stahmann et al. (1993)
endo-β-1,6-glucanase	Rapp (1989)
β-1,4-galactanase	Urbanek and Zalewska-Sobczak (1986)
α-galactosidase	Cooper (1984)

[a] Only one reference is given for each enzyme.

and cellulases (Jones et al. 1972; Mankarios and Friend 1980). The action of pectic enzymes and, in particular, of endopolygalacturonase on cell walls appears to be a prerequisite for wall degradation by other CWDEs (Karr and Albersheim 1970). The sequential appearance of the CWDE has been explained by a better accessibility of the pectic polymers in the primary cell wall to enzymatic attack and subsequent rapid release of pectic enzyme inducers (Mankarios and Friend 1980; Collmer and Keen 1986). Only after pectic enzymes have acted on their substrates, does the cellulose-xyloglucan framework, which is normally embedded in the pectin matrix (Carpita and Gibeaut 1993), become accessible; only then are inducers for cellulase and hemicellulase released. Recent evidence suggests that the easier access of pectic enzymes through the undamaged cell walls may depend on a peculiar structure of these enzymes that facilitates the function of binding to and cleaving buried galacturonate polymers. The three-dimensional structure of pectate lyase from *Erwinia chrysanthemi* has been determined. The enzyme folds into a unique motif of parallel β-strands coiled into a large helix. Since sequence similarities are shared between bacterial pectate lyase and fungal pectin lyase (González-Candelas and Kolattukudy 1992), it has been suggested that the parallel β-helix motif may also occur in fungal pectin lyases and perhaps in other pectic enzymes (Yoder et al. 1993).

Pectic enzymes exhibit a high degree of polymorphism (Cervone et al. 1986b). The molecular basis of this polymorphism has been elucidated for only a few enzymes. The endopolygalacturonase secreted by the phytopathogenic fungus *Fusarium moniliforme* consists of four molecular mass glycoforms, which derive from a single endopolygalacturonase gene product (Caprari et al. 1993a). In *Aspergillus niger* and *Sclerotinia sclerotiorum*, endopolygalacturonase isoenzymes are encoded by different genes (Bussink et al. 1992; Fraissinet-Tachet et al. 1995), and a similar situation has been described for pectin and pectate lyase isoenzymes (Harmsen et al. 1990; Templeton et al. 1994; Guo et al. 1995b).

The occurrence of multiple isoforms, each of which may in turn comprise multiple glycoforms, may have a physiological significance. Redundancy in components of the offense arsenal may allow invasion in a variety of different conditions and hosts, as well as protecting the fungus from losses of pathogenicity functions, such as the cell wall-degrading enzymes. Different forms of the same enzyme may differ in terms of stability, specific activity, pH optimum, substrate preference or degradation kinetics, and types of oligosaccharides released. They may also differ in the mechanisms or extent of their induction in planta. Unfortunately, insufficient information is available about in planta expression of pectic enzymes, which may differ greatly from in vitro expression (Fraissinet-Tachet et al. 1995). It has also been suggested that different forms of the same enzyme are related to the host range and specificity of a fungus (see Sect. II.A.4 below: Pectic Enzymes as Determinants of Specificity).

Activities of pectic enzymes and, in general, of CWDE may be related to the strategy of host colonization by a fungus. Enzymes with a high specific activity are expected to be produced by fungi that invade the tissue in an aggressive manner, while less efficient enzymes with lower specific activity are likely to be utilized by fungi that behave as biotrophic parasites or mutualists. For example, *Venturia inaequalis* establishes a kind of biotrophic relationship with the host and grows in intimate contact with cells of the upper epidermis, without penetrating or invading the lower tissue. This fungus produces a polygalacturonase with an "exo" mode of action that is far less destructive than a polygalacturonase with an "endo" mode of action (Valsangiacomo and Gessler 1992).

Regulation of enzyme production in planta may also contribute to the strategy of fungal invasion, with massive amounts of enzymes associated with rapid and disruptive invasion of the host tissue, and low and strictly regulated amounts of enzymes produced by nondestructive biotrophic or symbiotic fungi. For example, mycorrhizal ericoid fungi have been reported to produce very limited quantities of polygalacturonase both in vitro and in vivo (Cervone et al. 1988). However, the same fungi, when growing saprophytically on nonhost plants, produce much higher quantities of polygalacturonase (P. Bonfante, pers. comm.).

The pH status of the walls at the onset of and during pathogenesis can restrain in planta the ability of fungal enzymes to degrade the plant cell walls. Under physiological conditions, the pH of cell walls is between 5 and 6, but may rise during the late stages of the infection process or even decrease in infections characterized by breakdown of cells with extensive release of the vacuolar contents. Pectic enzymes vary widely in their pH optima, ranging from 5 or less for polygalacturonases, through 6–8 for pectin lyases, to 8 or above for pectate lyases. Each class of enzyme is likely to operate at specific stages of pathogenesis and the relative importance of each may differ in different host-fungus combinations. Together, these three classes of enzymes cover a broad range of pH and guarantee that a fungus can cope with the physiological and pathological variations possible in most plant tissues subject to invasion. Pectic enzyme production and variation of pH during the infection of *Phaseolus vulgaris* by *Colletotrichum lindemuthianum* have been analyzed in detail. The action of endopolygalacturonase may be crucial during the initial biotrophic phase of the infection process, when wall pH is low and degradation is highly localized around invading hyphae. At a later stage of infection (4 days after inoculation), cell wall pH rises and exceeds 6.5. At this value, polygalacturonase activity is greatly reduced, while activity of pectin lyase increases, concomitantly with a switch from a biotrophic to a highly destructive necrotrophic growth of the fungus (Wijesundera et al. 1989).

Purified pectic enzymes capable of cleaving the α-1,4-glycosidic bonds of homogalacturonan in an endo manner cause plant tissue maceration (cell separation). These enzymes also cause injury and death of unplasmolyzed plant cells (Fig. 2). It has been proposed that cell death results from a physical weakening of the wall caused by degradation of pectic polysaccharides, such that the wall can no longer resist the pressure exerted by the protoplast (Bateman and Basham 1976). The most convincing evidence supporting this theory is that plasmolyzed cells are not killed when treated with purified pectic enzymes, and that cell wall degradation, maceration, and cell death appear simultaneously with no spatial and temporal gaps (Gardner and Kado 1976). However, it has been argued that cellulose and hemicellulose, and not

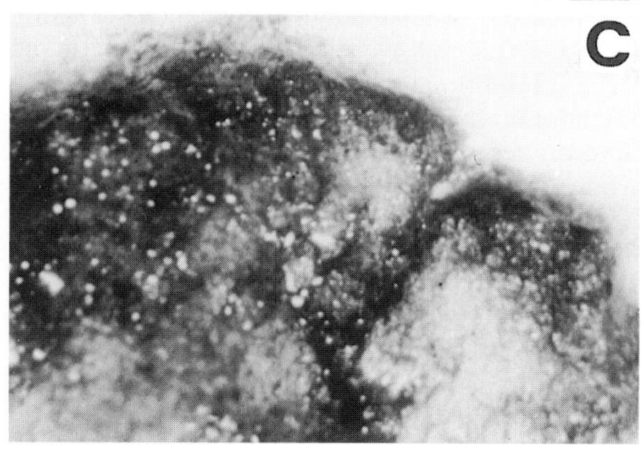

Fig. 2A–C. Effect of *Fusarium moniliforme* endopolygalacturonase on medullary tissue of potato. **A** Buffer control. **B** Tissue treated with endopolygalacturonase in the presence of excess PGIP of *P. vulgaris*. **C** Tissue treated with endopolygalacturonase. After treatment tissue was stained with Evans blue. *Staining* indicates dead cells

the pectic polymers, are the load-bearing components of primary cell walls (Hahn et al. 1989), and therefore that degradation of the pectic polysaccharides is unlikely to weaken the wall to such an extent that protoplasts would burst. Some observations support the hypothesis that cell death may not be a mere consequence of maceration. Homogeneous fungal endopolygalacturonases exhibited both macerating and killing activity when applied to potato tuber disks; however, when an excess of the inhibitor protein PGIP was added, killing activity in the absence of detectable macerating activity was observed (Fig. 2). This suggests that maceration and killing are two different functions of the endopolygalacturonase (Cervone et al. 1989a).

2. Pectic Enzymes as Elicitors of Plant Defense Responses

Active plant defense responses include the rapid death of the plant cells which first come into contact with the pathogen (hypersensitive response), the induction of the phenylpropanoid pathway and the synthesis of lignin, the synthesis and the accumulation of antimicrobial compounds named phytoalexins, the synthesis of hydroxyproline-rich glycoproteins (HRGPs) and fungal wall-degrading enzymes (chitinases, glucanases), and the production of ethylene. Many different molecules, termed elicitors, have been shown to induce defense responses when applied to plant tissues (Hahn et al. 1989; Darvill et al. 1992; Chaps. 5, 6, this Vol.).

Pectic enzymes with an endo mode of action behave as elicitors. The ability of a *Rhizopus stolonifer* endopolygalacturonase to elicit a phytoalexin biosynthesis-related enzyme (casbene synthase) in castor bean seedlings was described by Lee and West (1981). The authors suggested that the elicitor activity of the enzyme was mediated by the products released from the plant cell wall, and demonstrated that the catalytic activity of the endopolygalacturonase was necessary for its eliciting activity (Bruce and West 1982). Since these initial observations, the number of reports on the elicitor activity of fungal endopolygalacturonases in different systems has steadily increased. Endopolygalacturonase from *Rhizopus stolonifer* elicited the synthesis of the phytoalexin pisatin in pea and of the proteinase inhibitor I in tomato leaves (Walker-Simmons et al. 1984). The same enzyme elicited the synthesis of lignin and three forms of peroxidase in castor bean (Bruce and West 1989). The endopolygalacturonase and endopectate lyase of *Cladosporium cucumerinum* elicited the synthesis of lignin in cucumber hypocotyls (Robertsen 1987, 1989). Endopolygalacturonase from *Fusarium moniliforme* elicited the production of the phytoalexins 6-methoxymellein and glyceollin in, respectively, carrot suspension-cultured cells (Marinelli et al. 1991) and soybean cotyledons (C. Caprari et al., unpubl. results of our laboratory). Two endopolygalacturonases from *Sclerotinia sclerotiorum* elicited the synthesis of glyceollin in soybean hypocotyls (Favaron et al. 1992). Two endopolygalacturonases of *Aspergillus japonicus* elicited ethylene production in tomato and orange fruits (Baldwin and Pressey 1989). Endopolygalacturonase purified from *Colletotrichum lindemuthianum* elicited production of β-1,3-glucanase in French bean (Lafitte et al. 1993).

Elicitor activity of endopolygalacturonases suggests that during an attempted invasion the enzymes may function, simultaneously, but antithetically, as efficient agents of fungal aggression, or as potential signal molecules (elicitors). The early timing of endopolygalacturonase production is compatible with both functions. The ability of endopolygalacturonase to act as a signal molecule raises the possibility that it may function as a determinant of specificity in the plant-fungus interactions (see below).

3. Oligogalacturonides Released by Endopolygalacturonases Are the Elicitors of Plant Defense Responses

The available evidence indicates that fungal endopolygalacturonases are not directly responsible for the induction of plant defense responses, but are rather preelicitors that release from the plant cell wall the true elicitors (Anderson 1989), the oligogalacturonides. The first evidence that a galacturonide-rich fraction released from the walls induces accumulation of the phytoalexin glyceollin in soybean was given by Hahn et al. (1981), who named the active component of the fraction endogenous elicitor. Since then, oligogalacturonides with degree of polymerization between 10 and 15 have been reported to elicit a variety of defense responses, including the accumulation of phytoalexins (Davis et al. 1986), the synthesis of endo-β-1,3-glucanase (Davis and Hahlbrock 1987) and chitinase (Broekaert and Peumans 1988), the synthesis of lignin (Robertsen 1986; Bruce and West 1989), and elicitation of necrosis (Marinelli et al. 1991). In carrot suspension-cultured cells, oligogalacturonides induced the activation of PAL, a key enzyme in phenylpropanoid and lignin biosynthesis (De Lorenzo et al. 1987). Responses elicited in carrot cells by oligogalacturonides appeared to be a consequence of a transcriptional activation of several defense genes (Messiaen and Van Cutsem 1993).

At present, only a single report has suggested that oligogalacturonides are released in planta at strategic sites during the infection by a fungal pathogen (Benhamou et al. 1990). The mechanisms by which oligogalacturonides act as signals for activation of defense responses remain unknown. Several rapid responses occurring at the plant cell surface may be part of the transduction of the oligogalacturonide signal (Chap. 5, this Vol.). Oligogalacturonides have been shown to be internalized through a rapid receptor-mediated endocytosis in soybean suspension-cultured cells (Horn et al. 1989) and to induce, within 5 min, transient stimulation of cytoplasmic Ca^{2+} influx, K^+ efflux, cytoplasmic acidification, and depolarization of plasma membrane of tobacco cultured cells (Mathieu et al. 1991; Thain et al. 1995). The elevation of cytosolic free Ca^{2+} levels induced by oligogalacturonides has been also observed in carrot protoplasts (Messiaen et al. 1993). It is of interest that the increase of external pH may represent, in vivo, a stabilization factor of the elicitor-active oligogalacturonides, since the lower activity exhibited by endopolygalacturonase at higher nonoptimal pH values would prevent their degradation to shorter inactive products. Oligogalacturonides have been shown to generate H_2O_2 production in cucumber (Svalheim and Robertsen 1993), soybean (Apostol et al. 1989; Legendre et al. 1993; Chandra and Low 1995), and castor bean (Bruce and West 1989). In plants, production of reactive oxygen species and resulting biochemical changes, such as the oxidative cross-linking of the cell wall and lipid peroxidation, have been proposed to contribute to both programmed cell death and rapid activation of defense responses (Levine et al. 1994). Oligogalacturonides also enhance the in vitro phosphorylation of a 34-kDa protein associated with plasma membranes of potato and tomato (pp34; Jacinto et al. 1993; Chandra and Low 1995; Reymond et al. 1995).

Oligogalacturonides bind Ca^{2+}-forming intermolecular complexes named egg boxes. Conformational analysis has established that the egg box conformation requires oligopectate fragments with a degree of polymerization higher than 10 (Kohn 1975). The correlation between the degree of polymerization required for these intermolecular conformations and the elicitor activity suggests that oligogalacturonide-Ca^{2+} complexes are the active molecular signals (Liners et al. 1989; Messiaen and Van Cutsem 1994).

Oligogalacturonides have been also shown to regulate growth and morphogenesis (Branca et al. 1988; Marfà et al. 1991; Bellincampi et al. 1993, 1996); this aspect of the oligogalacturonide function is, however, beyond the purview of this chapter. The fact that the same molecules act as signals for either plant development or plant defense responses raises the question of how a plant cell can distinguish a physiological situation from a pathological one, and activate distinct and appropriate transduction pathways. It has been observed that the concentration at which oligogalacturonides regulate growth and morphogenesis (Bellincampi et al. 1993) is one to two orders of magnitude lower than that required to induce plant defense responses. In healthy plants, micromolar levels of oligogalacturonides, probably released by a developmentally regulated plant polygalacturonase (Peretto et al. 1992), may regulate growth and morphogenesis. In diseased plant tissues, much higher levels of oligogalacturonides, released by pathogen-derived polygalacturonase and enhanced by PGIP-endopolygalacturonase interaction (see below), may be responsible for the activation of defense. Different classes of receptors with different affinities for the same oligogalacturonides may discriminate signaling for growth and morphogenesis from signaling for defense.

4. Pectic Enzymes as Determinants of Specificity

Specificity in plant-fungus interactions is likely to be determined by recognition steps involving pathogen-derived signals and complementary sensor (receptor) molecules of plant origin. Both signals and receptors are thought to play their roles at the contact surfaces between the two organisms.

On the basis of the observation that *P. vulgaris* tissue absorbed endopolygalacturonase of *C. lindemuthianum* but not endopolygalacturonases from several fungi nonpathogenic to French bean, it was proposed that binding of endopolygalacturonases to cell walls might be important for the establishment of basic compatibility between a fungus and its host plant (Cervone et al. 1981). Later, it was also suggested that, during plant-phytopathogenic fungus coevolution, a clever defense strategy of a plant would be to exploit the ability to recognize polygalacturonase, a factor that is required for basic compatibility and therefore must be maintained by the microorganism during evolution for successful parasitism (Cervone et al. 1986b).

Although endopolygalacturonases behave as aspecific elicitors, a role of these enzymes in race-cultivar specificity is still conceivable. A differential binding of endopolygalacturonase from different races of *C. lindemuthianum* to cell walls of different cultivars of *P. vulgaris* was observed: the rate of enzyme absorption was faster in the incompatible combinations (i.e., resistant cultivars) and slower in the compatible ones (i.e., susceptible cultivars) (De Lorenzo and Cervone 1986). In a recent report, the endopolygalacturonase of the *C. lindemuthianum* race β was able to discriminate between two near-isogenic lines of *P. vulgaris*, one susceptible and one resistant towards race β of the fungus, and differentially elicited β-1,3-glucanase activity. Induction of β-1,3-glucanase occurred earlier in the resistant line than in the susceptible one. The endopolygalacturonase mimicked the effect of the fungus itself, since during infection, glucanase activity in the resistant plant was stimulated earlier than in the susceptible one (Lafitte et al. 1993).

If polygalacturonases are implicated in race-cultivar specificity, the mechanism underlying their differential recognition by plant cultivars is still unclear. Differences in biochemical properties of the fungal enzymes have not been found (De Lorenzo et al. 1990; Keon et al. 1990). A detailed analysis showed that polygalacturonase-inhibiting proteins (PGIPs) purified from four different bean cultivars were equally inhibitory in vitro to polygalacturonases purified from three different races of *C. lindemuthianum*. Furthermore, the inhibitors had similar effects on the release of oligogalacturonides from polygalacturonic acid (De Lorenzo et al. 1990).

More recently, it has been found that levels of PGIP in bean tissue increase upon infection with *C. lindemuthianum* (Bergmann et al. 1994) and accumulation of *pgip* transcripts in response to infection with the gamma race of *C. lindemuthianum* is faster and more abundant in resistant near-isogenic lines of bean than in susceptible lines (Nuss et al. 1996). These observations suggest that the interaction between endopolygalacturonase and PGIP and the way the two proteins are regulated in planta (kinetics of induction and levels of expression) are associated with race-cultivar specificity.

B. Cellulases

Cellulolytic enzymes may be classified into three different groups: (1) exo-β-1,4-glucan cellobiohydrolases that split the β-1,4-glycosidic bonds at the nonreducing end of the cellulose molecule to yield cellobiose; (2) endo-β-1,4-glucanases that attack internal glycosidic bonds, and (3) β-glucosidases that cleave short cello-oligosaccharides to glucose. Each group contains several isoenzymes. Cellulolytic enzymes show synergism (references in Knowles et al. 1987; Køj and Fincher 1995): all three classes of enzymes are required for the complete depolymeryzation of native cellulose. Cellulose hydrolysis is generally a slow process because of the crystalline nature of the polymer. Although cellulases have been described in several phytopathogenic fungi (Mullen and Bateman 1975; Anderson 1978; Bodenmann et al. 1985; Cooper et al. 1988; Katoh et al. 1988; Christakopoulos et al. 1990; Kollar 1994), studies on the role of these enzymes in pathogenesis are still limited (Cooper 1984; Sposato et al. 1995).

Recently, cellulase genes have been cloned from several filamentous fungi (Table 2). Several fungal cellulase genes share the characteristics of lacking the cellulose-binding domain. These include the endoglucanase genes of *Macrophomina phaseolina* (Wang and Jones 1995); the cellobiohydrolase gene, *cbh-1*, from *Phanerochaete chrysosporium* (Covert et al. 1992) and *Cryphonectria parasitica* (Wang and Nuss 1995); and *CEL1* from *Cochliobolus carbonum* (Sposato et al. 1995). In this regard, deletion of the cellulose-binding domain and adjacent hinge from the *Trichoderma reesei* CBH-1 enzyme by proteolytic cleavage has been shown to alter substrate specificity so that the enzyme is active on small soluble cellulose substrates but no longer active on larger insoluble substrates (Tomme et al. 1988). It is likely that several cellulase genes exist that encode a complex of enzymes exhibiting a range of different cellulose substrate specificities.

Induction of expression of the *C. parasitica cbh-1* gene is suppressed in an isogenic strain rendered hypovirulent due to hypovirus infection. Significantly, virus-free *C. parasitica* strains rendered hypovirulent by transgenic cosuppression of a GTP-binding protein α subunit are deficient in the induction of *cbh-1* transcript accumulation (Wang and Nuss 1995).

C. Depolymerases for Neutral Matrix Polysaccharides

A large number of different enzymes that have been found to be present in infected tissues are included in this class. Evidence suggesting a role in pathogenesis for some of these enzymes has been summarized by Cooper (1987). Most of these enzymes do not cause maceration, protoplast damage, or even major changes to the wall. In disease of monocotyledonous plants, where arabinoxylan is the main matrix component of the plant cell wall, arabinases and xylanases may play a role analogous to that of pectic enzymes in dicotyledonous plants. For example, xylanase is the first cell wall-degrading enzyme secreted in *Rhizoctonia cerealis*-infected wheat seedlings (Cooper et al. 1988).

Fungal xylanases also behave as elicitors of plant defense responses such as phytoalexin induction (Farmer and Helgeson 1987), cell death (Bucheli et al. 1990), and ethylene production (Fuchs et al. 1989). A β-1,4-xylanase purified from *Trichoderma viride* with an endo mode of action induced ethylene in tobacco leaf disks. Antibodies against this enzyme, as well as heat treatment or chemical inactivation, inhibited both xylanase- and ethylene-inducing activities (Dean et al. 1991). Induction of ethylene appeared to be due to the stimulation of 1-aminocyclopropane-1-carboxylic acid synthase activity (Bailey et al. 1990). Xylanase of *T. viride* is synthesized as a 25-kDa precursor and subsequently processed to a 22-kDa secreted polypeptide. By gel permeation chromatography, the enzyme appeared to undertake a compact conformation with an apparent molecular mass of about 9.2 kDa, a size that is likely to allow the protein to pass through cell wall pores and interact with the plasmalemma (Dean and Anderson 1991). Also the endoxylanase of the maize pathogen *Cochliobolus carbonum* exhibited a low apparent molecular weight in nondenaturing conditions and passed through a 10-kDa cut-off ultrafiltration membrane (Holden and Walton 1992). Protoplasts could respond to elicitation by xylanase, further suggesting that the protein may act directly at the level of the plasmalemma (Sharon et al. 1993). Genetic analyses of two cultivars with different sensitivities to xylanase showed that elicitation of ethylene production was controlled by a single dominant gene (Bailey et al. 1993). This gene might encode the putative plant receptor of xylanase.

Table 2. Genes for cell wall-degrading enzymes cloned from fungi

Enzyme	Fungus	Reference
Pectic enzymes		
Hydrolases		
*endo*polygalacturonase		
pgaI, pgaII, pgaC	*Aspergillus niger*	Bussink et al. (1990, 1991a, 1992)
pgaII	*A. tubigensis*	Ruttkowski et al. (1990); Bussink et al. (1991b)
pg	*A. oryzae*	Kitamoto et al. (1993)
pecA	*A. parasiticus*	Cary et al. (1995)
PGN1	*Cochliobolus carbonum*	Scott-Craig et al. (1990)
pg	*Fusarium moniliforme*	Caprari et al. (1993b)
pg1, pg2, pg3	*Sclerotinia sclerotiorum*	Reymond et al. (1994); Fraissinet-Tachet et al. (1995)
Lyases		
Pectate lyase		
pelA	*A. nidulans*	Ho et al. (1995)
pelA, pelB, pelC	*Nectria haematococca*	González-Candelas and Kolattukudy (1992); Guo et al. (1995a,b)
Pectin lyase		
pelA, pelB, pelD	*A. niger*	Harmsen et al. (1990); Kester and Visser (1994); Gysler et al. (1990)
plnA	*Glomerella cingulata*	Templeton et al. (1994)
Esterases		
Pectinmethylesterase		
pmeA	*A. niger*	Khanh et al. (1991)
Rhamnogalacturonan acetylesterase		
rha1	*A. aculeatus*	Kauppinen et al. (1995)
Cellulase		
Glucanase		
egl	*Penicillium funiculosum*	Sahasrabudhe and Ranjekar (1990)
egl1, egl2, egl3, egl4, egl5	*Trichoderma reesei*	Van Arsdell et al. (1987); Penttilä et al. (1986); Saloheimo et al. (1988); Gilkes et al. (1991); Saloheimo et al. (1994)
egl2	*Macrophomina phaseolina*	Wang and Jones (1995)
egl	*Fusarium oxysporum*	Sheppard et al. (1994)
Cellobiohydrolase		
cbh1.1, cbh1.2, cbh1.3, cbhII	*Phanerochaete chrysosporium*	Covert et al. (1992); Sims et al. (1988); Tempelaars et al. (1994)
cbh1, cbh2	*T. reesei*	Teeri et al. (1987)
CEL1	*C. carbonum*	Sposato et al. (1995)
cbh1	*Cryphonectria parasitica*	Wang and Nuss (1995)
Depolymerases for neutral matrix polisaccharides		
Arabinase		
abnA	*A. niger*	Flipphi et al. (1993)
Xylanase		
XYL1	*C. carbonum*	Apel et al. (1993)
Xyn22, Xyn33	*Magnaporthe grisea*	Wu et al. (1995)
XynA	*Pyromyces*	Fanutti et al. (1995)
β-1,3 glucanase		
EXG1	*C. carbonum*	Schaeffer et al. (1994)
Mannanase		
ManA	*Pyromyces*	Fanutti et al. (1995)

Although a xylanase from *Magnaporthe grisea* (Chap. 3, Vol. V, Part B) has been shown to release cell wall fragments that kill plant cells (Bucheli et al. 1990), the ability of *T. viride* xylanase to induce ethylene does not appear to involve release of biologically active cell wall fragments (Sharon et al. 1993). The response of tobacco tissues to xylanase is very similar to the response of plant tissues to other proteinaceous elicitors for which no known enzyme activity has been identified (de Wit et al. 1985; Ricci et al. 1989; Blein et al. 1991). Interestingly, a compact conformation and a high isolectric point similar to that observed for xylanase (pI = 9.1) have been also found in the proteinaceous elicitor crytogein (Ricci et al. 1989).

Xylanase from *T. viride* also elicited PR proteins in tobacco leaves through an ethylene-independent pathway (Lotan and Fluhr 1990). In tobacco plants this enzyme induced typical defense responses such as necrosis and electrolyte leakage not only at the site of application but also at distant sites. This long-distance effect was associated with the translocation of the enzyme in the plant xylem (Bailey et al. 1991).

III. Molecular Genetics of Cell Wall-Degrading Enzymes

Whether cell wall-degrading enzymes (CWDE) are indispensable for pathogenicity, rather than for the saprophytic life of a fungus, remains unclear. Biochemical and genetic approaches have proved inadequate to clarify the role of CWDE in invasion and pathogenesis: for example, enzyme multiplicity has presented considerable problems for isolation of mutants lacking a particular enzyme activity. The increasing ability to isolate genes, to modify their sequences in highly directed and nearly unlimited ways, and to introduce them back into fungi, may in the near future provide definite answers about the function of CWDE. Expression in heterologous systems such as *Saccharomyces cerevisiae* and *Aspergillus niger* (Fincham 1989; Punt and Van den Hondel 1992) offers the possibility to investigate the structure-function relationship of a particular enzyme by site-directed mutagenesis.

The manipulation of the genome of several phytopathogenic fungi, including *Cochliobolus carbonum* (Scott-Craig et al. 1990; Chap. 8, this Vol.), *Magnaporthe grisea* (Parsons et al. 1987; Chap. 3, Vol. V, Part B), *Ustilago maydis* (Wang et al. 1988), *Colletotrichum trifolii* (Dickman 1988), *Colletotrichum lindemuthianum* (Rodriguez and Yoder 1987), *Fusarium oxysporum* (Diolez et al. 1993), *Fusarium solani* f. sp. *pisi* state of *Nectria haematococca* (Stahl and Schäfer 1992), *Phytophthora* spp (Bailey et al. 1993; Chap. 2, Vol. V, Part B), and *Cladosporium fulvum* (Marmeisse et al. 1993; Chap. 1, Vol. V, Part B) can be accomplished to determine the importance of CWDE in pathogenicity, virulence, and specificity. Once a gene has been cloned, the importance in vivo can be assessed by its specific disruption in a fungal genome. If the gene corresponds to a function essential for pathogenicity, the disrupted or replaced mutant will be impaired in its virulence. Also, by expressing the cloned gene in heterologous fungi, pathogens expressing a novel function can be obtained to assess the contribution of this function to the pathogenicity phenotype. In bacteria, similar approaches have led to the discovery of enzymes that had been phenotypically masked (Kelemu and Collmer 1993) or of factors that confer specificity (Keen et al. 1993).

The molecular approach may also be valuable to identify novel forms of CWDE. Several laboratories are preparing subtracted cDNA libraries representing mRNAs present when a pathogenic fungus is grown on plant cell walls or single components of the cell wall (pectin, hemicellulose, etc.) as opposed to sucrose or glucose (D'Ovidio et al. 1990; J.D. Walton, pers. comm.). Genes thus identified can be compared to the DNA data banks in order to identify their possible function, and can be expressed in heterologous systems to produce high amounts of the corresponding proteins for biochemical studies. Sequences of unknown function can also be used for gene disruption in the fungus from which they were isolated to assess their importance in invasion and pathogenesis.

Several genes encoding CWDE have been cloned (Table 2) and reports on the disruption of CWDE genes have been recently published. The first was by Scott-Craig et al. (1990), who described the generation of a specific endopolygalacturonase mutant of *C. carbonum* by homologous integration of a plasmid containing an internal fragment of the gene. Pathogenicity on maize of the mutant was qualitatively indistinguishable from that of the wild-type strain, indicating that in this compatible interaction the endopolygalacturonase is not required. Similarly, pathogenicity of *C. carbonum* on maize was not altered upon disruption of the *XYL1* gene encoding a β-1,4-xylanase (Apel et al. 1993) and of the *EXG1* gene encoding an exo-β-1,3-glucanase (Schaeffer et al. 1994). However, whether degradation of pectin, xylan, and β-1,3 glucan is important for the colonization of maize tissues by *C. carbonum* remains an open question, because of the possibility that other enzymes (some of which may be produced by the fungus only in planta) may compensate for the loss of the products of the disrupted genes. In fact, the disrupted mutants still presented residual polygalacturonase, xylanase, and β-1,3-glucanase activity, respectively. The endopolygalacturonase mutant still exhibited 10 or 35% (based on viscometric or reducing sugar assays, respectively) of the polygalacturonase activity observed in vitro for the wild type, and was still able to grow on pectin. End-product analysis indicated that the residual activity in the mutant was due to an exopolygalacturonase (Scott-Craig et al. 1990). Residual xylan-degrading activity (6–15% of the total extracellular activity of the wild type when grown on corn cell walls) was observed in the xylanase mutant and was apparently sufficient to support wild-type growth rates on xylan (Apel et al. 1993). Extracellular β-1,3-glucanase activity produced in vitro declined by only 56% in the *EXG1* mutant (Schaeffer et al. 1994). No analyses as to whether these mutants have altered abilities to induce plant defense responses have yet been carried out.

Gene disruption experiments have also been performed for the cellobiohydrolase gene *CEL1* of *C. carbonum*, and exclude an important role for *CEL1* in pathogenicity. However, no enzyme activity has been associated with the *CEL1* gene: attempts to find a cellobiohydrolase activity that was specifically lacking in culture filtrates of the *CEL1* mutant were unsuccessful (Sposato et al. 1995). The case of *CEL1* demonstrates that, although isolating genes for cell wall-degrading enzymes of pathogenic fungi (using, for example, heterologous probes) and drawing conclusions on their importance for pathogenesis can be straightforward, detailed biochemical analysis of the gene products can be difficult.

In our laboratory, a site-directed mutagenesis approach has been adopted to study the structure and function of the endopolygalacturonase of *Fusarium moniliforme*. The sequence encoding the mature endopolygalacturonase of *F. moniliforme* has been cloned into a yeast secretion vector, YEpsec1, and the recombinant plasmid was introduced into *Saccharomyces cerevisiae*. The protein produced by the yeast was hyperglycosylated, but possessed biochemical properties similar to the enzyme produced by *F. moniliforme*, including the capability to be inhibited by the bean PGIP, to macerate and kill plant tissue, and to elicit phytoalexin accumulation in soybean. A detailed analysis is being carried out to identify residues involved in enzymatic activity, substrate binding, recognition by PGIP, and elicitor activity: preliminary results indicate that replacement of His-234 with Lys abolishes enzymatic activity, confirming the biochemical evidence obtained by Cooke et al. (1976) and Rexová-Benková and Mracková (1978). Interestingly, the inactive enzyme still exhibits elicitor activity, suggesting that elicitation by polygalacturonase may not be due solely to the ability of the enzyme to release oligogalacturonides from the plant cell wall. Replacement of either Ser-237 or Ser-240 with Gly reduces enzyme activity to 48 and 6%, respectively, of the wild-type enzyme (C. Caprari et al., unpubl. results from our laboratory).

IV. Inhibitors

A. Polygalacturonase-Inhibiting Proteins (PGIPs)

The first evidence of a proteinaceous inhibitor of fungal endopolygalacturonases was obtained by Uritani and Stahmann (1961) in water extract of sweet potato tissue. The isolation of a polygalacturonase inhibitor of *P. vulgaris* was reported in the early 1970s (Albersheim and Anderson 1971). Since these initial reports, the occurrence of inhibitor proteins (PGIP = **P**oly**G**alacturonase-**I**nhibiting **P**roteins) has been reported in a variety of dicotyledonous plants, including bean, cucumber, pea, green pepper, tomato, apple, pear, peach, oranges, alfalfa (Hahn et al. 1989 and references therein) and, more recently, soybean (Favaron et al. 1994) and raspberry (Johnston et al. 1993). PGIPs have also been identified in the pectin-rich monocotyledonous plants *Allium cepa* and *Allium porrum* (Favaron et al. 1993). The characteristics of a few purified PGIPs are reported in Table 3.

Some of these inhibitors are predominantly ionically bound to the plant cell wall, while others are readily extracted from it with dilute buffers. PGIPs are typically effective against the endopolygalacturonases of fungi and ineffective against other pectic enzymes of either microbial or fungal origin (Cervone et al. 1990). PGIP from bean hypocotyls in vitro protected bean cell walls against degradation by endopolygalacturonase of *C. lindemuthianum* (Lafitte et al. 1984). Similarly, PGIP prepared from tomato cell walls in vitro protected these walls from degradation by a complex CWDE mixture produced in culture by *Fusarium oxysporum* f.sp. *lycopersici* (Jones et al. 1972).

Table 3. Characteristics of purified PGIPs

Plant (tissues)	Molecular mass (kDa)	Isoelectric point	Mode of inhibition	Reference
P. vulgaris (hypocotyls)	41	9.5	Noncompetitive	Cervone et al. (1987) Lafitte et al. (1984)
Pear (fruits)	43	6.6	Competitive	Abu-Goukh et al. (1983)
Apple (fruits)	36	9.3	Competitive and noncompetitive	Müller and Gessler (1993)
	44–54	4.6	Competitive and noncompetitive	Yao et al. (1995)
Raspberry (fruits)	38.5	n.d.	Noncompetitive	Johnston et al. (1993)

Some observations suggest a correlation between presence of PGIP and resistance of plants to fungi. In a study on the distribution of PGIP in various tissues of *P. vulgaris*, levels of PGIP in hypocotyls were shown to increase six- to ninefold during seedling growth (Salvi et al. 1990). Although susceptibility tests were not performed, the increase in PGIP levels may be correlated with the increase in resistance occurring in bean hypocotyls when primary leaves begin to develop (Bailey et al. 1992). Increasing susceptibility of ripening pear fruits to *Dothiorella gregaria* and *Botrytis cinerea* correlated with a decline in the concentration of PGIP (Abu-Goukh et al. 1983). In raspberry fruits, the level of PGIP was maximal in immature green fruit, which are more resistant to fungal attack, and decreased in mature, more susceptible fruits (Johnston et al. 1993). Tissues of Chinese chestnut resistant to a virulent strain of *Cryphonectria parasitica* contain PGIP levels considerably higher than the susceptible American chestnut (Gao and Shain 1995).

B. Inhibitors of Pectin Lyases

The first evidence of a proteinaceous inhibitor of fungal endopectin lyase was obtained in extracts of onion, French bean, sweet paprika, white cabbage, and cucumber (Bock et al. 1975). A protein able to inhibit pectin lyases of *Rhizoctonia solani*, *Phoma betae* and *Aspergillus japonicus* was purified from sugar beet walls. This protein exhibited an apparent molecular mass of 57.5 kDa and a pI of 9.9; its mode of inhibition was uncompetitive (Bugbee 1993). No investigations have yet been performed on the physiological role of this inhibitor.

C. Polygalacturonase-Inhibiting Proteins Favor the Accumulation of Elicitor-Active Oligogalacturonides

Oligogalacturonides of chain length varying between 10 and 15, which are transiently produced by the action of the endopolygalacturonase on homogalacturonan, are elicitors of defense responses; shorter oligomers have little or no elicitor activity (Hahn et al. 1989; Darvill et al. 1992). Thus, endopolygalacturonases release elicitor-active oligogalacturonides, but also degrade them into inactive oligomers. This implies that an extensive degradation of the cell wall homogalacturonan negatively affects the eliciting activity of the endopolygalacturonase. PGIP modulates the endopolygalacturonase activity in vitro in such a way that the balance between release of elicitor-active oligogalacturonides and depolymerization of the active oligogalacturonides into inactive molecules is altered and the accumulation of elicitor-active molecules is favored (Cervone et al. 1986a). Endopolygalacturonase and PGIP form a specific, reversible, saturable, high affinity complex with a dissociation constant of 1.1 ± 10^{-9} M (Cervone et al. 1989b). Recognition in animal cells often involves binding of signals originating from one cell to a specific receptor on another cell. Such signal molecules bind specifically, reversibly, saturably, and with high affinity to a receptor protein often located on the surface of the target cell. The interaction between the signal and the receptor initiates the accumulation of other signals that lead to altered gene expression in the target cells (Hahn 1989). Endopolygalacturonase and PGIP may function respectively as a signal molecule and its receptor protein in the recognition that must occur between plants and their potential patho-

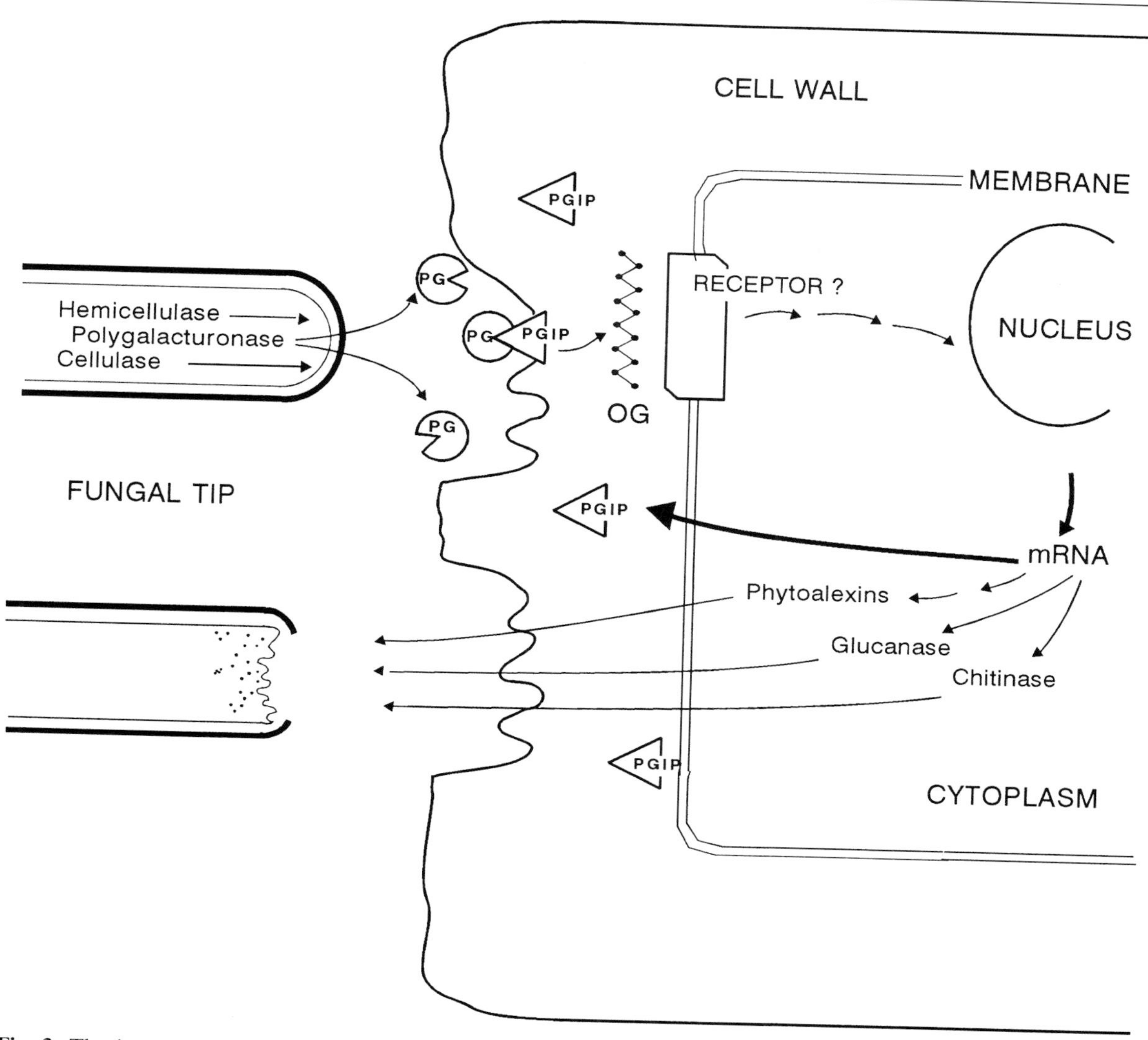

Fig. 3. The interaction of endopolygalacturonase, PGIP, and oligogalacturonides as a signaling system in communication between plants and fungi. *PG* Endopolygalacturonase; *PGIP* endopolygalacturonase-inhibiting protein; *OG* oligogalacturonides

gens. A model of PGIP as a component of the cell surface signaling system that leads to the formation of elicitor-active oligogalacturonides is presented in Fig. 3.

V. Molecular Genetics of Polygalacturonase-Inhibiting Proteins

Genes encoding PGIP have been cloned from *Phaseolus vulgaris* (Toubart et al. 1992), pear (Stotz et al. 1993), soybean (Favaron et al. 1994), and tomato (Stotz et al. 1994).

The *pgip* gene of *P. vulgaris* predicted a 342-amino-acid polypeptide including a 29-amino-acid signal peptide for secretion and four potential glycosylation sites. The polypeptide was rich in Leu, Asn, Ser, Gly, and Thr. By Northern blot analysis, a 1.2-kb transcript was detected in suspension-cultured cells and to a lesser extent, in leaves, hypocotyls, and flowers (Toubart et al. 1992).

The *pgip* genes from pear and tomato predicted 330 and 327 amino acid polypeptides, respectively, both including a putative signal sequence for secretion (Stotz et al. 1993, 1994). The expression of pear PGIP was regulated in a tissue-specific manner. Levels of *pgip* transcripts were much higher in fruits than in flowers of leaves. The deduced pear PGIP was approximately 50% identical and 65% similar to *P. vulgaris* PGIP at the amino acid level (Stotz et al. 1993). Tomato PGIP shared 68 and 50% amino acid sequence identity with pear and bean PGIP, respectively (Stotz et al. 1994). The cloned soybean PGIP was 77.6 and 60.3% identical to bean and pear PGIP, respectively (Favaron et al. 1994). A gene, *FIL2*, isolated from *Antirrhinum* encodes a protein showing a very high degree of similarity with PGIPs (Steinmayr et al. 1994); however, the possible polygalacturonase-inhibiting activity of the FIL2 protein has not yet been investigated.

Accumulation of *pgip* transcripts was induced in suspension-cultured bean cells following addition of elicitor-active oligogalacturonides and fungal glucan to the medium; *pgip* mRNA accumulated in *P. vulgaris* hypocotyls in response to wounding or treatment with salicylic acid (Bergmann et al. 1994). Accumulation of *pgip* mRNA has also been followed in different race-cultivar interactions (either compatible or incompatible) between *Colletotrichum lindemuthianum* and *P. vulgaris* by Northern blot and in situ hybridization analyses. Rapid accumulation of *pgip* mRNA correlated with the appearance of the hypersensitive response in incompatible interactions, while a more delayed increase, coincident with the onset of lesion formation, occurred in compatible interactions (Nuss et al. 1996). In incompatible interactions, the accumulation of *pgip* mRNA was localized around the site of infection (Fig. 4). These data indicate that PGIP expression is regulated upon the early race-specific recognition event in a manner similar to that observed for the known defense-related genes, and demonstrates that induction of *pgip* expression represents an active defense mechanism of plants against fungal pathogens.

The induction of PGIP expression during infection may be considered as indirect evidence of the importance of fungal endopolygalacturonase in pathogenesis. Direct evidence may be provided by analysis of transgenic plants that overexpress PGIP. Transgenic tomato plants expressing high levels of pear PGIP have been already obtained. A chimeric gene has been constructed possessing the structural pear *pgip* gene under the control of the CaMV 35S promoter, that allows constitutive and high-level expression in most plant tissues. The chimeric gene was introduced into tomato plants by *Agrobacterium tumefaciens*-mediated transformation. Ripe fruits from the transgenic plants obtained exhibited an enhanced resistance to colonization by *Botrytis cinerea* (Powell et al. 1994).

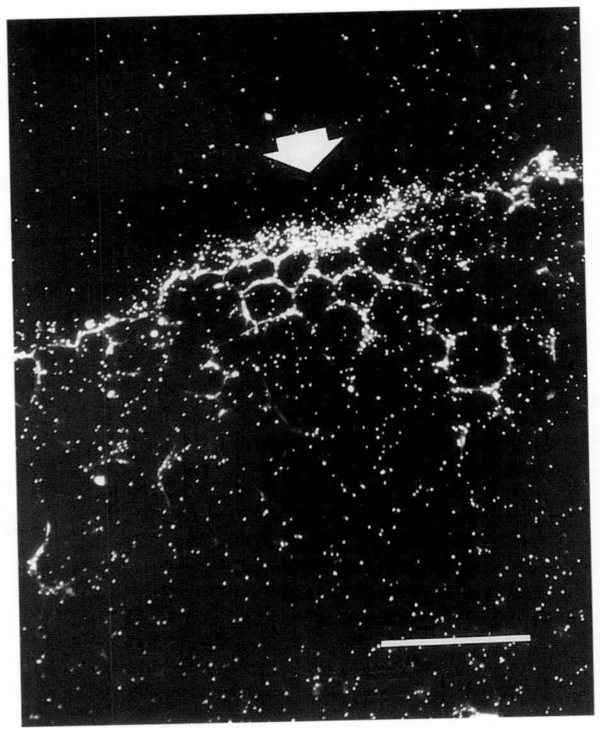

Fig. 4. In situ localization of *pgip* transcripts in hypocotyls of *P. vulgaris* infected with an avirulent race of *C. lindemuthianum*. *Arrow* indicates the infection site. *Bar* 100 μm. (Original photograph Dr. A.J. Clark, Department of Plant Sciences, University of Cambridge, Cambridge, UK)

A. The Structure of PGIP Is Specialized for Interaction with Other Macromolecules

The amino acid sequence of the PGIP exhibits characteristics of significant internal homology (De Lorenzo et al. 1994). The internal homology domain spans about 258 amino acids, from position 69 to 326, and exhibits a modular structure: it can be divided into a set of 10.5 tandemly repeating units, each derived from modifications of a 24-amino-acid peptide. The alignment between the 10.5 segments shows a periodic distribution of a few amino acids rather than a recurrent sequence of several different amino acids. The repeat element has regularly spaced Leu residues (often replaced by the bulky hydrophobic amino acids Val, Phe, Ile) and conserved Gly, Pro, and Asn at the 1st, 4th, and 21st position, respectively. A

```
69     G.L..T.T..  ..VN.LDLSG ..L...
95     .PIPS.L.NL  ..LN.L.I.G.N.L.
120    GPIP..I..L  ..L..L.IT.  ..VS
144    G.IP..LS.I  K.L..LDFS.  N.LS
168    G.LP.SIS.L  .NL..I.F.G  N.IS
192    G.IP.S...F.K.F..M.IS.   N.LT
217    G.IP.TF.NL   NL..VDLS.  N.L.
240    G.....F...  KN...I.L..  N.L.
264    .........   KNLN.LDL..  N.I.
287    G.LP..LT.L  K.L..L.VSF  NNL.
311    G.IP.. .NL  K.   FDVS.................*

Cons   G.IP..L..L  KnL..LdLS.  N.L.
```

Fig. 5. Structure of the *Phaseolus vulgaris* PGIP periodic repeat. The sequence for positions 69–342 is indicated. *Numbers on the left* are coordinates for the amino acid position at the left of each line. *Dots* replace nonconserved residues, *blanks* indicate the placements of gaps to maximize alignment, and *asterisk* indicates the end of the protein. The consensus sequence (*Cons*) for the periodic repeat is indicated at the *bottom*. Residues conserved in at least four repeats are reported in the consensus sequence. *Small letters* indicate residues identical in five or less repeats

hydrophilic residue preferentially appears at positions 6, 19, and 24. Negatively and positively charged residues are prominent at positions 17 and 11, respectively (Fig. 5).

Over 30 proteins, from bacteria to human, contain leucine-rich tandem repeats similar to those found in the PGIP: human leucine-rich α_2-glycoprotein (LRG), a putative membrane-derived serum protein (Takahashi et al. 1985); the regulatory domains or subunits of the enzymes adenylyl cyclase and carboxypeptidase N; the cell adhesion protein *Toll* and the photoreceptor membrane glycoprotein chaoptin of *Drosophila*; the human extracellular matrix-binding glycoprotein decorin; the alpha chain of human platelet glycoprotein (GpIb), a transmembrane receptor; receptors for gonadotrophins and neurotrophins; the *sds22+* nuclear protein; bacterial virulence factors; molecules involved in splicing and DNA repair; and a porcine RNAse inhibitor (Kobe and Deisenhofer 1993 and references therein). The consensus sequences for the repeating units are surprisingly similar, considering the range of organisms and the widely divergent functions of the different proteins. The great similarity in the structure of these proteins indicates a strong selection pressure for conservation of this structure, which is likely to have been attained by a series of unequal cross-overs of an oligonucleotide sequence coding for a prototype leucine-rich building block. The similarity may also indicate an evolutionary conservation between the proteins, or may reflect the coincidental and convergent evolution of a protein domain. Since a common feature among these proteins appears to be that of membrane association and/or protein binding, a domain composed of tandem leucine-rich repeats may represent a structure specialized to achieve strong interactions between macromolecules.

The crystal structure of the RNAse inhibitor has revealed that leucine-rich repeats correspond to β-alpha structural units. These units are arranged for a parallel β-sheet with one surface exposed to solvent so that the protein acquires an unusual nonglobular shape, which may be responsible for protein-binding functions (Kobe and Deisenhofer 1995).

B. The Structure of PGIP Is Similar to that of Proteins Involved in Signal Transduction

Many of the proteins with leucine-rich repeats are involved in signal transduction pathways. In this regard, the similarity between PGIP and a cloned *Arabidopsis thaliana* receptor-like protein kinase (RLK5; Walker 1993) is of particular interest. The RLK5 protein exhibits a three-domain molecular topology similar to all of the known receptor tyrosine kinases, with an N-terminal putative extracellular domain, a central putative transmembrane domain, and a C-terminal domain homologous to the catalytic core of serine-threonine kinases. The homology of RLK5 with PGIP is, however, limited to the putative extracellular receptor-like domain (Fig. 6).

The catalytic domain of RLK5 is homologous to that present in the deduced protein of other cloned genes (*ZmPK1*, *RLK1*, *RLK4*; Walker 1993). However, the putative extracellular domains of the ZmPK1, RLK1, and RLK4 deduced proteins differ from that of RLK5 or PGIP and appear to be related to the products of *Brassica SLG* (S Locus Glycoprotein) and *SRK* (S Locus Receptor Kinase) genes (Nasrallah and Nasrallah 1993). These two genes are physically linked to each other and linked to the S locus controlling self-incompatibility in *Brassica*. The SLG product is a secreted glycoprotein, while SRK is a transmembrane receptor protein kinase; SLG and the extracellular receptor domain of SRK isolated

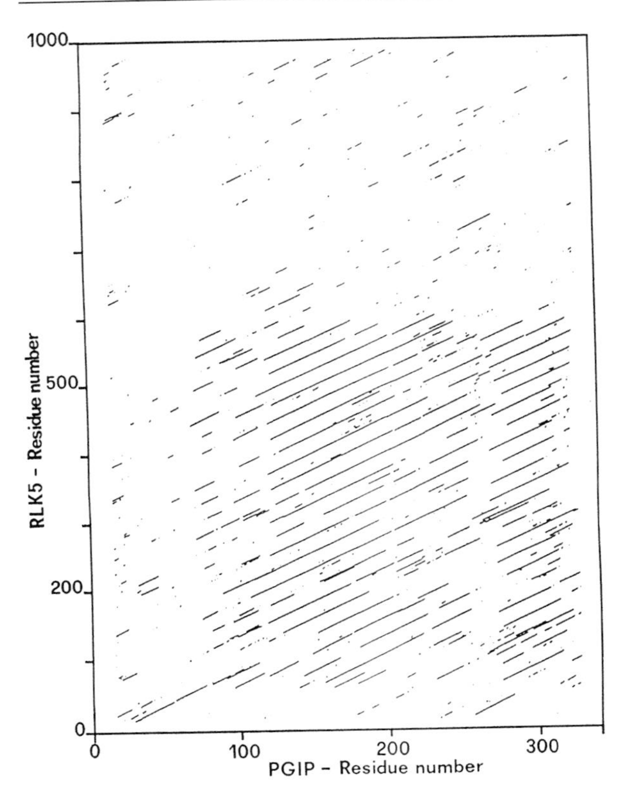

Fig. 6. Homology between *Phaseolus vulgaris* PGIP and the *Arabidopsis* receptor-like protein kinase RLK5. Homologies were analyzed according to Maizel and Lenk (1981) using the programs COMPARE and DOTPLOT of the University of Wisconsin Genetic Computer Group (UWGCG; Devereaux et al. 1984). The program COMPARE was set at a window of 30.0 and a stringency of 12. *Points* are plotted at positions of homology equal or greater than this value. *Diagonal lines* indicate regions of extensive homology

from the same S haplotype share a high level (>80%) of sequence identity (Nasrallah and Nasrallah 1993). Since both the SLG and SRK proteins are required for the self-incompatibility response, a model has been proposed in which SRK, acting in combination with SLG, couples the recognition event at the pollen-stigma interface to a cytoplasmic phosphorylation cascade that leads to pollen rejection (Nasrallah and Nasrallah 1993; Dickinson 1994).

C. PGIP Is Related to the Products of a Resistance Gene

Recently, the molecular characterization of several resistance genes has been reported (Dangl 1995; Staskawicz et al. 1995): four (the tomato *Cf-9*, the tobacco *N*, the *Arabidopsis RPS2*, and the flax L^6) encode proteins containing leucine-rich repeat domains. A fifth gene, the tomato *PTO*, encodes a serine/threonine kinase; *PTO*-mediated resistance, however, requires an additional gene, *PRF*, tightly linked to *PTO*, that encodes a leucine-rich repeat protein. The *Cf-9* product was found to be homologous to PGIPs, not only in the leucine-rich region but also outside. Cf-9 clearly shares an evolutionary relation with PGIPs (Jones et al. 1994).

VI. Conclusions

Our knowledge of the role of fungal cell wall-degrading enzymes (CWDE) in invasion is still limited. Molecular genetics is proving to be a useful tool in this regard: cloning of genes encoding different CWDE, site-directed mutagenesis, and gene disruption studies are currently being applied to a limited number of fungi, but this number is expected to increase steadily in the near future. New CWDE are expected to be discovered; the structure-function relationship of some CWDE is being elucidated and, most interestingly, the role of these enzymes in invasion and pathogenesis is being clarified. On the other hand, more inhibitors of CWDE are expected to be discovered and their role in resistance and in cell-cell signaling to be elucidated.

Among the fungal CWDE, endopolygalacturonase exhibits some interesting features: this enzyme is not only a pathogenicity factor of fungi but also activates the plant defense responses. The fact that plants possess a recognition system for endopolygalacturonases and are able to respond to them suggests that these enzymes are important in pathogenesis. Pathogens cannot evolve to nonproduction of a pathogenicity factor and therefore are unlikely to elude recognition. Moreover, since the enzyme products, the oligogalacturonides, rather than the enzyme itself, are the true elicitors of the plant defense response, evolution away from recognition is rendered even more complicated for a pathogen. Recently, it has been shown that endopolygalacturonase of *Colletotrichum lindemuthianum* race β differentially elicits defense responses in near-isogenic lines of *Phaseolus vulgaris* resistant and susceptible to the β race. It has also been found that, upon infection with *C. lindemuthianum* race 9, polygalacturonase-inhibiting protein (PGIP) is differentially induced in near-isogenic lines resistant and susceptible to race 9. Thus, a role of endopolygalacturonase and PGIP in race-cultivar specificity is conceivable.

The interaction between PGIP and fungal endopolygalacturonases has the requisites for functioning in a perception mechanism that leads to incompatibility. The analysis of the *pgip* gene evidenced that the PGIP has a leucine-rich repeat structure, similar to that found in the products of several resistance genes, and in particular, is related to the *Cf-9* product.

The analogy between the recognition system involved in pollen-stigma interactions and that involved in plant-pathogen interactions at the cellular and genetic levels (De Lorenzo et al. 1994; Lamb 1994) and the relation between PGIP and the extracellular domain of a receptor-like protein kinase of *Arabidopsis*, RLK5, raises the possibility that PGIP acts as a secreted "receptor" involved in recognition between plants and fungi. A PGIP-related class of two-component (secreted receptor/transmembrane receptor-kinase) signaling systems, similar to that controlling self-incompatibility in *Brassica*, may be present on the plant cell surface. This class would differ from the *Brassica* system in the features of the receptor component and/or the type of catalytic kinase domain, and might be implicated in different aspects of plant-fungus recognition, such as that leading to nonhost resistance or to race-cultivar specificity. PGIP may therefore represent an S-locus glycoprotein counterpart in a class of receptor/receptor kinase complex of the kind involved in *Brassica* self-incompatibility (De Lorenzo et al. 1994).

In the attempt to elucidate the role of PGIP in plant resistance to fungi, different *pgip*-related genes are being characterized. It has been shown that a small family of *pgip* genes, most likely clustered on chromosome 10, is present in the genome of *Phaseolus vulgaris* (Frediani et al. 1993) and several *pgip*-related clones have already been isolated. The structural and functional analyses of these clones are in progress. Also the possible presence of membrane-bound PGIP-related proteins is being investigated. In this connection, it is of interest that endopolygalacturonase of *Colletrotrichum lindemuthianum* has been shown to absorb to protoplasts of *P. vulgaris*, suggesting the presence of a polygalacturonase-binding protein (a PGIP?) at the level of plasmalemma (Cervone and De Lorenzo 1985).

Acknowledgments. Research in the authors' laboratory is supported by the EC Grants CHRX-CT93-0244 and AIR 3-CT94-2215 and by the Ministero delle Risorse Agroindustriali e Forestali (MIRAAF) and the Istituto Pasteur-Fondazione Cenci Bolognetti.

References

Abu-Goukh AA, Strand LL, Labavitch JM (1983) Development-related changes in decay susceptibility and polygalacturonase inhibitor content of Bartlett pear fruit. Physiol Plant Pathol 23:101–109

Albersheim P, Anderson AJ (1971) Proteins from plant cell walls inhibit polygalacturonases secreted by plant pathogens. Proc Natl Acad Sci USA 68:1815–1819

Anderson AJ (1978) Extracellular enzymes produced by *Colletotrichum lindemuthianum* and *Helminthosporium maydis* during growth on isolated bean and corn cell walls. Phytopathology 68:1585–1589

Anderson AJ (1989) The biology of glycoproteins as elicitors. In: Kosuge T, Nester E (eds) Plant-microbe interactions. Molecular and genetic perspectives, vol 3. McGraw Hill, New York, pp 87–130

Apel PC, Panaccione DG, Holden FR, Walton JD (1993) Cloning and targeted gene disruption of *XYL1*, a β-1,4-xylanase gene from the maize pathogen *Cochliobolus carbonum*. Mol Plant-Microbe Interact 6:467–473

Apostol I, Heinstein PF, Low PS (1989) Rapid stimulation of an oxidative burst during elicitation of cultured plant cells. Role in defense and signal transduction. Plant Physiol 90:109–116

Bailey AM, Mena GL, Herrera-Estrella L (1993) Transformation of four pathogenic *Phytophthora* spp. by microprojectile bombardment on intact mycelia. Curr Genet 23:42–46

Bailey BA, Dean JFD, Anderson JD (1990) An ethylene biosynthesis-inducing endoxylanase elicits electrolyte leakage and necrosis in *Nicotiana tabacum* cv. Xanthi leaves. Plant Physiol 94:1849–1854

Bailey BA, Taylor R, Dean JFD, Anderson JD (1991) Ethylene biosynthesis-inducing endoxylanase is translocated through the xylem of *Nicotiana tabacum* cv. Xanthi plants. Plant Physiol 97:1181–1186

Bailey BA, Korcak RF, Anderson JD (1993) Sensitivity to an ethylene biosynthesis-inducing endoxylanase in *Nicotiana tabacum* L. cv. Xanthi is controlled by a single dominant gene. Plant Physiol 101:1081–1088

Bailey JA, O'Connell RJ, Pring RJ, Nash C (1992) Infection strategies of *Colletotrichum* species. In: Bailey JA, Jeger MJ (eds) *Colletotrichum* biology, pathology and control. CAB International, Wallingford, pp 88–120

Baldwin EA, Pressey R (1989) Pectic enzymes in pectolyase. Separation, characterization, and induction of ethylene in fruits. Plant Physiol 90:191–196

Bateman DF, Basham HG (1976) Degradation of plant cell walls and membranes by microbial enzymes. In: Heitefuss R, Williams PH (eds) Encyclopedia of plant physiology, new series, vol 4. Physiological plant pathology. Springer, Berlin Heidelberg New York, pp 316–355

Bellincampi D, Salvi G, De Lorenzo G, Cervone F, Marfà V, Eberhard S, Darvill A, Albersheim P (1993) Oligogalacturonides inhibit the formation of roots on tobacco explants. Plant J 4:207–213

Bellincampi D, Cardarelli M, Zaghi D, Serino G, Salvi G, Gatz C, Cervone F, Altamura MM, Costantino P, De Lorenzo G (1996) Oligogalacturonides prevent rhizogenesis in *rol* B transformed tobacco explants by inhibiting auxin-induced expression of the *rol B* gene. Plant Cell 8:477–487

Benhamou N, Chamberland H, Pauzé FJ (1990) Implication of pectic components in cell surface interactions between tomato root cells and *Fusarium oxysporum* f. sp. *radicis-lycopersici*. A cytochemical study by means of a lectin with polygalacturonic acid-binding specificity. Plant Physiol 92:995–1003

Bergmann C, Ito Y, Singer D, Albersheim P, Darvill AG, Benhamou N, Nuss L, Salvi G, Cervone F, De Lorenzo G (1994) Polygalacturonase-inhibiting protein accumulates in *Phaseolus vulgaris* L. in response to wounding, elicitors, and fungal infection. Plant J 5:625–634

Blanchette RA, Abad AR, Cease KR, Lovrien RE, Leathers TD (1989) Collodial gold cytochemistry of endo-1,4-β-glucanase, 1,4-β-D-glucan cellobiohydrolase, and endo-1,4-β-xylanase: ultrastructure of sound and decayed birch wood. Appl Environ Microbiol 55:2293–2301

Blein J-P, Milat M-L, Ricci P (1991) Responses of cultured tobacco cells to cryptogein, a proteinaceous elicitor from *Phytophthora cryptogea*. Possible plasmalemma involvement. Plant Physiol 95:486–491

Bock W, Dongowski G, Göbel H, Krause M (1975) Nachweis der Hemmung mikrobieller Pektin- und Pektat Lyase durch Pflanzeneigene Inhibitoren. Nahrung 19: 411–416

Bodenmann J, Heiniger U, Hohl HR (1985) Extracellular enzymes of *Phytophthora infestans*: endo-cellulase, β-glucosidases, and 1,3-β-glucanases. Can J Microbiol 31:75–82

Branca C, De Lorenzo G, Cervone F (1988) Competitive inhibition of the auxin-induced elongation by α-D-oligogalacturonides in pea stem segments. Physiol Plant 72:499–504

Broekaert WF, Peumans WJ (1988) Pectic polysaccharides elicit chitinase accumulation in tobacco. Physiol Plant 74:740–744

Bruce RJ, West CA (1982) Elicitation of casbene synthetase activity in castor bean. The role of pectic fragments of the plant cell wall in elicitation by a fungal endopolygalacturonase. Plant Physiol 69:1181–1188

Bruce RJ, West CA (1989) Elicitation of lignin biosynthesis and isoperoxidase activity by pectic fragments in suspension-cultures of castor bean. Plant Physiol 91:889–897

Bucheli P, Doares SH, Albersheim P, Darvill A (1990) Host-pathogen interactions XXXVI. Partial purification and characterization of heat-labile molecules secreted by the rice blast pathogen that solubilize plant cell wall fragments that kill plant cells. Physiol Mol Plant Pathol 36:159–173

Bugbee WM (1993) A pectin lyase inhibitor protein from cell walls of sugar beet. Phytopathology 83:63–68

Bussink HJD, Kester HCM, Visser J (1990) Molecular cloning, nucleotide sequence and expression of the gene encoding prepro-polygalacturonase II of *Aspergillus niger*. FEBS Lett 273:127–130

Bussink HJD, Brouwer KB, De Graaff LH, Kester HCM, Visser J (1991a) Identification and characterization of a second polygalacturonase gene of *Aspergillus niger*. Curr Genet 20:301–307

Bussink HJD, Buxton FP, Visser J (1991b) Expression and sequence comparison of the *Aspergillus niger* and *Aspergillus tubigensis* genes encoding polygalacturonase II. Curr Genet 19:467–474

Bussink HJD, Buxton FP, Fraaye BA, De Graaff LH, Visser J (1992) The polygalacturonases of *Aspergillus niger* are encoded by a family of diverged genes. Eur J Biochem 208:83–90

Caprari C, Bergmann C, Migheli Q, Salvi G, Albersheim P, Darvill A, Cervone F, De Lorenzo G (1993a) *Fusarium moniliforme* secretes four *endo*polygalacturonases derived from a single gene product. Physiol Mol Plant Pathol 43:453–462

Caprari C, Richter A, Bergmann C, Lo Cicero S, Salvi G, Cervone F, De Lorenzo G (1993b) Cloning and characterization of the gene encoding the endopolygalacturonase of *Fusarium moniliforme*. Mycol Res 97:497–505

Carpita NC, Gibeaut DM (1993) Structural models of primary cell walls in flowering plants: consistency of molecular structure with the physical properties of the walls during growth. Plant J 3:1–30

Cary JW, Brown R, Cleveland TE, Whitehead M, Dean RA (1995) Cloning and characterization of a novel polygalacturonase-encoding gene from *Aspergillus parasiticus*. Gene 153:129–133

Cervone F, De Lorenzo G (1985) Pectic enzymes as phytotoxins: absorption of polygalacturonase from *Colletotrichum lindemuthianum* to French bean protoplasts. Phytopathol Mediterr 24:322–324

Cervone F, Andebrhan T, Coutts RHA, Wood RKS (1981) Effects of French bean tissue and leaf protoplasts on *Colletotrichum lindemuthianum* polygalacturonase. Phytopathol Z 102:238–246

Cervone F, De Lorenzo G, Degrà L, Salvi G (1986a) Interaction of fungal polygalacturonase with plant proteins in relation to specificity and regulation of plant defense response. In: Lugtenberg B (ed) Recognition in microbe-plant symbiotic and pathogenic interactions. NATO ASI Series, vol H4. Springer, Berlin Heidelberg New York, pp 253–258

Cervone F, De Lorenzo G, Salvi G, Camardella L (1986b) Molecular evolution of fungal polygalacturonase. In: Bailey J (ed) Biology and molecular biology of plant-pathogenic interactions. NATO ASI Series, vol H1. Springer, Berlin Heidelberg New York, pp 385–392

Cervone F, De Lorenzo G, Degrà L, Salvi G, Bergami M (1987) Purification and characterization of a polygalacturonase-inhibiting protein from *Phaseolus vulgaris* L. Plant Physiol 85:631–637

Cervone F, Castoria R, Spanu P, Bonfante-Fasolo P (1988) Pectinolytic activity in some ericoid mycorrhizal fungi. Trans Br Mycol Soc 91:537–539

Cervone F, De Lorenzo G, D'Ovidio R, Hahn MG, Ito Y, Darvill A, Albersheim P (1989a) Phytotoxic effects and phytoalexin-elicitor activity of microbial pectic enzymes. In: Graniti A, Durbin RD, Ballio A (eds) Phytotoxins and plant pathogenesis. NATO ASI Series, vol H27. Springer, Berlin Heidelberg New York, pp 473–477

Cervone F, Hahn MG, De Lorenzo G, Darvill A, Albersheim P (1989b) Host-pathogen interactions. XXXIII. A plant protein converts a fungal pathogenesis factor into an elicitor of plant defense responses. Plant Physiol 90:542–548

Cervone F, De Lorenzo G, Pressey R, Darvill AG, Albersheim P (1990) Can *Phaseolus* PGIP inhibit pectic enzymes from microbes and plants? Phytochemistry 29:447–449

Chandra S, Low PS (1995) Role of phosphorylation in elicitation of the oxidative burst in cultured soybean cells. Proc Natl Acad Sci USA 92:4120–4123

Christakopoulos P, Macris BJ, Kekos D (1990) On the mechanism of direct conversion of cellulose to ethanol by *Fusarium oxysporum*: effects of cellulase and glucosidase. Appl Microbiol Biotechnol 33:18–20

Collings A, Davis B, Mills J (1988) Endo-β-1,4-glucanase, exo-β-1,4-glucanase, β-glucosidase and related enzyme activity in culture filtrates of thermophilic, thermotolerant and mesophilic filamentous fungi. Microbios 56:131–147

Collmer A, Keen NT (1986) The role of pectic enzymes in plant pathogenesis. Annu Rev Phytopathol 24:383–409

Cooke RD, Ferber CEM, Kanagasabapathy L (1976) Purification and characterization of polygalacturonases from a commercial *Aspergillus niger* preparation. Biochim Biophys Acta 452:440–451

Cooper RM (1984) The role of cell wall degrading enzymes in infection and damage. In: Wood RKS, Jellis GJ (eds) Plant disease: infection, damage and loss. Blackwell, Oxford, pp 13–27

Cooper RM (1987) The use of mutants in exploring depolymerases as determinants of pathogenicity. In: Day PR, Jellis GJ (eds) Genetics and plant pathogenesis. Blackwell, Oxford, pp 261–281

Cooper RM, Longman D, Campbell A, Henry M, Lees PE (1988) Enzymic adaptation of cereal pathogens to the monocotyledonous primary wall. Physiol Mol Plant Pathol 32:33–47

Covert SF, Vanden Wymelenberg A, Cullen D (1992) Structure, organization and transcription of a cellobiohydrolase gene cluster from *Phanerochaete chrysosporium*. Appl Environ Microbiol 58:2168–2175

Dangl JL (1995) Pièce de résistance: novel classes of plant disease resistance genes. Cell 80:363–366

Darvill A, Augur C, Bergmann C, Carlson RW, Cheong J-J, Eberhard S, Hahn MG, Ló V-M, Marfà V, Meyer B, Mohnen D, O'Neill MA, Spiro MD, van Halbeek H, York WS, Albersheim P (1992) Oligosaccharins – oligosaccharides that regulate growth, development and defence responses in plants. Glycobiology 2:181–198

Davis KR, Hahlbrock K (1987) Induction of defense responses in cultured parsley cells by plant cell wall fragments. Plant Physiol 85:1286–1290

Davis KR, Darvill AG, Albersheim P, Dell A (1986) Host-pathogen interactions. XXIX. Oligogalacturonides released from sodium polypectate by endopolygalacturonic acid lyase are elicitors of phytoalexins in soybean. Plant Physiol 80:568–577

Dean JFD, Anderson JD (1991) Ethylene biosynthesis-inducing xylanase. II. Purification and physical characterization of the enzyme produced by *Trichoderma viride*. Plant Physiol 95:316–323

Dean JFD, Gross KC, Anderson JD (1991) Ethylene biosynthesis-inducing xylanase. III. Product characterization. Plant Physiol 96:571–576

De Lorenzo G, Cervone F (1986) Differential absorption rate of polygalacturonase from two races of *Colletotrichum lindemuthianum* to resistant and susceptible cultivars of *Phaseolus vulgaris* L. Ann Bot (Rome) 44: 147–154

De Lorenzo G, Ranucci A, Bellincampi D, Salvi G, Cervone F (1987) Elicitation of phenylalanine ammonia-lyase in *Daucus carota* by oligogalacturonides released from sodium polypectate by homogenous polygalacturonase. Plant Sci 51:147–150

De Lorenzo G, Ito Y, D'Ovidio R, Cervone F, Albersheim P, Darvill AG (1990) Host-pathogen interactions. XXXVII. Abilities of the polygalacturonase-inhibiting proteins from four cultivars of *Phaseolus vulgaris* to inhibit the *endo*polygalacturonases from three races of *Colletrichum lindemuthianum*. Physiol Mol Plant Pathol 36:421–435

De Lorenzo G, Cervone F, Bellincampi D, Caprari C, Clark AJ, Desiderio A, Devoto A, Forrest R, Leckie F, Nuss L, Salvi G (1994) Polygalacturonase, PGIP and oligogalacturonides in cell-cell communication. Biochem Soc Trans 22:396–399

Devereaux J, Haekerli P, Smithies O (1984) A comprehensive set of sequence programs for the VAX. Nucleic Acids Res 11:1645–1655

de Wit PJGM, Hofman AE, Velthuis GCM, Kuc JA (1985) Isolation and characterization of an elicitor of necrosis isolated from intercellular fluids of compatible interactions of *Cladosporium fulvum* (syn. *Fulvia fulva*) and tomato. Plant Physiol 77:642–647

Dickinson H (1994) Self-pollination: simply a social disease? Nature 367:517–518

Dickman MB (1988) Whole-cell transformation of the alfalfa fungal pathogen *Colletotrichum trifolii*. Curr Genet 14:241–246

Diolez A, Langin T, Gerlinger C, Brygoo Y, Daboussi M-J (1993) The *nia* gene of *Fusarium oxysporum*: isolation, sequence and development of a homologous transformation system. Gene 131:61–67

D'Ovidio R, De Lorenzo G, Cervone F (1990) Isolation and characterization of pectin-inducible cDNA clones from the phytopathogenic fungus *Fusarium moniliforme*. Mycol Res 94:635–640

Fanutti C, Ponyi T, Black GW, Hazlewood GP, Gilbert HJ (1995) The conserved noncatalytic 40-residue sequence in cellulases and hemicellulases from anaerobic fungi functions as a protein docking domain. J Biol Chem 270:29314–29322

Farmer EE, Helgeson JP (1987) An extracellular protein from *Phytophthora parasitica* var. *nicotianae* is associated with stress metabolite accumulation in tobacco callus. Plant Physiol 85:733–740

Favaron F, Alghisi P, Marciano P (1992) Characterization of two *Sclerotinia sclerotiorum* polygalacturonases with different abilities to elicit glyceollin in soybean. Plant Sci 83:7–13

Favaron F, Castiglioni C, Di Lenna P (1993) Inhibition of some rot fungi polygalacturonases by *Allium cepa* L. and *Allium porrum* L. extracts. J Phytopathol 139:201–206

Favaron F, D'Ovidio R, Porceddu E, Alghisi P (1994) Purification and molecular characterization of a soybean polygalacturonase-inhibiting protein. Planta 195:80–87

Fincham JRS (1989) Transformation of fungi. Microbiol Rev 53:148–170

Flipphi MJA, Panneman H, Van der Veen P, Visser J, De Graaff LH (1993) Molecular cloning, expression and structure of the endo-1,5-α-L-arabinase gene of *Aspergillus niger*. Appl Microbiol Biotechnol 40:318–326

Fraissinet-Tachet L, Reymond-Cotton P, Fèvre M (1995) Characterization of a multigene family encoding an endopolygalacturonase in *Sclerotinia sclerotiorum*. Curr Genet 29:96–99

Frediani M, Cremonini R, Salvi G, Caprari C, Desiderio A, D'Ovidio R, Cervone F, De Lorenzo G (1993) Cytological localization of the *pgip* genes in the embryo suspensor cells of *Phaseolus vulgaris* L. Theor Appl Genet 87:369–373

Fuchs Y, Saxena A, Gamble HR, Anderson JD (1989) Ethylene biosynthesis-inducing protein from cellulysin is an endoxylanase. Plant Physiol 89:139–143

Gao S, Shain L (1995) Activity of polygalacturonase produced by *Cryphonectria parasitica* in chestnut bark and its inhibition by extracts from American and Chinese chestnut. Physiol Mol Plant Pathol 46:199–213

Gardner JM, Kado CI (1976) Polygalacturonic acid trans-eliminase in the osmotic shock fluid of *Erwinia rubrifaciens*: characterization of the purified enzyme and its effect on plant cells. J Bacteriol 127:451–460

Gilkes NR, Henrissat B, Kilburn DG, Miller RC Jr, Warren RAJ (1991) Domains in microbial β-1,4-glycanases: sequence conservation, function, and enzyme families. Microbiol Rev 55:303–315

González-Candelas L, Kolattukudy PE (1992) Isolation and analysis of a novel inducible pectate lyase gene from the phytopathogenic fungus *Fusarium solani* f. sp. *pisi* (*Nectria haematococca*, mating population VI). J Bacteriol 174:6343–6349

Guo WJ, Gonzalez-Candelas L, Kolattukudy PE (1995a) Cloning of a new pectate lyase gene *pelC* from *Fusarium solani* f. sp. *pisi* (*Nectria haematococca*, mating type VI) and characterization of the gene product expressed in *Pichia pastoris*. Arch Biochem Biophys 323:352–360

Guo WJ, González-Candelas L, Kolattukudy PE (1995b) Cloning of a novel constitutively expressed pectate lyase gene pelB from *Fusarium solani* f. sp. *pisi* (*Nectria haematococca*, mating type VI) and characterization of the gene product expressed in *Pichia pastoris*. J Bacteriol 177:7070–7077

Gysler C, Harmsen JAM, Kester HCM, Visser J, Heim J (1990) Isolation and structure of the pectin lyase D-encoding gene from *Aspergillus niger*. Gene 89:101–108

Hahn MG (1989) Animal receptors – examples of cellular signal perception molecules. In: Lugtenberg BJJ (ed) Signal molecules in plants and plant-microbe interactions. NATO ASI Series, vol H36. Springer, Berlin Heidelberg New York, pp 1–26

Hahn MG, Bucheli P, Cervone F, Doares SH, O'Neill RA, Darvill A, Albersheim P (1989) Roles of cell wall constituents in plant-pathogen interactions. In: Kosuge T, Nester EW (eds) Plant-microbe interactions. Molecular and genetic perspectives, vol 3. McGraw Hill, New York, pp 131–181

Hahn MG, Darvill AG, Albersheim P (1981) Host-pathogen interactions. XIX. The endogenous elicitor, a fragment of a plant cell wall polysaccharide that elicits phytoalexin accumulation in soybeans. Plant Physiol 68:1161–1169

Harmsen JAM, Kusters-van Someren MA, Visser J (1990) Cloning and expression of a second *Aspergillus niger* pectin lyase gene (*pel*A): indications of a pectin lyase gene family in *A. niger*. Curr Genet 18:161–166

Ho M-C, Whitehead MP, Cleveland TE, Dean RA (1995) Sequence analysis of the *Aspergillus nidulans* pectate lyase *pelA* gene and evidence for binding of promoter regions to CREA, a regulator of carbon catabolite repression. Curr Genet 27:142–149

Høj PB, Fincher GB (1995) Molecular evolution of plant β-glucan endohydrolases. Plant J 7:367–379

Holden FR, Walton JD (1992) Xylanases from the fungal maize pathogen *Cochliobolus carbonum*. Physiol Mol Plant Pathol 40:39–47

Horn MA, Heinstein PF, Low PS (1989) Receptor-mediated endocytosis in plant cells. Plant Cell 1:1003–1009

Howard RJ, Ferrari MA, Roach DH, Money NP (1991) Penetration of hard substrates by a fungus employing enormous turgor pressures. Proc Natl Acad Sci USA 88:11281–11284

Jacinto T, Farmer EE, Ryan CA (1993) Purification of potato leaf plasma membrane protein pp34, a protein phosphorylated in response to oligogalacturonide signals for defense and development. Plant Physiol 103: 1393–1397

Johnston DJ, Ramanathan V, Williamson B (1993) A protein from immature raspberry fruits which inhibits endopolygalacturonases from *Botrytis cinerea* and other micro-organisms. J Exp Bot 44:971–976

Jones DA, Thomas CM, Hammond-Kosack KE, Balint-Kurti PJ, Jones JDG (1994) Isolation of the tomato *Cf-9* gene for resistance to *Cladosporium fulvum* by transposon tagging. Science 266:789–793

Jones TM, Anderson AJ, Albersheim P (1972) Host-pathogen interactions. IV. Studies on the polysaccharide-degrading enzymes secreted by *Fusarium oxysporum* f. sp. *lycopersici*. Physiol Plant Pathol 2: 153–166

Karr AL, Albersheim P (1970) Polysaccharide-degrading enzymes are unable to attack plant cell walls without prior action by a "wall-modifying enzyme". Plant Physiol 46:69–80

Katoh M, Hirose I, Kubo Y, Hikichi Y, Kunoh H, Furusawa I, Shishiyama J (1988) Use of mutants to indicate factors prerequisite for penetration of *Colletotrichum lagenarium* by appressoria. Physiol Mol Plant Pathol 32:177–184

Kauppinen S, Christgau S, Kofod LV, Halkier T, Dörreich K, Dalboge H (1995) Molecular cloning and characterization of a rhamnogalacturonan acetylesterase from *Aspergillus aculeatus* – synergism between rhamnogalacturonan-degrading enzymes. J Biol Chem 270: 27172–27178

Keen NT, Sims JJ, Midland S, Yoder M, Jurnak F, Shen H, Boyd C, Yucel I, Lorang J, Murillo J (1993) Determinants of specificity in the interaction of plants with bacterial pathogens. In: Nester EW, Verma DPS (eds) Advances in molecular genetics of plant-microbe interactions, vol 2. Kluwer, Dordrecht, pp 211–220

Kelemu S, Collmer A (1993) *Erwinia chrysanthemi* EC16 produces a second set of plant-inducible pectate lyase isozymes. Appl Environ Microbiol 59:1756–1761

Keon JPR, Waksman G, Bailey JA (1990) A comparison of the biochemical and physiological properties of a polygalacturonase from two races of *Colletotrichum lindemuthianum*. Physiol Mol Plant Pathol 37:193–206

Kester HCM, Visser J (1994) Purification and characterization of pectin lyase B, a novel pectinolytic enzyme from *Aspergillus niger*. FEMS Microbiol Lett 120:63–68

Khanh NQ, Ruttkowski E, Leidinger K, Albrecht H, Gottschalk M (1991) Characterization and expression of a genomic pectin methyl esterase-encoding gene in *Aspergillus niger*. Gene 106:71–77

Kitamoto N, Kimura T, Kito Y, Ohmiya K, Tsukagoshi N (1993) Structural features of a polygalacturonase gene cloned from *Aspergillus oryzae* KBN616. FEMS Microbiol Lett 111:37–42

Knowles J, Lehtovaara P, Teeri T, Penttilä M, Salovuori I, André L (1987) The application of recombinant-DNA technology to cellulase and lignocellulosic wastes. Philos Trans R Soc Lond A 321:449–454

Kobe B, Deisenhofer J (1993) Crystal structure of porcine ribonuclease inhibitor, a protein with leucine-rich repeats. Nature 366:751–756

Kobe B, Deisenhofer J (1995) A structural basis of the interactions between leucine-rich repeats and protein ligands. Nature 374:183–186

Kohn R (1975) Ion binding on polyuronates-alginate and pectin. Pure Appl Chem 42:371–397

Kollar A (1994) Characterization of specific induction, activity, and isozyme polymorphism of extracellular cellulases from *Venturia inequalis* detected in vitro and on the host plant. Mol Plant-Microbe Interact 7:603–611

Lafitte C, Barthe JP, Montillet JL, Touzé A (1984) Glycoprotein inhibitors of *Colletotrichum lindemuthianum* endopolygalacturonase in near isogenic lines of *Phaseolus vulgaris* resistant and susceptible to anthracnose. Physiol Plant Pathol 25:39–53

Lafitte C, Barthe J-P, Gansel X, Dechamp-Guillaume G, Faucher C, Mazau D, Esquerré-Tugayé M-T (1993) Differential induction by endopolygalacturonase of β-1,3-glucanases in *Phaseolus vulgaris* isoline susceptible and resistant to *Colletotrichum lindemuthianum* race β. Mol Plant-Microbe Interact 6:628–634

Lamb CJ (1994) Plant disease resistance genes in signal perception and transduction. Cell 76:419–422

Lee SC, West CA (1981) Polygalacturonase from *Rhizopus stolonifer*, an elicitor of casbene synthetase activity in castor bean (*Ricinus communis* L.) seedlings. Plant Physiol 67:633–639

Legendre L, Rueter S, Heinstein PF, Low PS (1993) Characterization of the oligogalacturonide-induced oxidative burst in cultured soybean (*Glycine max*) cells. Plant Physiol 102:233–240

Levine A, Tenhaken R, Dixon R, Lamb C (1994) H_2O_2 from the oxidative burst orchestrates the plant hypersensitive disease resistance response. Cell 79:583–593

Liners F, Letesson J-J, Didembourg C, Van Cutsem P (1989) Monoclonal antibodies against pectin. Recognition of a conformation induced by calcium. Plant Physiol 91:1419–1424

Linhardt RJ, Galliher PM, Cooney CL (1986) Polysaccharide lyases. Appl Biochem Biotechnol 12:135–176

Lotan T, Fluhr R (1990) Xylanase, a novel elicitor of pathogenesis-related proteins in tobacco, uses a non-ethylene pathway for induction. Plant Physiol 93:811–817

Maizel J, Lenk R (1981) Enhanced graphic matrix analysis of nucleic acid and protein sequences. Proc Natl Acad Sci USA 78:7665–7669

Mankarios AT, Friend J (1980) Polysaccharide degrading enzymes of *Botrytis allii* and *Sclerotium cepivorum*: enzyme production in culture and the effect of the enzymes on isolated onion cell walls. Physiol Plant Pathol 17:93–104

Marfà V, Gollin DJ, Eberhard S, Mohnen D, Darvill A, Albersheim P (1991) Oligogalacturonides are able to induce flowers to form on tobacco explants. Plant J 1:217–225

Marinelli F, Di Gregorio S, Nuti Ronchi V (1991) Phytoalexin production and cell death in elicited carrot cell suspension cultures. Plant Sci 77:261–266

Marmeisse R, van den Ackerveken GFJM, Goosen T, de Wit PJGM, van den Broek HWJ (1993) Disruption of the avirulence gene *avr9* in two races of the tomato pathogen *Cladosporium fulvum* causes virulence on tomato genotypes with the complementary resistance gene *Cf9*. Mol Plant-Microbe Interact 6:412–417

Mathieu Y, Kurkdijan A, Xia H, Guern J, Koller A, Spiro M, O'Neill M, Albersheim P, Darvill A (1991) Membrane responses induced by oligogalacturonides in suspension-cultured tobacco cells. Plant J 1:333–343

Messiaen J, Van Cutsem P (1993) Defense gene transcription in carrot cells treated with oligogalacturonides. Plant Cell Physiol 34:1117–1123

Messiaen J, Van Cutsem P (1994) Pectic signal transduction in carrot cells: membrane, cytosolic and nuclear responses induced by oligogalacturonides. Plant Cell Physiol 35:677–689

Messiaen J, Read ND, Van Cutsem P, Trewavas AJ (1993) Cell wall oligogalacturonides increase cytosolic free calcium in carrot protoplasts. J Cell Sci 104:365–371

Mullen JM, Bateman DF (1975) Polysaccharide-degrading enzymes produced by *Fusarium roseum* "avenaceum" in culture and during pathogenesis. Physiol Plant Pathol 6:233–246

Müller M, Gessler C (1993) A protein from apple leaves inhibits pectinolytic activity of *Venturia inaequalis* in vitro. In: Fritig B, Legrand M (eds) Mechanisms of plant defense responses. Kluwer, Dordrecht, pp 68–71

Nasrallah JB, Nasrallah ME (1993) Pollen-stigma signaling in the sporophytic self-incompatibility response. Plant Cell 5:1325–1335

Nuss L, Mahé A, Clark AJ, Grisvard J, Dron M, Cervone F, De Lorenzo G (1996) Differential accumulation of polygalacturonase-inhibiting protein (PGIP) mRNA in two near-isogenic lines of *Phaseolus vulgaris* L. upon infection with *Colletotrichum lindemuthianum*. Physiol Mol Plant Pathol 48:83–89

Parsons KA, Chumley FG, Valent B (1987) Genetic transformation of the fungal pathogen responsible for rice blast disease. Proc Natl Acad Sci USA 84:4161–4165

Penttilä M, Lehtovaara P, Nevalainen H, Bhikhabhai R, Knowles J (1986) Homology between cellulase genes of *Trichoderma reesei*: complete nucleotide sequence of the endoglucanase I gene. Gene 45:253–263

Peretto R, Favaron F, Bettini V, De Lorenzo G, Marini S, Alghisi P, Cervone F, Bonfante P (1992) Expression and localization of polygalacturonase during the outgrowth of lateral roots in *Allium porrum* L. Planta 188:164–172

Powell ALT, Stotz HU, Labavitch JM, Bennett AB (1994) Glycoprotein inhibitors of fungal polygalacturonases. In: Daniels MJ, Downie JA, Osbourn AE (eds) Advances in molecular genetics of plant-microbe interactions. Kluwer, Dordrecht, pp 399–402

Punt PJ, Van den Hondel CAMJJ (1992) Transformation of filamentous fungi based on hygromycin B and phleomycin resistance markers. Methods Enzymol 216:447–457

Rapp P (1989) 1,3-β-Glucanase, 1,6-β-glucanase and β-glucosidase activities of *Sclerotium glucanicum*: synthesis and properties. J Gen Microbiol 135:2847–2858

Rexová-Benková L, Mracková M (1978) Active groups of extracellular endo-D-galacturonanase of *Aspergillus niger* derived from pH effect on kinetic data. Biochim Biophys Acta 523:162–169

Reymond P, Deléage G, Rascle C, Fèvre M (1994) Cloning and sequence analysis of a polygalacturonase-encoding gene from the phytopathogenic fungus *Sclerotinia sclerotiorum*. Gene 146:233–237

Reymond P, Grünberger S, Paul K, Müller M, Farmer EE (1995) Oligogalacturonide defense signals in plants: Large fragments interact with the plasma membrane in vitro. Proc Natl Acad Sci USA 92:4145–4149

Ricci P, Bonnet P, Huet J-C, Sallantin M, Beauvais-Cante F, Bruneteau M, Billard V, Michel G, Pernollet J-C (1989) Structure and activity of proteins from pathogenic fungi *Phytophthora* eliciting necrosis and acquired resistance in tobacco. Eur J Biochem 183:555–563

Robertsen B (1986) Elicitors of the production of lignin-like compounds in cucumber hypocotyls. Physiol Mol Plant Pathol 28:137–148

Robertsen B (1987) *Endo*polygalacturonase from *Cladosporium cucumerinum* elicits lignification in cucumber hypocotyls. Physiol Mol Plant Pathol 31:361–374

Robertsen B (1989) Pectate lyase from *Cladosporium cucumerinum*, purification, biochemical properties and ability to induce lignification in cucumber hypocotyls. Mycol Res 94:595–602

Rodriguez RJ, Yoder OC (1987) Selectable genes for transformation of the fungal plant pathogen *Glomerella cingulata* f. sp. *phaseoli* (*Colletotrichum lindemuthianum*). Gene 54:73–81

Ruttkowski E, Labitzke R, Khanh NQ, Löffler F, Gottschalk M, Jany K-D (1990) Cloning and DNA sequence analysis of a polygalacturonase cDNA from *Aspergillus niger* RH5344. Biochim Biophys Acta 1087:104–106

Sahasrabudhe NA, Ranjekar PK (1990) Cloning of the cellulase gene from *Penicillium funiculosum* and its expression in *Escherichia coli*. FEMS Microbiol Lett 66:291–294

Saloheimo A, Henrissat B, Hoffren AM, Teleman O, Penttila M (1994) A novel, small endoglucanase gene, *egl5*, from *Trichoderma reesei* isolated by expression in yeast. Mol Microbiol 13:219–228

Saloheimo M, Lehtovaara P, Penttilä M, Teeri TT, Stahlberg J, Johansson G, Pettersson G, Claeyssens M, Tomme P, Knowles JK (1988) EGIII, a new endoglucanase from *Trichoderma reesei*: the characterization of both gene and enzyme. Gene 63:11–21

Salvi G, Giarrizzo F, De Lorenzo G, Cervone F (1990) A polygalacturonase-inhibiting protein in the flowers of *Phaseolus vulgaris* L. J Plant Physiol 136:513–518

Schaeffer HJ, Leykam J, Walton JD (1994) Cloning and targeted gene disruption of *EXG1*, encoding exo-β1,3-glucanase, in the phytopathogenic fungus *Cochliobolus carbonum*. Appl Environ Microbiol 60:594–598

Schejter A, Marcus L (1988) Isozymes of pectinesterase and polygalacturonase from *Botrytis cinerea* Pers. Methods Enzymol 161:366–373

Schols HA, Geraeds CCJM, Searle-van Leeuwen MF, Kormelink FJM, Voragen AGJ (1990) Rhamnogalacturonase: a novel enzyme that degrades the hairy regions of pectins. Carbohydr Res 206:105–115

Scott-Craig JS, Panaccione DG, Cervone F, Walton JD (1990) Endopolygalacturonase is not required for pathogenicity of *Cochliobolus carbonum* on maize. Plant Cell 2:1191–1200

Searle-van Leeuwen MJF, Van den Broek LAM, Schols HA, Beldman G, Voragen AGJ (1992) Rhamnogalacturonan acetylesterase: a novel enzyme from *Aspergillus aculeatus*, specific for the deacetylation of hairy (ramified) regions of pectins. Appl Microbiol Biotechnol 38:347–349

Sharon A, Fuchs Y, Anderson JD (1993) The elicitation of ethylene biosynthesis by a *Trichoderma* xylanase is not related to the cell wall degradation activity of the enzyme. Plant Physiol 102:1325–1329

Sheppard PO, Grant FJ, Oort PJ, Sprecher CA, Foster DC, Hagen FS, Upshall A, McKnight GL, O'Hara PJ (1994) The use of conserved cellulase family-specific sequences to clone cellulase homologue cDNAs from *Fusarium oxysporum*. Gene 150:163–167

Sims P, James C, Broda P (1988) The identification, molecular cloning and characterisation of a gene from *Phanerochaete chrysosporium* that shows strong homology to the exo-cellobiohydrolase I gene from *Trichoderma reesei*. Gene 74:411–422

Sposato P, Ahn J-H, Walton JD (1995) Characterization and disruption of a gene in the maize pathogen *Cochliobolus carbonum* encoding a cellulase lacking a cellulose-binding domain and hinge region. Mol Plant-Microbe Interact 8:602–609

Stahl DJ, Schäfer W (1992) Cutinase is not required for fungal pathogenicity on pea. Plant Cell 4:621–629

Stahmann K-P, Schimz K-L, Sahm H (1993) Purification and characterization of four extracellular 1,3-β-glucanases of *Botrytis cinerea*. J Gen Microbiol 139:2833–2840

Staskawicz BJ, Ausubel FM, Baker BJ, Ellis JG, Jones JDG (1995) Molecular genetics of plant disease resistance. Science 268:661–667

Steinmayr M, Motte P, Sommer H, Saedler H, Schwarz-Sommer Z (1994) FIL2, an extracellular leucine-rich repeat protein, is specifically expressed in *Antirrhinum* flowers. Plant J 5:459–467

Stotz HU, Powell ALT, Damon SE, Greve LC, Bennett AB, Labavitch JM (1993) Molecular characterization of a polygalacturonase inhibitor from *Pyrus communis* L. cv. Bartlett. Plant Physiol 102:133–138

Stotz HU, Contos JJA, Powell ALT, Bennett AB, Labavitch JM (1994) Structure and expression of an inhibitor of fungal polygalacturonases from tomato. Plant Mol Biol 25:607–617

Stratilová E, Markovic O, Skrovinová D, Rexová-Benková L, Jörnvall H (1993) Pectinase *Aspergillus* sp. polygalacturonase: multiplicity, divergence, and structural patterns linking fungal, bacterial, and plant polygalacturonases. J Protein Chem 12:15–22

Svalheim O, Robertsen B (1993) Elicitation of H_2O_2 production in cucumber hypocotyl segments by oligo-1,4-α-D-galacturonides and an oligo-β-glucan preparation from cell walls of *Phytophthora megasperma* f.sp. *glycinea*. Physiol Plant 88:675–681

Takahashi N, Takahashi Y, Putnam FW (1985) Periodicity of leucin and tandem repetition of a 24-amino acid segment in the primary structure of leucin-rich α_2-glycoprotein of human serum. Proc Natl Acad Sci USA 82:1906–1910

Tan LUL, Mayers P, Illing M, Saddler JN (1987) The copurification of beta-glucosidase, β-xylosidase, and 1,3-β-glucanase in two separate enzyme complexes isolated from *Trichoderma harzianum* E58. Biochem Cell Biol 65:822–832

Teeri TT, Lehtovaara P, Kauppinen S, Salovuori I, Knowles J (1987) Homologous domain in *Trichoderma reesei* cellulolytic enzymes: gene sequence and expression of cellobiohydrolase II. Gene 51:43–52

Tempelaars CA, Birch PR, Sims PF, Broda P (1994) Isolation, characterization, and analysis of the expression of the *cbhII* gene of *Phanerochaete chrysosporium*. Appl Environ Microbiol 60:4387–4393

Templeton MD, Sharrock KR, Bowen JK, Crowhurst RN, Rikkerink EHA (1994) The pectin lyase-encoding gene (*pnl*) family from *Glomerella cingulata*: characterization of *pnlA* and its expression in yeast. Gene 142:141–146

Thain JF, Gubb IR, Wildon DC (1995) Depolarization of tomato leaf cells by oligogalacturonide elicitors. Plant Cell Environ 18:211–214

Tomme P, Van Tilbeurgh H, Pettersson G, van Damme J, Vandekerckhove J, Knowles J, Teeri T, Claeyssens M (1988) Studies on the cellulolytic system of *Trichoderma reesei* QM 9414: analysis of domain function in two cellobiohydrolases by limited proteolysis. Eur J Biochem 170:575–581

Toubart P, Desiderio A, Salvi G, Cervone F, Daroda L, De Lorenzo G, Bergmann C, Darvill AG, Albersheim P (1992) Cloning and characterization of the gene encoding the endopolygalacturonase-inhibiting protein (PGIP) of *Phaseolus vulgaris* L. Plant J 2:367–373

Urbanek H, Zalewska-Sobczak J (1986) 1.4-β-galactanases and 1.3-β-glucanases of *Botrytis cinerea* isolate infecting apple. Biochem Physiol Pflanz 181:321–329

Uritani I, Stahmann MA (1961) Pectolytic enzymes of *Ceratocystis fimbriata*. Phytopathology 51:277–285

Valsangiacomo C, Gessler C (1992) Purification and characterization of an exo-polygalacturonase produced by *Venturia inaequalis*, the causal agent of apple scab. Physiol Mol Plant Pathol 40:63–77

Van Arsdell JN, Kwok S, Schweickart VL, Ladner MB, Gelfand DH, Innis MA (1987) Cloning, characterization, and expression in *Saccharomyces cerevisiae* of endoglucanase I from *Trichoderma reesei*. Bio/Technology 5:60–64

Vincken J-P, Beldman G, Voragen AGJ (1994) The effect of xyloglucans on the degradation of cell-wall-embedded cellulose by the combined action of cellobiohydrolase and endoglucanases from *Trichoderma viride*. Plant Physiol 104:99–107

Walker JC (1993) Receptor-like protein kinase genes of *Arabidopsis thaliana*. Plant J 3:451–456

Walker-Simmons M, Jin D, West CA, Hadwiger L, Ryan CA (1984) Comparison of proteinase inhibitor-inducing activities and phytoalexin elicitor activities of a pure fungal endopolygalacturonase, pectic fragments, and chitosan. Plant Physiol 76:833–836

Wang H, Jones RW (1995) Cloning, characterization and functional expression of an endoglucanase-encoding gene from the phytopathogenic fungus *Macrophomina phaseolina*. Gene 158:125–128

Wang J, Holden DW, Leong SA (1988) Gene-transfer system for the phytopathogenic fungus *Ustilago maydis*. Proc Natl Acad Sci USA 85:865–869

Wang P, Nuss DL (1995) Induction of a *Cryphonectria parasitica* cellobiohydrolase I gene is suppressed by hypovirus infection and regulated by a GTP-binding-protein-linked signaling pathway involved in fungal pathogenesis. Proc Natl Acad Sci USA 92:11529–11533

Wijesundera RLC, Bailey JA, Byrde RJW, Fielding AH (1989) Cell wall degrading enzymes of *Colletotrichum lindemuthianum*: their role in the development of bean anthracnose. Physiol Mol Plant Pathol 34:403–413

Wu S-C, Kauffmann S, Darvill AG, Albersheim P (1995) Purification, cloning and characterization of two xylanases from *Magnaporthe grisea*, the rice blast fungus. Mol Plant-Microbe Interact 8:506–514

Yao CL, Conway WS, Sams CE (1995) Purification and characterization of a polygalacturonase-inhibiting protein from apple fruit. Phytopathology 85:1373–1377

Yoder MD, Keen NT, Jurnak F (1993) New domain motif: the structure of pectate lyase C, a secreted plant virulence factor. Science 260:1503–1507

5 Signals in Host-Parasite Interactions

J. Ebel[1] and D. Scheel[2]

CONTENTS

I. Introduction

Plants are resistant to most potential pathogens in their environment; in fact, relatively few true host-pathogen pairs exist in which the plant is susceptible and the pathogen virulent. Within a susceptible plant species, specific cultivars carry genes which give them resistance to specific pathogen races harboring the corresponding avirulence genes (Chaps. 15, 16, Vol. V, Part B). Several models have been proposed to explain the biochemical basis for gene-for-gene complementarity in such plant-pathogen interactions. One model envisions a ligand-receptor-like interaction between the products of a fungal avirulence gene (elicitor) and the corresponding plant resistance gene (receptor) that triggers the initiation of a multicomponent defense response in the plant. The same response can also be activated by other types of elicitors that are not encoded by avirulence genes, including compounds released from fungal or plant cell walls during early phases of pathogen attack or substances secreted by the pathogen. Exogenous elicitors are derived from the pathogen, whereas endogenous elicitors are released from the plant cell wall during pathogen attack. Whatever the source of the elicitor, the mechanism of signal perception appears to rely on the presence of specific receptors on the plant cell surface (Fig. 1) which initiate signaling processes that activate plant defenses. Typical elements of the multicomponent defense response include the hypersensitive reaction (HR), the production of reactive oxygen species (oxidative burst), the activation of defense-related genes, structural changes of the cell wall, and the synthesis of phytoalexins.

In addition to these localized defenses, systemic acquired resistance (SAR), which transiently increases the plant's resistance systemically to subsequent pathogen attack, is induced in many plants. SAR development is accompanied by systemic gene activation and is inducible by certain elicitors. The initiation of the individual defense reactions, which occurs in different cellular compartments or even in distant plant tissues, requires intra- and intercellular signal transduction events that are originally triggered by pathogen recognition, i.e., the interaction of a relevant elicitor with its receptor (Fig. 1).

The discussion in this chapter will focus on signaling processes in plant-fungus pathogenic interactions and will not cover details that are described in recent reviews on similar topics (Atkinson 1993; Ebel and Cosio 1994; Dixon et al. 1994; Côté and Hahn 1994; Ricci et al. 1993).

[1] Botanisches Institut, Universität München, Menzinger Straße 67, 80638 München, Germany
[2] Institut für Pflanzenbiochemie, Abteilung Stress- und Entwicklungsbiologie, Weinberg 3, 06120 Halle (Saale), Germany

The Mycota V Part A
Plant Relationships
Carroll/Tudzynski (Eds.)

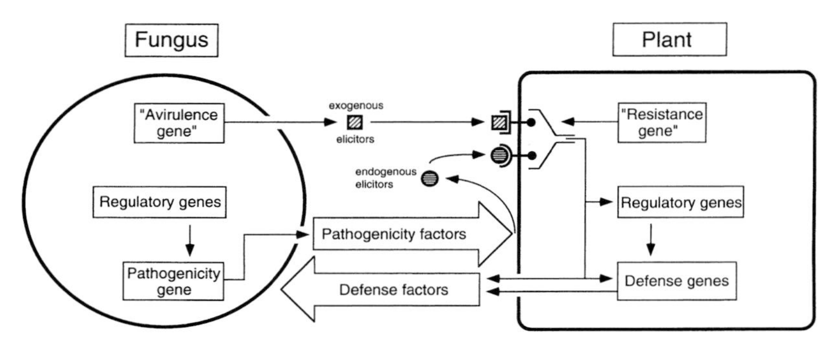

Fig. 1. Hypothetical model of the interaction between plants and potential pathogenic fungi describing signaling events possibly occurring within and between the interacting organisms. Exogenous elicitors may be encoded by avirulence genes or originate from fungal surface structures. In this latter case, the genes encoding the corresponding receptor would not fit the definition of resistance genes. Since otherwise the physiological model is identical for both cases, the terms, avirulence and resistance gene, have been marked with inverted commas

II. Elicitors of Defense Responses

Several elicitors of fungal origin have been purified to homogeneity (Table 1). Specific components of the defense response have been used in the purification procedures of different elicitors and it is, therefore, often unknown whether they activate the entire defense response or only part of it. Further, given the structural diversity of the purified fungal elicitors (Table 1), it appears that no universal elicitor exists which can serve as a general signal for pathogen attack. Rather, single fungal elicitors appear to act in only a few plant species, a single plant species, or even in specific cultivars of one species.

A. Protein Elicitors

1. Cultivar-Specific Elicitors

Two proteinaceous elicitors of fungal origin have been described to date that display cultivar-specific activity (De Wit and Spikman 1982; Scholtens-Toma and De Wit 1988; Hahn et al. 1993). Both are encoded by avirulence genes of the fungal plant pathogens, one in *Cladosporium fulvum* and the other in *Rhynchosporium secalis* (Van Kan et al. 1991; De Wit 1992; Knogge et al. 1994). The *avr4* and *avr9* genes of *C. fulvum* encode preproproteins that are synthesized, secreted, and processed during fungal growth in planta (Van Kan et al. 1991; Van den Ackerveken et al. 1993; Joosten et al. 1994; Chap. 1, Vol. V, Part B). The proteins, AVR4 and AVR9, specifically elicit an HR in leaves of tomato cultivars carrying the corresponding resistance genes, *Cf4* and *Cf9*, respectively (De Wit and Spikman 1982; Scholtens-Toma and De Wit 1988).

The *nip1* gene of *R. secalis* encodes a peptide elicitor that is secreted by the fungus during growth in plants and in liquid culture (Hahn et al. 1993; Wevelsiep et al. 1991; Knogge et al. 1994). The 82 amino-acid product of the *nip1* gene is processed to yield a 60 amino-acid mature protein that is cysteine-rich (Rohe et al. 1995). It elicits the activation of two genes encoding the defense-related proteins, peroxidase and PRHv-1, in a cultivar-specific manner in barley cultivars carrying the resistance gene, *Rrs1* (Hahn et al. 1993). In addition, exogenously applied NIP1 protein completely protected a barley cultivar carrying *Rrs1* against a virulent fungal race lacking the *nip1* gene, whereas virulence of the same fungal races on *rrs1* plants was unaffected (Knogge et al. 1994). Interestingly, NIP1 also appears to function as a virulence or pathogenicity factor that stimulates the plant plasma membrane H^+-ATPase (Wevelsiep et al. 1993). Although similar functions have not yet been demonstrated for other fungal avirulence gene products, the dual activity of the gene products would offer an attractive explanation for the evolutionary stability of avirulence genes.

It has recently been demonstrated for avirulence genes of both *C. fulvum* and *R. secalis* that point mutations leading to single amino acid exchanges, as well as complete loss of avirulence genes occur in nature and result in virulence of the respective fungal race on the corresponding plant cultivar (Van den Ackerveken et al. 1992; Van Kan et al. 1991; Joosten et al. 1994; Knogge et al.

Table 1. Selected fungus- or plant-derived elicitors of plant defense responses

Source	Elicitor	Defense response	Reference[a]
Fungi			
Cladosporium fulvum	Protein (M_r 3186)	Necrosis in tomato leaves	1
Fusarium moniliforme	Oligo-1,4-β-N-acetylglucosamine (chitin fragments)	Phytoalexins in rice	2
Fusarium solani f. sp. *phaseoli*	Oligo-1,4-β-glucosamine (chitosan fragments)	Phytoalexins in pea	3
Monilinia fructicola	Protein (M_r 8000)	Phytoalexins in bean	4
Phytophthora spp.	Proteins (M_r ~10000)	Necrosis and SAR in tobacco	5, 6
Phytophthora infestans	Eicosapentaenoic and arachidonic acids	Phytoalexins in potato tubers	7
Phytophthora megasperma H20	Glycoprotein (M_r 32000)	Necrosis and SAR in tobacco	8
Phytophthora parasitica var. nicotianae	Glycoprotein (M_r 46000)	Phytoalexins in tobacco	9
Phytophthora sojae	Hepta-1,3–1,6-β-glucoside	Phytoalexins in soybean	10, 11
	Glycoprotein (M_r 42000)	Phytoalexins in parsley	12
	Oligopeptide (M_r 1622)	Phytoalexins in parsley	13
Puccinia graminis f. sp. *tritici*	Peptidoglycan (M_r 67000)	Lignin-like compounds in wheat leaves	14
Rhynchosporium secalis	Protein (M_r 6440)	PR proteins in barley	15
Plants			
Glycine max	Oligo-1,4-α-galacturonide	Phytoalexins in soybean	16, 17
Ricinus communis	Oligo-1,4-α-galacturonide	Casbene synthetase and lignin-like compounds in castor bean	18, 19
Vigna unguiculata	Oligo-1,4-α-galacturonide	Necrosis in cowpea pods	20

[a] 1, Scholtens-Toma and De Wit (1988); 2, Ren and West (1992); 3, Hadwiger and Beckman (1980); 4, Cruickshank and Perrin (1968); 5, Pernollet et al. (1993); 6, Ricci et al. (1989); 7, Preisig and Kuc (1985); 8, Baillieul et al. (1995); 9, Farmer and Helgeson (1987); 10, Sharp et al. (1984a); 11, Sharp et al. (1984b); 12, Parker et al. (1991); 13, Nürnberger et al. (1994a); 14, Kogel et al. (1988); 15, Hahn et al. (1993); 16, Davis et al. (1986); 17, Nothnagel et al. (1983); 18, Bruce and West (1989); 19, Jin and West (1984); 20, Cervone et al. (1987).

1994). The mutated avirulence proteins were found to be inactive as elicitors on resistant cultivars (Joosten et al. 1994; Knogge et al. 1994). After transfer of the avirulence genes into virulent fungal races, transformants were isolated that are avirulent on plants carrying the respective resistance genes (Joosten et al. 1994; Rohe et al. 1995; Van den Ackerveken et al. 1992).

2. Elicitins

Another important class of fungal protein elicitors includes the elicitins, small extracellular proteins that are secreted by most *Phytophthora* species (Ricci et al. 1992; Kamoun et al. 1994; Chap. 2, Vol. V, Part B). These proteins, having a molecular mass of around 10 kDa, induce an HR as well as SAR in tobacco and certain radish and turnip cultivars (Ricci et al. 1989; Kamoun et al. 1993b). In addition, the synthesis of ethylene, sesquiterpenoid phytoalexins, and PR proteins is stimulated in tobacco upon treatment with the elicitin cryptogein (Milat et al. 1991; Ricci et al. 1993). The different elicitins share 80–90% sequence identity at the amino acid level and are encoded by multigene families (Kamoun et al. 1993a; Ricci et al. 1993). Six conserved cysteine residues form three disulfide bridges (Nespoulos et al. 1992) that appear to be important for elicitor activity (Ricci et al. 1989).

The production of elicitins is almost ubiquitous in *Phytophthora* species except for isolates of *P. parasitica* from tobacco, which either synthesize no or low amounts of elicitins (Kamoun et al.

1994) or lack elicitin-encoding genes altogether (Kamoun et al. 1993a). At least in this interaction, lack of elicitin production appears to be required for high virulence, and production of increasing levels of elicitins appears to confer a quantitative decrease in virulence rather than complete avirulence (Kamoun et al. 1994). As with NIP1 protein of *R. secalis* (Wevelsiep et al. 1993), elicitins appear to be required for pathogenicity and sporulation of *Phytophthora*; a *P. parasitica* isolate from tomato completely lacking genes encoding elicitins (Kamoun et al. 1993a) was found to be unable to sporulate and to attack tomato as well as tobacco (Kamoun et al. 1994).

A 32-kDa glycoprotein elicitor from *Phytophthora megasperma* H20, a pathogen of Douglas fir nonpathogenic on tobacco, was shown to cause lesion formation and defense gene expression when infiltrated into tobacco leaves that closely resembled hypersensitive response lesions (Baillieul et al. 1995). After several days of treatment acquired resistance was also induced in tobacco.

B. Glycoprotein Elicitors

Three glycoproteins secreted by phytopathogenic fungi have been purified on the basis of their elicitor activity (Coleman et al. 1992; Kogel et al. 1988; Parker et al. 1991). In the glycoprotein elicitors from *Colletotrichum lindemuthianum* (Coleman et al. 1992) and *Puccinia graminis* f. sp. *tritici*, the carbohydrate moieties were found to be responsible for elicitor activity (Kogel et al. 1988), whereas in *Phytophthora sojae* (formerly *Phytophthora megasperma* f. sp. *glycinea*) on parsley the protein portion of a 42-kDa glycoprotein elicited phytoalexin production (Parker et al. 1991).

In this last case, a detailed analysis of biochemical structure and function proved possible. An oligopeptide fragment of 13 amino acids (Pep-13) in the C-terminal half of the latter glycoprotein was found to be necessary and sufficient for elicitor activity (Nürnberger et al. 1994a; Sacks et al. 1995). Removal of the N- and C-terminal as well as replacement of most remaining amino acid residues of the oligopeptide by alanine only slightly reduced the elicitor activity of the peptide (Nürnberger et al. 1994a). Beyond this, it was shown that only the change of tryptophan in position two or proline in position five lowered the elicitor activity of the corresponding peptide significantly in the target cells, indicating that both specific structural elements and a minimal length are essential. Similar results have been reported for systemin, an oligopeptide apparently involved in systemic wound signaling in potato and tomato (Pearce et al. 1993). Interestingly, Pep-13 activated in cultured parsley cells all defense reactions observed in infected leaves except callose accumulation and hypersensitive cell death (Nürnberger et al. 1994). The initiation of these two typical elements of the multicomponent defense response, therefore, either requires intact plant tissue rather than cultured cells or is triggered by different elicitors.

In glycopeptides derived from yeast invertase, both oligopeptide and N-linked high-mannose oligoglycan moieties proved necessary to elicit ethylene biosynthesis and increases in phenylalanine ammonia-lyase activity in cultured tomato cells (Basse et al. 1992). Again, detailed analyses of structure and function have been carried out. Reducing the peptide part of a glycopeptide to the dipeptide arginine-asparagine did not affect elicitor activity. Glycans released enzymatically from these purified elicitors suppressed the elicitor activity of the intact glycopeptides when both were applied to tomato cells simultaneously. Glycan moieties with 10 to 12 mannosyl residues were most effective; those with 9 residues were threefold and those with 8 residues 100-fold less effective as elicitors or suppressors, indicating that the same recognition site is used by the glycan suppressors and the corresponding glycopeptide elicitors. It is unknown, however, whether such elicitors, derived from fungi non-pathogenic to plants, induce typical elements of the multicomponent defense response in plants.

C. Oligosaccharide Elicitors

Three major classes of elicitor-active oligosaccharides of fungal origin have been identified (for a most recent review see Côté and Hahn 1994): oligoglucan (Ayers et al. 1976), oligochitosan (Hadwiger and Beckman 1980), and oligochitin fragments (Barber et al. 1989; Ren and West 1992; Kuchitsu et al. 1993). The structural requirements for oligoglucan elicitors have been most extensively studied for soybean by Albersheim and coworkers, who found oligosaccharides with 3-, 6-, and

3,6-β-linked glucosyl residues to be responsible for elicitor activity and one specific hepta-β-glucoside to be the smallest elicitor-active compound within a complex mixture of oligoglucan fragments chemically derived from mycelial walls of *P. sojae* (Sharp et al. 1984a,b).

The minimal degree of polymerization of oligochitosan and oligochitin molecules required for biological activity appears to be four (Walker-Simmons et al. 1983, 1984). Beyond that, the structural requirements differ greatly from plant to plant and for different responses analyzed (for review see Côté and Hahn 1994).

Oligogalacturonides possibly released from plant cell walls during pathogen attack may act as endogenous elicitors (for review see Côté and Hahn 1994; Chap. 4, this Vol.). Many plants efficiently recognize oligogalacturonides with 10 to 15 uronic acid residues as signals for pathogen attack and respond with an array of defense responses. Oligogalacturonides of similar or different degree of polymerization also initiate wound responses or developmental processes in plants, suggesting that these compounds may play a role as internal systemic signals as well. A similar function has also been suggested for chitin derivatives (Spaink and Lugtenberg 1994; Röhrig et al. 1995) and has been reported for the oligopeptide systemin (Narváez-Vásquez et al. 1995). The elicitor perception machinery might therefore be part of a general intercellular communication system of plants, whose features and components remain to be elucidated in detail.

III. Elicitor Perception

Initiation of plant defense reactions requires perception of either exogenous or endogenous elicitors. The cultivar or species specificity of purified elicitors, their diverse chemical nature, their ability to induce various plant responses at low (nanomolar) concentrations, elicitor dose/response relationships, and the high degree of signal specificity observed for elicitor activity all strongly suggest the involvement of receptors in elicitor recognition and subsequent intracellular signal propagation (Ebel and Cosio 1994). As in animal cells, different classes of membrane-localized receptors may operate as activators of ion channels, protein kinases, or pathways generating intracellular signal compounds.

An important general principle in biochemical signaling suggests that a given cell type will respond in specific ways only to a particular subset of a group of external signal compounds. Consequently, a detailed investigation of primary events in perception and signaling during pathogenesis in plants requires that putative perception and signaling events for a pure elicitor be related to at least one typical defense reaction of the plant cell. Recent evidence indicates that high-affinity, receptor-like binding proteins for fungal elicitors do reside in the plasma membranes of several plants. Moreover, elicitor-induced changes in ion fluxes across the plant plasma membrane, in protein phosphorylation, and in the production of suspected intracellular signaling compounds have been described that precede the activation of defense responses (see Sect. IV). Evidence for the existence of elicitor-binding sites has been obtained for carbohydrate, protein, and glycoprotein elicitors.

A. Glucan-Binding Sites

In the initial work with β-glucan elicitors derived from the cell walls of *Phytophthora* spp., the use of mixtures of structural isomers of the fungal β-glucans strongly indicated the existence of specific β-glucan-binding sites in soybean cell membranes (Cosio et al. 1988; Schmidt and Ebel 1987; Yoshikawa et al. 1983). These earlier results were confirmed when high-affinity binding sites for the (1–3,1–6)-linked hepta-β-glucoside elicitor conjugated with radioiodinated aminophenylethylamine (Cosio et al. 1990b) or tyramine (Cheong and Hahn 1991) were demonstrated. The binding was saturable, reversible, and β-glucan-specific (Cosio et al. 1990b; Cheong and Hahn 1991). Affinity measurements gave apparent K_d values of about 1 to 3 nM and of 10 to 40 nM for the hepta-β-glucoside and the fungal β-glucan fraction, respectively. Competition for binding of the radiolabeled β-glucan fraction (average degree of polymerization = 18) by the hepta-β-glucoside was uniform and complete, showing that the heptaglucoside had access, with similar affinity, to all sites available to the radioligand (Cosio et al. 1990b). Specific glucan binding was detected in membrane fractions of all major organs of soybean seedlings and in soybean cell cultures, but not in soluble protein extracts from these tissues. The binding sites resided in the plasma membrane and were accessible at the external surface of protoplasts isolated from soybean cell cultures (Cheong et al. 1993; Cosio et al. 1988; Schmidt and Ebel 1987).

Analysis of the ligand specificity of the soybean β-glucan-binding sites with a variety of fungal (Cosio et al. 1988, 1990b; Schmidt and Ebel 1987) and synthetic (Cheong and Hahn 1991; Cosio et al. 1990b) oligo- and polyglucosides demonstrated that the hepta-β-glucoside identified earlier by Sharp et al. (1984a,b) was bound with the highest affinity. Furthermore, the binding affinities of the binding sites for the various glucan ligands correlated well with their abilities to stimulate phytoalexin production in a bioassay utilizing soybean cotyledon tissue (Cheong and Hahn 1991; Cosio et al. 1990b). These results showed that oligo-β-glucosides with high elicitor activity are efficient ligands for the binding sites. In contrast, some divergence in binding affinity and elicitor activity was observed when β-glucans naturally released from germinating zoospores of *P. sojae* (Waldmüller et al. 1992) or released by enzyme treatment of hyphal cell walls (Yoshikawa and Sugimoto 1993) were compared with the hepta-β-glucoside. This apparent contradiction deserves further analysis.

Although (1-3,1-6)-β-linked glucans are the most representative structural components of cell surfaces in the Oomycota, their role in the induction of defense responses in plants other than soybean has only rarely been reported (Ebel and Cosio 1994). A screening for β-glucan-binding sites in membranes of other legumes revealed high-affinity binding sites in the four legume species, *Pisum sativum*, *Phaseolus vulgaris*, *Medicago sativa*, and *Lupinus albus* (Cosio et al. 1996). Competition assays with unlabeled hepta-β-glucoside showed that the binding sites of *P. sativum*, *M. sativa*, and *L. albus* displayed the same apparent affinity towards the heptaglucoside as soybean, while the sites of *P. vulgaris* showed a significantly lower affinity. Low binding activity for the β-glucan fraction of relatively weak affinity was detected in membrane fractions of *Cicer arietinum* and *Vicia faba*. Overall, the binding activity observed for the β-glucans in the several legumes examined correlated well with their ability to stimulate phytoalexin accumulation in these plants.

Extracellular polysaccharides may also play critical roles in establishing structural and metabolic cooperation in mutualistic associations between microorganisms and plants. Studies on the symbiotic interactions between *Rhizobium* or *Bradyrhizobium* and their host legumes point the way to similar work on fungal mutualists. In these bacteria some of the extracellular carbohydrates may be active as signal compounds to the plant, not merely as a passive coat for protection (Long and Staskawicz 1993). One hypothesis suggests that extracellular compounds may be important for avoiding host defense (Long and Staskawicz 1993). Cyclic (1-3,1-6)-β-glucans synthesized by *B. japonicum* USDA 110 (Miller et al. 1990; Rolin et al. 1992) not only inhibited the binding of the elicitor-active β-glucans from *P. sojae* to soybean membranes but they also prevented the stimulation of phytoalexin production in a soybean bioassay when tested in combination with the fungal β-glucans (Mithöfer et al. 1996a). It was concluded (Mithöfer et al. 1996a) that the bradyrhizobial cyclic glucans cannot be considered as potent elicitors of phytoalexin production in soybean, but rather they could act to prevent stimulation of plant defenses.

The combined results of the biological assays (Cheong et al. 1991; Cosio et al. 1988, 1990b) and the binding studies (Cheong and Hahn 1991; Cosio et al. 1988, 1990b) provide evidence that the binding sites are the functional receptors of the hepta-β-glucoside and β-glucan elicitors. Unambiguous experimental proof of this hypothesis requires molecular characterization, including functional reconstitution of the isolated binding sites. Investigations towards this goal disclosed several properties of the binding sites. Glucan-binding proteins were identified by photoaffinity labeling of detergent-solubilized proteins from soybean root membranes using a photoreactive conjugate of the hepta-β-glucoside (Cosio et al. 1992). The major component of the solubilized β-glucan-binding sites appeared to be a protein of 70 to 75 kDa together with two proteins of 100 and 150 to 170 kDa labeled to a lesser extent by the photoaffinity ligand. The labeling characteristics suggested that these are elicitor-binding proteins (Cosio et al. 1992; Frey et al. 1993).

β-Glucan-binding proteins were solubilized from root microsomal membranes using detergents (Cheong et al. 1993; Cosio et al. 1990a; Ebel and Cosio 1994) and retained their binding affinity for the hepta-β-glucoside and the specificity for differentially elicitor-active oligoglucosides (Cheong et al. 1993; Cosio et al. 1990a; Frey et al. 1993). A one-step purification procedure using an affinity adsorbent consisting of a *P. sojae* glucan fraction conjugated to controlled-pore glass beads (Mithöfer et al. 1996b) gave a purified fraction containing one major protein of 75 kDa. Electrophoretic analyses of the purified and photaffinity-labeled binding protein showed that the native protein was an oligomer with apparent molecular mass of about 240 kDa (Mithöfer et al. 1996b). A polyclonal antipeptide antiserum raised against a synthetic 15-mer internal oligopeptide sequence derived from the 75-kDa protein recognized the purified binding protein in immunoblotting experiments. Ligand saturation studies and the kinetics of ligand interaction demonstrated that the hepta-β-glucoside-binding characteristics of the enriched solubilized protein fractions were similar to those of the membrane-bound binding sites (Frey et al. 1993).

B. Chitin-Binding Sites

Recent investigations with elicitor-active chitin fragments have demonstrated the existence of high-affinity binding sites for these oligoglycosides in two plant tissues, tomato (Baureithel et al. 1994) and rice (Shibuya et al. 1993). By using a radiolabeled tyramine conjugate of N-acetylchitooctaose a single class of binding sites with a K_d of 5.4 nM was identified in a microsomal fraction of suspension-cultured rice (*Oryza sativa* L.) cells (Shibuya et al. 1993). Ligand binding was inhibited by unlabeled N-acetylchitoheptaose, but not by its deacetylated form; this differential affinity correlates well with the ability of these compounds to stimulate phytoalexin production in rice cells (Yamada et al. 1993).

Specific high affinity binding of chitin fragments was also found with suspension-cultured cells and membranes of tomato (Baureithel et al. 1994). An N-acetylchitopentaose conjugate, which stimulated a rapid alkalinization of the culture medium of tomato cells, was synthesized and shown to bind to high affinity, saturable binding sites on tomato cells and in a membrane fraction isolated from the cells (K_d values of 1.4 and 23 nM, respectively). The concentrations of different chitin fragments required to inhibit binding of the radioligand at the 50% level closely resembled those that stimulated comparable alkalinization responses. Deacetylated chitooligosaccharides and N-propanoylchitooligosaccharides only weakly stimulated alkalinization or influenced radioligand binding. Conversely, a lipooligosaccharide from *Rhizobium leguminosarum* (Nod factor) effectively stimulated alkalinization of the cell culture medium and inhibited binding of chitin fragments to the tomato cells (Baureithel et al. 1994). It will be interesting to learn about the relationship between this perception system for the chitin fragments in tomato cells and perception mechanisms underlying some of the characterized defense responses of tomato against pathogen attack.

C. Peptidoglycan/Glycopeptide-Binding Sites

Binding sites for peptidoglycans or glycopeptides have been identified in wheat plasma membranes (Kogel et al. 1991; Langen et al., unpubl. results) and in microsomal membranes of tomato cells (Basse et al. 1993). The carbohydrate moiety has been found to be sufficient for binding in both systems. In wheat, a peptidoglycan elicitor from germtube walls of *Puccinia graminis* f. sp. *tritici* binds to a plasma membrane protein fraction (apparent K_d value of 0.33 μM). A direct correlation was reported between the elicitor activity of chemically and enzymatically modified peptidoglycan and its ability to bind to the membrane fraction. Removal of the oligogalactan moiety abolished both binding and elicitor activity. Elicitor-binding sites were present in plasma membranes from wheat cultivars irrespective of the presence of the *Sr5* gene for rust resistance. This result is in agreement with the observation that the elicitor acts in a cultivar-nonspecific manner.

A glycopeptide (gp 8c), effective in stimulating ethylene biosynthesis and increasing activity levels of phenylalanine ammonia-lyase, was isolated from yeast invertase and chemically conjugated to yield a suitable ligand for binding studies (Basse et al. 1993). Binding of this ligand to both intact tomato cells and to microsomal fractions derived from these cells was specific, saturable, reversible, and of high affinity (K_d = 3.3 nM for the microsomal membrane fraction). Several glycopeptides and preparations from yeast invertase were compared for their abilities to compete for binding of radiolabeled gp 8c to tomato membranes and to induce ethylene biosynthesis in tomato cells. A high degree of correlation was observed between the elicitor activities of the compounds in vivo and their radioligand displacement potency in vitro. High activity glycopeptides contained a glycan chain consisting of more than eight mannosyl residues. Interestingly, high mannose oligosaccharides released from glycopeptide elicitors by enzymatic cleavage not only acted as suppressors of elicitor activity in vivo (Basse et al. 1992) but also competed for membrane binding of radiolabeled gp 8c (Basse et al. 1993). The results suggest that agonistic and antagonistic effects of the glycan compounds are transmitted by their interaction with the same target site (Basse et al. 1993).

D. Peptide-Binding Sites

A high affinity oligopeptide-binding site has recently been identified in parsley plasma membranes for an oligopeptide of 13 amino acids (Pep-13) derived from a 42-kDa glycoprotein elicitor of *P. sojae* (Nürnberger et al. 1994a). Binding of radiolabeled Pep-13 to parsley microsomes and protoplasts was specific, saturable, and reversible (Nürnberger et al. 1994a), and identical structural elements within Pep-13 were required for binding to parsley membranes and for induction of defense responses (see above). Further, similar concentrations of the individual peptides were required to activate phytoalexin production and to inhibit binding of radiolabeled Pep-13 half-maximally. In a saturation analysis with Pep-13 as ligand, apparent K_d values of 2.4 nM for parsley microsomal membranes and of 11.4 nM for parsley protoplasts were determined (Nürnberger et al. 1994a).

This same binding site was apparently visualized after silver-enhanced immunogold labeling at the surface of parsley protoplasts by epipolarization microscopy (Diekmann et al. 1994). The significantly lower number of binding sites per protoplast detected by this method compared to results obtained from saturation analysis of binding (Nürnberger et al. 1994a) suggested the ligand-induced clustering of receptors. Further, a 91-kDa protein was labeled by chemical cross-linking of radioiodinated Pep-13 to parsley microsomal and plasma membrane preparations (Nürnberger et al. 1995). The amount of unlabeled ligand required to reduce cross-linked labeling by 50% was similar to the IC_{50} value obtained from binding assays, indicating that the Pep-13 binding site is identical to the 91-kDa protein. Furthermore, the same structural elements within the ligand were found to be essential for successful cross-linking (Nürnberger et al. 1995), activation of a Ca^{2+} channel (Zimmermann et al. 1997), stimulation of an oxidative burst, and induction of phytoalexin production (Nürnberger et al. 1994a). These results suggest that the 91-kDa plasma membrane protein represents the oligopeptide elicitor receptor of parsley plasma membranes which mediates activation of a multicomponent defense response. This peptide-binding site may represent a member of a new class of receptors in plants involved in peptide signaling (Nürnberger et al. 1994a).

For the protein elicitor, cryptogein, from *P. cryptogea*, evidence for the existence of specific binding sites (K_d of 2 nM) in tobacco plasma membranes has been reported (Blein et al. 1991; Wendehenne et al. 1995).

Recently, binding sites for the race-specific peptide elicitor AVR9 from *Cladosporium fulvum* have been identified in the tomato host plant (K_d value of 0.07 nM) (Kooman-Gersmann et al. 1996; Chap. 1, Vol. V, Part B). Very similar binding activity was present in plasma membranes from both leaves of the tomato cultivar Moneymaker without *Cf* resistance genes and from a near-isogenic genotype with the *Cf9* resistance gene. Furthermore, AVR9 peptide-binding sites were present in all solanaceous plant species tested but not detected in all other species analyzed. The *Cf9* resistance gene of tomato, which is complementary to the *Avr9* avirulence gene of *C. fulvum*, was cloned by transposon tagging and sequenced (Jones et al. 1994). It belongs to a clustered gene family and encodes a putative membrane-anchored extracytoplasmic glycoprotein harboring regions with homology to the receptor domain of several receptor-like protein kinases of *Arabidopsis*, to polygalacturonase-inhibiting proteins of plants with antifungal activity, and to proteins of the leucine-rich repeat family. Although these structural features may be consistent with a function of the Cf9 protein as receptor for the AVR9 peptide, the role of the Cf9 protein in elicitor signaling has yet to be defined.

A second resistance gene, the L^6 gene conferring rust resistance to flax, has also been isolated by transposon tagging (Lawrence et al. 1994). In contrast to the *Cf9* gene of tomato, no obvious membrane-spanning domains and regions homologous to protein kinases were detected in the deduced amino acid sequence of the L^6 gene, whereas a nucleotide-binding site motif and two leucine-rich repeats were present.

IV. Signal Transduction

For several of the elicitor (ligand) interactions discussed above, identical structural elements were required for membrane binding and for induction of plant defense. This common observation may indicate that the binding sites analyzed indeed function as receptors in the transmission of extracellular elicitor signals leading to the activation of the defense response, e.g., nuclear gene activation in the case of phytoalexin production or of PR protein expression. The mechanisms underlying transmembrane signaling in plants in response to environmental stimuli, including elicitors, and subsequent intracellular signal transmission to target sites are not well understood (Ebel and Cosio 1994). Considerable evidence exists for Ca^{2+}-based transduction pathways in plants (Bush 1993; Webb et al. 1996). Other compounds that have been associated with signaling events in plant defense responses are jasmonate and salicylate (Enyedi et al. 1992; Farmer and Ryan 1992; Malamy and Klessig 1992).

A. The Role of Ca^{2+} Uptake and Correlated Ion Fluxes

To place Ca^{2+} in signal transduction pathways requires the identification of (1) stimulus-mediated changes in intracellular (cytosolic) Ca^{2+} levels and (2) the specific target proteins that are regulated in vivo by Ca^{2+}, proteins whose action is necessary for the cellular response. While the first requirement may be fulfilled in several elicitor-plant cell systems, less information exists about the second one. Activation of defense responses in several systems depends on the presence of extracellular Ca^{2+}. Omission of Ca^{2+} from the culture medium blocked defense-related phytoalexin formation in soybean (Stäb and Ebel 1987), carrot (Kurosaki et al. 1987a), parsley (Scheel et al. 1989), and tobacco cells (Tavernier et al. 1995). Depletion of Ca^{2+} also prevented elicitor-mediated gene activation at the level of transcript accumulation (Dietrich et al. 1990; Nürnberger et al. 1994a) and transcript synthesis (Ebel and Scheel 1992). The Ca^{2+} ionophore A 23187 stimulated phytoalexin synthesis in soybean (Stäb and Ebel 1987) and carrot (Kurosaki et al. 1987a), but not in parsley (Scheel et al. 1991), and tobacco (Tavernier et al. 1995), whereas the ionophore in the presence of EGTA inhibited the response in soybean cells.

Ca^{2+} influx has been correlated with the activation of plant defenses in cells of a number of plant species. Thus, recent work with elicitor-treated parsley cells has revealed a complex pattern of ion fluxes involving uptake of Ca^{2+}, efflux of Cl^- and K^+, and rapid alkanization of the culture medium, changes detectable within 2 min of elicitor addition (Scheel et al. 1991). Rapid elicitor-stimulated Ca^{2+} uptake and alkalinization of the culture medium were also observed in soybean (Ebel et al. 1995) and tobacco cells (Tavernier et al. 1995). Similar changes in Ca^{2+}, K^+, and proton fluxes were found upon treatment of plant cells with elicitors of callose formation, such as chitosan (Kauss et al. 1989), and after challenge with oligogalacturonides, which activate defense response genes as well as a number of developmental changes (Mathieu et al. 1991). Other cell responses to elicitor treatment, such as 4-hydroxybenzoic acid production in carrot cells and protoplasts and the release of active oxygen species from cultured spruce cells were also correlated with increased Ca^{2+} influx (Bach et al. 1993; Schwacke and Hager 1992). Dimeric polygalacturonic acid and oligogalacturonides (DP > 9) both increased internal Ca^{2+} concentration as well as phenylalanine ammonia-lyase activity in carrot protoplasts (Messiaen et al. 1993).

pH changes in the medium which occur concomitantly with Ca^{2+} influxes are also correlated with the activation of plant defenses. Thus, treatment of tomato cells with chitin fragments induced a rapid, transient alkalinization of the growth medium (Felix et al. 1993). In several other plant systems, changes in extracellular and/or intracellular pH (Strasser et al. 1983; Osswald et al. 1985; Ojalvo et al. 1987) or changes in membrane potential (Pelissier et al. 1986; Mayer and Ziegler 1988; Thain et al. 1990) have been reported to precede the expression of defense reactions. The change in external pH associated with K^+ efflux occurred at a H^+/K^+ ratio of 0.83, suggesting that K^+ and H^+ exchange accounted for most of the observed external pH alteration in tobacco cells (Mathieu et al. 1991). Horn et al. (1992), however, reported that suspension-cultured soybean cells did not respond to elicitation by significantly changing the pH of their vacuolar or cytoplasmic compartments. It is not yet known whether the observed ion flux changes are a general consequence of elicitor-membrane interactions and whether the transient permeability changes are primary events in the transduction of the elicitor signals by the challenged plant cells. Precise measurements of ion fluxes in response to different types of elicitor may, therefore, reveal whether quantitative, temporal, and spatial differences in release or uptake of specific ions lead to qualitative and quantitative differences in defense responses. Further studies are also required to identify elicitor-responsive ion channels and to study their possible involvement in elicitor signal transduction.

Several experiments have shown that pharmacological agents which block ion channels in animal cells can also inhibit Ca^{2+} influx as well as fluxes of other ions across the plasma membrane in plants. In several cases such agents could also block elicitation of plant defenses (Kurosaki et al. 1987a; Nürnberger et al. 1994b; Ebel et al. 1993, 1995). However, such experiments often yielded variable results, depending on the pharmacological agent and the plant experimental system, indicating that the use of these inhibitors alone in physiological studies is not sufficient to demonstrate the involvement of calcium channels in elicitor signal transduction in plants.

Patch-clamp analysis of parsley protoplasts revealed a novel Ca^{2+} channel of large conductance in the plasma membrane that is activated within 2 to 5 min upon treatment with an oligopeptide elicitor (Pep-13) from *Phytophthora sojae* (Zimmermann et al. 1997). This channel was sensitive to La^{3+} and Gd^{3+}, two Ca^{2+} channel blockers that also inhibited elicitor-stimulated phytoalexin synthesis in parsley. The same structural elements of the peptide elicitor previously found to be essential for receptor binding, defense gene activation, and phytoalexin formation are also required for activation of this ion channel, indicating its causal involvement in elicitor signaling (Nürnberger et al. 1994a; Zimmermann et al. 1997).

If Ca^{2+} were an element of elicitor signal transduction, one would assume that rapid changes in cytosolic Ca^{2+} from low basal levels should occur in response to exogenous elicitor treatment. Measurements of cytosolic Ca^{2+} levels in plant cells have now been made possible using a variety of techniques (Bush 1993; Read et al. 1992). Knight et al. (1991) reported that Ca^{2+} levels in transgenic tobacco plants expressing aequorin increased upon treatment of the seedlings with a yeast elicitor preparation.

B. Other Components in Signaling Systems

A number of components of a Ca^{2+}-based messenger system have been proposed to exist in plant cells (Bush 1993; Drøbak 1993). Among these are the phosphoinositide cycle, calmodulin, protein kinases, protein phosphatases, ion transporters, and lipases. Indirect evidence is available that at least some of these elements may be involved in Ca^{2+}-mediated signaling during plant cell elicitation.

1. Phosphoinositides

In animal cells, interaction between one class of receptors and a stimulus leads to the activation of phosphoinositidase C (PIC) and subsequent cleavage of phosphatidylinositol 4,5-bisphosphate to soluble inositol trisphosphate (IP_3) and membrane-bound diacylglycerol (DG). IP_3 induces a rapid and transient release of Ca^{2+} from the endoplasmic reticulum, whereas DG modulates the activity of protein kinase C and thereby initiates a protein phosphorylation cascade.

Two main types of plant PIC have so far been identified. One type is associated predominantly with the plasma membrane, has a high affinity for polyphosphoinositides, is fully activated at micromolar Ca^{2+} levels (Melin et al. 1987; Tate et al. 1989; Yotsushima et al. 1993), and thus is likely to resemble most closely mammalian PIC with regard to cellular function (Drøbak 1993). The antibiotic neomycin, an inhibitor of PIC which is known to inhibit the generation of second messengers, IP_x and DG from animal membrane phosphoinositides (Berridge and Irvine 1989; Gabev et al. 1989), was found to block elicitor-mediated phytoalexin production in parsley protoplasts (Renelt et al. 1993). Both IP_3 and the metabolically stable analog IP_3S_3 enhanced the elicitation of phytoalexin accumulation when introduced into parsley cells by electroporation, whereas they had no effect in untreated control cells (Renelt et al. 1993). This could indicate that elevated concentrations of cytosolic Ca^{2+} could arise not only through an increase in the permeability of the plasma membrane but also through Ca^{2+} release from internal stores by IP_3. Neomycin also prevented the oxidative burst usually elicited upon treatment of cultured soybean cells with polygalacturonic acid (Legendre et al. 1993). No general agreement, however, exists about possible changes in IP_3 levels in response to elicitor treatment in different plant systems (Strasser et al. 1986; Kurosaki et al. 1987b; Legendre et al. 1993; Kamada and Muto 1994). In addition, Li^+, which inhibits the enzymatic hydrolysis of inositol phosphates and thereby reduces the supply of inositol in animal and plant cells (Berridge et al. 1989; Gillaspy et al. 1995), had no effect on elicitor-stimulated phytoalexin accumulation in parsley cells (Renelt et al. 1993).

2. Protein Phosphorylation

Yet another mode of short-term regulation of cytosolic Ca^{2+} might involve protein phosphorylation or dephosphorylation. Indeed, it is possible that Ca^{2+}-dependent protein phosphorylation may serve to integrate intracellular Ca^{2+} in the transduction pathway linking elicitor perception to defense gene activation (Ebel and Cosio 1994). Changes in the level of phosphorylation of cellular proteins have been observed upon elicitor treatment of different cell cultures (Grab et al. 1989; Dietrich et al. 1990; Felix et al. 1991; Viard et al. 1994) or membrane preparations (Farmer et al. 1989, 1991). These changes preceded the onset on defense responses and, at least in parsley cells, were dependent on the availability of extracellular Ca^{2+} (Dietrich et al. 1990). In parsley cells, the phosphorylation of a 45-kDa microsomal and cytoplasmic protein and of a 26-kDa nuclear protein could be observed within 1 min of elicitation (Dietrich et al. 1990). In soybean cells, changes in protein phosphorylation/dephosphorylation were found within 5 min of elicitation (Grab et al. 1989). In tomato cells, treatment with chitin oligomers or fungal xylanase elicited similar extracellular alkalinization responses and the same pattern of rapid protein phosphorylation (Felix et al. 1993). Elicitation with chitin oligomers caused a refractory state of about 8 h during which the cells did not respond to a second dose of chitin fragments, but remained fully responsive to xylanase.

While several of the protein kinases identified to date in plant tissues depend strictly on Ca^{2+} for activity (Ranjeva and Boudet 1987; Budde and Chollet 1988; Roberts and Harmon 1992), no protein kinase has as yet been biochemically characterized as a target protein in the putative Ca^{2+}-dependent signaling pathway for defense response activation. Indirect evidence for a role for protein kinases in elicitation was obtained from pharmacological studies. The protein kinase inhibitors K-252a and staurosporine inhibited alkalinization of the culture medium, ethylene synthesis, and increases in phenylalanine ammonia-lyase activity, as well as changes in protein phosphorylation during elicitation of tomato cells (Felix et al. 1991, 1993; Grosskopf et al. 1990). The correlation between both the time course and the dose response for the effects of K-252a on protein phosphorylation and the other cellular responses has been used to argue in favor of continuous phosphorylation/dephosphorylation of proteins in the maintenance of the elicited state (Felix et al. 1991). Consistent with the above results, staurosporine also blocked the oxidative burst in cultured spruce and tobacco cells (Schwacke and Hager 1992; Viard et al. 1994) and defense gene activation in suspension-cultured tobacco cells after treatment with a crude cell wall elicitor from *Phytophthora infestans* (Suzuki et al. 1995).

Other inhibitors of protein kinases also serve to suppress plant defense responses. Thus, for example, an inhibitor of Ca^{2+} phospholipid-dependent protein kinase C suppressed the accumulation of the phytoalexin, 6-methoxymellein, in carrot (Kurosaki et al. 1987b). The phosphatase inhibitor okadaic acid prevented elicitor-enhanced phytoalexin production in parsley protoplasts (Renelt et al. 1993) and decreased inducible Ca^{2+} uptake and callose synthesis in cells of *Catharanthus roseus* (Kauss and Jeblick 1991). Defense gene activation as well as the transient activation of a 47-kDa kinase, a suspected mitogen-activated protein kinase, in tobacco cells were inhibited by staurosporine and the Ca^{2+} channel blocker Gd^{3+}, whereas the protein phosphatase inhibitor, calyculin A, sustained its elicitor activation (Suzuki and Shinshi 1995; Suzuki et al. 1995).

While these results suggest that protein kinases are directly and uniformly involved in elicitor signaling, a more general view of the literature reveals a welter of less clear and even contradictory results. Protein kinase inhibitors do not always block plant defense responses. For example, staurosporine and K-252a did not block elicitor enhancement of phytoalexin production in parsley protoplasts (Renelt et al. 1993). Further, some protein phosphatase inhibitors may act as elicitors themselves. Thus, the protein phosphatase inhibitor, calyculin A, mimicked elicitor action in cultured tomato cells and induced rapid hyperphosphorylation of certain proteins whose phosphorylation state may also be affected by the fungal elicitor, xylanase (Felix et al. 1994). Another protein phosphatase inhibitor, okadaic acid, induced accumulation of pathogenesis-related proteins in tobacco in a manner similar to that of the elicitor xylanase and the stress hormone ethylene (Raz and Fluhr 1993). The application of a variety of structurally different protein phosphatase inhibitors, including okadaic acid, acanthifolicin, microcystins, nodularin, tautomycin, calyculin A, cantharidin, and endothall to cut surfaces of soybean cotyledons resulted in the production of isoflavonoid compounds such as daidzein (Mac Kintosh et al. 1994). The addition of some of the inhibitory compounds to cultured soybean cells elicited expression of phenylalanine ammonia-lyase activity and alkalinization of the cell culture medium. Paradoxically, such responses can themselves be prevented by K-252a in some situations (MacKintosh et al. 1994). Often disparate results are obtained with the same plant system in different laboratories, a situation which may arise because of slight variations in experimental conditions, elicitor fractions, responses analyzed, and concentrations of pharmacological agents (Conrath et al. 1991; Ebel et al. 1993; Renelt et al. 1993; Waldmüller et al. 1993; MacKintosh et al. 1994). These results do suggest, however, that defense signaling pathways in plants are no less complex than signaling pathways in animal cell. At present, none of the studies confirmed causal links between protein phosphorylation/dephosphorylation and elicitation of plant defenses. A clearer understanding of the role of these posttranslational protein modifications in elicitor signaling awaits identification of the enzymes involved and their substrates in vivo.

Even less consistent evidence exists for a role of target proteins other than protein kinases or phosphoprotein phosphatases in Ca^{2+}-mediated signaling during elicitation. Inhibitors of the Ca^{2+}-binding protein calmodulin prevented elicitor-induced phytoalexin accumulation in cell cultures of tobacco (Vögeli et al. 1992) and carrot (Kurosaki and Nishi 1993), but not in soybean (Stäb and Ebel 1987). A number of studies have been carried out to identify the mammalian equivalents of protein kinase C in plant cells. The enzymes so far characterized all differ from mammalian protein kinase C (Drøbak 1993), a finding which might explain the inefficiency of the kinase inhibitor phorbolester to interfere with elicitation of phytoalexin synthesis in parsley protoplasts (Renelt et al. 1993). The observation that no direct equivalent to protein kinase C seems to exist in plants and that one type of plant protein kinase contains a calmodulin-like domain (Roberts and Harmon 1992) indicates that Ca^{2+}-dependent signaling in plant cells may require protein kinases other than those in mammalian cells.

3. Jasmonic Acid

Jasmonic acid (3-oxo-2-[2′-*cis*-pentyl]-cyclopentane-1-acetate) and its methyl ester occur naturally in many plant species (Sembdner and Parthier 1993). Various growth-regulating effects of these compounds have been observed, but only recently have they been attracting broad interest as potentially important signal compounds mediating gene regulation (Staswick 1992; Reinbothe et al. 1994; Chap. 9, this Vol.). It has not yet been demonstrated that jasmonic acid or its methyl ester induced systemic pest resistance when applied to plants (Choi et al. 1994). These compounds are, however, associated with pathogen- and pest-induced wounding and trigger the expression of a family of wound-inducible proteinase inhibitor genes (Creelman et al. 1992; Farmer and Ryan 1992; Staswick 1992; Xu et al. 1993), of genes for vegetative storage proteins (Mason and Mullet 1990; Staswick 1990; Berger et al. 1995), of genes encoding enzymes of phenylpropanoid metabolism (Creelman et al. 1992; Gundlach et al. 1992), and of a gene encoding a leaf thionin (Andresen et al. 1992; Epple et al. 1995; Kogel et al. 1995). Jasmonates may not always be involved with signaling for plant defense, however. In barley, the thionin-encoding gene was not activated upon powdery mildew infection, nor was accumulation of jasmonate detectable, suggesting that jasmonate is not involved in signaling processes associated with resistance of barley against *Erysiphe graminis* f. sp. *hordei* (Kogel et al. 1995).

It has been proposed (Farmer and Ryan 1992) that wounding initiates a lipid-based intracellular signaling pathway by causing the release of a precursor linolenic acid and its subsequent conversion into jasmonic acid. Jasmonate synthesis in plants includes the action of a lipoxygenase, a dehydrase, and enzymes involved in β-oxidation (Vick and Zimmerman 1984). Observations in support of a role for jasmonate as an endogenous signal compound include its rapid accumulation in plant tissues upon wounding (Creelman et al. 1992) or elicitor treatment (Gundlach et al. 1992; Mueller et al. 1993; Nojiri et al. 1996), the effects of wounding on plant membrane biosynthesis and degradation (for review see Ryan 1992), and the finding that lipoxygenase is induced upon wounding (Galliard 1978), treatment with fungal elicitors (Fournier et al. 1986), or fungal infection (Koch et al. 1992). Wound induction of vegetative storage protein accumulation was prevented by treatment with lipoxygenase inhibitors that block jasmonic acid synthesis, whereas exogenous methyl jasmonate effectively induced the synthesis of these proteins even in the presence of the inhibitors (Ryan 1992; Staswick et al. 1991). Further, the synthesis of defensive proteinase inhibitors could be stimulated by precursors of jasmonic acid biosynthesis, whereas closely related compounds that are not precursors were ineffective (Farmer and Ryan 1992).

These latter data suggest that the pathway of jasmonic acid biosynthesis is active in unstressed cells and that its production may be limited through the supply of precursors such as linolenic acid. The rapid phospholipase-mediated release of the precursor linolenic acid (Mueller et al. 1993) from a membrane lipid pool as an early event in signal transduction could then constitute a critical step in wound-induced jasmonic acid synthesis. The time course of inducible lipoxygenase expression is relatively slow when compared to jasmonic acid accumulation (Creelman et al. 1992; Gundlach et al. 1992) and defense gene activation (Esquerré-Tugayé et al. 1993; Meier et al. 1993). Thus, regulation of lipoxygenase at the level of expression might not be a prerequisite for increased jasmonic acid production, but might be involved in other processes during plant attack (Slusarenko et al. 1993), while direct activation of lipoxygenases initiate the rapid elicitor-stimulated accumulation of jasmonate. A plasma membrane-associated lipoxygenase with a pH optimum around pH 6.0, which is stimulated by nanomolar levels of H_2O_2 has recently been purified from soybean (Macrì et al. 1994). Since acidification of the cytosol (Strasser et al. 1983; Ojalvo et al. 1987) and an oxidative burst (Apostol et al. 1989; Schwacke and Hager 1992; Nürnberger et al. 1994a; Levine et al. 1994) are among the most rapid responses of plant cells to elicitors, this enzyme may catalyze increased formation of jasmonate precursors in response to elicitor treatment.

It has been suggested that jasmonic acid may also function as a signal transducer for elicitor stimulation of phytoalexin accumulation and other defense responses involved in plant disease resistance. Thus, exogenous jasmonic acid induced the expression of phenylalanine ammonia-lyase, chalcone synthase, and proline-rich cell wall protein transcripts (but not the pterocarpan glyceollin, a phytoalexin typically produced in abundance by soybean as a defense compound) (Creelman et al. 1992; Gundlach et al. 1992). In a similar study, jasmonic acid or its methyl ester stimulated the accumulation of a variety of secondary metabolites in cell cultures of a wide range of plant species (Gundlach et al. 1992), although at relatively high concentrations. In suspension-cultured rice cells, jasmonate induced the formation of the phytoalexin, momilactone A; inhibition of elicitor-stimulated increases in endogenous jasmonate by ibuprofen reduced phytoalexin production, suggesting the involvement of jasmonate in elicitor signaling in rice (Nojiri et al. 1996). Interestingly, preincubation of parsley cells at low, ineffective concentrations of methyl jasmonate potentiated the response to fungal elicitors, as measured by the accumulation of coumarin derivatives and of esterified hydroxycinnamic acids and of lignin-like polymers into the cell walls (Kauss et al. 1992).

Jasmonates or their precursors have been shown to stimulate the production of flavonoids whose role in plant defense is as yet uncertain. Thus, exogenous jasmonate induced the accumulation of the isoflavonoid genistein, a minor isoflavonoid in the spectrum of soybean defense compounds (Morris et al. 1991), in different tissues of soybean (Gundlach et al. 1992). The jasmonate precursor, 12-oxo-phytodienoic acid, elicited the accumulation in parsley cells of the flavonoid glycoside, apiin, whose function in plant defense is unknown (Dittrich et al. 1992).

Recently, it has been demonstrated that methyl jasmonate and the fungal elicitor arachidonic acid induced different 3-hydroxy-3-methylglutaryl-CoA reductase genes and different antimicrobial isoprenoids in potato tuber disks (Choi et al. 1994). The results suggested that methyl jasmonate can function as a signal compound for the steroidglycoalkaloid pathway following wounding, but not for the sesquiterpenoid phytoalexin pathway challenged after infection.

A better understanding of the proposed lipid-based signaling pathway requires answers to a number of questions. Central questions relate to the nature of the primary (systemic) signal and to the nature of the putative receptors for that signal compound. The mode of activation of the signaling pathway resulting in the release of precursor fatty acids for jasmonate synthesis needs further clarification. Further questions concern the specificity of jasmonic acid signals in relation to the multiplicity of responses stimulated by exogenous application, and whether the endogenous jasmonic acid pathway is causally related to defense activation.

4. Active Oxygen Production and the Role of Salicylic Acid

The production of active oxygen species, the oxidative burst, is one of the earliest reactions of plants to infection or elicitor treatment (for review see Sutherland 1991; Mehdy 1994). It resembles the oxidative burst occurring during phagocyte activation (Babior 1992). Antisera raised against subunits of human phagocyte NADPH oxidase recognize soybean microsomal or soluble proteins of similar apparent molecular masses (Tenhaken et al. 1995; Dwyer et al. 1996). Recently, a rice homologue of the mammalian gene, encoding the subunit of the respiratory burst oxidase, *gp91phox*, has been isolated (Groom et al. 1996). In addition, elicitor-stimulated H_2O_2 production by soybean cells is efficiently inhibited by diphenylene iodonium, a suicide substrate inhibitor of the mammalian NADPH oxidase (Levine et al. 1994). Microsomal preparations from cultured rose and *Arabidopsis* cells were found to harbor enzymes catalyzing the generation of superoxide anions in a way similar to the mammalian NADPH oxidase (Desikan et al. 1996; Murphy and Auh 1996). Active oxygen species were shown to be directly toxic to plant pathogens (Peng and Kuc 1992) and to be involved in cell-wall stiffening by cross-linking of cell wall structural proteins (Bradley et al. 1992; Brisson et al. 1994). Accumulating evidence now suggests that they also play a role in plant defense signaling.

Cultured soybean cells respond rapidly to treatment with a fungal elicitor or an avirulent strain of *Pseudomonas syringae* pv. *glycinea* with the production of H_2O_2, while a virulent strain did not stimulate this response (Levine et al. 1994). Plant cell death was only induced by avirulent bacteria or by millimolar levels of H_2O_2. Cell death caused by bacterial infection was inhibited by preincubation of the plant cells with the oxidase inhibitor, diphenylene iodonium. Lower levels of H_2O_2 activated genes encoding glutathione S-transferase and glutathione peroxidase which may protect the cells from oxidative stress. H_2O_2 from the oxidative burst is therefore believed to "orchestrate" the localized hypersensitive response (Levine et al. 1994).

In addition, H_2O_2 also appears to be involved in SAR induction via salicylic acid, which was found to be part of the systemic signaling machinery in this defense response (for review see Kessmann et al. 1994). The soluble salicylic acid-binding protein in tobacco (Chen and Klessig 1991) was found to copurify with a 280-kDa protein (Chen et al. 1993a). The sequence of the protein, as deduced from a cDNA encoding it, was similar to that of catalases and the protein exhibited catalase activity (Chen et al. 1993b). Salicylic acid inhibited catalase activity in vitro and induced a slight increase in H_2O_2 concentration in vivo, which was suggested to serve as a signal in the activation of genes encoding PR proteins. The ability of a variety of phenolic compounds to compete for salicylic acid binding was correlated with their biological activity in inducing defense-related genes (Chen et al. 1993a). Recent results from different laboratories do not support the involvement of H_2O_2 in SAR signaling, but rather suggest the importance of active oxygen species in local defense gene activation and hypersensitive cell death (Bi et al. 1995; Neuenschwander et al. 1995; Levine et al. 1996; Jabs et al. 1997; Jones and Dangl 1996). Future research is expected to unequivocally demonstrate if active oxygen species do play a triple role in plant defense as toxic agents, intracellular signals, and systemic signals.

V. Conclusions and Perspectives

A combination of genetic, molecular, and biochemical approaches is beginning to unravel the mechanisms by which plants recognize pathogenic fungi and initiate local as well as systemic signal transduction chains that finally activate the many components of the defense response. Elicitors of diverse structure serve as signals for non-self, species- and race-specific recognition by plants. High-affinity receptors appear to be involved in the perception of all three types of elicitors. Several binding sites of the non-self and species-specific recognition machinery have been detected by biochemical means. However, no structural data on these putative receptors or the corresponding genes are available. For the few resistance genes that have been cloned, the biochemical function remains to be elucidated. It will be exciting to learn if structural and/or functional similarities exist between proteins encoded by resistance genes and the plasma membrane-located binding sites of the elicitors not encoded by classical avirulence genes. The structural analysis of these binding sites will probably provide insight into their possible function in plant metabolism and development. In addition, the availability of the corresponding genes is expected to allow the determination of the role these elicitor/receptor pairs play in non-host or basic resistance and thereby possibly open a new dimension for the transfer of this stable type of resistance to unrelated plant species using the tools of gene technology.

The bulk of primarily pharmacological data available on signal transduction from elicitor perception to initiation of individual defense responses is confusing rather than enlightening. Genetic and molecular approaches need to be employed to throw light on causal relationships and to identify genes encoding components of these signal chains. Identification of such genes and their combination with appropriate promoters will allow the generation of transgenic crops with artificially activatable pathogen defense. In addition, knowledge of the structure of elicitor receptors will allow elements further downstream of the corresponding signal transduction chain to be biochemically identified and analyzed. Interesting results may therefore be expected from this area of research with respect to intraorganismic, intercellular, and intracellular communication.

References

Andresen I, Becker W, Schlüter K, Burges J, Parthier B, Apel K (1992) The identification of leaf thionin as one of the main jasmonate-induced proteins of barley (*Hordeum vulgare*). Plant Mol Biol 19:193–204

Apostol I, Heinstein PF, Low PS (1989) Rapid stimulation of an oxidative burst during elicitation of cultured plant cells. Plant Physiol 90:109–116

Atkinson MM (1993) Molecular mechanisms of pathogen recognition by plants. Adv Plant Pathol 10:35–64

Ayers AR, Ebel J, Finelli F, Berger N, Albersheim P (1976) Host-pathogen interactions. IX. Quantitative assays of elicitor activity and characterization of the elicitor present in the extracellular medium of *Phytophthora megasperma* var. *sojae*. Plant Physiol 57:751–754

Babior BM (1992) The respiratory burst oxidase. Adv Enzymol Relat Areas Mol Biol 65:49–95

Bach M, Schnitzler J-P, Seitz HU (1993) Elicitor-induced changes in Ca^{2+} influx, K^+ efflux, and 4-hydroxybenzoic acid synthesis in protoplasts of *Daucus carota* L. Plant Physiol 103:407–412

Baillieul F, Genetet I, Kopp M, Saindrenan P, Fritig B, Kauffmann S (1995) A new elicitor of the hypersensitive response in tobacco: a fungal glycoprotein elicits cell death, expression of defence genes, production of salicylic acid, and induction of systemic acquired resistance. Plant J 8:551–560

Barber MS, Bertram RE, Ride JP (1989) Chitin oligosaccharides elicit lignification in wounded wheat leaves. Physiol Mol Plant Pathol 34:3–12

Basse CW, Bock K, Boller T (1992) Elicitors and suppressors of the defense response in tomato cells. Purification and characterization of glycopeptide elicitors and glycan suppressors generated by enzymatic cleavage of yeast invertase. J Biol Chem 267:10258–10265

Basse CW, Fath A, Boller T (1993) High affinity binding of a glycopeptide elicitor to tomato cells and microsomal membranes and displacement by specific glycan suppressors. J Biol Chem 268:14724–14731

Baureithel K, Felix G, Boller T (1994) Specific, high affinity binding of chitin fragments to tomato cells and membranes. J Biol Chem 269:17931–17938

Berger S, Bell E, Sadka A, Mullet JE (1995) *Arabidopsis thaliana Atvsp* is homologous to soybean *VspA* and *VspB*, genes encoding vegetative storage protein acid phosphatases, and is regulated similarly by methyl jasmonate, wounding, sugars, light and phosphate. Plant Mol Biol 27:933–942

Berridge MJ, Irvine RF (1989) Inositol phosphates and cell signalling. Nature 341:197–205

Berridge MJ, Downes CP, Hanley MR (1989) Neural and developmental actions of lithium: a unifying hypothesis. Cell 59:411–419

Bi Y-M, Kenton P, Mur L, Darby R, Draper J (1995) Hydrogen peroxide does not function downstream of salicylic acid in the induction of PR protein expression. Plant J 8:235–245

Blein J-P, Milat M-L, Ricci P (1991) Responses of cultured tobacco cells to cryptogein, a proteinaceous elicitor from *Phytophthora cryptogea*. Plant Physiol 95:486–491

Bradley DJ, Kjellbom P, Lamb CJ (1992) Elicitor- and wound-induced oxidative cross-linking of a proline-rich plant cell wall protein: a novel, rapid defense response. Cell 70:21–30

Brisson LF, Tenhaken R, Lamb C (1994) Function of oxidative cross-linking of cell wall structural proteins in plant disease resistance. Plant Cell 6:1703–1712

Bruce RJ, West CA (1989) Elicitation of lignin biosynthesis and isoperoxidase activity by pectic fragments in suspension cultures of castor bean. Plant Physiol 91:889–897

Budde RJA, Chollet R (1988) Regulation of enzyme activity in plants by reversible phosphorylation. Physiol Plant 72:435–439

Bush DS (1993) Regulation of cytosolic calcium in plants. Plant Physiol 103:7–13

Cervone F, De Lorenzo G, Degrá L, Salvi G (1987) Elicitation of necrosis in *Vigna unguiculata* Walp. by homogenous *Aspergillus niger* endopolygalacturonase and by α-D-galacturonate oligomers. Plant Physiol 85:626–630

Chen Z, Klessig DF (1991) Identification of a soluble salicylic acid-binding protein that may function in signal transduction in the plant disease-resistance response. Proc Natl Acad Sci USA 88:8179–8183

Chen Z, Ricigliano JW, Klessig DF (1993a) Purification and characterization of a soluble salicylic-binding protein from tobacco. Proc Natl Acad Sci USA 90:9533–9537

Chen Z, Silva H, Klessig DF (1993b) Active oxygen species in the induction of plant systemic acquired resistance by salicylic acid. Science 262:1883–1886

Cheong J-J, Hahn MG (1991) A specific, high affinity binding site for the hepta-β-glucoside elicitor exists in soybean membranes. Plant Cell 3:137–147

Cheong J-J, Birberg W, Fügedi P, Pilotti Å, Garegg PJ, Hong N, Ogawa T, Hahn MG (1991) Structure-activity relationships of oligo-β-glucoside elicitors of phytoalexin accumulation in soybean. Plant Cell 3:127–136

Cheong J-J, Alba R, Côté F, Enkerli J, Hahn MG (1993) Solubilization of functional plasma membrane-localized hepta-β-glucoside elicitor-binding proteins from soybean. Plant Physiol 103:1173–1182

Choi D, Bostock RM, Avdiushko, Hildebrand DF (1994) Lipid-derived signals that discriminate wound- and pathogen-responsive isoprenoid pathways in plants: methyl jasmonate and the fungal elicitor arachidonic acid induce different 3-hydroxy-3-methylglutaryl-coenzyme A reductase genes and antimicrobial isoprenoids in *Solanum tuberosum*. Proc Natl Acad Sci USA 91:2329–2333

Coleman MJ, Mainzer J, Dickerson AG (1992) Characterization of a fungal glycoprotein that elicits a defence response in French bean. Physiol Mol Plant Pathol 40:333–351

Conrath U, Jeblick W, Kauss H (1991) The protein kinase inhibitor, K-252a, decreases elicitor-induced Ca^{2+} uptake and K^+ release, and increases coumarin synthesis in parsley cells. FEBS Lett 279:141–144

Cosio EG, Pöpperl H, Schmidt WE, Ebel J (1988) High-affinity binding of fungal β-glucan fragments to soybean (*Glycine max* L.) microsomal fractions and protoplasts. Eur J Biochem 175:309–315

Cosio EG, Frey T, Ebel J (1990a) Solubilization of soybean membrane binding sites for fungal β-glucans that elicit phytoalexin accumulation. FEBS Lett 264:235–238

Cosio EG, Frey T, Verduyn R, van Boom J, Ebel J (1990b) High-affinity binding of a synthetic heptaglucoside and fungal glucan phytoalexin elicitors to soybean membranes. FEBS Lett 271:223–226

Cosio EG, Frey T, Ebel J (1992) Identification of a high-affinity binding protein for a hepta-β-glucoside phytoalexin elicitor in soybean. Eur J Biochem 204:1115–1123

Cosio EG, Feger M, Miller CJ, Antelo L, Ebel J (1996) High-affinity binding of fungal β-glucan elicitors to cell membranes of species of the plant family Fabaceae. Planta 200:92–99

Côté F, Hahn MG (1994) Oligosaccharins: structures and signal transduction. Plant Mol Biol 26:1379–1411

Creelmann RA, Tierney ML, Mullet JE (1992) Jasmonic acid/methyl jasmonate accumulate in wounded soybean hypocotyls and modulate wound gene expression. Proc Natl Acad Sci USA 89:4938–4941

Cruickshank IAM, Perrin DR (1968) The isolation and partial characterization of monilicolin A, a polypeptide with phaseollin-inducing activity from *Monilinia fructicola*. Life Sci 7:449–458

Davis KR, Darvill AG, Albersheim P, Dell A (1986) Host-pathogen interactions. XXX. Characterization of elicitors of phytoalexin accumulation in soybean released from soybean cell walls by endopolygalacturonide acid lyase. Z Naturforsch 41c:39–48

De Wit PJGM (1992) Molecular characterization of gene-for-gene systems in plant-fungus interactions and the application of avirulence genes in control of plant pathogens. Annu Rev Phytopathol 30:391–418

De Wit PJGM, Spikman G (1982) Evidence for the occurrence of race- and cultivar-specific elicitors of necrosis in intercellular fluids of compatible interactions between *Cladosporium fulvum* and tomato. Physiol Plant Pathol 21:1–11

Desikan R, Hancock JT, Coffey MJ, Neill SJ (1996) Generation of active oxygen in elicited cells of *Arabidopsis thaliana* is mediated by a NADPH oxidase-like enzyme. FEBS Lett 382:213–217

Diekmann W, Kerkt B, Low PS, Nürnberger T, Scheel D, Terschüren C, Robinson DG (1994) Visualization of elicitor-binding loci at the plant cell surface. Planta 195:126–137

Dietrich A, Mayer JE, Hahlbrock K (1990) Fungal elicitor triggers rapid, transient and specific protein phosphorylation in parsley cell suspension cultures. J Biol Chem 265:6360–6368

Dittrich H, Kutchan TM, Zenk MH (1992) The jasmonate precursor, 12-oxo-phytodienoic acid, induces phytoalexin synthesis in *Petroselinum crispum* cell cultures. FEBS Lett 309:33–36

Dixon RA, Harrison MJ, Lamb CJ (1994) Early events in the activation of plant defense responses. Annu Rev Phytopathol 32:479–501

Drøbak BK (1993) Plant phosphoinositides and intracellular signalling. Plant Physiol 102:705–709

Dwyer SC, Legendre L, Low PS, Leto TL (1996) Plant and human neutrophil oxidative burst complexes contain immunologically related proteins. Biochim Biophys Acta 1289:231–237

Ebel J, Cosio EG (1994) Elicitors of plant defense responses. Int Rev Cytol 148:1–36

Ebel J, Scheel D (1992) Elicitor recognition and signal transduction. In: Boller T, Meins F (eds) Plant gene research. Genes involved in plant defense, vol 8. Springer, Berlin Heidelberg New York, pp 183–205

Ebel J, Cosio EG, Feger M, Frey T, Kissel U, Reinold S, Waldmüller T (1993) Glucan elicitor-binding proteins and signal transduction in the activation of plant defence. In: Nester EW, Verma DPS (eds) Advances in molecular genetics of plant-microbe interactions, vol 2. Kluwer, Dordrecht, pp 477–484

Ebel J, Bhagwat AA, Cosio EG, Feger M, Kissel U, Mithöfer A, Waldmüller T (1995) Elicitor-binding proteins and signal transduction in the activation of a phytoalexin defense response. Can J Bot 73:506–510

Enyedi AJ, Yalpani N, Silverman P, Raskin I (1992) Signal molecules in systemic plant resistance to pathogens and pests. Cell 70:879–886

Epple P, Apel K, Bohlmann H (1995) An *Arabidopsis thaliana* gene is inducible via a signal transduction pathway different from that for pathogenesis-related proteins. Plant Physiol 109:813–820

Esquerré-Tugayé MT, Fournier J, Pouenat ML, Veronesi C, Rickauer M, Bottin A (1993) Lipoxygenases in plant signalling. In: Fritig B, Legrand M (eds) Mechanisms of plant defense responses. Kluwer, Dordrecht, pp 202–210

Farmer EE, Helgeson JP (1987) An extracellular protein from *Phytophthora parasitica* var. *nicotianae* is associated with stress metabolite accumulation in tobacco callus. Plant Physiol 85:733–740

Farmer EE, Ryan CA (1992) Octadecanoic precursors of jasmonic acid activate the synthesis of wound-inducible proteinase inhibitors. Plant Cell 4:129–134

Farmer EE, Pearce G, Ryan CA (1989) In vitro phosphorylation of plant plasma membrane proteins in response to the proteinase inhibitor inducing factor. Proc Natl Acad Sci USA 86:1539–1542
Farmer EE, Moloshok TD, Saxton MJ, Ryan CA (1991) Oligosaccharide signaling in plants. Specificity of oligouronide-enhanced plasma membrane protein phosphorylation. J Biol Chem 266:3140–3145
Felix G, Grosskopf DG, Regenass M, Boller T (1991) Rapid changes of protein phosphorylation are involved in transduction of the elicitor signal in plant cells. Proc Natl Acad Sci USA 88:8831–8834
Felix G, Regenass M, Boller T (1993) Specific perception of subnanomolar concentrations of chitin fragments by tomato cells: induction of extracellular alkalinization, changes in protein phosphorylation, and establishment of a refractory state. Plant J 4:307–316
Felix G, Regenass M, Spanu P, Boller T (1994) The protein phosphatase inhibitor calyculin A mimics elicitor action in plant cells and induces rapid hyperphosphorylation of specific proteins as revealed by pulse labeling with [^{33}P]phosphate. Proc Natl Acad Sci USA 91:952–956
Fournier J, Pelissier B, Esquerré-Tugayé MT (1986) Induction d'une activité lipoxygénase dans les cellules de tabac *Nicotiana tabacum* en culture, par des éliciteurs d'éthylène de *Phytophthora parasitica* var. *nicotianae*. CR Acad Sci 303:651–656
Frey T, Cosio EG, Ebel J (1993) Affinity purification and characterization of a binding protein for a hepta-β-glucoside phytoalexin elicitor in soybean. Phytochemistry 32:543–550
Gabev E, Kasianovicz J, Abbot T, McLaughlin S (1989) Binding of neomycin to phosphatidylinositol 4,5-bisphosphate (PIP_2). Biochim Biophys Acta 979:105–112
Galliard T (1978) Lipolytic and lipoxygenase enzymes in plants and their action in wounded tissues. In: Kahl G (ed) Biochemistry of wounded plant tissues. Walter de Gruyter, Berlin, pp 155–201
Gillaspy GE, Keddie JS, Oda K, Gruissem W (1995) Plant inositol monophosphatase is a lithium-sensitive enzyme encoded by a multigene family. Plant Cell 7:2175–2185
Grab D, Feger M, Ebel J (1989) An endogenous factor from soybean (*Glycine max* L.) cell cultures activates phosphorylation of a protein which is dephosphorylated in vivo in elicitor-challenged cells. Planta 179:340–348
Groom QJ, Torres MA, Fordham-Skelton AP, Hammond-Kosack KE, Robinson NJ, Jones JDG (1996) *rbohA*, a rice homologue of the mammalian *gp91phox* respiratory burst oxidase gene. Plant J 10:515–522
Grosskopf DG, Felix G, Boller T (1990) K-252a inhibits the response of tomato cells to fungal elicitors in vivo and their microsomal protein kinase in vitro. FEBS Lett 275:177–180
Gundlach H, Müller MJ, Kutchan TM, Zenk MH (1992) Jasmonic acid is a signal transducer in elicitor-induced plant cell cultures. Proc Natl Acad Sci USA 89:2389–2393
Hadwiger LA, Beckman JM (1980) Chitosan as a component of pea-*Fusarium solani* interactions. Plant Physiol 66:205–211
Hahn M, Jüngling S, Knogge W (1993) Cultivar-specific elicitation of barley defense reactions by the phytotoxic peptide NIP1 from *Rhynchosporium secalis*. Mol Plant-Microbe Interact 6:745–754
Horn MA, Meadows RP, Apostol I, Jones CR, Gorenstein DG, Heinstein PF, Low PS (1992) Effect of elicitation and changes in extracellular pH on the cytoplasmic and vacuolar pH of suspension-cultured soybean cells. Plant Physiol 98:680–686
Jabs T, Tschöpe M, Colling C, Hahlbrock K, Scheel D (1997) Elicitor-stimulated ion fluxes and O_2^- from the oxidative burst are essential components in triggering defense gene activation and phytoalexin synthesis in parsley. Proc Natl Acad Sci USA in press
Jin DF, West CA (1984) Characteristics of galacturonic acid oligomers as elicitors of casbene synthetase activity in castor bean seedlings. Plant Physiol 74:989–992
Jones AM, Dangl JL (1996) Logjam at the styx: programmed cell death in plants. Trends Plant Sci 1:114–119
Jones DA, Thomas CM, Hammond-Kosack KE, Balint-Kurti PJ, Jones JDG (1994) Isolation of the tomato *Cf-9* gene for resistance to *Cladosporium fulvum* by transposon tagging. Science 266:789–793
Joosten MHAJ, Cozijnsen TJ, De Wit PJGM (1994) Host resistance to a fungal tomato pathogen lost by a single base-pair change in an avirulence gene. Nature 367:384–386
Kamada Y, Muto S (1994) Stimulation by fungal elicitor of inositol phospholipid turnover in tobacco suspension culture cells. Plant Cell Physiol 35:397–404
Kamoun S, Klucher KM, Coffey MD, Tyler BM (1993a) A gene encoding a host-specific elicitor protein of *Phytophthora parasitica*. Mol Plant-Microbe Interact 6:573–581
Kamoun S, Young M, Glascock CB, Tyler BM (1993b) Extracellular protein elicitors from *Phytophthora*: host-specificity and induction of resistance to bacterial and fungal phyto-pathogens. Mol Plant-Microbe Interact 6:15–25
Kamoun S, Young M, Förster H, Coffey MD, Tyler BM (1994) Potential role of elicitins in the interaction between *Phytophthora* species and tobacco. Appl Environ Microbiol 60:1593–1598
Kauss H, Jeblick W (1991) Induced Ca^{2+} uptake and callose synthesis in suspension-cultured cells of *Catharanthus roseus* are decreased by the protein phosphatase inhibitor okadaic acid. Physiol Plant 81:309–312
Kauss H, Waldmann T, Jeblick W, Euler G, Ranjeva R, Domard A (1989) Ca^{2+} is an important but not the only signal in callose synthesis induced by chitosan, saponins and polyene antibiotics. In: Lugtenberg BJJ (ed) Signal molecules in plants and plant-microbe interactions. Springer, Berlin Heidelberg New York, pp 107–116
Kauss H, Krause K, Jeblick W (1992) Methyl jasmonate conditions parsley suspension cells for increased elicitation of phenylpropanoid defense responses. Biochem Biophys Res Commun 189:304–308
Kessmann H, Staub T, Oostendorp M, Ryals J (1994) Activation of systemic acquired disease resistance in plants. Eur J Plant Pathol 100:359–369
Knight MR, Campbell AK, Smith SM, Trewavas AJ (1991) Transgenic plant aequorin reports the effects of touch and cold-shock and elicitors on cytoplasmic calcium. Nature 352:524–526
Knogge W, Gierlich A, Hermann H, Wernert P, Rohe M (1994) Molecular identification and characterization of the *nip1* gene, an avirulence gene from the barley pathogen *Rhynchosporium secalis*. In: Daniels MJ, Downie JA, Osbourn AE (eds) Advances in molecular genetics of plant-microbe interactions, vol 3. Kluwer,

Dordrecht, pp 207–214
Koch E, Meier BM, Eiben H-G, Slusarenko AJ (1992) A lipoxygenase from leaves of tomato (*Lycopersicon esculentum* Mill.) is induced in response to plant pathogenic pseudomonads. Plant Physiol 99:571–576
Kogel G, Beissmann B, Reisener HJ, Kogel KH (1988) A single glycoprotein from *Puccinia graminis* f. sp. *tritici* cell walls elicits the hypersensitive lignification response in wheat. Physiol Mol Plant Pathol 33:173–185
Kogel G, Beissmann B, Reisener HJ, Kogel KH (1991) Specific binding of a hypersensitive lignification elicitor of *Puccinia graminis* f. sp. *tritici* to the plasma membrane from wheat (*Triticum aestivum* L.). Planta 183:164–169
Kogel K-H, Ortel B, Jarosch B, Atzorn R, Schiffer R, Wasternack C (1995) Resistance in barley against the powdery mildew fungus (*Erysiphe graminis* f. sp. *hordei*) is not associated with enhanced levels of endogenous jasmonates. Eur J Plant Pathol 101:319–332
Kooman-Gersmann M, Honée G, Bonnema G, De Wit PJGM (1996) A high-affinity binding site for the AVR9 peptide elicitor of *Cladosporium fulvum* is present on plasma membranes of tomato and other solanaceous plants. Plant Cell 8:929–938
Kuchitsu K, Kikugama M, Shibuya N (1993) N-Acetylchitooligosaccharides, biotic elicitor for phytoalexin production, induce transient membrane depolarization in suspension-cultured rice cells. Protoplasma 174:79–81
Kurosaki F, Nishi A (1993) Stimulation of calcium influx and calcium cascade by cyclic AMP in cultured carrot cells. Arch Biochem Biophys 302:144–151
Kurosaki F, Tsurusawa Y, Nishi A (1987a) The elicitation of phytoalexins by Ca^{2+} and cyclic AMP in carrot cells. Phytochemistry 26:1919–1923
Kurosaki F, Tsurusawa Y, Nishi A (1987b) Breakdown of phosphatidylinositol during the elicitation of phytoalexin production in cultured carrot cells. Plant Physiol 85:601–604
Lawrence GJ, Ellis JG, Finnegan EJ (1994) Cloning a rust-resistance gene in flax. In: Daniels MJ, Downie JA, Osbourne AE (eds) Advances in molecular genetics of plant-microbe interactions, vol 3. Kluwer, Dordrecht, pp 303–306
Legendre L, Yueh YG, Crain R, Haddock N, Heinstein PF, Low PS (1993) Phospholipase C activation during elicitation of the oxidative burst in cultured plant cells. J Biol Chem 268:24559–24563
Levine A, Tenhaken R, Dixon R, Lamb C (1994) H_2O_2 from the oxidative burst orchestrates the plant hypersensitive disease resistance response. Cell 79:583–593
Levine A, Pennell RI, Alvarez ME, Palmer R, Lamb C (1996) Calcium-mediated apoptosis in a plant hypersensitive disease resistance response. Curr Biol 6:427–437
Long SR, Staskawicz BJ (1993) Procaryotic plant parasites. Cell 73:921–935
MacKintosh C, Lyon GD, MacKintosh RW (1994) Protein phosphatase inhibitors activate anti-fungal defence responses of soybean cotyledons and cell cultures. Plant J 5:137–147
Macrí F, Braidot E, Petrussa E, Vianello A (1994) Lipoxygenase activity associated to isolated soybean plasma membranes. Biochim Biophys Acta 1215:109–114
Malamy J, Klessig DF (1992) Salicylic acid and plant disease resistance. Plant J 2:643–654
Mason HS, Mullet JE (1990) Expression of two soybean vegetative storage protein genes during development and in response to water deficit, wounding, and jasmonic acid. Plant Cell 2:569–579
Mathieu Y, Kurkdjian A, Xia H, Guern J, Koller A, Spiro MD, O'Neill M, Albersheim P, Darvill A (1991) Membrane responses induced by oligogalacturonides in suspension cultured tobacco cells. Plant J 1:333–343
Mayer MG, Ziegler E (1988) An elicitor of *Phytophthora megasperma* f. sp. *glycinea* influences the membrane potential of soybean cotyledonary cells. Physiol Mol Plant Pathol 33:397–407
Mehdy MC (1994) Active oxygen species in plant defense against pathogens. Plant Physiol 105:467–472
Meier BM, Shaw N, Slusarenko AJ (1993) Spatial and temporal accumulation of defense gene transcripts in bean (*Phaseolus vulgaris* L.) leaves in relation to bacteria-induced hypersensitive cell death. Mol Plant-Microbe Interact 6:453–466
Melin P-M, Sommarin M, Sandelius AS, Jergil B (1987) Identification of Ca^{2+}-stimulated polyphosphoinositide phospholipase C in isolated plant plasma membranes. FEB Lett 223:87–91
Messiaen J, Read ND, Van Cutsem P, Trewavas AJ (1993) Cell wall oligogalacturonides increase cytosolic free calcium in carrot protoplasts. J Cell Sci 104:365–371
Milat M-L, Ducruet J-M, Ricci P, Marty F, Blein J-P (1991) Physiological and structural changes in tobacco leaves treated with crytogein, a proteinaceous elicitor from *Phytophthora cryptogea*. Phytopathology 81:1364–1368
Miller KJ, Gore RS, Johnson R, Benesi AJ, Reinhold N (1990) Cell-associated oligosaccharides of *Bradyrhizobium* spp. J Bacteriol 172:136–142
Mithöfer A, Bhagwat AA, Feger M, Ebel J (1996a) Suppression of fungal β-glucan-induced plant defence in soybean (*Glycine max* L.) by cyclic 1,3-1,6-β-glucans from the symbiont *Bradyrhizobium japonicum*. Planta 199:270–275
Mithöfer A, Lottspeich F, Ebel J (1996b) One-step purification of the β-glucan elicitor-binding protein from soybean (*Glycine max* L.) roots and characterization of an anti-peptide antiserum. FEBS Lett 381:203–207
Morris PF, Savard ME, Ward EWB (1991) Identification and accumulation of isoflavonoids and isoflavone glucosides in soybean leaves and hypocotyls in resistance responses to *Phytophthora megasperma* f. sp. glycinea. Physiol Mol Plant Pathol 39:229–244
Mueller MJ, Brodschelm W, Spannagl E, Zenk MH (1993) Signaling in the elicitation process is mediated through the octadecanoid pathway leading to jasmonic acid. Proc Natl Acad Sci USA 90:7490–7494
Murphy TM, Auh C-K (1996) The superoxide synthases of plasma membrane preparations from cultured rose cells. Plant Physiol 110:621–629
Narváez-Vásquez J, Pearce G, Orozco-Cardenas ML, Franceschi VR, Ryan CA (1995) Autoradiographic and biochemical evidence for the systemic translocation of systemin in tomato plants. Planta 195:593–600
Nespoulos C, Huet J-C, Pernollet J-C (1992) Structure-function relationships of a and b elicitins, signal proteins involved in the plant-*Phytophthora* interaction. Planta 186:551–557
Neuenschwander U, Vernooij B, Friedrich L, Uknes S, Kessmann H, Ryals J (1995) Is hydrogen peroxide a second messenger of salicylic acid in systemic acquired resistance? Plant J 8:227–233
Nojiri H, Sugimori M, Yamane H, Nishimura Y, Yamada A, Shibuya N, Kodama O, Murofushi N, Omori T

(1996) Involvement of jasmonic acid in elicitor-induced phytoalexin production in suspension-cultured rice cells. Plant Physiol 110:387–392

Nothnagel EA, McNeil M, Albersheim P, Dell A (1983) Host-pathogen interactions. XXII. A galacturonic acid oligosaccharide from plant cell walls elicits phytoalexins. Plant Physiol 71:916–926

Nürnberger T, Nennstiel D, Jabs T, Sacks WR, Hahlbrock K, Scheel D (1994a) High affinity binding of a fungal oligopeptide elicitor to parsley plasma membranes triggers multiple defense responses. Cell 78:449–460

Nürnberger T, Colling C, Hahlbrock K, Jabs T, Renelt A, Sacks WR, Scheel D (1994b) Perception and transduction of an elicitor signal in cultured parsley cells. Biochem Soc Symp 60:173–182

Nürnberger T, Nennstiel D, Hahlbrock K, Scheel D (1995) Covalent cross-linking of the *Phytophthora megasperma* oligopeptide elicitor to its receptor in parsley membranes. Proc Natl Acad Sci USA 92:2338–2342

Ojalvo I, Rokem JS, G. Goldberg I (1987) ^{31}P NMR study of elicitor treated *Phaseolus vulgaris* cell suspension cultures. Plant Physiol 85:716–719

Osswald WF, Zieboll S, Elstner EF (1985) Comparison of pH changes and elicitor induced production of glyceollin isomers in soybean cotyledons. Z Naturforsch 40c:477–481

Parker JE, Schulte W, Hahlbrock K, Scheel D (1991) An extracellular glycoprotein from *Phytophthora megasperma* f. sp. *glycinea* elicits phytoalexin synthesis in cultured parsley cells and protoplasts. Mol Plant-Microbe Interact 4:19–27

Pearce G, Johnson S, Ryan CA (1993) Structure-activity of deleted and substituted systemin, an 18-amino acid polypeptide inducer of plant defensive genes. J Biol Chem 268:212–216

Pelissier B, Thibaud JB, Grignon C, Esquerré-Tugayé M-T (1986) Cell surfaces in plant-microorganism interactions. VII. Elicitor preparations from two fungal pathogens depolarize plant membranes. Plant Sci 46:103–109

Peng M, Kuc J (1992) Peroxidase-generated hydrogen peroxide as a source of antifungal activity in vitro and on tobacco leaf disks. Phytopathology 82:696–699

Pernollt J-C, Sallantin M, Sallé-Tourne M, Huet J-C (1993) Elicitin isoforms from seven *Phytophthora* species: comparison of their physico-chemical properties and toxicity to tobacco and other plant species. Physiol Mol Plant Pathol 42:53–67

Preisig CL, Kuc JA (1985) Arachidonic acid-related elicitors of the hypersensitive response in potato and enhancement of their activities by glucans from *Phytophthora infestans* (Mont.) de Bary. Arch Biochem Biophys 236:379–389

Ranjeva R, Boudet AM (1987) Phosphorylation of proteins in plants: regulatory effects and potential involvement in stimulus/response coupling. Annu Rev Plant Physiol 38:73–93

Raz V, Fluhr R (1993) Ethylene signal is transduced via protein phosphorylation events in plants. Plant Cell 5:523–530

Read ND, Allan WTG, Knight H, Knight MR, Malhó R, Russell A, Shacklock PS, Trewavas AJ (1992) Imaging and measurement of cytosolic free calcium in plant and fungal cells. J Microsc 166:57–86

Reinbothe S, Mollenhauer B, Reinbothe C (1994) JIPs and RIPs: the regulation of plant gene expression by jasmonate in response to environmental cues and pathogens. Plant Cell 6:1197–1209

Ren Y-Y, West CA (1992) Elicitation of diterpene biosynthesis in rice (*Oryza sativa* L.) by chitin. Plant Physiol 99:1169–1178

Renelt A, Colling C, Hahlbrock K, Nürnberger T, Parker JE, Sacks WR, Scheel D (1993) Studies on elicitor recognition and signal transduction in plant defence. J Exp Bot 44:257–268

Ricci P, Bonnet P, Huet J-C, Sallantin M, Beauvais-Cante F, Bruneteau M, Billard V, Michel G, Pernollet J-C (1989) Structure and activity of proteins from pathogenic fungi *Phytophthora* eliciting necrosis and acquired resistance in tobacco. Eur J Biochem 183:555–563

Ricci P, Trentin F, Bonnet P, Venard P, Mouton-Perronnet F, Bruneteau M (1992) Differential production of parasiticein, an elicitor of necrosis and resistance in tobacco, by isolates of *Phytophthora parasitica*. Plant Pathol 41:298–307

Ricci P, Panabieres F, Bonnet P, Maia N, Ponchet M, Devergne J-C, Marais A, Cardin L, Milat ML, Blein JP (1993) Proteinaceous elicitors of plant defense responses. In: Fritig B, Legrand M (eds) Mechanisms of plant defense responses. Kluwer, Dordrecht, pp 121–135

Roberts DM, Harmon AC (1992) Calcium-modulated proteins: targets for intracellular calcium signals in higher plants. Annu Rev Plant Physiol Plant Mol Biol 43:375–414

Rohe M, Gierlich A, Hermann H, Hahn M, Schmidt B, Rosahl S, Knogge W (1995) The race-specific elicitor, NIP1, from the barley pathogen, *Rhynchosporium secalis*, determines avirulence on host plants of the *Rrs1* resistance genotype. EMBO J 14:4168–4177

Röhrig H, Schmidt J, Walden R, Czaja I, Miklasevics E, Wieneke U, Schell J, John M (1995) Growth of tobacco protoplasts stimulated by synthetic lipochitooligosaccharides. Science 269:841–843

Rolin DB, Pfeffer PE, Osmann SF, Szwergold BS, Kappler F, Benesi AJ (1992) Structural studies of a phosphocholine substituted β-(1,3);(1,6) macrocyclic glucan from *Bradyrhizobium japonicum* USDA 110. Biochim Biophys Acta 1116:215–225

Ryan CA (1992) The search for the proteinase inhibitor-inducing factor, PIIF. Plant Mol Biol 19:123–133

Sacks W, Nürnberger T, Hahlbrock K, Scheel D (1995) Molecular characterization of nucleotide sequences encoding the extracellular glycoprotein elicitor from *Phytophthora megasperma*. Mol Gen Genet 246:45–55

Scheel D, Colling C, Keller H, Parker J, Schulte W, Hahlbrock K (1989) Studies on elicitor recognition and signal transduction in host and non-host plant/fungus pathogenic interactions. In: Lugtenberg BJJ (ed) Signal molecules in plants and plant-microbe interactions. Springer, Berlin Heidelberg New York, pp 211–218

Scheel D, Colling C, Hedrich R, Kawalleck P, Parker E, Sacks WR, Somssich IE, Hahlbrock K (1991) Signals in plant defense gene activation. In: Hennecke H, Verma DPS (eds) Advances in molecular genetics of plant-microbe interactions, vol 1. Kluwer, Dordrecht, pp 373–380

Schmidt WE, Ebel J (1987) Specific binding of a fungal phytoalexin elicitor to membrane fractions from soybean *Glycine max*. Proc Natl Acad Sci USA 84:4117–4121

Scholtens-Toma IMJ, De Wit PJGM (1988) Purification and primary structure of a necrosis-inducing peptide from the apoplastic fluids of tomato infected with *Cladosporium fulvum* (syn. *Fulvia Fulva*). Physiol Mol

Plant Pathol 33:59–67

Schwacke R, Hager A (1992) Fungal elicitors induce a transient release of active oxygen species from cultured spruce cells that is dependent on Ca^{2+} and protein-kinase activity. Planta 187:136–141

Sembdner G, Parthier B (1993) The biochemistry and the physiological and molecular actions of jasmonates. Annu Rev Plant Physiol Plant Mol Biol 44:569–589

Sharp JK, Valent B, Albersheim P (1984a) Purification and partial characterization of a β-glucan fragment that elicits phytoalexin accumulation in soybean. J Biol Chem 259:11312–11320

Sharp JK, McNeil M, Albersheim P (1984b) The primary structures of one elicitor-active and seven elicitor-inactive hexa (β-D-glucopyranosyl)-D-glucitols isolated from the mycelial walls of *Phytophthora megasperma* f. sp. *glycinea*. J Biol Chem 259:11321–11336

Shibuya N, Kaku K, Kuchitsu K, Maliarik MJ (1993) Identification of a novel high-affinity binding site for N-acetylchitooligosaccharide elicitor in the membrane fraction from suspension-cultured rice cells. FEBS Lett 329:75–78

Slusarenko AJ, Meier BM, Croft KPC, Eiben HG (1993) Lipoxygenase in plant disease. In: Fritig B, Legrand M (eds) Mechanisms of plant defense responses. Kluwer, Dordrecht, pp 211–220

Spaink HP, Lugtenberg BJJ (1994) Role of rhizobial lipochitin oligosaccharide signal molecules in root nodule organogenesis. Plant Mol Biol 26:1413–1422

Stäb MR, Ebel J (1987) Effects of Ca^{2+} on phytoalexin induction by fungal elicitor in soybean cells. Arch Biochem Biophys 257:416–423

Staswick PE (1990) Novel regulation of vegetative storage protein genes. Plant Cell 2:1–6

Staswick PE (1992) Jasmonate, genes, and fragrant signals. Plant Physiol 99:804–807

Staswick PE, Huang J, Rhee Y (1991) Nitrogen and methyl jasmonate induction of soybean vegetative storage protein genes. Plant Physiol 96:130–136

Strasser H, Tietjen KG, Himmelspach K, Matern U (1983) Rapid effect of an elicitor on uptake and intracellular distribution of phosphate in cultured parsley cells. Plant Cell Rep 2:140–143

Strasser H, Hoffmann C, Grisebach H, Matern U (1986) Are polyphosphoinositides involved in signal transduction of elicitor-induced phytoalexin synthesis in cultured plant cells? Z Naturforsch 41c:717–724

Sutherland MW (1991) The generation of oxygen radicals during host plant responses to infection. Physiol Mol Plant Pathol 39:79–93

Suzuki K, Shinshi H (1995) Transient activation and tyrosine phosphorylation of a protein kinase in tobacco cells treated with a fungal elicitor. Plant Cell 7:639–647

Suzuki K, Fukuda Y, Shinshi H (1995) Studies on elicitor-signal transduction leading to differential expression of defense genes in cultured tobacco cells. Plant Cell Physiol 36:281–289

Tate BF, Schaller GE, Sussman MR, Crain RC (1989) Characterization of a polyphosphoinositide phospholipase C from the plasma membrane of *Avena sativa*. Plant Physiol 91:1275–1279

Tavernier E, Wendehenne D, Blein J-P, Pugin A (1995) Involvement of free calcium in action of cryptogein, a proteinaceous elicitor of hypersensitive reaction in tobacco cells. Plant Physiol 109:1025–1031

Tenhaken R, Levine A, Brisson LF, Dixon RA, Lamb C (1995) Function of the oxidative burst in hypersensitive disease resistance. Proc Natl Acad Sci USA 92:4158–4163

Thain JF, Doherty HM, Bowles DJ, Wildon DC (1990) Oligosaccharides that induce proteinase inhibitor activity in tomato plants cause depolarization of tomato leaf cells. Plant Cell Environ 13:569–574

Van den Ackerveken GFJM, Van Kan JAL, De Wit PJGM (1992) Molecular analysis of the avirulence gene *avr9* of the fungal tomato pathogen *Cladosporium fulvum* fully supports the gene-for-gene hypothesis. Plant J 2:359–366

Van den Ackerveken GFJM, Vossen P, De Wit PJGM (1993) The AVR9 race-specific elicitor of *Cladosporium fulvum* is processed by endogenous and plant proteases. Plant Physiol 103:91–96

Van Kan JAL, Van den Ackerveken GFJM, De Wit PJGM (1991) Cloning and characterization of cDNA of avirulence gene Avr9 of the fungal pathogen *Cladosporium fulvum*, causal agent of tomato leaf mold. Mol Plant-Microbe Interact 4:52–59

Viard M-P, Martin F, Pugin A, Ricci P, Blein J-P (1994) Protein phosphorylation is induced in tobacco cells by the elicitor cryptogein. Plant Physiol 104:1245–1249

Vick BA, Zimmerman DC (1984) Biosynthesis of jasmonic acid by several plant species. Plant Physiol 75:458–461

Vögeli U, Vögeli-Lange R, Chappel J (1992) Inhibition of phytoalexin biosynthesis in elicitor-treated tobacco cell suspension cultures by calcium/calmodulin antagonists. Plant Physiol 100:1369–1376

Waldmüller T, Cosio EG, Grisebach H, Ebel J (1992) Release of highly elicitor-active glucans by germinating zoospores of *Phytophthora megasperma* f. sp. glycinea. Planta 188:498–505

Waldmüller T, Feger M, Ebel J (1993) Enhancement of β-glucan and hepta-β-glucoside elicitor activity in soybean by protein kinase inhibitor K-252a. In: Fritig B, Legrand M (eds) Mechanisms of plant defense responses. Kluwer, Dordrecht, pp 117–120

Walker-Simmons M, Hadwiger L, Ryan CA (1983) Chitosans and pectic polysaccharides both induce the accumulation of the antifungal phytoalexin pisatin in pea pods and antinutrient proteinase inhibitors in tomato leaves. Biochem Biophys Res Commun 110:194–199

Walker-Simmons M, Jin D, West CA, Hadwiger L, Ryan CA (1984) Comparison of proteinase inhibitor-inducing activities and phytoalexin elicitor activities and phytoalexin elicitor activities of a pure fungal endopolygalacturonase, pectic fragments, and chitosan. Plant Physiol 76:833–836

Webb AAR, McAinsh MR, Taylor JE, Hetherington AM (1996) Calcium ions as intracellular second messengers in higher plants. Adv Bot Res 22:45–96

Wendehenne D, Binet M-N, Blein J-P, Ricci P, Pugin A (1995) Evidence for specific, high-affinity binding sites for a proteinaceous elicitor in tobacco plasma membrane. FEBS Lett 374:203–207

Wevelsiep L, Kogel KH, Knogge W (1991) Purification and characterization of peptides from *Rhynchosporium secalis* inducing necrosis in barley. Physiol Mol Plant Pathol 39:471–482

Wevelsiep L, Rüpping E, Knogge W (1993) Stimulation of barley plasmalemma H^+-ATPase by phytotoxic peptides from the fungal pathogen, *Rhynchosporium secalis*. Plant Physiol 101:297–301

Xu D, McElroy D, Thornburg RW, Wu R (1993) Systemic

induction of a potato pin2 promoter by wounding, methyl jasmonate, and abscisic acid in transgenic rice plants. Plant Mol Biol 22:573–588
Yamada A, Shibuya N, Kodama O, Akatsuka T (1993) Induction of phytoalexin formation in suspension-cultured rice cells by *N*-acetylchitooligosaccharides. Biosci Biotech Biochem 57:405–409
Yoshikawa M, Sugimoto K (1993) A specific binding site in soybean membranes for a phytoalexin elicitor released from fungal cell walls by β-1,3-endoglucanase. Plant Cell Physiol 34:1229–1237
Yoshikawa M, Keen NT, Wang M-C (1983) A receptor on soybean membranes for a fungal elicitor of phytoalexin accumulation. Plant Physiol 73:497–506
Yotsushima K, Mitsui T, Takaoka T, Hayakawa T, Igaue I (1993) Purification and characterization of membrane-bound inositol phospholipid-specific phospholipase C from suspension-cultured rice (*Oryza sativa* L.) cells. Plant Physiol 102:165–172
Zimmermann S, Nürnberger T, Frachisse J-M, Wirtz W, Guern J, Hedrich R, Scheel D (1997) Receptor-mediated activation of a plant Ca^{2+}-permeable ion channel involved in pathogen defense. Proc Nate Acad Sci USA in press

6 Pathogenesis-Related Proteins and Plant Defense

E. Kombrink and I.E. Somssich

CONTENTS

I. Introduction

A. Pathogen-Induced Plant Defense Responses

Plants represent the largest and most important group of autotrophic organisms. Their abundant organic material serves as nutritional source for all heterotrophic organisms, including animals, insects, and microorganisms, and, as a consequence, they have developed effective mechanisms to protect themselves against herbivores and pathogens. Despite the large number of fungi and bacteria that are actively involved in the decomposition of dead plant material, very few of these potentially pathogenic microorganisms have acquired the ability to colonize living plants. Thus, plants exhibit natural resistance to microbial attack (non-host resistance, incompatible interaction), and disease is the exception rather than the rule (Chap. 16, Vol. V, Part B).

Usually, the host range of a pathogen is very limited, often restricted to a single species; however, with compatible host-pathogen interactions the host may show severe symptoms. In extreme cases, when agriculturally and economically important crop plants are affected, such interactions may result in costly and even devastating epidemics. For example, the epidemic spread of the late blight fungus *Phytophthora infestans* caused destruction of the staple crop potato (*Solanum tuberosum*) in many European countries and culminated in the Irish potato famine of 1846 (Chap. 2, Vol. V, Part B). Selection and breeding for resistant cultivars and coevolution of pathogen races with altered virulence has led to the development of host/pathogen combinations in which only certain cultivars can be colonized by particular races of the pathogen, whereas others retain resistance (cultivar resistance).

The biochemical mechanisms that determine the resistant phenotype are still incompletely understood, although specificity and resistance are known to be under genetic control. Extensive studies of many different plant/pathogen interactions have shown that plants utilize a large arsenal of defense responses when attacked by pathogens, including rapid localized cell death (hypersensitive response), accumulation of antimicrobial phytoalexins, synthesis and deposition of phenolic compounds and proteins in the cell wall, and syn-

Max-Planck-Institut für Züchtungsforschung, Abteilung Biochemie, D-50829 Köln, Germany

The Mycota V Part A
Plant Relationships
Carroll/Tudzynski (Eds.)

thesis of pathogenesis-related proteins (PR proteins) (Dixon and Harrison 1990; Kombrink and Somssich 1995). The present chapter will provide only a general description of the structure and function of PR proteins, and the reader is referred to previously published comprehensive reviews for more detailed information (Linthorst 1991; Cutt and Klessig 1992; Stintzi et al. 1993).

II. Structure and Properties of Pathogenesis-Related Proteins

A. Definition

The term pathogenesis-related (PR) proteins was coined in 1980 to define a group of plant polypeptides that accumulate in pathological or related situations (Cutt and Klessig 1992). Such proteins were first described in 1970 in leaves of certain tobacco cultivars (*Nicotiana tabacum*) following infection by tobacco mosaic virus (TMV) (Cutt and Klessig 1992; Stintzi et al. 1993). Originally identified as additional bands in polyacrylamide gels, PR proteins were characterized as acidic, low-molecular-weight proteins, soluble at low pH and relatively resistant to endogenous plant and commercial proteases (Stintzi et al. 1993). Organic compounds such as polyacrylic acid, acetyl salicylic acid (aspirin), and salicylic acid could also induce the synthesis of PR proteins, and their presence has been correlated with increased disease resistance to various pathogens (Gianinazzi 1984). In the absence of any known biological function, the ten major tobacco PR proteins were grouped into five families, based solely on their electrophoretic mobility and immunological cross-reactivity (Cutt and Klessig 1992; Stintzi et al. 1993). Proteins with similar characteristics were subsequently detected in a large variety of plants, and thus, PR proteins are now assumed to be ubiquitous in the plant kingdom.

The application of sensitive molecular biological techniques to the study of pathogen-induced plant responses has led to the isolation of numerous genes from various plant species whose products share some sequence similarity with the five classical PR protein families of tobacco (Linthorst 1991; Somssich 1994). In addition, novel PR proteins continue to be isolated from many plants, and their discovery has complicated PR protein nomenclature. For this chapter, we have adopted the classification recommended by van Loon et al. (1994), but have also included certain proteins whose characterization is still too preliminary for them to be unequivocally considered bona fide PR proteins. No attempt has been made to include or describe in detail well-defined proteins such as peroxidases, superoxide dismutases, amylases, or enzymes of general phenyl propanoid metabolism, despite the fact that their synthesis may be induced by pathogen infection and that some consider them to be PR proteins (van Loon et al. 1994).

B. Family 1: Proteins of Type PR-1

Proteins of the PR-1 family have been extensively studied, but their function(s) remains elusive. In tobacco, three dominant and serologically closely related acidic polypeptides, PR-1a, -1b, -1c (M_r 15000), have been identified in TMV-infected leaves, and the corresponding cDNAs and genes have been isolated and characterized (Linthorst 1991; Cutt and Klessig 1992; Stintzi et al. 1993). The existence of additional, basic forms of PR-1 in tobacco has been inferred from genomic and cDNA clones (Linthorst 1991; Cutt and Klessig 1992; Eyal et al. 1992), although a corresponding protein product has only recently been identified (Stintzi et al. 1993). In tobacco, the acidic forms of PR-1 are found in the extracellular leaf space and can be readily extracted with the intercellular washing fluid. All PR-1 proteins contain N-terminal signal peptides which direct them into the endoplasmic reticulum for secretion; the basic forms also have peptide extension at their C-termini, presumed to function as retention signals directing the proteins to the vacuole. Related proteins and genes have been identified in numerous other dicotyledonous and monocotyledonous plants, including *Lycopersicon esculentum* (Linthorst 1991; van Kan et al. 1992; Tornero et al. 1994), *Arabidopsis thaliana* (Uknes et al. 1992), *Zea mays* (Casacuberta et al. 1991), and *Hordeum vulgare* (Bryngelsson et al. 1994).

C. Family 2: 1,3-β-Glucanases

The finding that several serologically related PR proteins of tobacco exhibit 1,3-β-glucanase activity can be considered a major breakthrough in the research on PR proteins (Kauffmann et al. 1987). Not only could a biochemical function be assigned to PR proteins, but suddenly they shared a functional relationship with an abundant and well-characterized group of plant proteins, the basic 1,3-β-glucanases, which were considered important plant defense factors (Boller 1988). This finding further supported the original idea that PR proteins function in the plant defense program. That certain PR proteins have 1,3-β-glucanase activities was independently shown for potato (Kombrink et al. 1988), tomato (Joosten and de Wit 1989), and numerous other plants. 1,3-β-glucanases are abundant proteins that occur in several isoforms in all plants analyzed. For reviews on antifungal glucanohydrolases and their genes see Boller (1988) and Meins et al. (1992).

Based on comparisons of amino acid sequences, tobacco contains at least three distinct classes of 1,3-β-glucanase (Meins et al. 1992). Class I enzymes include at least three related isoforms with basic pI values that undergo substantial post-translational modification (glycosylation, removal of N- and C-terminal peptides) before they accumulate in the vacuole (Sticher et al. 1992a). Class II enzymes, which include the proteins PR-2a, PR-2b, and PR-2c (formerly PR-2, PR-N, PR-O), are acidic isoforms that are secreted into the extracellular leaf space. They lack the C-terminal extension but otherwise share about 50% amino acid sequence similarity with class I enzymes. To date, only one class III enzyme has been identified, PR-Q′, an acidic, extracellular protein (Payne et al. 1990). The purified proteins differ significantly in their enzymatic and antifungal activities (Kauffmann et al. 1987; Sela-Buurlage et al. 1993). 1,3-β-glucanases corresponding to the three classes have also been cloned from tomato (van Kan et al. 1992), potato (Beerhues and Kombrink 1994), and numerous others plants (Meins et al. 1992). Tomato appears to contain at least four transcriptionally active 1,3-β-glucanase genes: in addition to class I and class II, two cDNAs encoding class III enzymes with acidic and basic pI values have been isolated, that are highly similar to tobacco PR-Q′ and likewise presumed to be extracellular (Domingo et al. 1994). Class I, II, and III 1,3-β-glucanases are strongly induced in various plants by pathogen infection, and are therefore considered genuine PR proteins.

D. Family 3: Chitinases

Chitinase activity is the second enzymatic function that could be assigned to a group of PR proteins, again, first in tobacco for PR-3a and PR-3b (formerly PR-P and PR-Q) (Legrand et al. 1987) and subsequently in potato (Kombrink et al. 1988), tomato (Joosten and de Wit 1989), and other plants. Numerous plant chitinases, mostly basic isoforms, have been characterized and implicated in the active defense response of plants (Boller 1988). Compared to 1,3-β-glucanases, chitinases represent an even more diverse group of ubiquitous plant proteins (Boller 1988; Meins et al. 1992; Collinge et al. 1993).

On the basis of primary structure, four classes of plant chitinases have been proposed (Meins et al. 1992; Mikkelsen et al. 1992; Collinge et al. 1993). Class I chitinases, basic isoforms localized in the vacuole, have a hydrophobic signal peptide at the N-terminus followed by a cysteine-rich domain of approximately 40 amino acid residues. This region, shown to be a chitin-binding domain, shows high similarity to hevein (Raikhel et al. 1993). A proline-glycine-rich region of variable length connects the hevein domain with the catalytic domain of approximately 300 amino acid residues and a C-terminal extension, which has been identified as a vacuolar targeting signal (Neuhaus et al. 1991b). Class II chitinases, acidic extracellular isoforms, show an overall similarity of about 60% to class I chitinases within the same species, but lack the N-terminal cysteine-rich domain and the C-terminal extension. Class II chitinases have considerably lower specific activity than class I isoforms (Legrand et al. 1987). Class III chitinases show no sequence similarity to those of class I and II and, because of this divergence, were separately grouped into family 8 of the PR proteins (van Loon et al. 1994). Class IV chitinases resemble class I chitinases in having a cysteine-rich domain, a hinge region, and a conserved main structure. However, due to four deletions, including one within the cysteine-rich domain and three within the catalytic domain, they are about 50 amino acid residues shorter than class I chitinases. Their overall similarity to class I is 40–50%, and even lower to class II. The establishment of class IV as a separate class is based on the sequences of proteins from *Beta vulgaris* (Mikkelsen et al. 1992; Nielsen et al. 1994), *Brassica napus* (Rasmussen et al. 1992a), *Phaseolus vulgaris* (Margis-Pinheiro et al. 1991), and *Zea mays* seeds (Huynh et al. 1992b). Recently, two basic chitinases from tobacco have been proposed to represent a novel class V (Melchers et al. 1994). These proteins show a similarity of about 30% to bacterial exochitinases, but none to other plant chitinases. Their specific activity is 200–500 times lower than that of class I chitinases.

The majority of plant chitinases are endochitinases; only in few cases have purified enzymes been shown to exhibit exochitinase activity (Roby and Esquerre-Tugaye 1987; Kurosaki et al. 1989; Kirsch et al. 1993; Kragh et al. 1993). Many chitinases also have lysozyme activity to varying degrees (Boller 1988; Meins et al. 1992; Kombrink et al. 1993b), and some also cleave chitosan (Bernasconi et al. 1987; Grenier and Asselin 1990; Osswald et al. 1993). Among plant chitinases some are found with unique structural features. Class I chitinases from tobacco and class IV chitinases from sugar beet contain hydroxyproline (Sticher et al. 1992b; Nielsen et al. 1994). Other chitinases are glycosylated, such as class IV chitinases from bean and sugar beet (Margis-Pinheiro et al. 1991; Nielsen et al. 1994), a class I chitinase from potato epidermal leaf cells (Kombrink et al. 1993a), and several of the exochitinases found in parsley cell cultures (Kirsch et al. 1993). The physiological functions of these modifications are unknown.

Most plants contain multiple chitinase isoforms from different classes. An extreme case is represented by sugar beet leaves, which contain 13 isoforms (Mikkelsen et al. 1992). Chitinases from all classes (I–V) are induced by infection or other types of stress and are therefore considered bona fide PR proteins.

E. Family 4: Win-Like Proteins (Chitin-Binding Proteins)

The PR-4 family consists of several proteins of M_r 13000–15000 which in tobacco have acidic pI values (Cutt and Klessig 1992; Stintzi et al. 1993), whereas the serologically related tomato protein P2 is basic (Joosten et al. 1990). cDNA clones for PR-4 proteins have been isolated from these two plant species and the deduced amino acid sequences revealed the presence of N-terminal signal peptides for secretion (Friedrich et al. 1991; Linthorst et al. 1991). In addition, the proteins share sequence similarity to the deduced amino acid sequences of two wound-induced genes, *win1* and *win2*, from potato (Stanford et al. 1989) and the precursor protein of hevein from rubber tree (*Hevea brasiliensis*; Broekaert et al. 1990), which have an amino-terminal lectin domain and a carboxy-terminal domain of unknown function. PR-4 is homologous to the carboxy-terminus of these proteins but does not contain the lectin domain. In contrast to pro-hevein and *win2*, but not *win1*, which have an extended C-terminal region, the PR-4 proteins lack such an extension, which is in accordance with their extracellular location.

Recently, an *Arabidopsis* PR-4-like cDNA was isolated encoding a protein which is similar not only to other PR-4 proteins (75%) but also to hevein itself (67%) (Potter et al. 1993). In this hevein-like protein, HEL, the hevein domain is connected to the carboxy-terminal PR-4 domain via a spacer region of eight amino acids, a structural feature reminiscent of class I chitinases.

The intracellular PR-4 counterpart, CBP20, a 20-kDa chitin-binding protein, was recently purified from tobacco (Ponstein et al. 1994). The protein represents the first class I PR-4 protein. Its amino acid sequence, as deduced from the cDNA, is 79% identical to the PR-4 protein and its overall structure similar to *win2* (potato) and HEL (*Arabidopsis*), including signal peptide, hevein domain, hinge region, main structure, and C-terminal extension. CBP20 has very low chitinase and lysozyme activity (about 1% of class I chitinase), but it is difficult to assess whether this is a true intrinsic feature of the protein (Ponstein et al. 1994). Its antimicrobial activity is shared by other chitin-binding proteins of the PR-4 family such as those isolated from barley leaves (Hejgaard et al. 1992). No function is known for any of the PR-4 proteins.

In conclusion, the hevein domain or chitin-binding function appears to be the common feature of the PR-4 family of proteins. Possibly, the original PR-4 proteins represent only a special case within this class.

F. Family 5: Thaumatin-Like Proteins (Osmotins)

A group of acidic tobacco proteins of M_r 22000–25000, formerly called PR-R and PR-S, in Samsun NN tobacco (Cutt and Klessig 1992; Stintzi et al. 1993) were grouped into the PR-5 family. The proteins share extensive sequence similarity (65%) with the sweet-tasting protein, thaumatin, of *Thaumatococcus daniellii* (Cornelissen et al. 1986), to a bifunctional trypsin/α-amylase inhibitor (57%) that occurs in maize seeds (Richardson 1991), as well as to the basic protein osmotin (65%) found in salt-adapted, cultured tobacco cells (Singh et al. 1987). Osmotin was shown to be the vacuolar counterpart of PR-R and PR-S which accumulate extracellularly, and accordingly, osmotin contains a C-terminal extension (Melchers et al. 1993; Stintzi et al. 1993). Both the intra- and extracellular PR-5 proteins are induced by viral and fungal infection in tobacco (Cornelissen et al. 1986; Woloshuk et al. 1991; Stintzi et al. 1993), and similar proteins have been identified in numerous other plants, including potato (Pierpoint et al. 1990), tomato (Rodrigo et al. 1993), *Arabidopsis* (Uknes et al. 1992), and barley (Bryngelsson and Gréen 1989; Hahn et al. 1993). In addition to these bona fide PR proteins of family 5, thaumatin-like proteins have been identified among the many antimicrobial seed storage proteins from maize, oats, sorghum, and wheat (Vigers et al. 1991). Except for antifungal activity (Vigers et al. 1991; Vigers et al. 1992), no other function of PR-5 proteins is known.

G. Family 6: Proteinase Inhibitors

Proteinase inhibitors (PIs) were not classically considered PR proteins. PIs are induced in vegetative organs, such as leaves, by wounding and insect bite, and are also found in large amounts together with other enzyme inhibitors in tubers and seed. The literature on these proteins is rather extensive, and the reader is referred to relevant reviews (Ryan 1990; Richardson 1991). Much of the work on plant PIs has focused on the inhibition of trypsin and chymotrypsin, two proteinases of the digestive tract of animals. Their function, along with that of other enzyme inhibitors, is thought to protect plants against foreign invaders, particularly insects. The most extensively studied wound-inducible PIs are those found in tomato and potato leaves. Based on their amino acid sequences, two groups are distinguished, PI-I (M_r 8000) and PI-II (M_r 12500), which are both synthesized as preproteins and accumulate in the vacuole. Interestingly, PI induction is not restricted to the wounded leaf area but occurs systemically throughout the plant (Ryan 1990; Farmer et al. 1992).

Induction of PIs by pathogens has also been demonstrated in melon infected by *Colletotrichum lagenarium* (Roby et al. 1987b), in tomato during bacterial pathogen invasion (Pautot et al. 1991), and during the hypersensitive reaction of tobacco to TMV infection (Geoffroy et al. 1990; Linthorst et al. 1993). Interestingly, the tobacco inhibitor (M_r 6000), which appears related to the potato PI-I, was shown to be highly active against different serine endoproteinases of fungal and bacterial origin, such as proteinase K of *Tritirachium album* or subtilisin of *Bacillus subtilis*, whereas it poorly inhibited the serine endoproteinases of animal origin, trypsin, and chymotrypsin (Geoffroy et al. 1990). Thus, it is reasonable to assume that PIs also function in inhibiting pathogen growth and development.

H. Family 7: Proteinases

In tomato leaves, the acid PR protein P69 (M_r 69000), which is induced by citrus exocortis viroid (CEV) infection and ethephon treatment, has been identified as an alkaline endoproteinase (Vera et al. 1989). A similar, if not identical glycoprotein, P70, has been isolated from tomato leaves following infection with *Phytophthora infestans* (Fischer et al. 1989). In the case of P70, partial N-terminal amino acid sequence information has been obtained. Apart from this, however, this family remains poorly characterized and nothing is known concerning the function of such proteinases in plant defense.

I. Family 8: Lysozymes/Chitinases (Class III)

The prototype of class III chitinases is a PR protein of cucumber (*Cucumis sativus*) which is induced locally and systemically in response to viral, bacterial, and fungal infection (Metraux et al. 1989). Purification and cloning of the extracellular, acidic endochitinase (M_r 28000) revealed no sequence similarity to other plant chitinases of class I and II, whereas 68% identity was found to a partial amino acid sequence of the bifunctional basic lysozyme/chitinase from *Parthenocissus quinquifolia* (Bernasconi et al. 1987; Metraux et al. 1989). However, for the cucumber chitinase no lysozyme activity could be demonstrated. The chromosomal region encoding the cucumber class III chitinase was recently isolated and sequenced (Lawton et al. 1994a). A 12-kb genomic fragment harbors three highly conserved open reading frames: the CHI2 gene, identical to the previously isolated cDNA, is flanked by two closely related genes, CHI1 and CHI3, with 93% similarity at the nucleotide level. CHI2 appears to be the only gene expressed under all the conditions tested.

Class III chitinases have also been purified and cloned from a number of other plant species, such as tobacco (Lawton et al. 1992), *Arabidopsis* (Samac et al. 1990), rubber tree latex (Jeckel et al. 1991), bean (Margis-Pinheiro et al. 1993), and chickpea (*Cicer arietinum*) (Vogelsang and Barz 1993). The acidic and basic class III chitinases identified in tobacco share 65% amino acid sequence identity among each other and about 60% with other class III chitinases (Lawton et al. 1992). Only the basic isoforms of tobacco, *Hevea*, and *Parthenocissus* display high lysozyme activity, whereas the acidic isoforms of cucumber, tobacco, *Arabidopsis*, bean, and chickpea are unable to hydrolyze the bacterial cell wall. All class III chitinases lack C-terminal extensions, in agreement with their extracellular localization.

J. Family 9: Peroxidases

Peroxidases catalyze the oxidation of various compounds in the presence of H_2O_2 and are found widely in animals, plants, and microbes (Tyson 1992). They constitute an enzyme superfamily displaying tremendous diversity in structure, function, and distribution within the plant kingdom. Peroxidases have been studied extensively because they are involved in a variety of physiological and biochemical processes, such as synthesis of lignin and suberin, cross-linking of cell-wall proteins (i.e., extensins) or phenolics with other substrates, processes which are not only important in growth and development but have also been associated with plant defense. Based on their inducibility upon pathogen attack in a variety of plants, the anionic, extracellular, lignin-forming peroxidases have very recently been suggested to be grouped into one family of PR proteins (van Loon et al. 1994). For the purpose of this chapter, however, they will be excluded from further discussion.

K. Family 10: Intracellular PR Proteins (IPRs)

The de novo synthesis upon pathogen attack, elicitor treatment, or wounding of a novel class of low-molecular-weight proteins (M_r 16000–18000) has been discovered in numerous plant species, including parsley (Somssich et al. 1988; van de Löcht et al. 1990), pea (Fristensky et al. 1988), potato (Matton and Brisson 1989), bean (Walter et al. 1990), and asparagus (*Asparagus officinalis*; Warner et al. 1992). IPR relationship is based strictly on their comparative amino acid sequences as deduced from isolated cDNA and genomic clones. They contain no signal peptides, suggesting that these "intracellular" PR proteins (IPR) are located in the cytosol. Interestingly, these proteins also share sequence similarity to pollen allergens of a number of tree species (Breiteneder et al. 1989). Nothing is known concerning the biological functions or antimicrobial activites of IPRs. However, a recent report that N- and C-terminal peptide sequences of a ribonuclease purified from ginseng callus show similarity to the parsley PR1 and PR2 proteins of 70 and 60%, respectively (Moiseyev et al. 1994) allows for the speculation that the specific biological functions of IPRs are intimately linked to a possible ribonuclease activity or RNA-binding capability.

L. Thionins

Thionins are a group of small, sulfur-rich plant proteins (M_r 5000) that are found mainly in cereals and mistletoes. Their three-dimensional structures are very compact, amphipathic, and stabilized by three or four disulfide bridges. Thionins are usually basic proteins and exert toxicity in various biological systems by altering membrane permeability. They are synthesized as preproproteins and deposited into vacuoles, protein bodies, and cell walls. Their antibacterial and antifungal activities towards plant pathogens point to a role as plant defense proteins, as first suggested in 1972 (Garcia-Olmedo et al. 1992), but otherwise their biological function remains obscure and they have not yet been classified as PR proteins. Structural and functional aspects of thionins and their possible role in plant protection are discussed in a number of recent reviews (Garcia-Olmedo et al. 1992; Bohlmann 1994). Their toxicity towards plant pathogens and the finding that the leaf-specific thionins of barley are induced upon infection with the powdery mildew fungus *Erysiphe graminis* f.sp. *hordei* (Bohlmann et al. 1988) permit their inclusion into the family of PR proteins.

M. Unclassified PR Proteins

A number of putative PR proteins which are induced by pathogen attack have been only partially characterized either as proteins or cDNAs that were obtained from differential screening procedures. These include the chitosanases of barley, cucumber, and tomato, that appear to be distinct from chitinases (Grenier and Asselin 1990), the PRP1 protein of potato, which shares considerable similarity to the soybean heat-shock protein HSP26 and was recently identified as glutathione S-transferease (Hahn and Strittmatter 1994), the PvPR3 protein from bean (Sharma et al. 1992) and a number of different protein families of parsley (Somssich et al. 1989).

Furthermore, plants contain a variety of mostly basic, low-molecular-weight proteins, which are particularly abundant in seeds but are also found in vegetative organs. Recently, numerous such proteins have been purified and shown to exhibit a broad spectrum of antimicrobial activity. These proteins include chitin-binding lectins and related proteins from stinging nettle (*Urtica dioica*), rubber tree and amaranth (*Amaranthus caudatus*) (Peumans and van Damme 1995), ribosome-inactivating proteins (RIPs) which catalytically inactivate nonself ribosomes (Stirpe et al. 1992), thaumatin-like proteins (Vigers et al. 1991; Huynh et al. 1992a), neurotoxin-like peptides from *Mirabilis jalapa* (Cammue et al. 1992), cysteine-rich proteins from radish (*Raphanus sativus*) and Brassicaceae, which are homologous to γ-thionins and α-amylase-inhibitor (Broekaert et al. 1995), and nonspecific lipid transfer proteins (Segura et al. 1993). Despite the fact that very little is known concerning the pathogen inducibility of such proteins, some do share certain common structural features with PR proteins. For example, the two very basic peptides, Ac-AMP1 and Ac-AMP2 (M_r 3100), purified from amaranth, share striking similarity with the cysteine/glycine-rich domain of chitin-binding proteins, such as chitinases and lectins (Raikhel et al. 1993; Peumans and van Damme 1995).

III. The Role of PR Proteins in Plant Defense

A. Inhibition of Pathogen Growth in Vitro

Since their identification in infected plants, PR proteins have been implicated in plant defense. The first experimental evidence of an antifungal function was obtained with purified bean class I chitinase, which strongly inhibited the growth of the saprophytic fungus *Trichoderma viride* at concentrations as low as $2\,\mu g\,ml^{-1}$ (Schlumbaum et al. 1986). In vitro inhibition of *Rhizoctonia solani* by bean chitinase was also demonstrated, but at much higher concentrations of $100–400\,\mu g\ ml^{-1}$ (Broglie et al. 1991). Using a whole series of test fungi, it was demonstrated that purified chitinase and 1,3-β-glucanase acted synergistically in degrading fungal cell walls and only in combination inhibited growth of most fungi, with the exception of *Phytophthora* and *Pythium* species, which are Oomycota, and thus lack chitin in their cell wall (Mauch et al. 1988b). Both enzymes, tested individually, were active only in exceptional cases. The enzyme concentrations required for inhibition of fungal growth correspond to only a fraction of the amount present in infected plant tissues, indicating that inhibition may well occur under physiological conditions.

Morphological studies showed that fungal hyphae are particularly susceptible at the tip and that exposure to chitinases and 1,3-β-glucanases causes extreme, balloon-like swelling and lysis of the hyphal tip, as shown in Fig. 1 (Mauch et al. 1988b; Broekaert et al. 1989; Arlorio et al. 1992). Apparently the hydrolases disturb the balance of cell wall synthesis and degradation required for tip growth of the hyphae. Thinning of the cell wall at the apex during tip growth exposes the chitin which is present in the inner parts of the wall, making it accessible to chitinase, and its hydrolysis leads to an imbalance of turgor pressure and wall tension which causes the tip to swell and to burst.

When different isoforms of chitinase and 1,3-β-glucanase of tobacco were tested for their antifungal activity, only vacuolar isoforms (class I) were active against *Fusarium solani* germlings, whereas extracellular isoforms (class II) of both hydrolases showed no or only limited antifungal activity (Sela-Buurlage et al. 1993). The biochemical basis for this difference is not yet known. In the case of the chitinases it may be related to the absence of the chitin-binding domain in the class II isoforms (Collinge et al. 1993). It is also absent from class III chitinases which also seem to lack antifungal activity, as demonstrated for the acid chitinases from chickpea, which was tested against *Trichoderma viride, T. harzianum*, *Helminthosporium tuberosum*, and *Rhizoctonia solani* at concentrations up to $30\,\mu g\,ml^{-1}$ (Vogelsang and Barz 1993). In contrast, the chickpea basic class I chitinase inhibited growth of the test fungi at $0.5\,\mu g\,ml^{-1}$ (Vogelsang and Barz 1993). A number of additional chitinases from a taxonomically di-

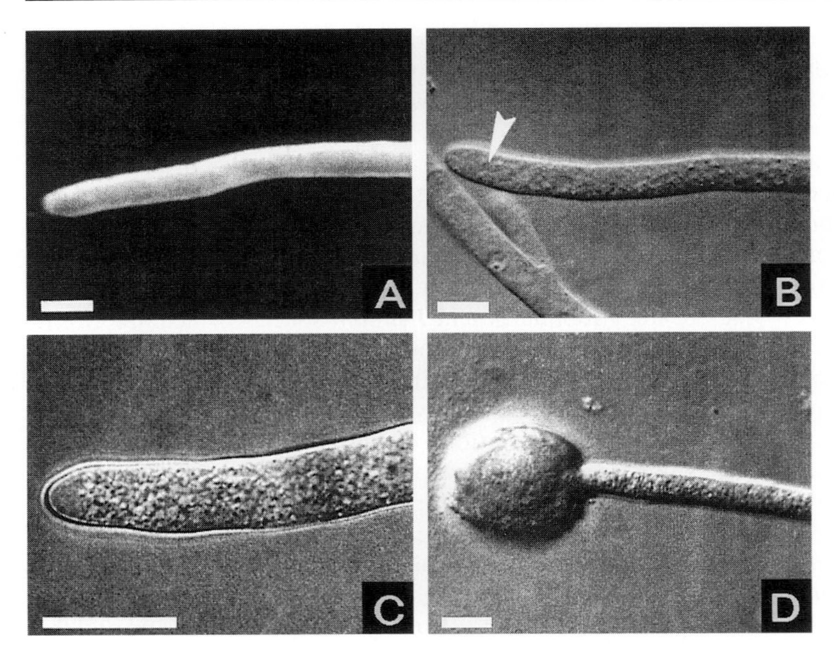

Fig. 1A–D. Influence of 1,3-β-glucanase and chitinase on hyphal morphology. *Trichoderma longibrachiatum* growing on malt extract agar was analyzed by scanning electron microscopy (**A**) and transmission electron microsopy (**B**), the *arrowhead* points to the subapical region. Treatment with 1,3-β-glucanase (**C**) did not cause morphological changes compared to control hyphae. In contrast, treatment with chitinase alone (**D**) or in combination with 1,3-β-glucanase caused marked swelling of the hyphal tip. *Bars* 10 μm. (After Arlorio et al. 1992)

verse set of plants have been reported to show in vitro antifungal activity (Broekaert et al. 1988; Roberts and Selitrennikoff 1988; Jacobsen et al. 1990; Leah et al. 1991; Verburg and Huynh 1991; Melchers et al. 1994).

The functional importance of the chitin-binding domain for antifungal activity is further supported by the fact that a number of proteins containing this domain such as hevein, stinging nettle lectin, and some peptides purified from amaranth seeds (Peumans and van Damme 1995) have been shown to strongly inhibit fungal growth. These chitin-binding proteins lack detectable levels of chitinase, chitosanase, N-acetylglucosaminidase, or 1,3-β-glucanase activities. In contrast to chitinase, inhibition of fungal growth by the stinging nettle lectin or hevein is not connected to swelling of hyphal tips but rather to the formation of thick budding hyphae by an unknown mechanism. However, other lectins such as wheat germ agglutinin, which contains four hevein domains in tandem, lack any antifungal activity (Schlumbaum et al. 1986; Raikhel et al. 1993). In contrast, some chitinases lacking this domain, such as barley chitinase C or a mutated tobacco class I chitinase, have been demonstrated to inhibit *Trichoderma viride* (Jacobsen et al. 1993; Iseli et al. 1993); therefore, the hevein domain appears not to be absolutely required for antifungal activity.

Additional chitin-binding proteins, belonging to the PR-4 family, were isolated from grains and infected leaves of barley and shown to inhibit growth of *Trichoderma harzianum* at concentrations of 10 μg ml^{-1} (Hejgaard et al. 1992). The proteins did not exhibit chitinase, lysozyme, or chitosanase activity. The pathogen- and wound-inducible chitin-binding protein, CBP20, exhibits antifungal activity towards *Trichoderma viride* and *Fusarium solani* by causing lysis of germ tubes and/or growth inhibition and acts synergistically with class I chitinase and 1,3-β-glucanase (Ponstein et al. 1994).

In vitro antifungal activity was also found to be associated with members of the rather diverse family of PR-5 proteins. The tobacco osmotin protein, AP24, strongly affects *Phytophthora infestans* by causing lysis of sporangia and severely inhibiting hyphal growth (Woloshuk et al. 1991). The related viroid-induced tomato protein, P23, showed antifungal activity against *Trichothecium roseum*, *Fusarium oxysporum* f.sp. *lycopersici*, and *Phytophthora citrophthora* (Rodrigo et al. 1993).

In contrast to osmotin, purified PR-5 of tobacco was not inhibitory towards *P. infestans* (Woloshuk et al. 1991). Thaumatin- or osmotin-like proteins are also found as constituents of plant seeds. Two immunochemically distinct proteins, R and S, purified from barley grains, inhibited growth of *Trichoderma viride* and *Candida albicans* (Hejgaard et al. 1991). Both proteins showed a synergistic effect in inhibiting fungal growth. Zeamatin, from corn, and similar proteins from seeds of other plants (oats, sorghum, wheat) inhibit *Candida albicans*, *Trichoderma reesei*, and *Neurospora crassa* (Vigers et al. 1991). All of these proteins apparently exert their antifungal activity by permeabilizing the plasma membrane of the fungi, hence they are called permatins, although they differ in their specificity for different fungal species (Vigers et al. 1992). Osmotin was found to be particularly effective in inhibiting the growth of *C. albicans*, *N. crassa*, and *T. reesei*, while tobacco PR-S showed no detectable activity against these fungi, but proved the most potent antifungal protein against the plant pathogen *Cercospora beticola* (Vigers et al. 1992).

Thionins have been shown to exhibit toxic effects on bacteria, fungi, yeast, animal, and plant cells (Garcia-Olmedo et al. 1992; Bohlmann 1994). The pathogen-induced leaf thionins from barley were specifically shown to inhibit two phytopathogenic fungi, *Thielaviopsis paradoxa*, a pathogen of sugar cane, and *Pyrenophora* (*Drechslera*) *teres*, a pathogen of barley, when tested in an agar diffusion assay (Bohlmann et al. 1988).

Because of their antimicrobial activity, numerous seed proteins have been purified from a variety of plants, including the four o'clock plant (*Mirabilis jalapa*), amaranth, and radish (Broekaert et al. 1995; Peumans and van Damme 1995). They all inhibit growth of a broad spectrum of fungal pathogens. The ribosome-inactivating protein (RIP) purified from barley seeds was demonstrated to inhibit the growth of fungi such as *Trichoderma reesei* and *Fusarium sporotrichioides*, and the inhibition was enhanced in the presence of chitinase and 1,3-β-glucanase (Leah et al. 1991). Finally, various molecular forms of stress-induced chitosanase from diverse plant species have the capacity to hydrolyze spores of different pathogens such as *Fusarium*, *Verticillium*, and *Ophiostoma* (Grenier and Asselin 1990).

In summary, various PR proteins tested individually or in combination with other PR proteins show in vitro antimicrobial activities. Although suggestive, such experiments do not prove that these proteins fulfill a similar biological function in vivo. One obvious problem with these results is a lack of correlation between pathogenicity and inhibition by PR proteins. Thus, frequently used test organisms such as *Trichoderma viride* which show inhibition are not pathogens, while recognized plant pathogens often show no inhibition. In addition, the spatial distribution and subcellular localization of particular PR proteins in plant/pathogen interactions is not taken into account, and thus it is difficult to assess whether a pathogen invading a plant is at any time in actual contact with critical concentrations of PR proteins. Finally, since PR proteins obviously differ in their biological function, only the combination of several proteins may result in effective pathogen inhibition.

B. Generation of Plant Defense Signals

Apart from direct antimicrobial activity, another proposed defense function, particularly of the hydrolytic enzymes 1,3-β-glucanase and chitinase, involves the release or degradation of chemical signals, so-called elicitors, from microbial cell walls (Chaps. 4, 5, this Vol.). Both chitinases and 1,3-β-glucanases have been shown to partially degrade isolated fungal cell walls (Boller 1988; Mauch et al. 1988b) and often the enzymatically released glucan and chitin fragments have been demonstrated to serve as active elicitors of other plant-defense responses. For example, 1,3-β-glucanases purified from soybean were able to release glucan elicitors of phytoalexin (glyceollin) accumulation from isolated cell walls of *Phytophthora megasperma* (Takeuchi et al. 1990; Ham et al. 1991). Soluble fragments produced from chitin or mycelial walls of the fungus *Chaetomium globosum* were able to induce phenylalanine ammonia-lyase (PAL) activity and accumulation of phenolic acid synthesis and lignification (Kurosaki et al. 1988). Similarly, chitin oligomers also induced rapid lignification in wheat leaves (Barber et al. 1989), chitinase activity in melon plants (Roby et al. 1987a), or synthesis of diterpene phytoalexins (momilactones, oryzalexins) in rice (*Oryza sativa*) cell cultures (Ren and West 1992).

Although diverse roles for oligosaccharides in the function of plants are well established (Fry et al. 1993), it remains to be proven that the observed patterns of enzyme induction account for the generation of signals which trigger plant defense. From the few systems which have been studied in detail, it appears that other defense responses are activated much more rapidly than 1,3-β-glucanase and chitinase (Kombrink and Somssich 1995). These rapid responses include the stimulation of general phenylpropanoid metabolism and induction of intracellular PR proteins (Somssich et al. 1989; Nicholson and Hammerschmidt 1992; Kombrink et al. 1993a), callose formation (Kauss 1987), the oxidative burst and oxidative cross-linking of cell wall proteins (Mehdy 1994; Tenhaken et al. 1995), cytoplasmic rearrangements (Freytag et al. 1994), and the very rapid alterations in membrane functions proposed to be part of the signaling events leading to defense gene activation (Atkinson 1993; Felix et al. 1993; Nürnberger et al. 1994). Therefore, a participation of the pathogen-induced 1,3-β-glucanases and chitinases in signaling events appears unlikely. Nevertheless, such function cannot be dismissed for the rather low levels of constitutively expressed enzymes, such as a specific chitinase, which is restricted to epidermal cells of potato leaves and stems (Kombrink et al. 1993a).

C. Expression Patterns in Infected Plants

In contrast to the antifungal activity demonstrated for some PR proteins in vitro, the participation of these proteins in restricting pathogen growth and development in planta is difficult to prove. In this respect, correlative evidence has been obtained from a number of plant/pathogen interactions in which race/cultivar-specific resistance is expressed. For instance, inoculation of two near-isogenic cultivars of bean with two races of *Colletotrichum lindemuthianum* resulted in more rapid increases of 1,3-β-glucanase and chitinase activities and mRNA amounts in incompatible than in compatible interactions (Hedrick et al. 1988; Daugrois et al. 1990). Similar differences were observed in oilseed rape cultivars upon infection with *Phoma lingam* (Rasmussen et al. 1992b), or in bean upon inoculation with virulent or avirulent races of the bacterial pathogen *Pseudomonas syringae* pv. *phaseolicola* (Voisey and Slusarenko 1989). The most detailed results have been obtained with tomato infected with *Cladosporium fulvum*. Several PR proteins, including chitinase, 1,3-β-glucanase, P4 and P6 of the PR-1 protein family, and P2 of the PR-4 family, are all induced more rapidly in incompatible interactions when compared to compatible ones (van Kan et al. 1992; Danhash et al. 1993; Chap. 14, this Vol.). Finally, inoculation of barley with *Rhynchosporium secalis* resulted in rapid accumulation of the transcript for a thaumatin-like protein (PR-5 family) only in incompatible interactions of those cultivars carrying the corresponding resistance gene to the appropriate fungal race (Hahn et al. 1993).

In contrast to results suggesting a causal relationship between PR protein induction and disease resistance, numerous other studies have failed to demonstrate such a correlation. Examples for this are the induction of 1,3-β-glucanase and chitinase activities in pea, potato, and chickpea cultivars upon inoculation with *Fusarium solani* (Mauch et al. 1988a), *Phytophthora infestans* (Schröder et al. 1992), and *Ascochyta rabiei* (Vogelsang and Barz 1993), respectively.

Since an understanding of specific plant defense responses requires that their expression within individual plant tissues be related to an invading pathogen, studies on enzymatic activities and transcript concentrations in whole extracts of infected plant organs are inadequate. Tissue-specific expression of defense responses has been studied by quantifying various defense responses, enzyme activities, and mRNA amounts, after localized tissue sampling (Bell et al. 1986; Schröder et al. 1992; Meier et al. 1993), by in situ localization of proteins and mRNA in sections of infected tissues (Somssich et al. 1988; Schmelzer et al. 1989; Taylor et al. 1990; Benhamou 1993; Kombrink et al. 1993a), or by histochemical analysis of the products of reporter genes driven by promoters of defense-related genes in transgenic plants (Castresana et al. 1990; Roby et al. 1990; Samac and Shah 1991; Hennig et al. 1993; Martini et al. 1993; Warner et al. 1993). In general, these studies show that the patterns of expression and localization of PR proteins and transcripts differ greatly. Induction can vary from rapid to slow and from local to systemic (Fig. 2). Proteins can accumulate intra- and extracellularly, and even organ- and cell-type-specific expression of distinct isoforms has been observed, e.g., restriction to epidermal cells of specific chitinases and 1,3-β-glucanases (Keefe et al. 1990; Mauch et al. 1992; Kombrink et al. 1993a). However, despite the wealth of accumulating information, a causal relationship be-

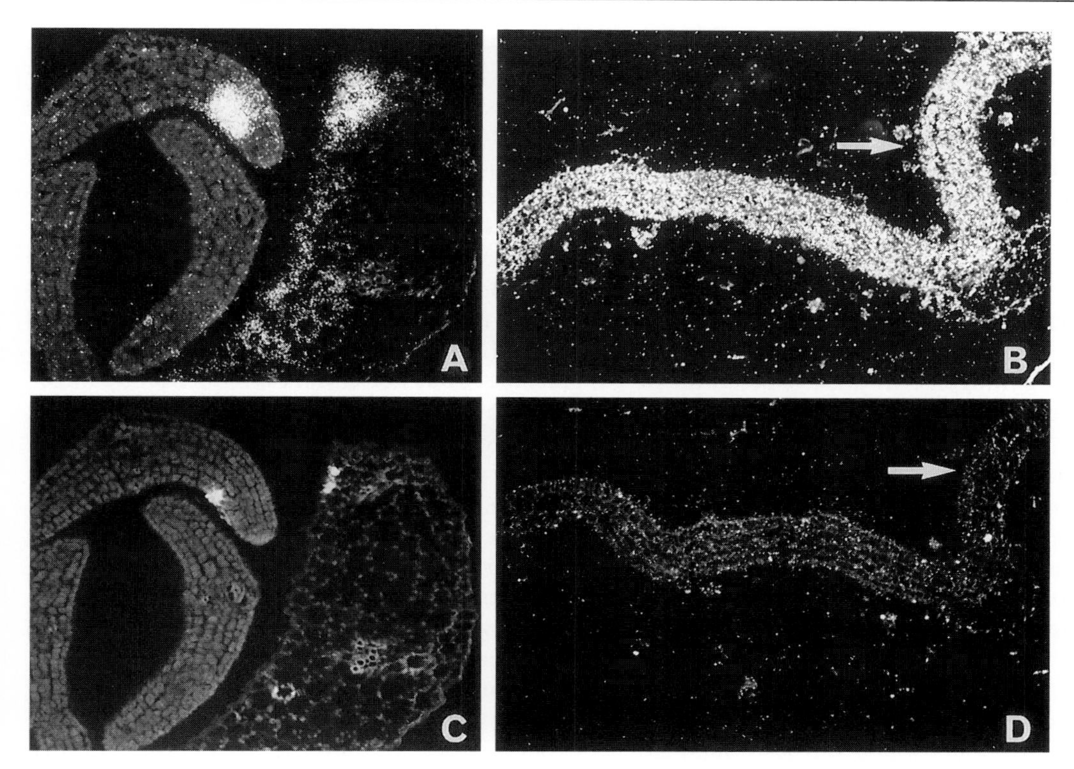

Fig. 2A–D. In situ localization of defense gene transcripts. Local accumulation of PR-10 mRNA at infection sites in parsley leaves inoculated with *Phytophthora megasperma* for 12 h (**A**). Two infection sites were identified by visualizing autofluorescing cells under UV-epifluorescent light (**C**). Systemic accumulation of 1,3-β-glucanase mRNA in potato leaves inoculated with *Phytophthora infestans* for 24 h (**B**). Serial sections were probed with either antisense (**A**, **B**) or, as control, sense RNA transcripts (**D**). *Arrows* point to an infection site. (**A** and **C** kindly provided by Elmon Schmelzer, Köln)

tween resistance and PR proteins cannot be deduced from such experiments.

D. Inhibition of Pathogen Growth in Vivo: Transgenic Plants Expressing PR Genes

The cloning of numerous PR genes enables us now to express them individually in transgenic plants and thereby to assess the role of their protein products in disease resistance in vivo. Although the number of such experiments is still too small to allow general conclusions, evidence has been obtained demonstrating the involvement of certain PR proteins in disease resistance.

The first experiment of this kind involved the constitutive expression of the tobacco PR-1a, -1b, and PR-S (PR-5 family) genes in tobacco under the control of the cauliflower mosaic virus (CaMV) 35S promoter which resulted in no apparent alteration in the susceptibility of the plants towards TMV or alfalfa mosaic virus (Cutt et al. 1989; Linthorst et al. 1989). No reduction in the number or size of local necrotic lesions was observed. When leaf-eating insects, *Spodoptera exigua* and *Heliothis virescens*, were kept during their larval stages on a diet of tobacco leaves constitutively expressing the tobacco PR-1a or PR-S genes, no detrimental effects on development, morphology, or reproduction were observed compared to the control group (Linthorst 1991). In contrast, constitutive high-level expression of the tobacco PR-la increased the tolerance of transgenic tobacco plants to infection by two oomycotan pathogens, *Peronospora tabacina*, the causal agent of blue mold disease and *Phytophthora parasitica* var. *nicotianae*, the causal agent of black shank disease (Alexander et al. 1993). Compared to control plants the reduction of blue mold disease symptoms varied from 27 to 42%, depending on whether homozygote or heterozygote PR-1a transgenic lines were analyzed. Transgenic tobacco lines expressing other PR genes, including different types of 1,3-β-glucanase and chitinase, showed no increased resistance to blue mold or black shank (Alexander et al. 1993). In accordance with previous investigations, no increased resistance to TMV, potato virus Y, *Cercospora nicotianae*, or *Pseudomonas syringae* pv. *tabaci* was observed.

Chitinases and 1,3-β-glucanases have been prime candidates for transformation experiments since single proteins were expected to confer resistance to pathogens. Indeed, seedlings of transgenic tobacco and canola (*Brassica napus*) constitutively expressing a bean class I chitinase gene under the control of the CaMV 35S promoter showed an increased ability to survive in soil infested with the fungal pathogen *Rhizoctonia solani*, an endemic, chitinous, soilborne fungus that is able to infect numerous plants (Broglie et al. 1991). When transgenic plants were challenged with *Pythium aphanidermatum*, a pathogen which lacks chitin in its cell wall, no difference in survival was detected compared to control plants. Protection against *Rhizoctonia solani* was also obtained in transgenic tobacco by transformation with a chimeric construct combining the CaMV 35S promoter with the bacterial chitinase gene ChiA derived from a selected strain of the soil bacterium *Serratia marcescens*, which among 203 different strains was the most efficient biocontrol agent for the fungal pathogen *Sclerotium rolfsii* (Ordentlich et al. 1988; Logemann et al. 1993). In contrast, high-level expression of a tobacco class I chitinase gene in *Nicotiana sylvestris* did not alter the susceptibility of the plants to *Cercospora nicotianae*, despite the fact that the fungus is sensitive to chitinases in vitro and the chitinase level was increased 120-fold in the transgenic plants (Neuhaus et al. 1991a). Lack of protection towards *Cercospora nicotianae* was also reported with transgenic tobacco (*Nicotiana benthamiana*) plants constitutively expressing the acidic class III chitinase SE2 from sugar beet (Nielsen et al. 1993). When class I chitinase and class I 1,3-β-glucanase genes were expressed simultaneously in tomato plants, partial resistance to *Fusarium oxysporum* f.sp. *lycopersici* could be obtained, whereas transgenic plants expressing either of the genes alone were not protected against infection (van den Elzen et al. 1993). Likewise, constitutive co-expression of chitinase and 1,3-β-glucanase genes yielded transgenic tobacco lines with enhanced resistance to *Cercospora nicotianae* (Zhu et al. 1994). These results are consistent with the observation that chitinases and 1,3-β-glucanases synergistically inhibit fungal growth in vitro, and provide the first experimental support that such synergy can contribute to enhanced fungal resistance in planta.

Expression of an antisense tobacco class I 1,3-β-glucanase construct in *Nicotiana sylvestris* suppressed the homologous 1,3-β-glucanases of class I, but not those of class II, and did not lead to increased susceptibility to *Cercospora nicotianae* infection (Neuhaus et al. 1992). In these plants 1,3-β-glucanase, but not chitinase induction by infection or by ethylene treatment, was efficiently blocked. Since the plants developed normally and were fully fertile, it was concluded that expression of the basic 1,3-β-glucanase is not necessary for "housekeeping" functions in the plant and that it is not important in the successful defense against this particular pathogen.

Transgenic plants with enhanced disease resistance have also been obtained by expression of different PR protein encoding genes. Tobacco plants containing a transcriptional fusion between the barley type I ribosome-inactivating protein (RIP) cDNA and the promoter of the wound- and pathogen-inducible potato *wun1* gene exhibited increased protection, as compared to untransformed plants, when grown in soil that was inoculated with *Rhizoctonia solani* mycelium (Logemann et al. 1993). Again, coexpression of different antifungal proteins (chitinase, 1,3-β-glucanase, RIP) resulted in enhanced quantitative resistance of transgenic tobacco against fungal attack (Jach et al. 1995). Expression of a barley endosperm α-thionin gene under the control of the CaMV 35S promoter conferred enhanced resistance to the bacterial plant pathogens *Pseudomonas syringae* pv. *tabaci* and *P. syringae* pv. *syringae* (Carmona et al. 1993). In different transgenic lines, the level of protection correlated with the level of thionin expression. Transgenic potato plants, expressing low levels of phage T4 lysozyme showed reduced maceration by *Erwinia carotovora atroseptica* of wounded tubers and increased sprouting capacity of inoculated tuber pieces when grown in soil (Düring et al. 1993). Expression of osmotin in potato plants resulted in delayed development of disease symptoms upon inoculation with *Phytophthora infestans* (Liu et al. 1994), confirming the observed in vitro inhibition of the fungus by osmotin. In contrast, transgenic potato plants expressing the STH-2 gene (PR-10 type) showed no altered susceptibility to a virulent race of *P. infestans* or to potato virus X (Constabel et al. 1993).

Despite the need for further investigations, some conclusions can already by drawn from such experiments. Individual PR proteins may play an important role in deterring certain pathogens, whereas they play only a minor or indeed no role at all in other plant/pathogen interactions. Therefore, future experiments should employ numerous diverse fungal and bacterial assay strains and viruses. In addition, since local and systemic resistance (see Sect. E, SAR) very likely involves multigene functions, there is a need for more studies in which combinations of two or more PR gene functions are assayed for in individual transgenic plants.

E. Systemic Acquired Resistance (SAR)

In addition to specific, genetically determined resistance, many plants also display nonspecific "immunity" to subsequent infection after initial inoculation by a necrotizing pathogen or upon treatment with certain chemicals. Induced resistance or systemically acquired resistance (SAR) are probably the most appropriate terms to refer to this well-documented phenomenon (Gianinazzi 1984; Kessmann et al. 1994; Ryals et al. 1994). Although SAR has been described in many plant species, its biochemical and molecular basis has been most extensively studied in tobacco, cucumber, and *Arabidopsis* (Kessmann et al. 1994). Typically, induced resistance is effective against a broad range of pathogens and can last for several weeks to months, or even for the whole life of annual plants. Furthermore, induced systemic resistance is graft-transmissible from root stock to scion, demonstrating that a mobile component is involved (Gianinazzi 1984; Kessmann et al. 1994; Ryals et al. 1995). From such studies the idea emerged that induced resistance mechanisms might provide new strategies for crop protection by (1) activation of natural disease resistance mechanisms, (2) sensitizing plants by application of new, inexpensive, harmless chemical compounds (Kessmann et al. 1994), or (3) the development of transgenic plants that constitutively express molecular components of induced resistance (Lamb et al. 1992; Lawton et al. 1993; Kessmann et al. 1994).

In tobacco, a close relationship between the accumulation of PR proteins induced by different infections or treatment with chemicals, such as polyacrylic acid and acetyl salicylic acid, and the subsequent level of resistance to TMV has been observed (Gianinazzi 1984). A further significant step towards demonstrating a possible implication of PR proteins in induced resistance has been made by the discovery that the reciprocal hybrids *Nicotiana glutinosa* × *Nicotiana debneyi* possess high constitutive levels of PR proteins and a much higher level of resistance to viral, bacterial, and fungal diseases than either parental species (Gianinazzi 1984; Ahl Goy et al. 1992). In addition to chitinase, 1,3-β-glucanase, peroxidase, and other PR proteins, also high constitutive levels of two phenolic compounds with antifungal activity, scopolin, and scopoletin, as well as an inhibitor of viral replication (IVR), were detected in the hybrid, and it is very likely that they all contribute to its high disease resistance (Ahl Goy et al. 1993).

Recent studies have shown that in tobacco a set of nine distinct classes of mRNAs encoding PR proteins are coordinately and systemically induced concomitantly with the onset of SAR following preinoculation with TMV (Ward et al. 1991). The same set of mRNAs was found to be coordinately induced after the exogenous application of two different chemicals: salicylic acid (SA), a putative resistance-inducing endogenous signal, and methyl-2,6-dichloroisonicotinic acid (INA), a synthetic regulator of induced resistance (Kessmann et al. 1994). From these and similar studies, it became apparent that in tobacco only the genes encoding acidic PR proteins (PR-1, PR-2, PR-3, PR-4, PR-5) strongly respond to SA and INA and are systemically expressed following local infection (Brederode et al. 1991; Ward et al. 1991). In contrast, the genes for the basic isoforms are highly induced by ethephon (ethylene), a treatment that does not lead to SAR, suggesting that the basic PR proteins are not involved in systemic acquired resistance to TMV (Brederode et al. 1991).

In cucumber, induced resistance can be triggered by viral, bacterial, and fungal preinoculation as well as treatment with SA and INA, and it has been correlated with a variety of biochemical defense responses, including increased levels of lignification, peroxidase, hydroxyproline-rich glycoproteins, chitinase, lipoxygenase, and SA itself (Hammerschmidt 1993; Kessmann et al. 1994).

SAR in *Arabidopsis* can be induced by treatment with chemicals (SA, INA) as well as with biological agents such as the necrotizing pathogen turnip crinkle virus (TCV) or *Pseudomonas syringae* (Uknes et al. 1993). Resistance is active against viral (TCV), bacterial (*Pseudomonas syringae*) and fungal (*Peronospora parasitica*) pathogens, and results in both a decrease in visible disease symptoms and a reduction in growth of the challenge pathogen. At least four PR proteins, related to PR-1, PR-2, and PR-5 of tobacco, have been identified, and their accumulation suggested to be responsible for restriction of the challenging pathogen (Uknes et al. 1993). *Arabidopsis* should prove an extremely useful experimental system allowing the genetic dissection of these resistance mechanisms. It should also allow us to determine the relative importance of single PR proteins in the resistance response of plants to particular types of pathogens as well as to unravel single components of the signal transduction pathways controlling resistance and PR gene expression.

IV. Regulation of PR Gene Expression

A. Coordinate Induction by Pathogens and Other Stimuli

Early experiments on the identification and characterization of PR proteins revealed that their synthesis and accumulation occurs in a coordinate fashion. For numerous cases it has been established that induction is regulated mainly at the transcriptional level (Somssich 1994). Coordinate induction is observed not only in response to different types of pathogens, but also to elicitors, plant growth regulators such as ethylene, auxin and cytokinin, metabolic compounds such as SA, arachidonic acid, methyl jasmonate (MeJa), as well as several other chemicals, including acetyl salicylic acid, benzoic acid, and metal ions (Cutt and Klessig 1992; Farmer and Ryan 1992; Raskin 1992; Staswick 1992; see Chap. 5, this Vol. for further information).

B. Signals Involved in PR Gene Expression

Ethylene is produced in most plant tissues in response to stress, including infection, mechanical wounding, or treatment with elicitors or chemicals (Boller 1988; Chap. 9, this Vol.). Infection of tobacco leaves with TMV resulted in a burst of ethylene which coincided with the hypersensitive response (HR). Treatment of leaves with exogenous ethylene, by application of ethephon, which decomposes into ethylene and phosphoric acid, resulted in induction or PR proteins and increased resistance (Enyedi et al. 1992; Lawton et al. 1994b). This suggested that ethylene is the natural mediator of PR protein induction, at least in tobacco. Further evidence for the involvement of ethylene was obtained by using supposedly specific inhibitors of ethylene production or action, although the existence of an ethylene-independent pathway of PR protein induction was also inferred from these experiments (Raz and Fluhr 1992). The contrary conclusion was drawn from similar experiments with pea pods, namely that ethylene is a symptom, not a signal, in the induction of a specific chitinase (Mauch et al. 1984). More recent experiments using norbornadien, an inhibitor of ethylene action, and ethylene-response mutants of *Arabidopsis* also indicate that ethylene is neither directly involved in signal transduction leading to PR-1 accumulation during the HR, nor does it cause salicylic acid accumulation (Silverman et al. 1993; Lawton et al. 1994b).

Despite the fact that only a limited number of investigations were performed in which the expression of all or at least many of the PR genes was analyzed simultaneously, the picture emerges that ethylene preferentially induces only the basic PR proteins, whereas the acidic isoforms, the classical PR proteins of tobacco, are preferentially induced by SA (Brederode et al. 1991; Linthorst 1991). Interestingly, more than 20 years ago it had already been demonstrated by White (1979) that SA, as well as benzoic acid and acetyl salicylic acid, induced resistance in tobacco (Gianinazzi 1984), a phenomenon which was later correlated with the accumulation of PR proteins, or more specifically, the coordinate induction of at least nine PR gene families (Ward et al. 1991).

It has now been shown that endogenous SA levels increase 20- to 180-fold in pathogen-infected tobacco, cucumber, and *Arabidopsis* plants (Hammerschmidt 1993; Uknes et al. 1993), an increase sufficient for the induction of PR proteins (Yalpani et al. 1993). Several reports supporting the role of SA in signaling PR gene expression and induction of SAR have been published (Enyedi et al. 1992; Raskin 1992; Shulaev et al. 1995). Particularly compelling evidence came from the analysis of transgenic tobacco plants expressing a bacterial salicylate hydroxylase, an enzyme converting SA to catechol and thus abolishing endogenous SA levels, in which resistance (SAR) can no longer be induced (Gaffney et al. 1993). However, other experiments suggest that SA is not the systemically transported signal molecule itself but that other signals must be involved (Hammerschmidt 1993; Vernooij et al. 1994; Ryals et al. 1995). Finally, other molecules such as MeJa, an 18-amino acid polypeptide named systemin, and active oxygen species such as H_2O_2 have been strongly implicated to function as local and systemic endogenous signal molecules in the expression of particular PR proteins (Enyedi et al. 1992; Hammerschmidt 1993; Chen et al. 1995; Kombrink and Somssich 1995). In view of the recent advances in identifying components of the signal transduction process in plants, further progress in this area of research can be expected within the next few years (Chap. 5, this Vol.).

C. Regulatory Elements in PR Gene Promoters

In nearly all cases studied to date, regulation of PR protein biosynthesis has been shown to occur primarily at the level of gene activation (Cutt and Klessig 1992; Somssich 1994). This implies that molecular analyses of the PR genes should reveal important *cis*-regulatory DNA elements involved in the transcriptional activation of PR genes as well as in modulating their expression. Such studies should provide extremely useful information not only for understanding the basic mechanisms involved in gene control but also for devising selective strategies aimed at inducing local and systemic resistance in agronomically important plants. Identification of important regulatory DNA elements is providing us with the means by which interacting transcription factors can be isolated, thereby identifying final components of participating signal transduction pathways. As delineation of the regulatory elements involved in PR gene expression is still at an early stage, no general conclusions on possible common mechanisms can be drawn (Somssich 1994). In most studies, rather large segments of the respective PR gene promoters fused to appropriate reporter genes were analyzed, utilizing transgenic plants or plant protoplasts. Only in a very few cases have the regions been sufficiently narrowed down to attempt pinpointing the *cis*-regulatory motifs (Matton et al. 1993; Meller et al. 1993; Somssich 1994). Even less is known on the *trans*-acting factors associated with such elements. DNA/protein interactions have been detected using gel mobility-shift assays and by in vivo DNA footprinting within the tobacco PR-1a (Hagiwara et al. 1993), the tobacco PRb-1b (Meller et al. 1993), the tobacco glucanase GLB (Hart et al. 1993), the tobacco chitinase CHN50 (Fukuda and Shinshi 1994), the potato STH-2 (Matton et al. 1993), the parsley PR1-1, and the parsley PR2 promoters (Meier et al. 1991; Korfhage et al. 1994). However, to date, only one report on the cloning of a putative transcription factor exists (Korfhage et al. 1994).

V. Conclusions and Prospects

Plant defense to pathogens comprises a highly sophisticated and complex pattern of constitutive and inducible reactions, often differentially regulated in a temporal, spatial, cell-type-, organ-, development- and species-specific manner. Although still mostly correlative, the increasing amount of experimental data strongly suggests that the various types of pathogenesis-related (PR) proteins represent an important part of the general, active defense response that has apparently evolved in all plant species. These studies, however, also indicate that the contribution of PR proteins to disease resistance can vary considerably, depending on the plant species and specific pathogen employed. Thus, a proper assessment of their function in future work will require appropriate bioassays including a broad range of different pathogens. In addition, the obvious synergistic effect of PR proteins must be further investigated. Exploitation of such synergism by expression of varying combinations of PR genes in transgenic plants may be an extremely useful, perhaps the only successful, strategy leading to a long-term and relatively stable improvement in crop protection. Furthermore, elucidation of the regulatory mechanisms involved in pathogen-induced expression of PR genes will undoubtedly be of utmost importance both for a basic understanding at the molecular level and for designing safe and efficient strategies to increase plant resistance by means of genetic engineering techniques. Use of inducible PR gene promoters or identified *cis*-regulatory elements thereof could allow limited and specific expression of antimicrobial compounds confined only to plant tissue within the vicinity of infection sites. Identification of the signal compounds involved in the induction of both local and systemic resistance and determination of their mode of action may help to design or use cheap and nontoxic substances that can serve to activate the plant's own defense response and thereby limit infections in the field without the use of toxic agrichemicals. In summary, the new insights gained by the continued study of the role of PR proteins both in defense and during developmental processes should prove profitable for basic as well as for applied scientific research.

Acknowledgments. We would like to thank our many colleagues who willingly contributed unpublished data for this chapter. Special thanks go to Prof. Klaus Hahlbrock for continuous support of our own work and helpful comments on the manuscript.

References

Ahl Goy P, Felix G, Métraux JP, Meins F Jr (1992) Resistance to disease in the hybrid *Nicotiana glutinosa* × *Nicotiana debneyi* is associated with high constitutive levels of β-1,3-glucanase, chitinase, peroxidase and polyphenoloxidase. Physiol Mol Plant Pathol 41:11–21

Ahl Goy P, Signer H, Reist R, Aichholz R, Blum W, Schmidt E, Kessmann H (1993) Accumulation of scopoletin is associated with the high disease resistance of the hybrid *Nicotiana glutinosa* x *Nicotiana debneyi*. Planta 191:200–206

Alexander D, Goodman RM, Gut-Rella M, Glascock C, KW, Friedrich L, Maddox D, Ahl-Goy P, Luntz T, Ward E, Ryals J (1993) Increased tolerance to two oomycete pathogens in transgenic tobacco expressing pathogenesis-related protein la. Proc Natl Acad Sci USA 90:7327–7331

Arlorio M, Ludwig A, Boller T, Bonfante P (1992) Inhibition of fungal growth by plant chitinases and β-1,3-glucanases. A morphological study. Protoplasma 171: 34–43

Atkinson MM (1993) Molecular mechanisms of pathogen recognition by plants. Adv Plant Pathol 10:35–64

Barber MS, Bertram RE, Ride JP (1989) Chitin oligosaccharides elicit lignification in wounded wheat leaves. Physiol Mol Plant 34:3–12

Beerhues L, Kombrink E (1994) Primary structure and expression of mRNAs encoding basic chitinase and 1,3-β-glucanase in potato. Plant Mol Biol 24:353–367

Bell JN, Ryder TB, Wingate VPM, Bailey JA, Lamb CJ (1986) Differential accumulation of plant defense gene transcripts in a compatible and an incompatible plant-pathogen interaction. Mol Cell Biol 6:1615–1623

Benhamou N (1993) Spatio-temporal regulation of defence genes: immunohistochemistry. In: Fritig B, Legrand M (eds) Mechanisms of plant defense responses. Kluwer, Dordrecht, pp 221–235

Bernasconi P, Locher R, Pilet PE, Jollès J, Jollès P (1987) Purification and N-terminal amino-acid sequence of a basic lysozyme from *Parthenocissus quinquifolia* cultured in vitro. Biochim Biophys Acta 915:254–260

Bohlmann H (1994) The role of thionins in plant protection. Crit Rev Plant Sci 13:1–16

Bohlmann H, Clausen S, Behnke S, Giese H, Hiller C, Reimann-Philipp U, Schrader G, Barkholt V, Apel K (1988) Leaf-specific thionins of barley – a novel class of cell wall proteins toxic to plant pathogenic fungi and possibly involved in the defense mechanism of plants. EMBO J 7:1559–1565

Boller T (1988) Ethylene and the regulation of antifungal hydrolases in plants. In: Miflin BJ (ed) Oxford surveys of plant molecular and cell biology, vol 5. Oxford University Press, Oxford, pp 145–174

Brederode FT, Linthorst HJM, Bol JF (1991) Differential induction of acquired resistance and PR gene expression in tobacco by virus infection, ethephon treatment, UV light and wounding. Plant Mol Biol 17:1117–1125

Breiteneder H, Pettenburger K, Bito A, Valenta R, Kraft D, Rumpold H, Schreiner O, Breitenbach M (1989) The gene coding for the major birch pollen allergen *BetvI*, is highly homologous to a pea disease resistance response gene. EMBO J 8:1935–1938

Broekaert W, Lee H-I, Kush A, Chua N-H, Raikhel N (1990) Wound-induced accumulation of mRNA containing a hevein sequence in laticifers of rubber tree (*Hevea brasiliensis*). Proc Natl Acad Sci USA 87:7633–7637

Broekaert WF, van Parijs J, Allen AK, Peumans WJ (1988) Comparison of some molecular, enzymatic and antifungal properties of chitinases from thorn-apple, tobacco and wheat. Physiol Mol Plant Pathol 33:319–331

Broekaert WF, van Parijs J, Leyns F, Joos H, Peumans WJ (1989) A chitin-binding lectin from stinging nettle rhizomes with antifungal properties. Science 245:1100–1102

Broekaert WF, Terras FRG, Cammue BPA, Osborn RW (1995) Plant defensins: novel antimicrobial peptides as components of the host defense system. Plant Physiol 108:1353–1358

Broglie K, Chet I, Holliday M, Cressman R, Biddle P, Knowlton S, Mauvais CJ, Broglie R (1991) Transgenic plants with enhanced resistance to the fungal pathogen *Rhizoctonia solani*. Science 254:1194–1197

Bryngelsson T, Gréen B (1989) Characterization of a pathogenesis-related, thaumatin-like protein isolated from barley challenged with an incompatible race of mildew. Physiol Mol Plant Pathol 35:45–52

Bryngelsson T, Sommer-Knudsen J, Gregersen PL, Collinge DB, Ek B, Thordal-Christensen H (1994) Purification, characterization, and molecular cloning of basic PR-1-type pathogenesis-related proteins from barley. Mol Plant-Microbe Interact 7:265–275

Cammue BPA, de Bolle MFC, Terras FRG, Proost P, van Damme J, Rees SB, JV, Broekaert WF (1992) Isolation and characterization of a novel class of plant antimicrobial peptides from *Mirabilis jalapa* L. seeds. J Biol Chem 267:2228–2233

Carmona MJ, Molina A, Fernández JA, López-Fando JJ, Garcia-Olmedo F (1993) Expression of the α-thionin gene from barley in tobacco confers enhanced resistance to bacterial pathogens. Plant J 3:457–462

Casacuberta JM, Puigdomènech P, San Segundo B (1991) A gene coding for a basic pathogenesis-related (PR-like) protein from *Zea mays*. Molecular cloning and induction by a fungus (*Fusarium moniliforme*) in germinating maize seeds. Plant Mol Biol 16:527–536

Castresana C, de Carvalho F, Gheysen G, Habets M, Inzé D, van Montagu M (1990) Tissue-specific and pathogen-induced regulation of a *Nicotiana plumbaginifolia* β-1,3-glucanase gene. Plant Cell 2:1131–1143

Chen Z, Malamy J, Henning, J, Conrath U, Sánchez-Casas P, Silva H, Ricigliano J, Klessig DF (1995) Induction, modification, and transduction of the salicylic acid signal in plant defense responses. Proc Natl Acad Sci USA 92:4134–4137

Collinge DB, Kragh KM, Mikkelsen JD, Nielsen KK, Rasmussen U, Vad K (1993) Plant chitinases. Plant J 3:31–40

Constabel CP, Bertrand C, Brisson N (1993) Transgenic potato plants overexpressing the pathogenesis-related STH-2 gene show unaltered susceptibility to *Phytophthora infestans* and potato virus X. Plant Mol Biol 22:775–782

Cornelissen BJC, Hooft van Huijsduijnen RAM, Bol JF (1986) A tobacco mosaic virus-induced tobacco protein is homologous to the sweet-tasting protein thaumatin. Nature 321:531–532

Cutt JR, Klessig DF (1992) Pathogenesis-related proteins. In: Boller T, Meins F Jr (eds) Genes involved in plant defense. Springer, Vienna New York, pp 209–243

Cutt JR, Harpster MH, Dixon DC, Carr JP, Dunsmuir P, Klessig DF (1989) Disease response to tobacco mosaic virus in transgenic tobacco plants that constitutively express the pathogenesis-related PR1b gene. Virology 173:89–97

Danhash N, Wagemakers CAM, van Kan JAL, de Wit PJGM (1993) Molecular characterization of four chitinase cDNAs obtained from *Cladosporium fulvum*-infected tomato. Plant Mol Biol 22:1017–1029

Daugrois JH, Lafitte C, Barthe JP, Touze A (1990) Induction of β-1,3-glucanase and chitinase activity in compatible and incompatible interactions between *Colletotrichum lindemuthianum* and bean cultivars. J Phytopathol 130:225–234

Dixon RA, Harrison MJ (1990) Activation, structure, and organization of genes involved in microbial defense in plants. Adv Genet 28:165–234

Domingo C, Conejero V, Vera P (1994) Genes encoding acidic and basic class III β-1,3-glucanases are expressed in tomato plants upon viroid infection. Plant Mol Biol 24:725–732

Düring K, Porsch P, Fladung M, Lörz H (1993) Transgenic potato plants resistant to the phytopathogenic bacterium *Erwinia carotovora*. Plant J 3:587–598

Enyedi AJ, Yalpani N, Silverman P, Raskin I (1992) Signal molecules in systemic plant resistance to pathogens and pests. Cell 70:879–886

Eyal Y, Sagee O, Fluhr R (1992) Dark-induced accumulation of a basic pathogenesis-related (PR-1) transcript and a light requirement for its induction by ethylene. Plant Mol Biol 19:589–599

Farmer EE, Ryan CA (1992) Octadecanoid precursors of jasmonic acid activate the synthesis of wound-inducible proteinase inhibitors. Plant Cell 4:129–134

Farmer EE, Johnson RR, Ryan CA (1992) Regulation of expression of proteinase inhibitor genes by methyl jasmonate and jasmonic acid. Plant Physiol 98:995–1002

Felix G, Regenass M, Boller T (1993) Specific perception of subnanomolar concentrations of chitin fragments by tomato cells: induction of extracellular alkalinization, changes in protein phosphorylation, and establishment of a refractory state. Plant J 4:307–316

Fischer W, Christ U, Baumgartner M, Erismann KH, Mösinger E (1989) Pathogenesis-related proteins of tomato: II. Biochemical and immunological characterization. Physiol Mol Plant Pathol 35:67–83

Freytag S, Arabatzis N, Hahlbrock K, Schmelzer E (1994) Reversible cytoplasmic rearrangements precede wall apposition, hypersensitive cell death and defense-related gene activation in potato/*Phytophthora infestans* interactions. Planta 194:123–135

Friedrich L, Moyer M, Ward E, Ryals J (1991) Pathogenesis-related protein 4 is structurally homologous to the carboxy-terminal domains of hevein, Win-1 and Win-2. Mol Gen Genet 230:113–119

Fristensky B, Horovitz D, Hadwiger LA (1988) cDNA sequences for pea disease resistance response genes. Plant Mol Biol 11:713–715

Fry SC, Aldington S, Hetherington PR, Aitken J (1993) Oligosaccharides as signals and substrates in the plant cell wall. Plant Physiol 103:1–5

Fukuda Y, Shinshi H (1994) Characterization of a novel *cis*-acting element that is responsive to a fungal elicitor in the promoter of a tobacco class I chitinase gene. Plant Mol Biol 24:485–493

Gaffney T, Friedrich L, Vernooij B, Negrotto D, Nye G, Uknes S, Ward E, Kessmann H, Ryals J (1993) Requirement of salicylic acid for the induction of systemic acquired resistance. Science 261:754–756

Garcia-Olmedo F, Carmona MJ, Lopez-Fando JJ, Fernandez JA, Castagnaro A, Molina A, Hernandez-Lucas C, Carbonero P (1992) Characterization and analysis of thionin genes. In: Boller T, Meins F Jr (eds) Genes involved in plant defense. Springer, Vienna New York, pp 283–302

Geoffroy P, Legrand M, Fritig B (1990) Isolation and characterization of a proteinaceous inhibitor of microbial proteinases induced during the hypersensitive reaction of tobacco to tobacco mosaic virus. Mol Plant-Microbe Interact 3:327–333

Gianinazzi S (1984) Genetic and molecular aspects of resistance induced by infections or chemicals. In: Kosuge T, Nester EW (eds) Plant-microbe interactions. Molecular and genetic berspectives, vol 1. Macmillan, New York, pp 321–342

Grenier J, Asselin A (1990) Some pathogenesis-related proteins are chitosanases with lytic activity against fungal spores. Mol Plant-Microbe Interact 3:401–407

Hagiwara H, Matsuoka M, Ohshima M, Watanabe M, Hosokawa D, Ohashi Y (1993) Sequence-specific binding of factors to two independent promoter regions of the acidic tobacco pathogenesis-related-1 protein (PR-1). Mol Gen Genet 240:197–205

Hahn K, Strittmatter G (1994) Pathogen-defence gene *prp1-1* from potato encodes an auxin-responsive glutathion *S*-transferease. Eur J Biochem 226:619–626

Hahn M, Jüngling S, Knogge W (1993) Cultivar-specific elicitation of barley defense reactions by the phytotoxic peptide NIP1 from *Rhynchosporium secalis*. Mol Plant-Microbe Interact 6:745–754

Ham K-S, Kauffmann S, Albersheim P, Darvill AG (1991) Host-pathogen interactions XXXIX. A soybean pathogenesis-related protein with β-1,3-glucanase activity releases phytoalexin elicitor-active heat-stable fragments from fungal walls. Mol Plant-Microbe Interact 4:545–552

Hammerschmidt R (1993) The nature and generation of systemic signals induced by pathogens, arthropod herbivores, and wounds. Adv Plant Pathol 10:307–337

Hart CM, Nagy F, Meins F Jr (1993) A 61 bp enhancer element of the tobacco β-1,3-glucanase B gene interacts with a regulated nuclear protein(s). Plant Mol Biol 21:121–131

Hedrick SA, Bell JN, Boller T, Lamb CJ (1988) Chitinase cDNA cloning and mRNA induction by fungal elicitor, wounding, and infection. Plant Physiol 86:182–186

Hejgaard J, Jacobsen S, Svendsen I (1991) Two antifungal thaumatin-like proteins from barley grain. FEBS Lett 291:127–131

Hejgaard J, Jacobsen S, Bjørn SE, Kragh KM (1992) Antifungal activity of chitin-binding PR-4 type proteins from barley grain and stressed leaf. FEBS Lett 307:389–392

Hennig J, Dewey RE, Cutt JR, Klessig DF (1993) Pathogen, salicylic acid and developmental dependent expression of a β-1,3-glucanase/GUS gene fusion in transgenic tobacco plants. Plant J 4:481–493

Huynh QK, Borgmeyer JR, Zobel JF (1992a) Isolation and characterization of a 22 kDa protein with antifungal properties from maize seeds. Biochem Biophys Res Commun 182:1–5

Huynh QK, Hironaka CM, Lemine DB, Smith CE, Borgmeyer IR, Shah DM (1992b) Antifungal proteins from plants. Purification, molecular cloning and antifungal properties of chitinases from maize seeds. J Biol Chem 267:6635–6640

Iseli B, Boller T, Neuhaus J-M (1993) The N-terminal cysteine-rich domain of tobacco class I chitinase is essential for chitin binding but not for catalytic or antifungal activity. Plant Physiol 103:221–226

Jach G, Görnhardt B, Mundy J, Logemann J, Pinsdorf E, Leah R, Schell J, Maas C (1995) Enhanced quantitative resistance against fungal disease by combinatorial expression of different barley antifungal proteins in transgenic tobacco. Plant J 8:97–109

Jacobsen S, Mikkelsen JD, Hejgaard J (1990) Characterization of two antifungal endochitinases from barley grain. Physiol Plant 79:554–562

Jeckel PA, Hartmann BH, Beintema JJ (1991) The primary structure of hevamine, an enzyme with lysozyme/chitinase activity from *Hevea basiliensis* latex. Eur J Biochem 200:123–130

Joosten MHAJ, de Wit PJGM (1989) Identification of several pathogenesis-related proteins in tomato leaves inoculated with *Cladosporium fulvum* (syn. *Fulvia fulva*) as 1,3-β-glucanases and chitinases. Plant Physiol 89:945–951

Joosten MHAJ, Bergmans CJB, Meulenhoff EJS, Cornelissen BJC, de Wit PJGM (1990) Purification and serological characterization of three basic 15-kilodalton pathogenesis-related proteins from tomato. Plant Physiol 94:585–591

Kauffmann S, Legrand M, Geoffroy P, Fritig B (1987) Biological function of "pathogenesis-related" proteins: four PR proteins of tobacco have 1,3-β-glucanase activity. EMBO J 6:3209–3212

Kauss H (1987) Some aspects of calcium-dependent regulation in plant metabolism. Annu Rev Plant Physiol 38:47–72

Keefe D, Hinz U, Meins F Jr (1990) The effect of ethylene on the cell-type-specific and intracellular localization of β-1,3-glucanase and chitinase in tobacco leaves. Planta 182:43–51

Kessmann H, Staub T, Hofmann C, Maetzke T, Herzog J, Ward E, Uknes S, Ryals J (1994) Induction of systemic acquired resistance in plants by chemicals. Annu Rev Phytopathol 32:439–459

Kirsch C, Hahlbrock K, Kombrink E (1993) Purification and characterization of extracellular, acidic chitinase isoenzymes from elicitor-stimulated parsley cells. Eur J Biochem 213:419–425

Kombrink E, Somssich IE (1995) Defense responses of plants to pathogens. In: Andrews JH, Tommerup IC (eds) Advances in botanical research, vol 21. Academic Press, London, pp 1–34

Kombrink E, Schröder M, Hahlbrock K (1988) Several "pathogenesis-related" proteins in potato are 1,3-β-glucanases and chitinases. Proc Natl Acad Sci USA 85:782–786

Kombrink E, Beerhues L, Garcia-Garcia F, Hahlbrock K, Müller M, Schröder M, Witte B, Schmelzer E (1993a) Expression patterns of defense-related genes in infected and uninfected plants. In: Fritig B, Legrand M (eds) Mechanisms of plant defense responses, vol 2. Kluwer, Dordrecht, pp 236–249

Kombrink E, Hahlbrock K, Kirsch C, Meyer R, Witte B (1993b) Properties and expression patterns of chitinases in potato and parsley. In: Muzzarelli RAA (ed) Chitin enzymology. European Chitin Society, Lyon, pp 245–256

Korfhage U, Trezzini GF, Meier I, Hahlbrock K, Somssich IE (1994) Plant homeodomain protein involved in transcriptional regulation of a pathogen defense-related gene. Plant Cell 6:695–708

Kragh KM, Jacobsen S, Mikkelsen JD, Nielsen KA (1993) Tissue specificity and induction of class I, class II and class III chitinases in barley (*Hordeum vulgare*). Physiol Plant 89:490–498

Kurosaki F, Tashiro N, Nishi A (1988) Role of chitinase and chitin oligosaccharides in lignification response of cultured carrot cells treated with mycelial walls. Plant Cell Physiol 29:527–531

Kurosaki F, Tashiro N, Gamou R, Nishi A (1989) Chitinase isoenzymes induced in carrot cell culture by treatment with ethylene. Phytochemistry 28:2989–2992

Lamb CJ, Ryals JA, Ward ER, Dixon RA (1992) Emerging strategies for enhancing crop resistance to microbial pathogens. Bio/Technology 10:1436–1445

Lawton K, Ward E, Payne G, Moyer M, Ryals J (1992) Acidic and basic class III chitinase mRNA accumulation in response to TMV infection of tobacco. Plant Mol Biol 19:735–743

Lawton K, Uknes S, Friedrich L, Gaffney T, Alexander D, Goodman R, Metraux JP, Kessmann H, Ahl Goy P, Gut Rella M, Ward E, Ryals J (1993) The molecular biology of systemic acquired resistance. In: Fritig B, Legrand M (eds) Mechanisms of plant defense responses. Kluwer, Dordrecht, pp 422–432

Lawton KA, Beck J, Potter S, Ward E, Ryals J (1994a) Regulation of cucumber class III chitinase gene expression. Mol Plant-Microbe Interact 7:48–57

Lawton KA, Potter SL, Uknes S, Ryals J (1994b) Acquired resistance signal transduction in *Arabidopsis* is ethylene independent. Plant Cell 6:581–588

Leah R, Tommerup H, Svendsen I, Mundy J (1991) Biochemical and molecular characterization of three barley seed proteins with antifungal properties. J Biol Chem 266:1564–1573

Legrand M, Kauffmann S, Geoffroy P, Fritig B (1987) Biological function of pathogenesis-related proteins: four tobacco pathogenesis-related proteins are chitinases. Proc Natl Acad Sci USA 84:6750–6754

Linthorst HJM (1991) Pathogenesis-related proteins of plants. Crit Rev Plant Sci 10:123–150
Linthorst HJM, Meuwissen RLJ, Kauffmann S, Bol JF (1989) Constitutive Expression of pathogenesis-related proteins PR-1, GRP, and PR-S in tobacco has no effect on virus infection. Plant Cell 1:285–291
Linthorst HJM, Danhash N, Brederode FT, van Kan JAL, de Wit PJGM, Bol JF (1991) Tobacco and tomato PR proteins homologous to *win* and pro-hevein lack the "hevein" domain. Mol Plant-Microbe Interact 4:586–592
Linthorst HJM, Brederode FT, van der Does C, Bol JF (1993) Tobacco proteinase inhibitor I genes are locally, but not systemically induced by stress. Plant Mol Biol 21:985–992
Liu D, Raghothama KG, Hasegawa PM, Bressan RA (1994) Osmotin overexpression in potato delays development of disease symptoms. Proc Natl Acad Sci USA 91:1888–1892
Logemann J, Jach G, Logemann S, Leah R, Wolf G, Mundy J, Oppenheim A, Chet I, Schell J (1993) Expression of a ribosome-inhibiting protein (RIP) or a bacterial chitinase leads to fungal resistance in transgenic plants. In: Fritig B, Legrand M (eds) Mechanisms of plant defense responses. Kluwer, Dordrecht, pp 446–448
Lucas J, Camacho Henriquez A, Lottspeich F, Henschen A, Sänger HL (1985) Amino acid sequence of the "pathogenesis-related" leaf protein p14 from viroid-infected tomato reveals a new type of structurally unfamiliar proteins. EMBO J 4:2745–2749
Margis-Pinheiro M, Metz-Boutigue MH, Awade A, de Tapia M, le Ret M, Burkard G (1991) Isolation of a complementary DNA encoding the bean PR4 chitinase: an acidic enzyme with an amino-terminus cysteine-rich domain. Plant Mol Biol 17:243–253
Margis-Pinheiro M, Martin C, Didierjean L, Burkard G (1993) Differential expression of bean chitinase genes by virus infection, chemical treatment and UV light. Plant Mol Biol 22:659–668
Martini N, Egen M, Rüntz I, Strittmatter G (1993) Promoter sequences of a potato pathogenesis-related gene mediate transcriptional activation selectively upon fungal infection. Mol Gen Genet 236:179–186
Matton DP, Brisson N (1989) Cloning, expression, and sequence conservation of pathogenesis-related gene transcripts of potato. Mol Plant-Microbe Interact 2:325–331
Matton DP, Prescott G, Bertrand C, Camirand A, Brisson N (1993) Identification of *cis*-acting elements involved in the regulation of the 17 kDa pathogenesis-related gene STH-2 in potato. Plant Mol Biol 22:279–291
Mauch F, Hadwiger LA, Boller T (1984) Ethylene: symptom, not signal for the induction of chitinase and β-1,3-glucanase in pea pods by pathogens and elicitors. Plant Physiol 76:607–611
Mauch F, Hadwiger LA, Boller T (1988a) Antifungal hydrolases in pea tissue. I. Purification and characterization of two chitinases and two 1,3-β-glucanases differentially regulated during development and in response to fungal infection. Plant Physiol 87:325–333
Mauch F, Mauch-Mani B, Boller T (1988b) Antifungal hydrolases in pea tissue. II. Inhibition of fungal growth by combinations of chitinase and 1,3-β-glucanase. Plant Physiol 88:936–942
Mauch F, Meehl JB, Staehelin LA (1992) Ethylene-induced chitinase and β-1,3-glucanase accumulate specifically in the lower epidermis and along vascular strands of bean leaves. Planta 186:367–375
Mehdy MC (1994) Active oxygen species in plant defense against pathogens. Plant Physiol 105:467–472
Meier BM, Shaw N, Slusarenko AJ (1993) Spatial and temporal accumulation of defense gene transcripts in bean (*Phaseolus vulgaris*) leaves in relation to bacteria-induced hypersensitive cell death. Mol Plant-Microbe Interact 6:453–466
Meier I, Hahlbrock K, Somssich IE (1991) Elicitor-inducible and constitutive in vivo DNA footprints indicate novel *cis*-acting elements in the promoter of a parsley gene encoding pathogenesis-related protein 1. Plant Cell 3:309–315
Meins F Jr, Neuhaus J-M, Sperisen C, Ryals J (1992) The primary structure of plant pathogenesis-related glucanohydrolases and their genes. In: Boller T, Meins F Jr (eds) Genes involved in plant defense. Springer, Vienna New York, pp 245–282
Melchers LS, Sela-Buurlage MB, Vloemans SA, Woloshuk CP, van Roekel JSC, Pen J, van den Elzen PJM, Cornelissen BJC (1993) Extracellular targeting of the vacuolar tobacco proteins AP24, chitinase and β-1,3-glucanase in transgenic tobacco. Plant Mol Biol 21:583–593
Melchers LS, Apotheker-de Groot M, van der Knaap JA, Ponstein AS, Sela-Buurlage MB, Bol JF, Cornelissen BJC, van den Elzen PJM, Linthorst HJM (1994) A new class of tobacco chitinase homologous to bacterial exo-chitinases displays antifungal activity. Plant J 5:469–480
Meller Y, Sessa G, Eyal Y, Fluhr R (1993) DNA-protein interactions on a *cis*-DNA element essential for ethylene regulation. Plant Mol Biol 23:453–463
Metraux JP, Burkhart W, Moyer M, Dincher S, Middlesteadt W, Williams S, Payne G, Carnes M, Ryals J (1989) Isolation of a complementary DNA encoding a chitinase with structural homology to a bifunctional lysozyme/chitinase. Proc Natl Acad Sci USA 86:896–900
Mikkelsen JD, Berglund L, Nielsen KK, Christiansen H, Bojsen K (1992) Structure of endochitinase genes from sugar beets. In: Brine CJ, Sandford PA, Zikakis JP (eds) Advances in chitin and chitosan. Elsevier, New York, pp 344–352
Moiseyev GP, Beintema JJ, Fedoreyeva LI, Yakovlev GI (1994) High sequence similarity between a ribonuclease from *ginseng* calluses and fungus-elicited proteins from parsley indicates that intracellular pathogenesis-related proteins are ribonucleases. Planta 193:470–472
Neuhaus J-M, Ahl-Goy P, Hinz U, Flores S, Meins F Jr (1991a) High-level expression of a tobacco chitinase gene in *Nicotiana sylvestris*. Susceptibility of transgenic plants to *Cercospora nicotianae* infection. Plant Mol Biol 16:141–151
Neuhaus J-M, Sticher L, Meins F Jr, Boller T (1991b) A short C-terminal sequence is necessary and sufficient for the targeting of chitinases to the plant vacuole. Proc Natl Acad Sci USA 88:10362–10366
Neuhaus J-M, Flores S, Keefe D, Ahl-Goy P, Meins F Jr (1992) The function of vacuolar β-1,3-glucanase investigated by antisense transformation. Susceptibility of transgenic *Nicotiana sylvestris* plants to *Cercospora nicotianae* infection. Plant Mol Biol 19:803–813
Nicholson RL, Hammerschmidt R (1992) Phenolic compounds and their role in disease resistance. Annu Rev Phytopathol 30:369–389

Nielsen KK, Mikkelsen JD, Kragh KM, Bojsen K (1993) An acidic class III chitinase in sugar beet: induction by *Cercospora beticola*, characterization, and expression in transgenic tobacco plants. Mol Plant-Microbe Interact 6:495–506

Nielsen KK, Bojsen K, Roepstorff P, Mikkelsen JD (1994) A hydroxyproline-containing class IV chitinase of sugar beet is glycosylated with xylose. Plant Mol Biol 25:241–257

Nürnberger T, Nennstiel D, Jabs T, Sacks WR, Hahlbrock K, Scheel D (1994) High affinity binding of a fungal oligopeptide elicitor to parsley plasma membranes triggers multiple defense responses. Cell 78:449–460

Ordentlich A, Elad Y, Chet I (1988) The role of chitinase of *Serratia marcescens* in biocontrol of *Sclerotium rolfsii*. Phytopathology 78:84–87

Osswald WF, Shapiro JP, McDonald RE, Niedz RP, Mayer RT (1993) Some citrus chitinases also possess chitosanase activities. Experientia 49:888–892

Pautot V, Holzer FM, Walling LL (1991) Differential expression of tomato proteinase inhibitor I and II genes during bacterial pathogen invasion and wounding. Mol Plant-Microbe Interact 4:284–292

Payne G, Ward E, Gaffney T, Ahl Goy P, Moyer M, Harper A, Meins F Jr, Ryals J (1990) Evidence for a third structural class of β-1,3-glucanase in tobacco. Plant Mol Biol 15:797–808

Peumans WJ, van Damme EJM (1995) Lectins as plant defense proteins. Plant Physiol 109:347–352

Pierpoint WS, Jackson PJ, Evans RM (1990) The presence of a thaumatin-like protein, a chitinase and a glucanase among the pathogenesis-related proteins of potato (*Solanum tuberosum*). Physiol Mol Plant Pathol 36:325–338

Ponstein AS, Bres-Vloemans SA, Sela-Buurlage MB, van den Elzen PJM, Melchers LS, Cornelissen BJC (1994) A novel pathogen- and wound-inducible tobacco (*Nicotiana tabacum*) protein with antifungal activity. Plant Physiol 104:109–118

Potter S, Uknes S, Lawton K, Winter AM, Chandler D, DiMaio J, Novitzki R, Ward E, Ryals J (1993) Regulation of a hevein-like gene in *Arabidopsis*. Mol Plant-Microbe Interact 6:680–685

Raikhel NV, Lee H-I, Broekaert WF (1993) Structure and function of chitin-binding proteins. Annu Rev Plant Physiol Plant Mol Biol 44:591–615

Raskin I (1992) Salicylate, a new plant hormone. Plant Physiol 99:799–803

Rasmussen U, Bojsen K, Collinge DB (1992a) Cloning and characterization of a pathogen-induced chitinase in *Brassica napus*. Plant Mol Biol 20:277–287

Rasmussen U, Giese H, Mikkelsen JD (1992b) Induction and purification of chitinase in *Brassica napus* L. ssp. *oleifera* infected with *Phoma lingam*. Planta 187:328–334

Raz V, Fluhr R (1992) Calcium requirement for ethylene-dependent responses. Plant Cell 4:1123–1130

Ren Y-Y, West CA (1992) Elicitation of diterpene biosynthesis in rice (*Oryza sativa* L.) by chitin. Plant Physiol 99:1169–1178

Richardson M (1991) Seed storage proteins: the enzyme inhibitors. In: Rogers LJ (ed) Methods in plant biochemistry, vol 5: Amino Acids, Proteins and Nucleic Acids. Academic Press, London, pp 259–305

Roberts WK, Selitrennikoff CP (1988) Plant and bacterial chitinases differ in antifungal activity. J Gen Microbiol 134:169–176

Roby D, Esquerre-Tugaye M-T (1987) Purification and some properties of chitinases from melon plants infected by *Colletotrichum lagenarium*. Carbohydr Res 165:93–104

Roby D, Gadelle A, Toppan A (1987a) Chitin oligosaccharides as elicitors of chitinase activity in melon plants. Biochem Biophys Res Commun 143:885–892

Roby D, Toppan A, Esquerré-Tugayé M-T (1987b) Cell surfaces in plant micro-organism interactions. VIII. Increased proteinase inhibitor activity in melon plants in response to infection by *Colletotrichum lagenarium* or to treatment with an elicitor fraction from this fungus. Physiol Mol Plant Pathol 30:453–460

Roby D, Broglie C, Cressman R, Biddle P, Chet I, Broglie R (1990) Activation of a bean chitinase promoter in transgenic tobacco plants by phytopathogenic fungi. Plant Cell 2:999–1007

Rodrigo I, Vera P, Tornero P, Hernández-Yago J, Conejero V (1993) cDNA cloning of viroid-induced tomato pathogenesis-related protein P23. Characterization as a vacuolar antifungal factor. Plant Physiol 102:939–945

Ryals J, Uknes S, Ward E (1994) Systemic acquired resistance. Plant Physiol 104:1109–1112

Ryals J, Lawton KA, Delaney TP, Friedrich L, Kessmann H, Neuenschwander U, Uknes S, Vernooij B, Weymann K (1995) Signal transduction in systemic acquired resistance. Proc Natl Acad Sci USA 92:4202–4205

Ryan CA (1990) Protease inhibitors in plants: genes for improving defense against insects and pathogens. Annu Rev Phytopathol 28:425–449

Samac DA, Shah DM (1991) Developmental and pathogen-induced activation of the *Arabidopsis* acidic chitinase promoter. Plant Cell 3:1063–1072

Samac DA, Hironaka CM, Yallaly PE, Shah DM (1990) Isolation and characterization of the genes encoding basic and acidic chitinase in *Arabidopsis thaliana*. Plant Physiol 93:907–914

Schlumbaum A, Mauch F, Vögeli U, Boller T (1986) Plant chitinases are potent inhibitors of fungal growth. Nature 324:365–367

Schmelzer E, Krüger-Lebus S, Hahlbrock K (1989) Temporal and spatial patterns of gene expression around sites of attempted fungal infection in parsley leaves. Plant Cell 1:993–1001

Schröder M, Hahlbrock K, Kombrink E (1992) Temporal and spatial patterns of 1,3-β-glucanase and chitinase induction in potato leaves infected by *Phytophthora infestans*. Plant J 2:161–172

Segura A, Moreno M, García-Olmedo F (1993) Purification and antipathogenic activity of lipid transfer proteins (LTPs) from the leaves of *Arabidopsis* and spinach. FEBS Lett 332:243–246

Sela-Buurlage MB, Ponstein AS, Bres-Vloemans SA, Melchers LS, van den Elzen PJM, Cornelissen BJC (1993) Only specific tobacco (*Nicotiana tabacum*) chitinases and β-1,3-glucanases exhibit antifungal activity. Plant Physiol 101:857–863

Sharma YK, Hinojos CM, Mehdy MC (1992) cDNA cloning, structure, and expression of a novel pathogenesis-related protein in bean. Mol Plant-Microbe Interact 5:89–95

Shulaev V, León J, Raskin I (1995) Is salicylic acid a translocated signal of systemic acquired resistance in tobacco? Plant Cell 7:1691–1701

Silverman P, Nuckles E, Ye YS, Kuc J, Raskin I (1993) Salicylic acid, ethylene, and pathogen resistance in tobacco. Mol Plant-Microbe Interact 6:775–781

Singh NK, Bracker CA, Hasegawa PM, Handa AK, Buckel S, Hermodson MA, Pfankoch E, Regnier FE, Bressan RA (1987) Characterization of osmotin. A thaumatin-like protein associated with osmotic adaptation in plant cells. Plant Physiol 85:529–536

Somssich IE (1994) Regulatory elements governing pathogenesis-related (PR) gene expression. In: Nover L (ed) Results and problems in cell differentiation, vol 20. Springer, Berlin Heidelberg New York, pp 163–179

Somssich IE, Schmelzer E, Kawalleck P, Hahlbrock K (1988) Gene structure and in situ transcript localization of pathogenesis-related protein 1 in parsley. Mol Gen Genet 213:93–98

Somssich IE, Bollmann J, Hahlbrock K, Kombrink E, Schulz W (1989) Differential early activation of defense-related genes in elicitor-treated parsley cells. Plant Mol Biol 12:227–234

Stanford A, Bevan M, Northcote D (1989) Differential expression within a family of novel wound-induced genes in potato. Mol Gen Genet 215:200–208

Staswick PE (1992) Jasmonate, genes, and fragrant signals. Plant Physiol 99:804–807

Sticher L, Hinz U, Meyer AD, Meins F Jr (1992a) Intracellular transport and processing of a tobacco vacuolar β-1,3-glucanase. Planta 188:559–565

Sticher L, Hofsteenge J, Milani A, Neuhaus J-M, Meins F Jr (1992b) Vacuolar chitinases of tobacco: a new class of hydroxyproline-containing proteins. Science 257:655–657

Stintzi A, heitz T, Prasad V, Wiedemann-Merdinoglu S, Kauffmann S, Geoffroy P, Legrand M, Fritig B (1993) Plant "pathogenesis-related" proteins and their role in defense against pathogens. Biochimie 75:687–706

Stirpe F, Barbieri L, Battelli MG, Soria M, Lappi DA (1992) Ribosome-inactivating proteins from plants: present status and future prospects. Bio/Technology 10:405–412

Takeuchi Y, Yoshikawa M, Takeba G, Tanaka K, Shibata D, Horino O (1990) Molecular cloning and ethylene induction of mRNA encoding a phytoalexin elicitor-releasing factor, β-1,3-endoglucanase, in soybean. Plant Physiol 93:673–682

Taylor JL, Fritzemeier K-H, Häuser I, Kombrink E, Rohwer F, Schröder M, Strittmatter G, Hahlbrock K (1990) Structural analysis and activation by fungal infection of a gene encoding a pathogenesis-related protein in potato. Mol Plant-Microbe Interact 3:72–77

Tenhaken R, Levine A, Brisson LF, Dixon RA, Lamb C (1995) Function of the oxidative burst in hypersensitive disease resistance. Proc Natl Acad Sci USA 92:4158–4163

Tornero P, Conejero V, Vera P (1994) A gene encoding a novel isoform of the PR-1 protein family from tomato is induced upon viroid infection. Mol Gen Genet 243:47–53

Tyson H (1992) Relationships among amino acid sequences of animal, microbial and plant peroxidases. Theor Appl Genet 84:643–655

Uknes S, Mauch-Mani B, Moyer M, Potter S, Williams S, Dincher S, Chandler D, Slusarenko A, Ward E, Ryals J (1992) Acquired resistance in *Arabidopsis*. Plant Cell 4:645–656

Uknes S, Winter AM, Delaney T, Vernooij B, Morse A, Friedrich L, Nye G, Potter S, Ward E, Ryals J (1993) Biological induction of systemic acquired resistance in *Arabidopsis*. Mol Plant-Microbe Interact 6:692–698

van de Löcht U, Meier I, hahlbrock K, Somssich IE (1990) A 125 bp promoter fragment is sufficient for strong elicitor-mediated gene activation in parsley. EMBO J 9:2945–2950

van den Elzen PJM, Jongedijk E, Melchers LS, Cornelissen BJC (1993) Virus and fungal resistance: from laboratory to field. Philos Trans R Soc Lond B 342:271–278

van Kan JAL, Joosten MHAJ, Wagemakers CAM, van den Berg-Velthuis GCM, de Wit PJGM (1992) Differential accumulation of mRNAs encoding extracellular and intracellular PR proteins in tomato induced by virulent and avirulent races of *Cladosporium fulvum*. Plant Mol Biol 20:513–527

van Loon LC, Pierpoint WS, Boller T, Conejero V (1994) Recommendations for naming plant pathogenesis-related proteins. Plant Mol Biol Rep 12:245–264

Vera P, Hernández Yago J, Conejero V (1989) Immunogold localization of the citrus exocortis viroid-induced pathogenesis-related proteinase P69 in tomato leaves. Plant Physiol 91:119–123

Verburg JG, Huynh QK (1991) Purification and characterization of an antifungal chitinase from *Arabidopsis thaliana*. Plant Physiol 95:450–455

Vernooij B, Friedrich L, Morse A, Reist R, Kolditz-Jawhar R, Ward E, Uknes S, Kessmann H, Ryals J (1994) Salicylic acid is not the translocated signal responsible for inducing systemic acquired resistance but is required in signal transduction. Plant Cell 6:959–965

Vigers AJ, Roberts WK, Selitrennikoff CP (1991) A new family of plant antifungal proteins. Mol Plant-Microbe Interact 4:315–323

Vigers AJ, Wiedemann S, Roberts WK, Legrand M, Selitrennikoff CP, Fritig B (1992) Thaumatin-like pathogenesis-related proteins are antifungal. Plant Sci 83:155–161

Vogelsang R, Barz W (1993) Purification, characterization and differential hormonal regulation of a β-1,3-glucanase and two chitinases from chickpea (*Cicer arietinum* L.). Planta 189:60–69

Voisey CR, Slusarenko AJ (1989) Chitinase mRNA and enzyme activity in *Phaseolus vulgaris* (L.) increase more rapidly in response to avirulent than to virulent cells of *Pseudomonas syringae* pv. *phaseolicola*. Physiol Mol Plant Pathol 35:403–412

Walter MH, Liu J-W, Grand C, Lamb CJ, Hess D (1990) Bean pathogenesis-related (PR) proteins deduced from elicitor-induced transcripts are members of a ubiquitous new class of conserved PR proteins including pollen allergens. Mol Gen Genet 222:353–360

Ward ER, Uknes SJ, Williams SC, Dincher SS, Wiederhold DL, Alexander DC, Ahl-Goy P, Métraux J-P, Ryals JA (1991) Coordinate gene activity in response to agents that induce systemic acquired resistance. Plant Cell 3:1085–1094

Warner SAJ, Scott R, Draper J (1992) Characterization of a wound-induced transcript from the monocot asparagus that shares similarity with a class of intracellular pathogenesis-related (PR) proteins. Plant Mol Biol 19:555–561

Warner SAJ, Scott R, Draper J (1993) Isolation of an asparagus intracellular *PR* gene (*AoPR1*) wound-responsive promoter by the inverse polymerase chain reaction and its characterization in transgenic tobacco. Plant J 3:191–201

White RF (1979) Acetylsalicylic acid (aspirin) induces resistance to tobacco mosaic virus in tobacco. Virology 99:410–412

Woloshuk CP, Meulenhoff JS, Sela-Buurlage M, van den Elzen PJM, Cornelissen BJC (1991) Pathogen-induced proteins with inhibitory activity toward *Phytophthora infestans*. Plant Cell 3:619–628

Yalpani N, Shulaev V, Raskin I (1993) Endogenous salicylic acid levels correlate with accumulation of pathogenesis-related proteins and virus resistance in tobacco. Phytopathology 83:702–708

Zhu Y, Maher EA, Masoud S, Dixon RA, Lamb CJ (1994) Enhanced protection against fungal attack by constitutive co-expression of chitinase and glucanase genes in transgenic tobacco. Bio/Technology 12:807–812

7 Fungal Phytotoxins: Biosynthesis and Activity

T.M. HOHN

CONTENTS

I. Introduction

Many filamentous fungi produce compounds that adversely affect the growth and development of plants. Determining the importance of these phytotoxic compounds in plant/fungal interactions has proven to be an extremely difficult task. Although research on fungal phytotoxins spans almost a century, it is only within the past few years that their role in plant disease has been unambiguously demonstrated. This chapter will focus on recent progress in characterizing the maromolecular aspects of fungal phytotoxin biosynthesis, which, together with advances in fungal molecular genetics, have provided powerful new approaches for investigating the significance of phytotoxins in plant/fungal interactions. A number of reviews covering phytotoxin biosynthesis, isolation, structure, and mechanism of action have been published over the past 15 years (Scheffer 1983; Mitchell 1984; Knoche and Duvick 1987). The reader is directed to the excellent review by Stoessl (1981) for a comprehensive account of fungal phytotoxin biosynthesis.

Mycotoxin Research Unit, National Center for Agricultural Utilization Research, USDA/ARS, 1815 North University Street, Peoria, Illinois, 61604, USA

A. Definitions

It is important to define the term phytotoxin, which is used in a variety of ways in the literature. Some definitions of phytotoxins take into consideration the possible involvement of phytotoxins in plant disease (Scheffer 1983). For the purposes of this chapter, phytotoxin will be defined in the broadest sense so as to include fungal molecules that adversely affect the growth and development of plants, regardless of whether or not they are presently acknowledged to play a role in plant disease. Macromolecules, such as fungal proteins and polysaccharides, that are either toxic to plants or thought to play a role in plant diseases are excluded from consideration. Unfortunately, the term phytotoxin is also used to describe toxic compounds produced by plants; however, these compounds are more frequently described as plant toxins. A further source of confusion arises from the fact that the definition of phytotoxin employed here is not parallel to the definition of another group of fungal toxins known as mycotoxins. Mycotoxins are narrowly defined as fungal molecules that are toxic to vertebrates. Because phytotoxins are frequently active against a wide spectrum of organisms, many phytotoxins are also mycotoxins.

Some fungal phytotoxins appear to have a very narrow range of activity with respect to plants. The terms host-specific toxin or host-selective toxin (HST) are used to describe phytotoxins that are typically active against a single host

The Mycota V Part A
Plant Relationships
Carroll/Tudzynski (Eds.)

plant or host plant genotype. This definition of phytotoxin host specificity is problematic, since it encourages the use of parallel terms such as host-nonspecific toxin to describe phytotoxins with a less restricted host range. While the degree of host specificity can vary dramatically among fungal phytotoxins, it is unlikely that any are truly "host-nonspecific," therefore this terminology is inaccurate and should be avoided. This chapter will be concerned primarily with those phytotoxins that are not typically classified as HSTs.

In discussing the role a phytotoxin plays in a specific plant/fungal interaction, it is also important to define the terms virulence and pathogenicity. Virulence is a quantitative measure of disease produced by a pathogen on a specific host, while pathogenicity refers to the ability of an organism to cause disease (Yoder 1980). Some HSTs are thought to be pathogenicity factors since they are required for disease to occur in the susceptible host. However, most phytotoxins involved in plant disease probably act to increase the severity or amount of disease that occurs and are best characterized as virulence factors.

B. Historical Perspective

Studies of fungal phytotoxins began as part of efforts to determine if fungal toxins were involved in plant diseases caused by fungi. Until recently, difficulties associated with their isolation and structural characterization presented major obstacles to investigations of phytotoxins. Progress in the area of phytotoxin research has resulted from the efforts of both plant pathologists to understand the role of phytotoxins in plant disease and of chemists to characterize the structure and biosynthetic pathways of natural products. Over the past 50 years, fungal phytotoxin research has been dominated by work on HSTs. Much of the attention directed toward HSTs is due to the fact that they present opportunities to study phytotoxins as molecular determinants of both disease and host specificity. A significant experimental advantage for some HSTs is that the producing organism has a known perfect state and, in contrast to most other phytotoxins, HST production segregates as single gene trait. Host plant susceptibility to some HSTs also appears to be monogenic, and this has facilitated the characterization of several HST molecular targets (see Chap. 8, this Vol. for a detailed discussion of HSTs).

Studies of fungal phytotoxins not characterized as HSTs have progressed more slowly. This has occurred partly because the fungi producing these phytotoxins appear less attractive as experimental systems. In many instances, fungi producing the wider host range phytotoxins are not amenable to genetic manipulation. In addition, the contribution of these phytotoxins to virulence can be relatively small and therefore difficult to quantify.

C. Involvement in Fungal/Plant Interactions

1. Determining Phytotoxin Significance

Fungi produce numerous phytotoxic metabolites in culture media. The types and amounts of the metabolites produced frequently depend on the growth conditions provided. Many of these phytotoxic metabolites are also active against a wide range of organisms including fungi, bacteria, and insects. This wide spectrum of activity can occur either because the phytotoxin molecular target is highly conserved or because the phytotoxin mechanism of action lacks specificity. For example, phytotoxins that are specific inhibitors of protein synthesis may have a wide range of activity because many of the molecular components of protein synthesis are highly conserved among eukaryotes. Likewise, photoactive phytotoxins would be expected to have a wide range of activity since the oxygen radicals produced following the exposure of these compounds to light are nonspecific in their interactions with host molecules.

The production of metabolites that are active against a wide spectrum of organisms may provide the producing fungus with a selective advantage since these toxins can be employed against a greater number of potential competitors. However, the wide host range observed for many fungal phytotoxins complicates efforts to understand their function. Determining which fungal phytotoxins play a role in the development of a specific plant disease has proven to be difficult. Attempts to establish experimental criteria for evaluating phytotoxins as determinants of plant diseases have generally been unsuccessful. Examples of such criteria include demonstrating that phytotoxin production in planta is correlated with the appearance of disease symptoms or that purified toxin is capable of producing symptoms typical of those observed in infected plants. While these types of criteria are useful for establishing the probable involvement of a phytotoxin in pathogenesis, as pointed out by Yoder (1980), they are inconclusive. The tremendous complexity of plant/fungal interactions makes it difficult or impossible to establish cause and effect relationships for specific events.

2. Chemical Analysis

Efforts to demonstrate the involvement of fungal phytotoxins in plant disease using approaches that depend upon the chemical analysis of phytotoxins are subject to a variety of experimental difficulties, the most significant of which is probably the molecular complexity of the plant/fungus matrix. The detection of phytotoxins is often obscured by the large background of host plant compounds, some of which are produced in response to fungal invasion. Detection of phytotoxins is further compounded by the fact that toxins may be produced in infected plant tissues in exceptionally small quantities, and their concentration can be further reduced through plant metabolism. Another problem arises from the production of multiple phytotoxins by the same fungus. The presence of multiple phytotoxins can result in complex synergistic effects that make it difficult to determine the role of individual phytotoxins.

3. Genetic/Molecular Genetic Approaches

The problem of phytotoxin relevance in plant disease would appear to be inaccessible to approaches requiring the chemical analysis of phytotoxins. For this reason, recent efforts have employed genetic/molecular genetic approaches (Bronson 1991). The advantage of such approaches is that they permit the isolation or construction of fungal strains that differ only in their ability to produce a specific phytotoxin. Genetic approaches depend on the identification of mutants deficient in phytotoxin production that have an altered virulence or pathogenicity phenotype. Genetic analysis can then be used to determine if the same genes that control phytotoxin production also control the changes in mutant virulence or pathogenicity (Yoder 1980). The validity of this analysis depends on the probability that the two strains used for the genetic cross differ only in their ability to produce the phytotoxin. Its major limitation is that there is no perfect state known for many phytotoxin-producing fungi. Obtaining mutants specifically altered in phytotoxin production can also be problematic because it is usually not possible to select for phytotoxin-deficient mutants. This means that the identification of mutants requires the use of time-consuming screening procedures (Beremand 1987).

Molecular genetic approaches involve the use of molecular gene disruption. Efforts to develop transformation vectors and transform filamentous fungi have been successful in nearly all fungi where such efforts have been attempted. The integration of transforming DNA via homologous recombination, a requirement for gene disruption and gene replacement techniques, has also been demonstrated in a large number of filamentous fungi (Fincham 1989). Therefore, it is now possible to genetically alter phytotoxin production in most fungi if the genes for phytotoxin biosynthetic pathway enzymes or pathway regulatory elements are available. Genetic alterations of this type permit the recognition of phytotoxin production as a factor in disease interactions. However, this approach also is not without potential problems. As pointed out by Bronson (1991), blocking a phytotoxin pathway at its branch point could perturb fungal metabolism in other ways that affect fungal virulence. An alternative approach is to truncate the phytotoxin pathway so that a pathway intermediate, thought to have minimal bioactivity, accumulates, although it is possible that this could also affect fungal virulence in unpredictable ways.

The major limitation to the use of molecular genetic approaches is that they require structural information about phytotoxin-related genes or enzymes. Phytotoxin and mycotoxin pathway genes have been isolated from filamentous fungi by both genetic (Skory et al. 1992; Payne et al. 1993) and biochemical (Hohn and Beremand 1989; Wang et al. 1991; Scott-Craig et al. 1992; Haese et al. 1993) approaches. Despite these potential problems, the application of genetic/molecular genetic techniques has provided the most convincing evidence to date that fungal phytotoxins can make important contributions to pathogen virulence (Walton and Panaccione 1993).

II. Biosynthesis and Activity

In this section, selected phytotoxins are discussed with respect to their biosynthesis and activity. The discussion is limited to those phytotoxins (1) about which something is known concerning the macromolecular aspects of their biosynthesis, and (2) which appear likely to function in plant/fungal interactions. While the biogenesis of fungal phytotoxins as a group is extremely diverse, only three biogenetic types will be discussed. These include cyclic terpenoids, polyketides, and peptides, which constitute the majority of fungal phytotoxins that have been characterized. The use of biogenetic concepts as the basis for classifying fungal natural products has several advantages and has been frequently used for this purpose (Stoessl 1981). One benefit of biogenetic classification, which has been realized only in recent years, is that enzymes involved in the biosynthesis of biogenetically related compounds are in some cases themselves structurally related. This has important implications for studies of phytotoxin biosynthesis, since it means that knowledge of biogenetic origins can sometimes provide insights into the structure and function of key biosynthetic enzymes. The possibility of obtaining structural information for some phytotoxin pathway enzymes without resorting to biochemical methods could speed progress toward the understanding of phytotoxin pathway regulation and function.

A. Cyclic Terpenoids

1. Terpene Synthases

Of the numerous cyclic terpenoids which have been isolated from fungi, a substantial number have been identified as phytotoxic. The biosynthetic pathways of cyclic terpenoids begin with cyclization of the prenyl diphosphate intermediates present in the isoprenoid pathway. Terpenoid pathway precursors are produced by condensation reactions initiated with dimethylallyl diphosphate (DMAP) and subsequently extended in a head-to-tail fashion with isopentenyl diphosphate (IPP) units (Fig. 1) by prenyltransferases (Ashby et al. 1990). Farnesyl diphosphate (C15) is the precursor of sesquiterpenoids, which appear to be the largest group of fungal terpenoids. Many phytotoxic diterpenoids (C20) and sesterterpenoids (C25) have also been identified from fungi.

Terpenoid cyclization reactions frequently involve a great variety of molecular rearrangements that lead to the formation of hundreds of structurally distinct products (Croteau 1987; Cane 1990).

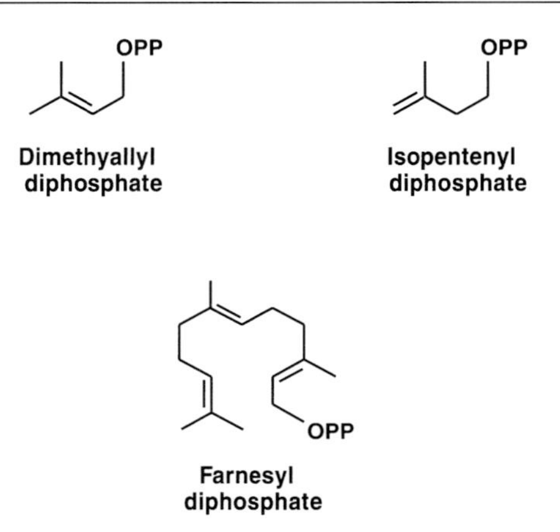

Fig. 1. Structures of the terpenoid precursors DMAP, IPP, and FPP

The enzymes that catalyze terpene cyclizations are called terpene synthases. They play a key role in terpenoid biosynthesis, since they are the first unique step in terpenoid pathways, and their products possess most of the structural information required for bioactivity. However, terpene synthase products themselves only rarely possess significant bioactivity. The phytotoxicity of fungal terpenoids is typically due to the modification of terpene synthase products by a variety of different types of oxygenations. There is limited information available on oxygenases involved in fungal terpenoid biosynthesis. Oxygen isotope labeling studies indicate that all six oxygens in the trichothecene, T2 toxin, are derived from molecular oxygen (Desjardins et al. 1986). The initial oxygenation step in fungal trichothecene biosynthesis has recently been shown to be catalyzed by a cytochrome P450 monooxygenase (Hohn et al. 1995a).

The sequences of two fungal terpene synthases involved in phytotoxin biosynthesis have been reported. Genes encoding trichodiene synthase, *Tri5* (*Tox5*), have been isolated from three different *Fusarium* species (Hohn and Beremand 1989; Hohn and Desjardins 1992; Proctor et al. 1995a) and *Myrothecium roridum* (S.C. Trapp et al., unpubl.), and a gene encoding aristolochene synthase (*Ari1*) has been isolated from *Penicillium roqueforti* (Proctor and Hohn 1993). Sequence comparisons among the trichodiene synthases revealed 70% sequence identity, but comparisons between the trichodiene synthase sequences and aristolochene synthase suggest that these polypeptides are distantly related. In addition, both of the fungal enzymes appear to be unrelated to pentalenene synthase (Cane et al. 1994), the only bacterial sesquiterpene synthase sequence that has been reported. They also appear to be unrelated to the sesquiterpene synthase sequences reported from plant sources (Facchini and Chappell 1992; Back and Chappell 1995). Interestingly, comparisons between the available sequences for the plant terpene synthases indicate that they are related (Colby et al. 1993). The sequence identity between these terpene synthases is variable, but sufficient to permit the design of degenerate primers capable of amplifying presently uncharacterized plant terpene synthase genes. Families of related terpene synthases may also exist in fungi, but sequence information for additional terpene synthase genes is required before this can be determined. Unlike the polyketide and peptide pathways discussed below, there is currently no apparent sequence-based approach for isolating genes that encode terpenoid pathway enzymes in fungi.

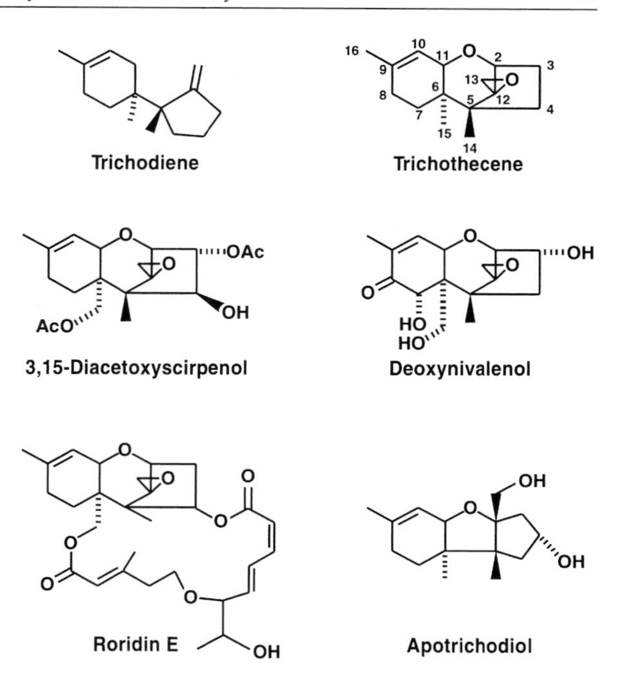

Fig. 2. Structures of trichodiene and trichodiene-derived phytotoxins

2. Trichodiene-Derived Toxins

Trichothecenes have been recognized as mycotoxins for over 30 years and are often found as contaminates of feed grains and other grain products. Because of their agricultural importance, considerable research attention has been directed toward an understanding of their biosynthesis, mechanism of action, and toxicology (Desjardins et al. 1993). Trichothecenes are produced by fungi belonging to several genera including *Fusarium*, *Myrothecium*, *Stachybotrys*, *Trichoderma*, *Trichothecium*, and *Cylindrocarpon*. Most of the trichothecene contamination found in agricultural products is due to *Fusarium* species.

The structures of over 60 trichothecenes have been reported. Two major structural types of trichothecenes are recognized. Simple trichothecenes such as diacetoxyscirpenol (DAS) and deoxynivalenol (DON) differ by the types and extent of oxygenation from the hypothetical core trichothecene structure (Fig. 2) (Desjardins et al. 1993). Complex or macrocyclic trichothecenes, such as the roridins (Fig. 2) and verrucarins, contain a macrocycle of variable structure esterified through C4 and C15 (Jarvis 1991). The production of a particular trichothecene structural type is characteristic of specific fungal genera. Simple trichothecenes are produced by *Fusarium* species, while complex trichothecenes are primarily associated with *Myrothecium* species but are also produced by members of *Stachybotrys* and *Cylindrocarpon*.

The trichothecene biosynthetic pathway has been extensively studied in several different fungi over the past 30 years. In most cases, it has been possible to identify the probable pathway intermediates, but little is known concerning pathway enzymes. Trichodiene synthase is the only pathway enzyme that has been purified and characterized (Hohn and Vanmiddlesworth 1986). Attempts to demonstrate the activities of other pathway enzymes have met with only limited success. As discussed above, the *Tri5* gene has been isolated from several sources. Recently, a single cosmid carrying the *Tri5* gene of *F. sporotrichioides* was found to contain at least three other genes involved in trichothecene biosynthesis (Hohn et al. 1993, 1995b). Two of the genes were shown to complement previously described trichothecene pathway mutants (Beremand 1987). Characterization of these closely linked genes has identified their products as a transacetylase (*Tri3*) responsible for acetylation of the 15-decalonectrin (McCormick et al. 1996), a cytochrome P450 (*Tri4*) responsible for the initial oxygenation of trichodiene (Hohn et al. 1995a), and a transcriptional activator (*Tri6*) that appears to regulate the expression of the entire trichothecene pathway

(Proctor et al. 1995b). Additional transcripts have been mapped on this cosmid but have not been characterized with respect to their role in trichothecene biosynthesis (T.M. Hohn et al., unpubl.). Cosmids carrying the *Tri5* gene have also been isolated from *M. roridum*, a macrocyclic trichothecene-producer, but it is unknown if they carry additional closely linked pathway genes (S.C. Trapp et al., unpubl.).

Trichothecenes are potent inhibitors of protein synthesis in most eukaryotic organisms. The 9,10 double bond and the 12,13 epoxide have both been identified as important for trichothecene toxicity although macrocyclic trichothecenes still retain considerable toxicity following the loss of the 12,13 epoxide (Jeker and Tamm 1988). Many trichothecenes are also potent phytotoxins that are able to exert effects such as wilting, chlorosis, and necrosis at very low concentrations (10^{-5} to 10^{-6}M) in a number of different plants (Cutler 1988). Examples of trichothecene phytotoxicity include the inhibition of protein synthesis in maize leaf disks and kernel sections by DON (Casale and Hart 1988) and the production of chlorotic and necrotic lesions by roridin E on muskmelon leaves (Kuti et al. 1989).

Because most trichothecene-producing fungi are plant pathogens, the possibility that trichothecenes may play a role in plant disease has been investigated. Several lines of evidence indicate that trichothecenes are virulence factors. The most convincing of such evidence is the demonstration that mutants lacking an intact trichothecene pathway are less virulent than their trichothecene-producing progenitors. UV-induced mutants of *F. sporotrichioides*, a T2 toxin producer, blocked at three different steps in the trichothecene pathway, exhibited different levels of virulence on parsnip root (Desjardins et al. 1989). The mutants accumulating the earliest pathway intermediates (trichodiene and diacetylcalonectrin) were significantly less virulent than a DAS-accumulating mutant. Coinoculation with the low-virulence mutants resulted in both the restoration of T2 toxin production in vitro and the partial restoration of virulence on parsnip root.

In a different experiment, mutants deficient in trichothecene biosynthesis have been constructed through molecular disruption of the *Tri5* (*Tox5*) gene in *F. sambucinum* (*Gibberella pulicaris*), a causative agent of various root and tuber dry rots (Hohn and Desjardins 1992). Of the five different *Tri5*$^-$ mutants analyzed, all were reduced in virulence on parsnip root (Desjardins et al. 1992). Genetic analysis revealed that loss of trichothecene production and reduced virulence cosegregated in crosses between a *Tri5*$^-$ mutant and an isogenic *Tri5*$^+$ strain (Fig. 3). In contrast, the virulence of *Tri5*$^-$ mutants on potato tuber slices was indistinguishable from that of the progenitor strain, suggesting that the importance of trichothecene production by a pathogen with respect to virulence can vary significantly, depending on the host. Recently, *Tri5*$^-$ mutants have also been constructed by gene disruption in *F. graminearum* (*Gibberella zeae*), an important pathogen of both maize and wheat (Proctor et al. 1995a). The virulence of two *Tri5*$^-$ mutants that no longer produced DON in either culture media or in diseased plant tissues was found to be unchanged from that of the progenitor strain for maize seedlings and detached ears. However, their virulence was reduced in wheat seedling and wheat head infec-

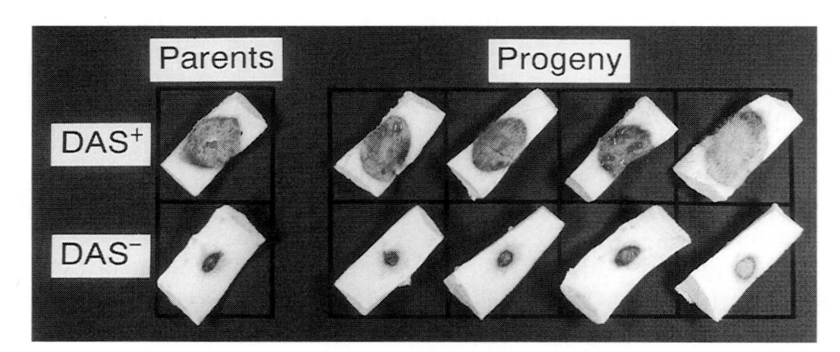

Fig. 3. Segregation of DAS (diacetoxyscirpenol) production and virulence among an eight-spored tetrad from a cross between a *Tri5*$^+$ parent and a *Tri5*$^-$ transformant. For each strain, inoculated parsnip root strips were incubated for 5 days at 25 °C in the dark. (Data and figure from Desjardins et al. 1992)

tions, again suggesting that the importance of trichothecenes in plant disease is host-specific.

Some trichothecene-producing fungi also produce low levels of trichodiene-derived metabolites that do not contain the core trichothecene structure. These metabolites are called apotrichothecenes and lack the C12, C13 epoxide that is characteristic of trichothecenes (Fig. 2). They also differ from trichothecenes by the presence of a *trans*-fused A/B ring system as opposed to the *cis*-fused system of trichothecenes (Greenhalgh et al. 1989). Although the toxicity of these compounds is considerably reduced over that of trichothecenes, they are also phytotoxic (Wang and Miller 1988). Apotrichothecenes appear to be shunt metabolites of the trichothecene pathway since they are found only in trichothecene-producing fungi. In addition, *F. sporotrichioides* mutants lacking a functional *Tri4* gene accumulate trichodiene but do not produce apotrichodiol (3a, 13-dihydroxy-epiapotrichothec-9-ene), suggesting that *Tri4* is involved in apotrichodiol biosynthesis (Hohn et al. 1995a). The importance of apotrichothecenes in fungal plant interactions is unknown. Unless the apotrichothecene biosynthetic pathway can be shown to employ enzymes not involved in trichothecene biosynthesis, it will be difficult to determine their significance in plant disease apart from that of trichothecenes.

3. Aristolochene-Derived Toxins

Sesquiterpenoids that appear to be derived from aristolochene constitute a large, widely distributed family of fungal sesquiterpenoid phytotoxins (Fig. 4). One of these, PR-toxin, is also considered to be a mycotoxin although it is found very infrequently in agricultural products. Interest in PR-toxin as a mycotoxin led to the isolation of the *Ari1* gene from *P. roqueforti* (Proctor and Hohn 1993). Because there is little evidence to indicate that *P. roqueforti* is a plant pathogen, it is doubtful that PR-toxin plays a role in plant disease. However, a number of aristolochene-derived phytotoxins are produced by fungi that are aggressive plant pathogens, and efforts have been made to determine the role of these phytotoxins in plant disease. Except for the apparent participation of aristolochene synthase, little is known concerning the biosynthesis of the aristolochene-derived toxins. It is not known if the aristolochene synthase found in *P. roqueforti* is related to other fungal aristolochene synthases. Southern blots employing stringent hy-

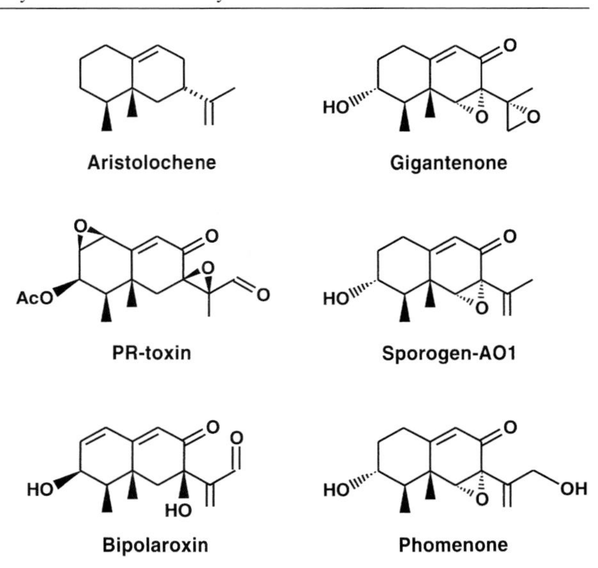

Fig. 4. Structures of aristolochene and apparent aristolochene-derived phytotoxins

bridization conditions and the *Ari1* gene as a probe failed to detect homologous sequences in fungi that produce aristolochene (Cane et al. 1987) or aristolochene-derived phytotoxins (R.H. Proctor and T.M. Hohn, unpubl.). Plants also make a number of sesquiterpenoids which are derived from sesquiterpene synthase products structurally related to aristolochene. For example, the sesquiterpenoid phytoalexin capsidiol, that is found in both pepper and tobacco, is derived from 5-epi-aristolochene (Whitehead et al. 1989).

Perhaps the most interesting of the apparent aristolochene-derived phytotoxins is gigantenone (Fig. 4) which is produced by the Bermuda grass and quack grass pathogen, *Drechslera gigantea* (Kenfield et al. 1989b). Application of gigantenone to the detached leaves of several monocot species results in the development of localized areas of chlorophyll retention called green islands, while the application of gigantenone to dicot species causes only necrotic lesions. The lesions induced by gigantenone application to the leaves of *Drechslera gigantea* host plants are similar in appearance to the lesions that result from natural infections. While it was initially thought that the green island effect caused by gigantenone was due to the stimulation of cytokinin production in plant tissues, further study revealed that this was not the case (Bunkers and Strobel 1991). The primary mode of action for gigantenone appears to be the inhibition of protein synthesis.

The phytotoxin bipolaroxin (Fig. 4) is produced by the Bermudagrass pathogen, *Bipolaris cynodonti* (Sugawara et al. 1985). This phytotoxin appears to exhibit more host selectivity than other aristolochene-derived toxins. It causes lesions on Bermudagrass at a concentration of 0.38 mM, but concentrations of 3.8 mM are required to produce detectable symptoms in other monocots such as wild oats, sugarcane, and maize. *B. cynodontis* also produces dihydrobipolaroxin, an inactive analog of bipolaroxin, that lacks the C12 aldehyde.

Other aristolochene-derived phytotoxins include phomenone and sporogen AO1 (13-deoxyphomenone; Fig. 4). Phomenone is produced by *Phoma destructiva* which causes a wilt disease of tomato. In a tomato cutting assay it was shown to cause wilting and necrosis. Sporogen AO1 (13-deoxyphomenone; Tanaka et al. 1984) was first characterized as a sporogenic substance from *Aspergillus oryzae* but is also produced by *Hansfordia pulvinata*, a hyperparasite of *Cladosporium fulvum*.

B. Polyketides

1. Polyketide Synthases

Polyketides also constitute a large group of fungal phytotoxins that can vary greatly in structure. The biosynthesis of polyketides occurs through the condensation of acetyl and malonyl coenzyme A units (Fig. 5) by either a multifunctional enzyme (multienzyme) or multienzyme complex employing a thiotemplate mechanism similar to that of fatty acid synthases (FAS; Katz and Donadio 1993). Like FASs, the enzymes responsible for the formation of the polyketide carbon skeleton, polyketide synthases (PKS), contain various catalytic domains that successively process enzyme-bound intermediates. Probably the most significant difference between FASs and PKSs is the ability of PKSs to process polyketide intermediates in a nonsystematic fashion. This results from the presence of domains within the multienzyme or multienzyme complex that catalyze selective reduction and dehydration of individual extender units (Donadio et al. 1991). In the case of multienzymes, the organization of these domains is colinear with the order in which they act upon the polyketide intermediates to yield keto, hydroxyl, enoyl, or alkyl functionalities at specific positions. It is this nonsystematic introduction of functionalities by PKSs that is responsible for much of the structural diversity observed in polyketides. The products of PKSs also frequently undergo additional modification by enzymes such as oxygenases and methylases.

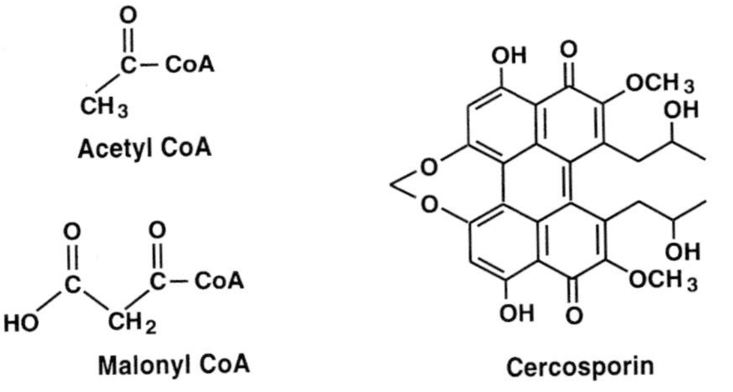

Fig. 5. Structures of the polyketide precursors acetyl and malonyl CoA, and cercosporin

The sequences of three fungal PKS genes have been reported. The gene encoding the 6-methylsalicylic acid synthase involved in patulin biosynthesis has been isolated from *P. patulum* (Beck et al. 1990) and *P. urticae* (Wang et al. 1991), while a gene encoding an apparent polyketide synthetase involved in the formation of a conidial wall pigment has been isolated from *A. nidulans* (Mayorga and Timberlake 1992). Recently, genes encoding the PKS responsible for aflatoxin/stergimaticystin biosynthesis have been reported from several sources (Change et al. 1995; Feng and Leonard 1995; Yu and Leonard 1995). Sequence comparisons between the fungal enzymes and the products of other PKS and FAS genes have revealed the existence of significant sequence homologies. The highest degree of amino acid sequence conservation occurs between the active sites of the various catalytic domains of PKSs and their FAS homologs (Hopwood and Sherman 1990). It is likely that this sequence conservation, together with polyketide structural information, can be used to develop molecular probes for isolating previously uncharacterized PKS genes.

Although less is known about the genes encoding other fungal polyketide pathway enzymes, it appears that in at least some pathways these genes are clustered. Evidence for aflatoxin pathway clustering was first reported for *A. parasiticus* (Skory et al. 1992). It has subsequently been shown that most aflatoxin/stergmaticystin pathway genes are present in a large gene cluster spanning 55 to 60 kb (Trail et al. 1995). Therefore, the isolation of a PKS gene or other pathway gene may lead to the identification of additional closely linked pathway genes.

2. Cercosporin

Many *Cercospora* species produce cercosporin (Fig. 5), a red perylenequinone phytotoxin that was first isolated from *C. kikuchii* in 1957 (Kuyama and Tamura 1957). Evidence suggesting that cercosporin is a virulence factor in plant diseases caused by *Cercospora* species includes the following: (1) the bioactivity of cercosporin is light-activated (Daub 1982) and light has been shown to be an important factor in the development of disease symptoms (Calpouzos and Stalknecht 1967); (2) application of cercosporin to plant tissue results in symptoms similar to those observed in infected plants; (3) cercosporin has been isolated from the necrotic lesions of diseased plants (Fajola 1978).

Cercosporin is a photosensitizing compound that produces superoxide radicals and singlet oxygen in the presence of light. These oxygen species then modify membrane lipids through the formation of fatty acid peroxides. The exposure of cells to cercosporin produces effects such as electrolyte leakage and decreased membrane fluidity, that eventually result in cell death. It appears that virtually all plants and fungi are sensitive to cercosporin, with the exception of organisms that themselves produce photosensitizing compounds. Cercosporin-producing fungi are resistant to cercosporin, but the mechanism of this resistance is presently unknown (Daub and Ehrenshaft 1993).

Little is known concerning the biosynthesis of cercosporin beyond the fact that it is an apparent heptaketide dimer derived from acetate (Okubo et al. 1975). No pathway intermediates have ever been observed, suggesting that, in contrast to fungal polyketides such as aflatoxin, cercosporin biosynthesis may occur primarily on either a multienzyme or multienzyme complex. In an effort to identify genes involved in cercosporin biosynthesis, cDNAs were isolated from *C. kikuchii* that showed enhanced expression in cultures exposed to light (Ehrenshaft and Upchurch 1991). The expression of one of these cDNAs (cLE6) was found to be closely correlated with cercosporin accumulation and could represent a gene involved in the cercosporin pathway. Efforts to disrupt the gene encoding cLE6 are reportedly in progress.

Recently, mutants of *C. kikuchii* with altered cercosporin production phenotypes have been reported (Upchurch et al. 1991). In one class of mutants that were found to occur spontaneously at low frequency (S2), cercosporin production is conditional on growth medium composition. The phenotype of these mutants must likely results from a mutation in a regulatory locus distinct from either cercosporin biosynthesis or the light-induced responses. The isolation of UV-induced mutants that produce less than 2% of the wild-type cercosporin levels independent of growth conditions has also been reported. While the pathogenicity of S2 is indistinguishable from the wild type, several of the cercosporin-deficient UV mutants are unable to produce lesions when inoculated on soybean leaves.

C. Peptides

1. Peptide Synthetases

Fungi produce a variety of phytotoxic peptides, several of which have been shown to play important roles in fungal/plant interactions. The biosynthesis of fungal peptide phytotoxins, like that of polyketides, involves a thiotemplate mechanism and the modular organization of distinct catalytic domains (Kleinkauf and von Döhren 1990b). Peptide phytotoxins are often cyclic and can contain unusual amino or hydroxy acids (depsipeptides). Peptide synthetases (PS) have been characterized from a number of prokaryotic sources and shown to consist of either a multienzyme or multienzyme complex (Kleinkauf and von Döhren 1990a).

The general mechanistic features of peptide synthetases include the following: (1) the multienzyme or multienzyme complex is composed of two or more amino acid-activating domains where amino acids undergo adenylation and subsequent thioester formation with 4′-phosphopantetheine, amino acid modifications such as methylation and epimerization also occur at this site; (2) the organization of the amino acid domains is colinear with the sequence of amino acids in the peptide product; (3) peptide (or ester) bonds are formed as the nascent peptide chain is transferred from one domain to the next.

The sequences of several fungal PS genes have been reported. Two of these are involved in the biosynthesis of phytotoxins, the enniatin synthetase (Haese et al. 1993) from *F. scirpi* and the HC toxin synthetase (HST-1) from *Cochliobolus carbonum* (Panaccione et al. 1992), while a third,

ACV synthetase from *P. chrysogenum*, participates in the biosynthesis of β-lactam compounds such as penicillin (Smith et al. 1990). All of these enzymes consist of a single multifunctional polypeptide. Comparisons between the functional domains of fungal and bacterial peptide synthetases have revealed significant sequence homologies. The two functional domains of enniatin synthetase show significant homology to each other (25% identity) but have even greater homology with the three domains of the ACV synthetase from *P. chrysogenum* (38–43% identity), and the four domains of the grsB gene product from *Bacillus brevis* (45–50% identity) which is involved in gramicidin S synthesis (Haese et al. 1993). The domain responsible for proline activation in HC-toxin synthetase exhibits significant homology to the proline-activating domain in the grsB gene product over a region of about 30 amino acids (Scott-Craig et al. 1992). Based on the amount of sequence homology observed between different PSs, it should be possible to develop molecular probes for the isolation of previously uncharacterized PS genes as discussed above for PKS genes.

2. Enniatins/Beauvericin

The enniatins are cyclic depsipeptides produced by several *Fusarium* species (Fig. 6). Beauvericin, an enniatin analog, is produced by a diverse group of fungi including *Fusarium* species (Plattner and Nelson 1994), *Beauveria bassiana* (Hamill et al. 1969), *Paecilomyces fumosoroseus* (Bernadini et al. 1975), and *Polyporus sulphureus* (Deol et al. 1978). Depsipeptides differ from other peptide antibiotics in that they contain ester bonds. Enniatins consist of a repeating dipeptidol unit made up of D-2-hydroxyisovaleric acid (D-HIV) and an N-methylated branched chain amino acid or N-methylated phenylalanine (beauvericin). Formation of enniatins involves the condensation of three dipeptidol units followed by cyclization. Enniatin synthetase has been shown to be a single polypeptide of 347 kDa that can synthesize enniatin in the presence of D-HIV, a branched chain amino acid (e.g., valine), ATP, and adenosylmethionine (Haese et al. 1993). N-methylation occurs while the branched chain amino acid is thioesterified to the multienzyme (Billich and Zocher 1991). Amino acid activation and N-methylation are

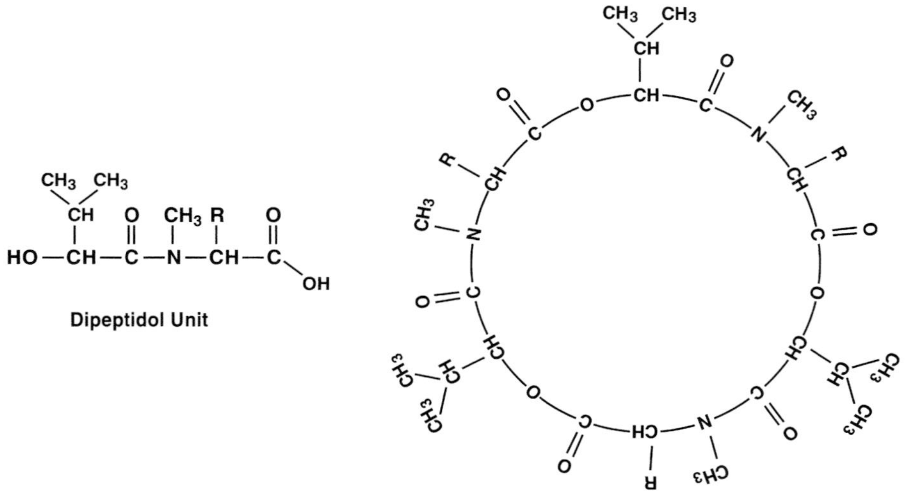

Dipeptidol Unit

	R
Enniatin A	$-CH(CH_3)CH_2CH_3$
Enniatin B	$-CH_2(CH_3)_2$
Enniatin C	$-CH_2CH(CH_3)_2$
Beauvericin	$-CH_2(C_6H_5)$

Fig. 6. Structures of dipeptidiol unit and enniatins/beauvericin

both thought to occur within the EB domain which is located at the C-terminal end of the multienzyme. The sequence of enniatin synthetase contains a 434-amino acid region in the EB domain that is similar to a conserved motif in S-adenosylmethionine-dependent methytransferases.

Beauvericin synthetase has also been shown to consist of a single polypeptide of about 250kDa (Peeters et al. 1988). As with enniatin synthetase, the methylation of phenylalanine occurs while it is thioesterified to the multienzyme. Enniatin and beauvericin synthetases may be closely related since beauvericin synthetase will also produce enniatin B if presented with valine as the amino acid substrate. Further evidence of their relatedness comes from immunochemical studies employing monoclonal antibodies generated against specific regions of enniatin synthetase. Cross-reactivity was observed with beauvericin synthetase for monoclonal antibodies directed against the D-HIV binding domain and the N-methyltransferase site of enniatin synthetase (Billich et al. 1987), but monoclonal antibodies directed against the valine-binding domain of enniatin synthetase did not cross-react.

While many of the fungi that produce enniatins/beauvericin are plant pathogens, relatively few studies have focused on the phytotoxicity of these compounds. Mixtures of enniatins were shown by Gäumann et al. (1960) to act synergistically in causing wilt symptoms of excised tomato shoots. Other symptoms included necrosis of leaves and loss of turgor. The effects of a mixture of enniatins on germinating wheat seeds were quite different. Wilting symptoms were not observed, but the growth of developing wheat plants was significantly reduced at concentrations as low as 10mg/ml (Burmeister and Plattner 1987). The inhibition of root elongation was found to be greater than the inhibition of leaf development. The mode of action for enniatins/beauvericins in plants has not been determined, but may involve the previously demonstrated ability of these compounds to disrupt membranes and to function as ionophors.

Recently, it has been shown that the application of a mixture of enniatins to potato tissue induces necrotic lesions (Herrman et al. 1996a). Enniatins were also isolated from potato tissue infected with several different *Fusarium* species, further suggesting a role for these metabolites in *Fusarium*/potato interactions. In addition, mutants of *Fusarium avenaceum* deficient in the production of enniatins have recently been generated by disruption of the enniatin synthetase gene (*esyn1*, Herrman et al. 1996b). Comparisons between wild-type and several *esyn1*$^{-}$ mutants grown on potato tuber slices indicated a correlation between enniatin production and higher levels of disease. This is the first genetic evidence that enniatins may serve as virulence factors in some plant diseases. The possible role of enniatins in *Fusarium* diseases of cereal crops remains unknown. Because enniatins/beavericin are thought to affect membrane function, it is possible that they synergize the activity of other phytotoxins produced by *Fusarium* species. The availability of the *esyn1* gene provides a means to address this question.

III. Practical Implications

Interest in fungal phytotoxins has primarily focused on the possible significance of these compounds in fungal ecology. Increasingly, efforts are being made to develop agricultural and research applications for fungal phytotoxins. In this section some promising phytotoxin applications are discussed.

A. Phytotoxins as Tools for Plant Research

Several fungal phytotoxins have proven useful as tools for basic studies of plant physiology. Probably the most widely used phytotoxin for research purposes is fusicoccin (Fig. 7), a diterpene glucoside produced by *Fusicoccum amygdali* (Marre 1979). The production of fusiccocin has been implicated as a factor in canker disease of almond and peach caused by *F. amygdali* (Aducci et al. 1983). Fusicoccin produces a number of physiological effects in plants including the stimulation of cell enlargement, stomatal cell opening, and seed germination. It can also act as an agonist or mimic of plant hormone responses. Extensive studies of fusicoccin-treated plants have determined that its primary mechanism of action is the stimulation of a plasmalemma H^{+}-ATPase. Because of its specificity for plasmalemma H^{+}-ATPase, fusicoccin has facilitated investigations of membrane transport processes and their relationship to other plant cellular responses.

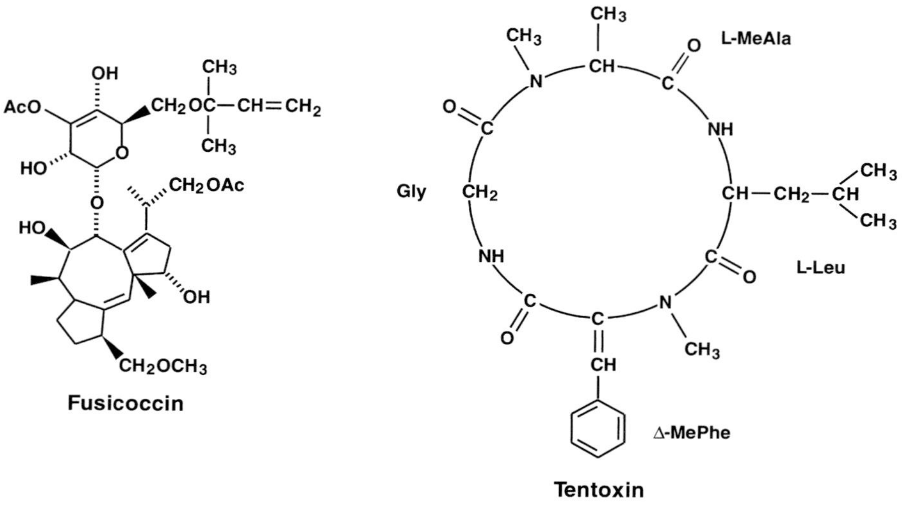

Fig. 7. Structures of fusicoccin and tentoxin

Another phytotoxin that has been frequently employed in plant research is tentoxin, a cyclic peptide produced by a number of *Alternaria* species (Fig. 7). Tentoxin consists of four amino acids (glycine, L-methylalanine, L-leucine, and methyldehyrophenylalanine). The N-methylation of alanine and dehydrophenylalanine appears to occur within the amino acid-activating domains of the PS involved in tentoxin biosynthesis (Liebermann and Ramm 1991). Tentoxin treatment results in chlorosis by preventing chlorophyll accumulation in the germinating seedlings of some, but not all, angiosperms (Durbin and Uchytil 1977).

Although tentoxin produces a variety of effects in plants (Klotz 1988), possibly involving multiple sites and modes of action, studies of this phytotoxin have focused on its potent inhibition of photophosphorylation (Arntzen 1972). Inhibition of photophosphorylation has been shown to result from the interaction of tentoxin with a specific site on a coupling factor (CF_1) of chloroplast H^+-ATPase (Steele et al. 1976). Recently, it has been shown that sensitivity to tentoxin in *Chlamydomonas reinhardtii* is determined by codon 83 of the b subunit of chloroplast H^+-ATPase (Avni et al. 1992). The presence of a glutamate residue at amino acid 83 was correlated with resistance, while the presence of an aspartate at this position was correlated with sensitivity. This result suggests that codon 83 may represent a critical site on the b subunit that is independent of the nucleotide-binding site or other catalytic activities. So, like fusicoccin, tentoxin may also prove valuable as a probe for investigations of membrane transport processes.

B. Novel Herbicides

One of the most promising applications for phytotoxins involves their use as herbicides (Kenfield et al. 1989a; Tanaka and Omura 1993). Many synthetic herbicides are currently viewed as undesirable since they degrade slowly in soil and the residual herbicide can cause adverse environmental effects. The widespread use of some synthetic herbicides has also resulted in the selection of resistant varieties of weeds. The major advantage of fungal herbicides is that they are likely to biodegrade rapidly, and therefore produce less stress on ecosystems. The diverse structures and activities of fungal herbicides also reduce the possibility of cross-resistance problems since, in most cases, they will be directed at molecular targets different from those recognized by synthetic herbicides.

The search for fungal metabolites that can protect crop plants from weeds or other pests is analogous to the exploitation of microbial antibiotics as chemotherapeutics. However, one major difference between the two applications is that the effective use of agricultural chemicals requires considerably larger quantities of material than is true for chemotherapeutics. This means that production costs play a much larger role in deciding their commercial feasibility. Difficulties in developing cost-effective production methods can represent a formidable barrier to the commercialization of potential biological herbicides.

Perhaps the most significant commercial contribution of fungal phytotoxins involves their use as models for the development of novel synthetic herbicides. Determining the mechanism of action for fungal phytotoxins can help to identify unique molecular targets in important weed species. Once a promising molecular target has been identified, structure/activity studies can then be used to facilitate the rational design of synthetic herbicides. Tentoxin is an example of a phytotoxin that has been considered for use as a herbicide because it shows differential activity toward some weed species (Lax et al. 1988). As discussed above, the primary molecular target of tentoxin is unique to plants. The synthesis of tentoxin and tentoxin analogs has permitted structure/activity studies with this phytotoxin (Edwards et al. 1987).

C. Improved Disease Resistance

There are several applications for fungal phytotoxins that are potentially useful for improving crop disease resistance. One of these is the use of phytotoxins suspected of being fungal virulence or pathogenicity factors to screen plants for increased resistance to the phytotoxin-producing pathogen. Phytotoxin screens for resistant individuals have been performed by analyzing the effects of toxins on various plant tissues, seedlings, tissue culture, or protoplasts (Durbin and Graniti 1989). The rationale for this approach is that exogenously added toxin may mimic the effect of toxin production by the fungal pathogen during infection. If this assumption is true, then increased resistance to the toxin as indicated by the assay is likely to reflect increased resistance of the plant to the toxin-producing pathogen. In the case of fungal phytotoxins, this has been applied to some toxins which are known to be pathogenicity factors and, to a lesser extent, to toxins considered virulence factors. With the increased availability of both purified phytotoxins and information concerning their role in plant diseases, this approach may be employed more frequently in the future.

Other phytotoxin-based applications include the genetic engineering of phytotoxin resistance in relevant crop plants. Because some phytotoxins are also antibiotics, the fungi that produce them have developed mechanisms to protect themselves. If genes specifying fungal resistance to phytotoxin/antibiotics can be isolated, then it may be possible to use these resistance genes to improve disease resistance. The approach should work for phytotoxins that have either a clearly defined mechanism of action or a specific molecular target. In *Actinomycetes*, antibiotic resistance can result from target site modification, antibiotic inactivation, or transport of the antibiotic out of the cell (Cundliffe 1989; Schoner et al. 1992). Antibiotic resistance genes are often closely linked to the genes specifying pathway biosynthetic enzymes. So far, no phytotoxin resistance genes have been reported from fungi.

Still another possible application for fungal phytotoxins is their use as plant-defensive chemicals through the introduction of phytotoxin biosynthetic pathway genes into plants. This is again suggested by the fact that some phytotoxins are also antibiotics. In addition, several examples are known of fungal phytotoxins with antibiotic activity that are also produced by plants. Probably the best example of this is the trichothecenes, which are produced by both fungi and at least two species of the plant genus *Baccharis* (Jarvis 1992). The macrocyclic trichothecenes that accumulate to high levels in the seeds and flowering parts of *B. cordifolia* are potent phytotoxins for most plants but appear to have little effect on this Brazilian shrub. Although the function of trichothecenes in *Baccharis* species is presently unknown, it is possible that their adaptation to permit the accumulation of trichothecenes is related to defense against either disease or herbivores. Examples of this sort blur the line between fungal phytotoxins and plant-defensive chemicals, suggesting that in some cases they are functionally identical. The use of phytotoxins as preformed defensive compounds would also require the introduction of phytotoxin resistance factors to protect the plant. As discussed above, such resistance factors have not yet been identified; however, it may be possible to use phytotoxins as part of plant-inducible defense systems. Many phytoalexins are known to possess significant phytotoxicity (Smith 1982), although the importance of this activity with respect to their defensive function has not been determined.

IV. Conclusions

Many plant pathogenic fungi produce phytotoxic metabolites that are active against a wide spectrum of organisms. Some of the most prolific phytotoxin producers are plant pathogens (e.g., *Fusarium* species) which exhibit a wide host range and against which it has proven difficult or impossible to develop effective resistance in crop plants. Identifying phytotoxins involved in disease has important practical implications for efforts to improve disease resistance. For example, introducing traits which increase plant defenses against a disease-related phytotoxin is likely to also increase disease resistance to the phytotoxin-producing pathogen. Available evidence suggests that many phytotoxins probably make minor contributions to pathogen virulence; however, for some pathogens of major crops even small improvements in resistance could be economically significant.

The use of molecular gene disruption to selectively block phytotoxin pathways has proven an effective approach for analyzing phytotoxin involvement in plant/fungal interactions. The application of this approach typically requires knowledge of both phytotoxin biosynthetic pathways and structural information for pathway genes. Studies of natural product biosynthesis in bacteria, fungi, and plants have recently made important contributions to our knowledge of phytotoxin biosynthetic enzymes which may speed efforts to develop molecular probes for some key phytotoxin pathway genes.

Acknowledgments. I thank Anne Desjardins, Susan McCormick, and Robert Proctor for helpful discussions and critically reading the manuscript.

References

Aducci P, Ballio A, Federico R, Marra M, Graniti A (1983) Translocation of fusicoccin within almond plants infected by *Fusicoccum amygdali* Del. Phytopathol Mediterr 22:100–102

Arntzen CJ (1972) Inhibition of photophosphorylation by tentoxin, a cyclic tetrapeptide. Biochim Biophys Acta 283:539

Ashby MN, Spear DH, Edwards PA (1990) Prenyltransferases: From yeast to man. In: Attie AD (ed) Molecular biology of atherosclerosis. Elsevier, Amsterdam, pp 27–34

Avni A, Anderson JD, Holland N, Rochaix J-D, Gromet-Elhanan Z, Edelman M (1992) Tentoxin sensitivity of chloroplasts determined by codon 83 of b subunit of proton-ATPase. Science 257:1245–1247

Back K, Chappell J (1995) Cloning and bacterial expression of a sesquiterpene cyclase from *Hyoscyamus muticus* and its molecular comparison to related terpene cyclases. J Biol Chem 270:7375–7381

Beck J, Pipka S, Siegner A, Schlitz E, Schweizer E (1990) The multifunctional 6-methylsalicylic acid synthase gene of *Penicillium patulum*: its gene structure relative to that of other polyketide synthases. Eur J Biochem 192:487–498

Beremand MN (1987) Isolation and characterization of mutants blocked in T-2 toxin biosynthesis. Appl Environ Microbiol 53:1855–1859

Bernadini M, Carilli A, Pacioni G, Santurbano B (1975) Isolation of beauvericin from *Paecilomyces fumosoroseus*. Phytochemistry 14:1865

Billich A, Zocher R (1991) Formation of N-methylated peptide bonds in peptides anf peptidols. In: Kleinkauf H, von Dohren H (eds) Biochemistry of peptide antibiotics. de Gruyter, Berlin, pp 57–79

Billich A, Zocher R, Kleinkauf H, Braun DG, Lavanchy D, Hochkeppel HK (1987) Monoclonal antibodies to the multienzyme enniatin synthetase. Production and use in structural studies. Biol Chem Hoppe-Seyler 368:521–529

Bronson CR (1991) The genetics of phytotoxin production by plant pathogenic fungi. Experientia 47:771–776

Bunkers GJ, Strobel GA (1991) A proposed mode of action for green island induction by the eremophilane phytotoxins produced by *Drechslera gigantea*. Physiol Mol Plant Pathol 38:313–323

Burmeister HR, Plattner RD (1987) Enniatin production by *Fusarium tricinctum* and its effect on germinating wheat seeds. Phytopathology 77:1483–1487

Calpouzos L, Stalknecht GF (1967) Symptomas of *Cercospora* leaf spot of sugar beets influenced by light intensity. Phytopathology 57:799–800

Cane DE (1990) Enzymatic formation of sesquiterpenes. Chem Rev 90:1089–1103

Cane DE, Rawlings BJ, Yang C-C (1987) Isolation of (−)-g-cadinene and aristolochene from *Aspergillus terreus*. J Antibiot 40:1331–1334

Cane DE, Sohng JK, Lamberson CR, Rudnicki SM, Wu Z, Lloyd MD, Oliver JS, Hubbard BR (1994) Pentalenene synthase. Purification, molecular cloning, sequencing, and high-level expression in *Escherichia coli* of a terpenoid cyclase from *Streptomyces* UC5319. Biochemistry 33:5846–5857

Casale WL, Hart LP (1988) Inhibition of ^{3}H-leucine incorporation by trichothecene mycotoxins in maize and wheat tissue. Phytopathology 78:1673–1677

Chang PK, Cary JW, Yu JJ, Bhatnagar D, Cleveland TE (1995) The *Aspergillus parasiticus* polyketide synthase gene pksA, a homolog of *Aspergillus nidulans* wA, is required for aflatoxin B-1 biosynthesis. Mol Gen Genet 248:270–277

Colby SM, Alonso WR, Katahira EJ, McGarvey DJ, Croteau R (1993) 4S-Limonene synthase from the oil glands of spearmint (*Mentha spicata*). cDNA isolation, characterization, and bacterial expression of the catalytically active monoterpene cyclase. J Biol Chem 268: 23016–23024

Croteau R (1987) Biosynthesis and catabolism of monoterpenoids. Chem Rev 87:929–954

Cundliffe E (1989) How antibiotic-producing organisms avoid suicide. Annu Rev Microbiol 43:207–233

Cutler HG (1988) Trichothecenes and their role in the expression of plant disease. Biotechnology of crop protection. ACS Symp Ser 379:50–72

Daub ME (1982) Cercosporin, a photosensitizing toxin from *Cercospora* species. Phytopathology 72:370–374
Daub ME, Ehrenshaft M (1993) The photoactivated toxin cercosporin as a tool in fungal photobiology. Physiol Plant 89:227–236
Deol BD, Ridley DD, Singh P (1978) Isolation of cyclodepsipeptides from plant pathogenic fungi. Aust J Chem 31:1397–1399
Desjardins AE, Plattner RD, VanMiddlesworth F (1986) Trichothecene biosynthesis in *Fusarium sporotrichioides*: origin of the oxygen atoms of T-2 toxin. Appl Environ Microbiol 51:493–497
Desjardins AE, Spencer GF, Plattner RD, Beremand MN (1989) Furanocoumarin phytoalexins, trichothecene toxins and infection of *Pastinaca sativa* by *Fusarium sporotrichioides*. Phytopathology 79:170–175
Desjardins AE, Hohn TM, McCormick SP (1992) The effect of gene disruption of trichodiene synthase (*Tox5*) on the virulence of *Gibberella pulicaris*. Mol Plant-Microbe Interact 5:214–222
Desjardins AE, Hohn TM, McCormick SP (1993) Trichothecene biosynthesis in *Fusarium* species: chemistry, genetics, and significance. Microbiol Rev 57:595–604
Donadio S, Staver MJ, McAlpine JB, Swanson SJ, Katz L (1991) Modular organization of genes required for complex polyketide biosynthsis. Science 252:675–679
Durbin RD, Graniti A (1989) Possible applications of phytotoxins. In: Graniti A, Durbin RD, Ballio A (eds) Phytotoxins and plan pathogenesis. Springer, Berlin Heidelberg New York, pp 337–355
Durbin RD, Uchytil TF (1977) A survey of plant insensitivity to tentoxin. Phytopathology 67:602–603
Edwards JV, Lax AR, Lillehoj EB, Boudreaux GJ (1987) Structure-activity relationships of cyclic and acyclic analogues of the phytotoxic peptide tentoxin. J Agric Food Chem 35:451–456
Ehrenshaft M, Upchurch RG (1991) Isolation of light-enhanced cDNAs of *Cercospora kikuchii*. Appl Environ Microbiol 57:2671–2676
Facchini PJ, Chappell J (1992) A gene family for an elicitor-induced sesquiterpene cyclase in tobacco. Proc Natl Acad Sci USA 89:11088–11092
Fajola AO (1978) Cercosporin, a phytotoxin of *Cercospora* spp. Physiol Plant Pathol 13:157–164
Feng GH, Leonard TJ (1995) Characterization of the polyketide synthase gene (pksL1) required for aflatoxin biosynthesis in *Aspergillus parasiticus*. J Bacteriol 177:6246–6254
Fincham JRS (1989) Transformation in fungi. Microbiol Rev 53:148–170
Gäumann E, Naef-Roth S, Kern H (1960) Zur phytotoxischen Wirksamkeit der Enniatine. Phytopathology 40:45–51
Greenhalgh R, Fielder DA, Morrison LA et al. (1989) Apotrichothecenes: Minor metabolites of the *Fusarium* species. In: Natori S, Hashimoto K, Ueno Y (eds) Mycotoxins and phycotoxins '88. Elsevier, Amsterdam, pp 223–232
Haese A, Schubert M, Herrmann M, Zocher R (1993) Molecular characterization of the enniatin synthetase gene encoding a multifunctional enzyme catalyzing N-methyldepsipeptide formation in *Fusarium scirpi*. Mol Microbiol 7:905–914
Hamill RL, Higgens CE, Boaz HE, Gorman M (1969) The structure of beauvericin, a new depsipeptide antibiotic toxic to *Artemia salina*. Tetrahedron Lett 4255–4258
Hermann M, Zocher R, Haese A (1996a) Enniatin production by *Fusarium* strains and its effect on potato tuber tissue. Appl Environ Microbiol 62:393–398
Hermann M, Zocher R, Haese A (1996b) Effect of disruption of the enniatin synthetase gene on the virulence of *Fusarium avenaceum*. Mol Plant-Microbe Interact 9:226–232
Hohn TM, Beremand PD (1989) Isolation and nucleotide sequence of a sesquiterpene cyclase gene from the trichothecene-producing fungus *Fusarium sporotrichioides*. Gene 79:131–138
Hohn TM, Desjardins AE (1992) Isolation and gene disruption of the *Tox5* gene encoding trichodiene synthase in *Gibberella pulicaris*. Mol Plant-Microbe Interact 5:249–256
Hohn TM, VanMiddlesworth F (1986) Purification and characterization of the sesquiterpene cyclase trichodiene synthetase from *Fusarium sporotrichioides*. Arch Biochem Biophys 251:756–761
Hohn TM, McCormick SP, Desjardins AE (1993) Evidence for a gene cluster involving trichothecene pathway biosynthetic genes in *Fusarium sporotrichioides*. Curr Genet 24:291–295
Hohn TM, Desjardins AE, McCormick SP (1995a) The *Tri4* gene of *Fusarium sporotrichioides* encodes a cytochrome P450 monooxygenase involved in trichothecene biosynthesis. Mol Gen Genet 248:95–102
Hohn TM, Desjardins AE, McCormick SP, Proctor RH (1995b) Biosynthesis of trichothecenes, genetic and molecuar aspects. In: Eklund M, Richard JL, Katsutoshi M (eds) Molecular approaches to food safety. Alaken, Fort Collins, Colorado, pp 239–247
Hopwood DA, Sherman DH (1990) Molecular genetics of polyketides and its comparison to fatty acid biosynthesis. Annu Rev Genet 24:37–60
Jarvis BB (1991) Macrocyclic trichothecenes. In: Sharma RP, Salunkhe DK (eds) Mycotoxins and phytoalexins. CRC Press, Boca Raton, pp 361–421
Jarvis BB (1992) Macrocyclic trichothecenes from Brazilian *Baccharis* species: from microanalysis to large scale isolation. Phytochem Anal 3:241–249
Jeker N, Tamm C (1988) Synthesis of new, unnatural macrocyclic trichothecenes: 4-epiverrucarin A. Helv Chim Acta 71:1904–1913
Katz L, Donadio S (1993) Polyketide synthesis: prospects for hybrid antibiotics. Annu Rev Microbiol 47:875–912
Kenfield D, Bunkers G, Strobel G et al. (1989a) Fungal phytotoxins: potential new herbicides. In: Graniti A, Durbin RD, Ballio A (eds) Phytotoxins and plant pathogenesis. Springer, Berlin Heidelberg New York, pp 319–335
Kenfield D, Bunkers G, Wu Y-H, Strobel G, Sugawara F, Hallock Y, Clardy J (1989b) Gigantenone, a novel sesquiterpene phytohormone mimic. Experientia 45:900–902
Kleinkauf H, von Döhren H (1990a) Bioactive peptides: Recent advances and trends. In: Kleinkauf H, von Döhren H (eds) Biochemistry of peptide antibiotics. deGruyter, Berlin, pp 1–31
Kleinkauf H, von Döhren H (1990b) Nonribosomal biosynthesis of peptide antibiotics. Eur J Biochem 192:1–15
Klotz MG (1988) The action of tentoxin on membrane processes in plants. Physiol Plant 74:575–582
Knoche HW, Duvick JP (1987) The role of fungal toxins in plant disease. In: Pegg GF, Ayers PG (eds) Fungal infection of plants. Cambridge University Press, Cambridge, pp 158–191

Kuti JO, Ng TJ, Bean GA (1989) Possible pathogen-produced trichothecene metabolite in *Myrothecium* leafspot of muskmelon. Physiol Mol Plant Pathol 34:41–54

Kuyama S, Tamura T (1957) A pigment of *Cercospora kikuchii* (Matsumoto et Tomoyasu). 1. Cultivation of the fungus, isolation and purification of pigment. J Am Chem Soc 79:5725–5726

Lax AR, Shepherd HS, Edwards JV (1988) Tentoxin, a chlorosis-inducing toxin from *Alternaria* as a potential herbicide. Weed Technol J Weed Sci Soc Am 2:540–544

Liebermann B, Ramm K (1991) N-methylation in the biosynthesis of the phytotoxin tentoxin. Phytochemistry 30:1815–1817

Marre E (1979) Fusicoccin: a tool in plant physiology. Annu Rev Plant Physiol 30:273–288

Mayorga ME, Timberlake WE (1992) The developmentally regulated *Aspergillus nidulans wA* gene encodes a polypeptide homologous to polyketide and fatty acid synthases. Mol Gen Genet 235:205–212

McCormick SP, Hohn TM, Desjardins AE (1996) Isolation and characterization of *Tri3*, a gene encoding 15-O-acetyltransferase from *Fusarium sporotrichioides*. Appl Environ Microbiol 62:353–359

Mitchell RE (1984) The relevance of non-host-specific toxins in the expression of virulence by pathogens. Annu Rev Phytopathol 22:215–245

Okubo A, Yamazaki S, Fuwa K (1975) Biosynthesis of cercosporin. Agric Biol Chem 39:1173–1175

Panaccione DG, Scott-Craig JS, Pocard J-A, Walton JD (1992) A cyclic peptide synthetase gene required for pathogenicity of the fungus *Cochliobolus carbonum* on maize. Proc Natl Acad Sci USA 89:6590–6594

Payne GA, Nystrom GJ, Bhatnagar D, Cleveland TE, Woloshuk CP (1993) Cloning of the *afl-2* gene involved in aflatoxin biosynthesis from *Aspergillus flavus*. Appl Environ Microbiol 59:156–162

Peeters H, Zocher R, Kleinkauf H (1988) Synthesis of beauvericin by a multifunctional enzyme. J Antibiol 41:352–359

Plattner RD, Nelson PE (1994) Production of beauvericin by a strain of *Fusarium proliferatum* isolated from corn fodder for swine. Appl Environ Microbiol 60:3894–3896

Proctor RH, Hohn TM (1993) Aristolochene synthase. Isolation, characteriztion, and bacterial expression of a sesquiterpenoid biosynthetic gene (*Ari1*) from *Penicillium roqueforti*. J Biol Chem 268:4543–4548

Proctor RH, Hohn TM, McCormick SP (1995a) Reduced virulence of *Gibberella zeae* caused by disruption of a trichothecene toxin biosynthetic gene. Mol Plant-Microbe Interact 8:593–601

Proctor RH, Hohn TM, McCormick SP, Desjardins AE (1995b) *Tri6* encodes an unsual zinc finger protein involved in the regulation of trichothecene biosynthesis in *Fusarium sporotrichioides*. Appl Environ Microbiol 61:1923–1930

Scheffer RP (1983) Toxins as chemical determinants of plant disease. In: Daly JM, Deverall BJ (eds) Toxins and plant pathogenesis. Academic Press, New York, pp 1–40

Schoner B, Geistlich M, Rosteck P Jr, Rao RN, Seno E, Reynolds P, Cox K, Burgett S, Hershberger C (1992) Sequence similarity between macrolide-resistance determinants and ATP-binding transport proteins. Gene 115:93–96

Scott-Craig JS, Panaccione DG, Pocard J-A, Walton JD (1992) The cyclic peptide synthetase catalyzing HC-toxin production in the filamentous fungus *Cochliobolus carbonum* is encoded by a 15.7-kilobase open reading frame. J Biol Chem 267:26044–26049

Skory CD, Chang P-K, Cary J, Linz JE (1992) Isolation and characterization of a gene from *Aspergillus parasiticus* associated with the converison of versicolorin A to sterigmastocystin in aflatoxin biosynthesis. Appl Environ Microbiol 58:3527–3537

Smith DA (1982) Toxicity of phytoalexins. In: Bailey JA, Mansfield JW (eds) Phytoalexins. Halstead Press, New York, pp 218–252

Smith DJ, Earl AJ, Turner G (1990) The multifunctional peptide synthetase performing the first step of penicilin biosythesis in *Penicillium chrysogenum* is a 421 073 dalton protein similar to *Bacillus brevis* peptide antibiotic sythetases. EMBO J 9:2743–2750

Steele JA, Uchytil TF, Durbin RD, Bhatnagar P, Rich DH (1976) Chloroplast coupling factor 1: a species-specific receptor for tentoxin. Proc Natl Acad Sci USA 73:2245–2248

Stoessl A (1981) Structure and biogentic relations: fungal nonhost-specific. In: Durbin RD (ed) Toxins in plant disease. Academic Press, New York, pp 109–219

Sugawara F, Strobel G, Fisher LE, Van Duyne GD, Clardy J (1985) Bipolaroxin, a selective phytotoxin produced by *Bipolaris cynodontis*. Proc Natl Acad Sci USA 82:8291–8294

Tanaka S, Wada K, Marumo S, Hattori H (1984) Structure of sporogen-AO 1, a sporogenic substance of *Aspergillus oryzae*. Tetrahedron Lett 25:5907–5910

Tanaka Y, Omura S (1993) Agroactive compounds of microbial origin. Annu Rev Microbiol 47:57–87

Trail F, Mahanti N, Linz J (1995) Molecular biology of aflatoxin biosynthesis. Microbiology-Uk 141:755–765

Upchurch RG, Walker DC, Rollins JA, Ehrenshaft M, Daub ME (1991) Mutants of *Cercospora kikuchii* altered in cercosporin synthesis and pathogenicity. Appl Environ Microbiol 57:2940–2945

Walton JD, Panaccione DG (1993) Host-selective toxins and disease specificity: perspectives and progress. Annu Rev Phytopathol 31:275–303

Wang I-K, Reeves C, Gaucher CM (1991) Isolation and sequencing of a genomic DNA clone containing the 3′ terminus of the 6-methylsalicylic acid polyketide synthetase gene of *Penicillium* urticae. Can J Microbiol 37:86–95

Wang YZ, Miller JD (1988) Effects of *Fusarium graminearum* metabolites of wheat tissue in relation to *Fusarium* head blight resistance. J Phytopathol 122: 118–125

Whitehead IM, Threlfall DR, Ewing DF (1989) 5-epi-aristolochene is a common precursor of the sesquiterpenoid phytoalexins capsidiol and debneyol. Phytochemistry 28:775–779

Yoder OC (1980) Toxins in pathogenesis. Annu Rev Phytopathol 18:103–129

Yu JH, Leonard TJ (1995) Sterigmatocystin biosynthesis in *Aspergillus nidulans* requires a novel type I polyketide synthase. J Bacteriol 177:4792–4800

8 *Cochliobolus* spp. and Their Host-Specific Toxins

O.C. Yoder[1], V. Macko[2], T. Wolpert[3], and B.G. Turgeon[1]

CONTENTS

I. Introduction

The genus *Cochliobolus* (class Ascomycetes, order Pleosporales, family Pleosporaceae) contains approximately 30 species (Luttrell 1973), nearly all of which are pathogens of wild grasses or cereal crops (Sivanesan 1984). The genus is predominately heterothallic, although four homothallic species have been described (Luttrell and Rogerson 1959; Waki et al. 1979). Two genera, *Bipolaris* and *Curvularia*, have been erected for the asexual (anamorph) stages of *Cochliobolus* (Alcorn 1983). Before teleomorphic states of these fungi were known, anamorphic *Cochliobolus* spp. were described under the name *Helminthosporium* (Alcorn 1988). The graminicolous *Helminthosporium* spp. were divided into three genera as their perfect states were discovered: *Bipolaris* for anamorphs of *Cochliobolus*, *Drechslera* for anamorphs of *Pyrenophora*, *Exserohilum* for anamorphs of *Setosphaeria*.

C. heterostrophus has been genetically domesticated for conventional and molecular laboratory analyses. Its features include: (1) ease of growth in culture, (2) an efficient sexual stage readily produced in the laboratory, (3) a collection of morphological and auxotrophic mutants of spontaneous and chemically induced origin, (4) routine genetic transformation and cotransformation (Turgeon et al. 1985, 1987), (5) high-frequency integration of transforming DNA by homologous recombination into the chromosome (Turgeon et al. 1987; Wirsel et al. 1996), (6) gene disruption and gene replacement (Mullin et al. 1993; Wirsel et al. 1996), (7) gene cloning by complementation (Turgeon et al. 1993a), (8) insertional mutagenesis by restriction enzyme-mediated integration (Lu et al. 1994), (9) an RFLP map (Tzeng et al. 1992), (10) availability of isogenic lines (Leach et al. 1982a; Klittich and Bronson 1986), (11) chromosome separation using pulsed-field gel electrophoresis (Tzeng et al. 1992). Genetic methods developed for *C. heterostrophus* are directly applicable to other *Cochliobolus* spp., e.g., *C. carbonum* and *C. victoriae* (Turgeon et al. 1993b, 1995b), and the molecular techniques can be readily adapted for use with asexual relatives such as *Bipolaris sacchari* (Sharon et al. 1996).

Prominent species of *Cochliobolus* in North America include *C. heterostrophus*, which causes Southern Corn Leaf Blight; *C. carbonum*, which

[1] Department of Plant Pathology, Cornell University, Ithaca, New York 14853, USA
[2] Boyce Thompson Institute, Cornell University, Ithaca, New York 14853, USA
[3] Department of Botany and Plant Pathology, Oregon State University, Corvallis, Oregon 97331, USA

The Mycota V Part A
Plant Relationships
Carroll/Tudzynski (Eds.)

causes northern leaf spot of corn; and *C. victoriae*, which causes Victoria blight of oats (Table 1). All three fungi produce chemically distinct host-specific toxins, considered to be pathogenicity factors in *C. carbonum* and *C. victoriae* and a virulence factor in *C. heterostrophus* (Yoder 1980). The three species have similar life cycles (Fig. 1), including a heterothallic (self-sterile) sexual stage that can be generated in the laboratory by pairing strains with alternate genes at the mating type (*MAT*) locus (Nelson 1957, 1959, 1960; Turgeon et al. 1993a). The sexual structures, commonly called perithecia, are more accurately designated pseudothecia in this genus.

Host-specific toxins produced by *Cochliobolus* spp. have been studied for nearly 50 years. Extensive data suggest roles for these unusual metabolites in the development of several diseases caused by *Cochliobolus* spp. (Yoder 1980). However, little is known about their biosynthetic pathways or the genetic events which have caused the sudden appearance of new toxin-producing races in certain species (Scheffer 1989a,b, 1991). Chemical structures have been determined for a number of the host-specific toxins, revealing a diverse group of secondary metabolites (Macko 1983; Macko et al. 1985; Wolpert et al. 1985). At first glance, this chemical diversity suggests that the genetic events leading to toxin production by different fungi have little or nothing in common with each other, which, in fact, may be true. However, it is now clear that, in at least some cases, the appearance of a new race was not due to a simple genetic change, such as a point mutation, but rather was associated with rearrangement of the genome. Is production of host-specific toxins (and perhaps other secondary metabolites) by fungi generally associated with genome rearrange-

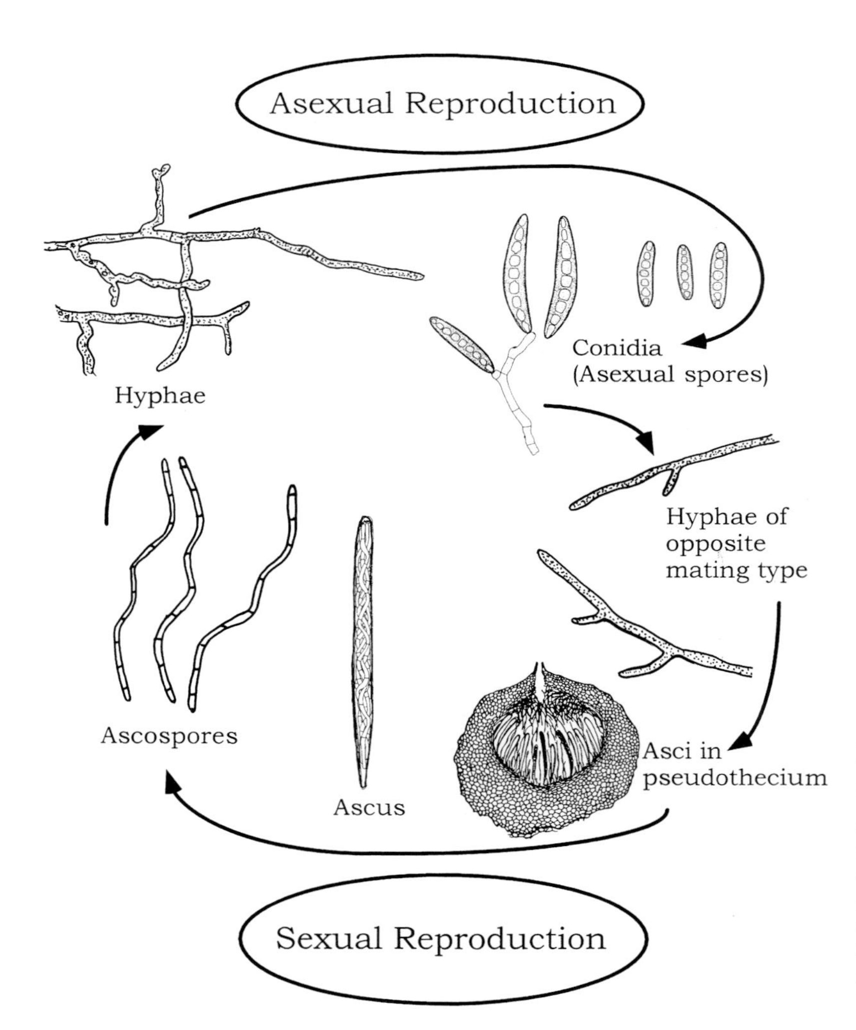

Fig. 1. Sexual (represented by ascospores) and asexual (represented by conidia) cycles of *Cochliobolus* spp. The *B. sacchari* asexual cycle and spore morphology are similar to those of *Cochliobolus* spp.; it has no known sexual cycle but appears closely related to *Cochliobolus* since it carries a functional homologue of the *Cochliobolus* mating type gene. (Sharon et al. 1996)

ments, which then give rise to new metabolic pathways or can simpler genetic events have the same effect? This question is now being addressed using the genetic and molecular technology which is available for host-specific toxin-producing fungi. As genes which control toxin biosynthesis and transport are cloned and analyzed, we will begin to understand the genetic bases of toxin production and the molecular mechanisms that lead to evolution of new fungal races.

This chapter focuses on the four most studied host-specific toxin-producing *Cochliobolus* spp. Their toxins (Fig. 2) differ strikingly in structure but may have common features in their evolution: T-toxin, a linear polyketide, produced by *C. heterostrophus* (a pathogen of corn), HC-toxin, a cyclic tetrapeptide, produced by *C. carbonum* (a pathogen of corn), victorin, a chlorinated cyclized peptide produced by *C. victoriae* (a pathogen of oats) and HS-toxin, a glycosylated sesquiterpene produced by *Bipolaris sacchari* (a pathogen of sugarcane), an asexual relative of *Cochliobolus* spp. (Table 1, Fig. 1).

II. *C. heterostrophus*, Corn, and T-Toxin

C. heterostrophus is a pathogen of corn found in nature as race T or race O. The two races are indistinguishable, except that race T is highly virulent toward Texas male sterile (T) cytoplasm corn, whereas race O is not. Both races are weakly virulent toward corn lacking T cytoplasm. The specificity of race T is associated with production of T-toxin (Kono and Daly 1979; Kono et al. 1981a,b).

A. Chemistry of T-Toxin

T-toxin is a mixture of several linear polyketols ranging in length from C_{35} to C_{41}, with each component possessing apparently identical specific toxicity toward T cytoplasm corn (Fig. 2). The C_{39} and C_{41} components comprise 60–90% of the native toxin (Kono and Daly 1979; Kono et al. 1981a). A number of shorter analogs (C_{15} to C_{26}) of these toxins were synthesized de novo (Suzuki et al. 1982b). Inhibition of dark CO_2 fixation by susceptible corn leaves was used as a bioassay to compare the relative toxicity of synthetic analogs with that of the native toxin. Analogs with C_{15}, C_{25}, and C_{26} chain lengths and 1,5-dioxo-3-hydroxy functions were only slightly less toxic ($2–6 \times 10^{-7}$M) than the individual components (C_{35}–C_{45} chain lengths) of the native T toxin (3×10^{-8}M) (Suzuki et al. 1982a,b). Like the native toxin, analogs were host-specific. The analogs did not inhibit dark CO_2 fixation in leaf tissue of resistant corn at concentrations 10^2 or 10^3 times greater than those effective with susceptible corn. A very close (C_{41}) analog of T-toxin, with almost the same distribution of carbonyl and hydroxyl groups as one of the natural T-toxins, was synthesized as well. Biological activity of this synthetic compound in the dark CO_2 fixation assay was indistinguishable from that of the native T-toxin complex (Suzuki et al. 1983). These findings support the validity of the structure proposed for native T-toxin.

Table 1. Genetics of fungal toxin production and host plant resistance/susceptibility

	Fungus			Host		
	Toxin	Nuclear locus[a]	Role in disease[b]	Relevant locus[c]	Genomic location	Dominant phenotype[d]
C. heterostrophus	T-toxin	*Tox1*	Virulence factor	*T-urf13*	Mitochondrion	Susceptibility
C. carbonum	HC-toxin	*Tox2*	Pathogenicity factor	*Hm1, Hm2*	Nucleus	Resistance
C. victoriae	victorin	*Tox3*	Pathogenicity factor	*Vb*	Nucleus	Susceptibility
B. sacchari	HS-toxin	?	?	?	?	?

[a] For each of the three sexual fungi conventional genetic analysis has defined a single genetic element controlling toxin production, but molecular analysis has revealed that both *TOX1* and *TOX2* are complex (see text).
[b] Null (toxin-less) mutants lose a level of virulence in *C. heterostrophus*, but become nonpathogenic in *C. carbonum* and *C. victoriae*. No mutants are available for *B. sacchari*.
[c] The host for each of the four fungi (top to bottom) is corn, corn, oats, sugarcane (Fig. 2).
[d] The dominant allele at the relevant host locus determines susceptibility to both fungus and toxin for *C. heterostrophus* and *C. victoriae*, but resistance for *C. carbonum*.

PATHOGEN	MAJOR HOST-SPECIFIC TOXIN	HOST	SUSCEPTIBLE GENOTYPE
Cochliobolus heterostrophus race T	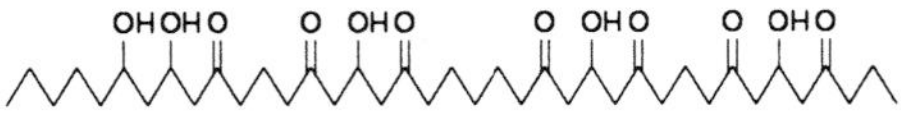 T-toxin 3	Maize (*Zea mays*)	*T-urf 13*
Mycosphaerella zeae-maydis	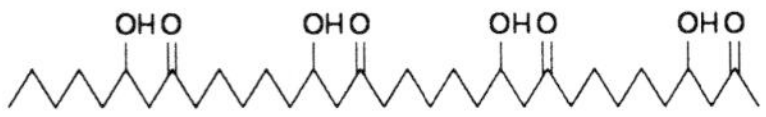 PM-toxin A	Maize (*Z. mays*)	*T-urf 13*
	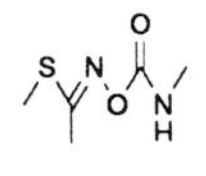 methomyl		
C. carbonum race 1	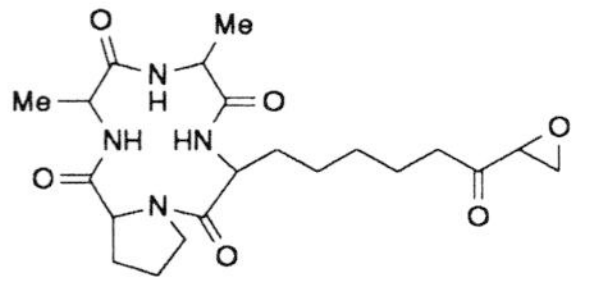 HC-toxin 1	Maize (*Z. mays*)	*hm1 hm1* *hm2 hm2*
C. victoriae	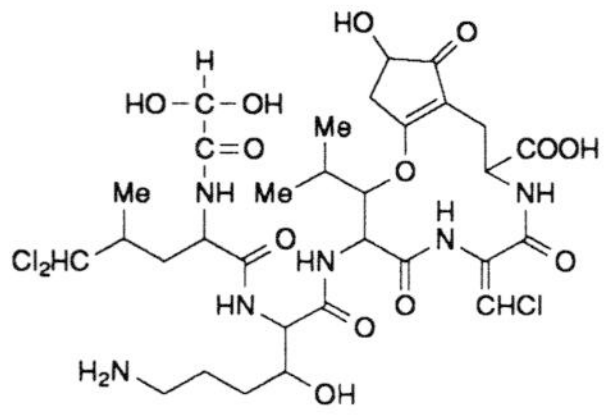 victorin C	Oats (*Avena sativa*)	*Vb Vb*
Bipolaris sacchari	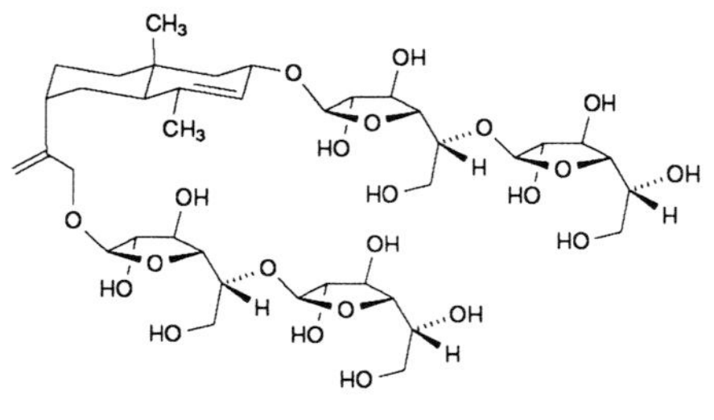 HS-toxin C	Sugarcane (*Saccharum officinarum*)	?

Fig. 2. Fungi, host plants, and structures of toxins discussed in this chapter; see text for details. Methomyl is included here because it has the same specificity toward T-cytoplasm corn as do T-toxin and PM-toxin (although at much higher concentration), yet it is structurally unrelated to the two fungal toxins

Some host selectivity may occur with chemically distinct compounds (Tegtmeier et al. 1982). Reduced toxin, a linear polyalcohol, possesses high activity and specificity without the carbonyl groups that are present in the native toxins and their synthetic analogs (Suzuki et al. 1982a,b), and binds to the URF13 protein (see below) in both corn mitochondria and *E. coli* (Braun et al. 1990). The insecticide methomyl (Fig. 2) is structurally quite different from T-toxin, but has the same host selectivity, although at much higher concentrations (10^{-3} M). The corn pathogen *Mycosphaerella zeae-maydis* (anamorph = *Phyllosticta maydis* = *Phoma zeae-maydis*) also produces a toxin with the same host selectivity (Comstock et al. 1973; Yoder 1973). In this last case, however, it was found that the toxin complex from *M. zeae-maydis* is structurally related (Fig. 2), though not identical, to T-toxin (Kono et al. 1983; Danko et al. 1984).

The number of chemical species in the T-toxin complex has been determined in acetone- or chloroform-insoluble precipitates, but activity also occurs in the soluble portion (Tegtmeier et al. 1982). It is therefore likely that additional species of T-toxin will emerge as newer separation and analytical HPLC methods are applied to extracts.

B. Genetics of T-Toxin Biosynthesis

The sudden appearance of *C. heterostrophus* race T on corn in 1969 in an endemic population of race O (Hooker 1974) suggests there may be an identifiable step in the molecular evolution of this plant-fungus interaction that can be defined genetically. Indeed, when race T, which produces T-toxin, is crossed with a naturally occurring nontoxin-producing strain (race O), only parental progeny are observed and they segregate in a 1:1 ratio (Tox^{+}:Tox^{-}). All progeny producing T-toxin are highly virulent on corn with T cytoplasm; all progeny not producing T-toxin have low virulence (Yoder and Gracen 1975; Bronson et al. 1990). Thus, both T-toxin and race T of the fungus are specific for T-cytoplasm corn and an apparent single genetic locus, designated *Tox1*, determines whether or not T-toxin is produced (Leach et al. 1982b). Because there is clear genetic definition underlying control of T-toxin production, a genetic (rather than biochemical) strategy has been used to investigate the molecular mechanism of toxin biosynthesis.

The simplicity suggested by the identification of *Tox1* was misleading. Genetic complexities associated with T-toxin production were masked by the existence of a reciprocal translocation breakpoint tightly linked to *Tox1* (Bronson 1988; Tzeng et al. 1992). These complexities have been revealed by molecular technology and knowledge of the genetics of polyketide biosynthesis, which have been used to produce chemically induced and tagged insertional Tox^{-} mutants. Although analyses of these two classes of mutant have not yet led to a complete understanding of the genetic event(s) which gave rise to race T, they have revealed several notable features of the *Tox1* locus (Tzeng et al. 1992; Yoder et al. 1993, 1994; Turgeon et al. 1995a): (1) it is inseparable from a translocation breakpoint, (2) it is associated with highly repeated DNA, some of which is known to be A + T-rich, (3) there is more DNA in the $Tox1^{+}$ genome than in a near-isogenic $Tox1^{-}$ genome, (4) the locus is probably very large, (5) it is not a single locus as previously thought, but at least two loci, and (6) it contains DNA that is missing or nonfunctional in $Tox1^{-}$ strains.

1. Enrichment for Nontagged Tox^{-} Mutants

Chemically induced mutants were obtained by constructing a $Tox1^{+}$ strain of *C. heterostrophus* that was conditionally sensitive to its own toxin (Yang et al. 1994). The *T-urf13* gene, which encodes a 13 kDa T-toxin-binding protein and is unique to the mitochondrial chromosome of T-cytoplasm corn (Levings and Siedow 1992), was placed under the control of a regulatable promoter and transformed into the genome of a $Tox1^{+}$ strain (Yang et al. 1994). Growth of the transformant was normal on medium that repressed the promoter, but retarded on medium that induced the promoter. Conidia of this transformant were treated with ethylmethanesulfonate and plated on inducing medium. Of 362 survivors, nine were Tox^{-}. Crosses of the nine mutants with $Tox1^{+}$ and $Tox1^{-}$ tester strains showed that each of them carried a single gene mutation and that each mutation mapped at the *Tox1* locus. One mutant, ctm45, produced no T-toxin at all, whereas the other eight mutants produced less toxin than the wild-type $Tox1^{+}$ strain. In all cases, reduced T-toxin-producing ability clearly segregated in progeny of crosses between mutant and wild type. Thus, one of the mutants has a tight Tox^{-} phenotype and eight of them have leaky phenotypes.

2. A Tagged *Tox*$^-$ Mutation

A strategy for placing a molecular tag at *Tox1* was based on the fact that T-toxin is a polyketide and should require a polyketide synthase (PKS) for its biosynthesis. Genes encoding PKSs have been cloned from several prokaryotes and a few fungi (Hopwood and Khosla 1992). Those found in certain Streptomycotina and most fungi tend to be very large, encode multifunctional polypeptides, and have domains which are conserved among various organisms. Primers were made to conserved regions of the PKS acyltransferase (AT) domain and used in a PCR reaction with *C. heterostrophus* race T genomic DNA as the template. A 300-bp PCR product was amplified, cloned, and by sequence, found to be similar to AT domains of PKSs from other organisms. The 300-bp fragment was inserted into a fungal transformation vector with *hygB* as the selectable marker and transformed into *C. heterostrophus* race T. Of 15 transformants recovered, 14 were Tox$^+$, and in each one *hygB* was unlinked to *Tox1*. One of the transformants, C4.PKS.13, was Tox$^-$ and, when crossed to race T, all Tox$^+$ progeny were *hygB*^{S} and all Tox$^-$ progeny were *hygB*^{R}, indicating that the *Tox*$^-$ mutation was tagged with *hygB*. When crossed with race O, all progeny were found to be Tox$^-$, indicating that the mutation maps at the *Tox1* locus. DNA flanking the plasmid insertion point was recovered and sequenced. It was found to be highly repeated, A + T-rich, and to have no resemblance to a PKS-encoding gene. Chromosomes of C4.PKS.13 were separated on a CHEF gel and probed with the transformation vector. The karyotype was identical to that of the progenitor race T strain, except that the chromosome bearing the plasmid insertion was about 100 kb smaller in the mutant than in wild type. When genomic DNAs were digested with the rare-cutting enzyme *Not*I, two large fragments (700 and 540 kb) were found in digests of the *Tox1*$^+$ strain which were missing in the wild-type *Tox1*$^-$ strain. In *Tox*$^-$ mutant C4.PKS.13, the 540-kb fragment was missing and replaced by a fragment of 440 kb, which also carried the tagged mutation site; the 700-kb fragment remained unchanged in the mutant. Taken together, the karyotype and *Not*I analyses indicate that insertion of the plasmid was associated with a large deletion. It is likely that at least some of the deleted sequences are required for the biosynthesis of T-toxin.

3. Tagged *Tox*$^-$ Mutants Using the REMI Procedure

A second strategy for obtaining tagged *Tox1*$^-$ mutants was based on a procedure, recently described for yeast (Schiestl and Petes 1991), to randomly mutagenize the genome and simultaneously mark the mutation with a selectable gene. This procedure, called REMI (Restriction Enzyme-Mediated Integration), has been used to mark developmental mutations in *Dictyostelium discoideum* (Kuspa and Loomis 1992; Dynes et al. 1994) and was adapted for use with *C. heterostrophus* (Lu et al. 1994). Transformants were screened for a Tox$^-$ phenotype. Among approximately 1300 *C. heterostrophus* transformants recovered after the REMI procedure, two (R.C4.186 and R.C4.350L) were Tox$^-$ when tested with a microbial assay (Ciuffetti et al. 1992). When either of the *Tox*$^-$ REMI transformants was crossed with race T, progeny segregated 1:1 (Tox$^+$:Tox$^-$), indicating mutation of a single gene in each case. All Tox$^-$ progeny were also *hygB*^{R}, providing evidence that the *Tox*$^-$ mutation in each transformant was tagged with *hygB*. When either of the *Tox*$^-$ mutants was crossed with race O, all progeny were Tox$^-$, indicating that both mutations mapped at the genetically defined *Tox1* locus. DNA flanking the plasmid insertion point was isolated from transformant R.C4.350L. Sequence analysis revealed a 7.6-kb ORF containing six enzymatic domains typical of a PKS: ketoacyl-ACP synthase (KS), acyltransferase (AT), dehydrogenase (DH), enoyl reductase (ER), ketoreductase (KR), and acyl-carrier protein (ACP). The gene, *PKS1*, contains one synthase unit, which occurs as a single copy in the genome of race T, but is completely lacking in race O. Disruption of *PKS1* in race T by site-specific mutagenesis causes loss of both T-toxin production and high virulence to T-cytoplasm corn.

4. Linkage Among *Tox*$^-$ Mutants

The tagged mutations R.C4.186, R.C4.350L, and C4.PKS.13 all map at the *Tox1* locus and therefore might be expected to show linkage to each other. Indeed, progeny of crosses between R.C4.186 and R.C4.350L were 100% *Tox*$^-$; *hygB*^{R}, indicating that these two mutations are linked. However, when either R.C4.186 or R.C4.350L was crossed to C4.PKS.13, progeny segregated 25% *Tox*$^+$, revealing that there is no detectable linkage between either of the two REMI mutations and the one sustained by C4.PKS.13. Physical evidence supporting the lack of linkage between these mutations can be seen upon probing separated chromosomes. The transformation plasmid inserted into two different chromosomes in the two mutants. These results at first seem inconsistent with the genetic data showing that all three insertional mutations are linked to *Tox1*. The resolution to this apparent dilemma lies in an earlier observation that the breakpoint of a reciprocal translocation is inseparably linked to *Tox1*; chromosomes 6 and 12 in race O exchanged arms to produce chromosomes 12;6 and 6;12 in race T (Tzeng et al. 1992). Considering the recently obtained molecular data along with the fact that race T and race O differ by a chromosomal rearrangement, it is clear that *Tox1* is not a single genetic locus, but rather two loci, one on chromosome 12;6 (designated *Tox1A*), the site of *PKS1*, and the other on chromosome 6;12 (designated *Tox1B*). Both *Tox1A* and *Tox1B* appear to be at or near the breakpoint of the translocation on each chromosome. The two loci always appear linked to each other (in fact appear inseparable) whenever the translocation breakpoint is heterozygous in a cross, i.e., in any cross between a race T and a race O strain. However, in a cross where the breakpoint is homozygous, as it is when either REMI mutant is crossed to C4.PKS.13, *Tox1A* and *Tox1B* segregate independently of each other.

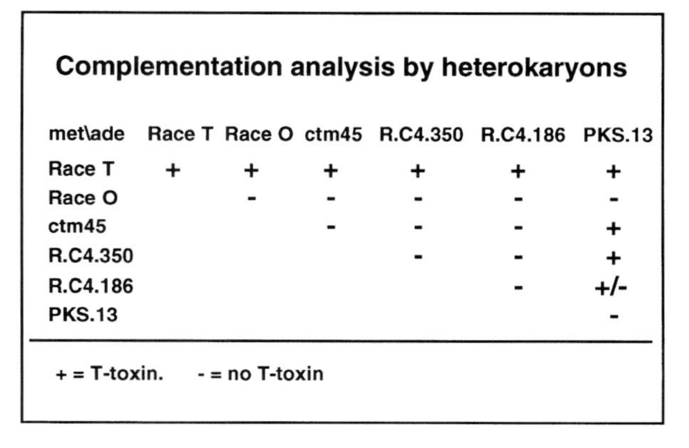

Complementation analysis by heterokaryons

met\ade	Race T	Race O	ctm45	R.C4.350	R.C4.186	PKS.13
Race T	+	+	+	+	+	+
Race O		-	-	-	-	-
ctm45			-	-	-	+
R.C4.350				-	-	+
R.C4.186					-	+/-
PKS.13						-

+ = T-toxin. - = no T-toxin

Fig. 3. Mutations that map at *Tox1A* (ctm45, R.C4.350, and R.C4.186) complement the one that maps at *Tox1B* (PKS.13), but not each other. Race O does not complement any $Tox1^-$ mutation, suggesting that it lacks functional DNA at both *Tox1A* and *Tox1B*; see text for details. The results shown here were also obtained when the forcing markers (*ade1* and *met2*) were reversed

5. Race O Lacks Both *Tox1A* and *Tox1B*

We have used heterokaryon analysis to investigate complementation among the various Tox^- mutants (Fig. 3). Auxotrophic markers *ade1* and *met2* were individually recombined into ctm45 (the chemically induced Tox^- mutant that has a tight phenotype; Sect. II.B.1), R.C4.186, and R.C4.350L, all of which map at *Tox1A*, and C4.PKS.13, which maps at *Tox1B* (Sect. II.B.4). Heterokaryons were forced on minimal medium between strains carrying *ade1* or *met2*, then assayed for ability to produce T-toxin. Heterokaryons carrying nuclei of C4.PKS.13 and either ctm45 or R.C4.186 or R.C4.350L produced T-toxin, (low levels were observed with R.C4.186). Complementation did not occur between any other pairs of mutants. These results are consistent with karyotype analysis, indicating that ctm45, R.C4.186 and R.C4.350L are at *Tox1A*, while C4.PKS.13 is at *Tox1B*. Interestingly, race O nuclei did not complement any of the induced Tox^- mutations, suggesting that the DNA necessary for T-toxin biosynthesis is missing or nonfunctional in race O or that race O has a mechanism for metabolizing T-toxin. However, race O/race T heterokaryons produced as much T-toxin in the microbial assay (Ciuffetti et al. 1992) as race T/race T heterokaryons, an observation which supports a previous report (Leach et al. 1982b) that the $Tox1^+$ gene is dominant, and is not consistent with the toxin metabolism hypothesis.

All meiotic and molecular genetic analyses of the ability of *C. heterostrophus* to attack corn have focused on production of T-toxin. Although T-toxin has been proven to be a virulence factor in southern corn leaf blight, it is clear that additional factors are required by *C. heterostrophus* for pathogenesis, since race O, which does not produce T-toxin, can be an effective pathogen (under appropriate conditions of environment and plant genotype). The existence of these hypothetical factors has been investigated (Yoder 1981) but so far none has been conclusively identified. A major impediment to analysis of pathogenicity in *C. heterostrophus* in the past has been the inability to produce nonpathogenic mutants. This problem may be rectified by the availability of the REMI procedure for producing tagged mutations. It can be predicted that several genes will be identified, since pathogenicity involves a complex process in the fungus and control of disease reaction in the host plant (corn) is multigenic (see Sect. II.C).

C. Genetics of Host Susceptibility

Susceptibility of corn to race T is inherited maternally, indicating non-nuclear genetic control of disease reaction (Hooker 1974). Consistent with this observation, the site of action of T-toxin in T-cytoplasm corn was determined to be a 13-kDa polypeptide (URF13) found only in the inner mitochondrial membrane (Levings and Siedow 1992). Plants with T-cytoplasm produce URF13, are male sterile, and are sensitive to T-toxin. URF13 is encoded by the gene *T-urf13*, which is unique to T-cytoplasm mitochondrial DNA and is the result of complex multiple rearrangements of the mitochondrial chromosome. After specific binding of URF13 to toxin (Braun et al. 1990), essential metabolites leak (Matthews et al. 1979) through enlarged pores created in the membrane by the T-toxin/URF13 complex (Korth et al. 1991; Kaspi and Siedow 1993).

The mitochondrial gene *T-urf13* appears to be the sole determinant of plant sensitivity to T-toxin. However, the level of plant sensitivity can be influenced by nuclear genes, since inhibition by T-toxin varies among corn inbreds and hybrids (Payne and Yoder 1978). The effect of the nuclear genome on plant sensitivity to T-toxin may be an artifact of the assay conditions, because the level of apparent sensitivity is a function of the type of assay used. Moreover, corn inbreds and hybrids vary widely in susceptibility to *C. heterostrophus* itself, yet there is no correlation between susceptibility to the fungus and sensitivity to T-toxin (Payne and Yoder 1978). These results indicate that increased levels of disease caused by race T (the producer of T-toxin) are due entirely to action of the mitochondrial *T-urf13* gene, while plant reaction to the fungus itself, independent of T-toxin, is controlled by nuclear genes.

The number of nuclear genes determining plant response to *C. heterostrophus* is not well established. Quantitative inheritance of plant reaction to *C. heterostrophus* was reported in 1954 (Pate and Harvey 1954) and observed later in additional sources of germplasm (Holley and Goodman 1989). A qualitative genetic effect was established with the discovery of chlorotic lesion resistance, which restricts lesion size and reduces sporulation by the fungus (Craig and Daniel-Kalio 1968). Genetic control of this resistance was first associated with two linked recessive genes (Craig and Fajemisin 1969), and later attributed to a single recessive gene called *rhm* (Smith and Hooker 1973; Smith 1975), which is located at the end of the short arm of chromosome 6 (Zaitlin et al. 1993). The relationship, if any, between *rhm* and two additional independent recessive genes for resistance (Thompson and Bergquist 1984) is unclear. In contrast to reports of recessive genes for resistance to *C. heterostrophus*, dominant resistance was found in inbred RD4515 (derived from parental strains Mo17 and W153R), which is controlled by either one or two nuclear genes (Ceballos and Gracen 1988). It is obvious that much confusion exists concerning nuclear gene control of resistance to *C. heterostrophus*, probably because several genes are involved and no study to date has related them to each other. There is no evidence that any of the nuclear genes described as controlling plant reaction to *C. heterostrophus* affect reaction of corn to T-toxin. Moreover, the nuclear genes *Rf1* and *Rf2*, which restore male fertility to T-cytoplasm corn (Levings and Siedow 1992), have little or no effect on reaction of corn to either *C. heterostrophus* or T-toxin.

D. Role of T-Toxin in Pathogenesis

The relative importance of various lines of evidence in evaluating the role of a toxin (or other fungal metabolite) in pathogenesis was discussed earlier (Yoder 1980). The strongest evidence is derived from both conventional genetics and cloning and site-specific mutation of genes involved in toxin biosynthesis. For *C. heterostrophus*, circumstantial evidence that T-toxin is a virulence factor was observed in segregating progenies of crosses between races T and O. Only parental types were found, indicating a close association between production of T-toxin and high virulence to T-cytoplasm corn (Yoder 1980). To more rigorously evaluate the role of T-toxin in pathogenesis, all induced *Tox*$^-$ mutants obtained to date (see above) have been tested on corn plants with T or N cytoplasm. All mutants with a tight Tox$^-$ phenotype (R.C4.186, R.C4.350L, C4.PKS.13, and ctm45) cause mild symptoms on corn that are indistinguishable from those caused by race O. Mutants with leaky Tox⁻ phenotypes cause weak race T symptoms on corn. These genetic correlations firmly support the hypothesis that T-toxin is required by *C. heterostrophus* for its high virulence on T-cytoplasm corn. It is interesting to note that in this case, specific interaction (binding) between a fungal virulence factor (T-toxin) and its host plant binding site (URF13) leads to a susceptible outcome of the plant/fungus interaction. Thus, the reason T-toxin is host-specific is because its binding site is unique to only one plant genotype, i.e., corn carrying *T-urf13*. All other living organisms apparently lack this gene; therefore, to them T-toxin is harmless. This molecule, in fact, should probably not be considered a toxin, but rather a virulence determinant which is active only in the presence of cells producing URF13. In contrast, for *C. carbonum* specific interaction leads to resistance and the fungal determinant of pathogenicity appears to have nonspecific toxic activity toward the ubiquitous enzyme histone deacetylase (see below).

III. *Cochliobolus carbonum*, Corn, and HC-Toxin

The *TOX2* gene of *C. carbonum* race 1 controls production of HC-toxin, which is specifically active against corn with the genotype *hmhm*, as is the fungus itself (Yoder 1980). Any corn with the genotype *Hm1* – is resistant to both HC-toxin and *C. carbonum*. In crosses between race 1 and race 2 (which makes no HC-toxin), the ability to produce toxin segregates as a single gene (*TOX2*; Yoder et al. 1989).

A. Chemistry of HC-Toxin

The structures of HC-toxins (and the victorins as well) resemble, in certain ways, those of many other bioactive peptides. They occur as families of closely related cyclic peptides that are resistant to proteases, are composed of various unusual amino acids and other non-protein constituents, and have a low molecular weight (250–4000 daltons), indicating a nonribosomal origin. For HC-toxin there is one prominent form (Fig. 2) called HC-toxin I, cyclo-(D-Pro-L-Ala-D-Ala-L-Aeo; Gross et al. 1982; Liesch et al. 1982; Walton et al. 1982; Kawai et al. 1983; Pope et al. 1983), the structure of which was confirmed by total synthesis (Kawai and Rich 1983). There are two closely related peptides, HC-toxins II and III (Kim et al. 1985; Tanis et al. 1986), which are less abundant in culture filtrates than is HC-toxin I. A fourth HC-toxin has been found in culture filtrates, but its structure has not been determined (Rasmussen and Scheffer 1988). All three of the characterized toxins contain Aeo (amino-oxo-epoxydecanoic acid). Both the epoxide (Ciuffetti et al. 1983; Walton and Earle 1983) and the keto group (Kim et al. 1987) of the Aeo residue (Fig. 2) are essential for toxicity. The epoxy function hydrolyzes readily in acidic solutions to a dihydroxy group, whereas the keto function can be reduced by sodium borohydride to form two epimeric hydroxy epoxides. All three reduction products are completely lacking in biological activity.

B. Genetics of HC-Toxin Biosynthesis

HC-toxin is synthesized by a 570-kDa multifunctional cyclic peptide synthase called HTS (Walton and Holden 1988). HTS has four amino acid activation domains, presumably corresponding to the four amino acids comprising the toxin. To understand *TOX2* genetically, it was assumed that the locus encodes a cyclic peptide synthetase and a biochemical approach employing reverse genetics was used to identify DNA at the *Tox2* locus (Walton and Panaccione 1993). Two enzyme fragments (called HTS-1 and HTS-2), released by proteolytic cleavage of HTS during purification of the enzyme and believed (prior to confirmation), to be involved in cyclic peptide synthesis, were purified from a *TOX2* strain. These fragments were not found in *tox2* strains. Antibody prepared against HTS-1 was used to isolate a portion of the corresponding gene (*HTS1*) from a cDNA expression library (Panaccione et al. 1992). The cDNA clone was used to identify a 16-kb lambda clone, which, in turn, was used to identify two additional overlapping lambda clones. Through a combination of restriction enzyme mapping of these clones, hybridization of restriction enzyme fragments of these clones to DNA from *Tox*⁺ and *Tox*⁻ strains, and sequencing, the *HTS1* gene was localized to a 22-kb genomic fragment. *HTS1* encodes a 15.7-kb ORF without introns, which has similarity to known cyclic peptide synthetase-encoding genes (Scott-Craig et al. 1992). There are two linked copies of the 22-kb fragment, embedded in highly repeated sequences, which are not found in *tox2* strains or in other fungi. Disruption of one or the other copy of *HTS1* resulted in reduced enzyme activity; however, the strain was still fully pathogenic on *hmhm* corn. When both copies of *HTS1* were disrupted in the genome, the fungus lost HTS activity, ability to produce HC-toxin, and pathogenicity to corn (Panaccione et al. 1992). Transfer of the 22-kb region into a *tox2* strain does not restore toxin production, suggesting that additional DNA beyond the 22-kb region is required for production of HC-toxin (Walton et al. 1994b).

Within the 22-kb unique fragment is an additional transcribed sequence, detected by probing a cDNA library with unique DNA from a *Tox*$^+$ strain. The gene (*TOXA*) encodes a 58-kDa protein with similarity to proteins which function as antibiotic pumps. Such pumps have been found in bacterial genomes adjacent to clusters of genes for antibiotic production (Walton et al. 1994a). Hydropathy plots of the amino acid sequence suggest that the protein has ten possible membrane-spanning domains. There are two linked copies of *TOXA*, and disruption of either of the copies has no detectable phenotype. It has not been possible, to date, to construct a strain with both copies to *TOXA* disrupted, suggesting that such a strain is nonviable due to intracellular accumulation of HC-toxin. Thus the function of *TOXA* appears to be HC-toxin efflux and/or auto-resistance (Walton et al. 1994a).

Recently, two additional genes, *TOXC* (Walton et al. 1994a) and *TOXD* (Scott-Craig et al. 1995), were identified on the same chromosome as *HTS1* in *TOX2* strains. *TOXC* is large (7 kb), present in three copies, and is thought to encode a fatty acid synthase involved in biosynthesis of Aeo, the fourth amino acid component of HC-toxin; *TOXC* is not found in *tox2* strains. The function of *TOXD* is unknown, but the gene is in triplicate at the *TOX2* locus, encodes a transcript of 1.4 kb, and is unique to *Tox*$^+$ strains (Scott-Craig et al. 1995).

Interestingly, *HTS1* is found on different chromosomes in different *TOX2* isolates. In isolate SB111 it is found on the largest chromosome (3.5 kb), but on chromosomes ranging in size from 1.8 to 3.5 Mb in other isolates (Walton et al. 1994a). Chromosome-specific probes are being used to determine if the chromosome on which *HTS1* resides can recombine with other chromosomes and if it is entirely dispensable.

C. Genetics of Host Resistance

Plant reaction to *C. carbonum* and to HC-toxin is determined by two genes, each of which has one or more dominant alleles for resistance; *Hm1* on chromosome 1 and *Hm2* on chromosome 9 (Nelson and Ullstrup 1964). Conventional genetic analysis defined two alleles at *Hm2* and four alleles at *Hm1*, with the following epistatic relationships: *Hm1* > *Hm1A* > *Hm1B* > *hm1*. The resistance conferred by *Hm1* is epistatic to that of all other alleles at either locus. Full susceptibility is achieved only if plants are homozygous recessive at both loci. Intermediate levels of resistance to the fungus and to HC-toxin are determined by various combinations of alleles at *Hm1* and *Hm2*.

The molecular basis for resistance of corn containing *Hm1* to HC-toxin and to *C. carbonum* race 1 was determined from two lines of evidence. First, radiolabeled HC-toxin was used to study the fate of HC-toxin in corn tissues and leaf extracts. *Hm1* corn was found to have an enzyme known as HC-toxin reductase (HCTR); corn with the *hm1hm1* genotype has little or no enzyme activity (Meeley et al. 1992). HCTR reduces the 8-carbonyl group of Aeo on the toxin molecule (Fig. 2) to a hydroxyl, thereby inactivating the toxin (Meeley and Walton 1991). Second, the *Hm1* gene was cloned by transposon tagging and sequenced, revealing homology to known carbonyl reductases and indicating that *Hm1* encodes HCTR (Johal and Briggs 1992). The *hm1* allele carries a transposon insertion which inactivates the gene, whereas the molecular nature of the *Hm1A* and *Hm1B* alleles is unknown. The cloning of *Hm2* is underway, with the prediction that it also encodes HCTR (S. Briggs, pers. comm.).

There are homologues of *Hm1* in all monocots and dicots tested to date, suggesting that they are all resistant to *C. carbonum* because they can detoxify HC-toxin, and predicting that mutation of the *Hm1* homologue of any plant would make it susceptible to *C. carbonum*. Natural susceptibility is caused by the rare case (in *hmhm* corn) of absence of the enzyme. Mutation of *Hm1* is thought to have occurred in a corn line in Kansas under little disease pressure (Briggs and Johal 1994). When the germplasm was moved to the cornbelt (where *C. carbonum* is endemic) in the 1930s, a new disease resulted.

Although HCTR interacts with HC-toxin to cause resistance in plants expressing *Hm1*, the mechanism by which the toxin interferes with plant metabolism has remained obscure (Yoder and Scheffer 1973a,b). Recent data, however, suggest that the site of toxin action may be histone deacetylase, a ubiquitous enzyme required for modification of histones (Walton et al. 1994a,b). This enzyme appears to be the site of action of trapoxin, a novel cyclotetrapeptide isolated from the fungus *Helicoma ambiens* and structurally related to HC-toxin (Kijima et al. 1993). Histone deacetylase is thought to facilitate binding of regulatory proteins to DNA, which then promotes induction of certain classes of genes. Walton et al. (Walton et al. 1994a) have proposed that HC-toxin inhibition of histone deacetylase prevents binding of regulatory proteins to promoters of genes involved in plant defense responses such as those that encode PR proteins and enzymes involved in biosynthesis of phytoalexins, etc. Since histone deacetylase is widely distributed among living things, it seems likely that most or all plants are sensitive to HC-toxin (but usually protected by HCTR), and therefore that this toxin is not really host-specific. Indeed, *C. carbonum* itself appears sensitive to HC-toxin, since circumstantial evidence suggests that null mutations in HC-toxin pump genes are lethal (see above).

D. Role of HC-Toxin in Pathogenesis

As with *C. heterostrophus* (Sect. II.D), segregating progenies of *C. carbonum* have shown an association between toxin production by the fungus and pathogenesis to susceptible corn (Yoder 1980). Availability of cloned DNA from the *TOX2* gene has made molecular analysis possible. When both copies of *HTS1* were disrupted in the genome, the fungus lost cyclic peptide synthetase activity, ability to produce HC-toxin, and pathogenicity on corn (Panaccione et al. 1992). This is unequivocal evidence that HC-toxin is required for pathogenicity of *C. carbonum* race 1. In contrast to the case of *C. heterostrophus* (Sect. II.D), specific interaction of the *C. carbonum* pathogenicity determinant (HC-toxin) with the product of the plant resistance gene (HC-toxin carbonyl reductase) leads to resistance rather than susceptibility. Another difference is that HC-toxin does not seem to be host-specific, since its putative target site, histone deacetylase, is widely distributed among living things. This toxin only *appears* to be host-specific, since among plant species generally, only mutant corn carrying the genotype *hm1hm1* lacks the carbonyl reductase encoded by *Hm1* which inactivates the toxin. Corn carrying *Hm1* and all other plant species which have homologues of it can inactivate HC-toxin and are therefore relatively insensitive to its effects.

IV. *Cochliobolus victoriae*, Oats, and Victorin

C. victoriae is highly virulent on oats which carry the dominant *Vb* allele (Litzenberger 1949), but causes no disease on any other genotype of oats or on other plant species (Yoder 1980). Specificity is associated with production, by the fungus, of a chlorinated cyclized peptide called victorin, which has the same specificity toward *Vb* oats as the fungus itself (Macko et al. 1985; Wolpert et al. 1985). The disease was of major economic importance in the early 1940s as a consequence of the introduction of commercial oat varieties carrying the *Pc* gene for resistance to crown rust caused by *Puccinia coronata* (Meehan and Murphy 1946, 1947). It was later discovered that *Pc* and *Vb* are either the same genetic locus or are tightly linked (Wallace et al. 1967; Scheffer 1976; Rines and Luke 1985). If they are the same, it suggests that a single plant protein can cause susceptibility to one fungus and resistance to another. On the other hand, if *Pc* and *Vb* are different, a cluster of genes involved in plant reaction to microbial attack would be indicated. For experimental purposes, the system has the advantage that the cloning of either gene should lead directly to the cloning of the other. The *Vb* gene is the more attractive of the two, since victorin can be used as a probe for the gene product (Wolpert and Macko 1989; Wolpert et al. 1994).

A. Chemistry of Victorin

The most abundant toxic component in the culture filtrate of *C. victoriae*, victorin C (Fig. 2), is composed of a cyclic combination of the six components: glyoxylic acid, 5,5-dichloroleucine, erythro-β-hydroxyleucine, 2-alanyl-3,5-dihydroxy-D^2-cyclopentenone-1 or victalanine (victala), threo-β-hydroxylysine, and α-amino-β-chloro-acrylic acid, (aclaa) and the complete structure of victorin C has been elucidated (Macko et al. 1985; Wolpert et al. 1985). Using this structure as a guide, related structures could be deduced for four congeners (Wolpert et al. 1986). Furthermore, structure-activity relationships were established for the natural victorins and their derivatives (Wolpert et al. 1988; Macko et al. 1989). Additionally, another victorin species of very low biological activity, victorin M, similar to victorin C, with glycine instead of glyoxylic acid, has been isolated (Kinoshita et al. 1989).

The structure of victorin suggests that several enzymes are required for its biosynthesis. At least one cyclic peptide synthetase can be predicted for activation of the constituent amino acids and their assembly in a cyclized structure. One or more additional enzymes is expected for chlorination of the molecule. There may also be unique enzymes required for production of the unusual amino acids found in victorin. The identity of each of these enzymes may be revealed by analysis of *Tox*⁻ REMI mutants (since this type of mutation should occur randomly in the genome), and by direct investigation of cyclic peptide synthetases, as described below.

B. Genetics of Victorin Biosynthesis

Genetic control of victorin production by *C. victoriae* has not yet been investigated in detail. The most convincing evidence that victorin production is controlled by a single genetic locus is from segregation of progeny in interspecific crosses between *C. victoriae*, which produces victorin, and its close relative, *C. carbonum*, which produces HC-toxin but not victorin (Scheffer et al. 1967). A ratio of 1:1:1:1 was observed for ability of progeny to produce victorin only, HC-toxin only, both toxins, or neither toxin (Table 2). This indicates segregation of two unlinked loci, one controlling production of HC-toxin and designated *Tox2* (described above), and one controlling production of victorin and designated *Tox3* (Yoder et al. 1989).

To investigate the molecular genetics of victorin biosynthesis, *Tox*⁻ mutants of *C. victoriae* have been produced using the REMI procedure described above for *C. heterostrophus* (S.W. Lu, A.C.L. Churchill et al., unpubl.). Culture fluids of REMI transformants are screened by automated HPLC to identify those with an altered peak for victorin C, the predominant form of the toxin. Among approximately 620 REMI transformants, eight have been found which are defective in victorin production. One is a null mutant which produces no detectable victorin and is no longer pathogenic to plants. Five are leaky, producing less than 0.5 to 5% as much victorin as wild type, two are overproducers of victorin, accumulating two to four times more toxin in culture than wild type. Several of the mutants with a small or absent victorin C peak in HPLC have new peaks not produced by wild type, and one has altered production of secondary metabolites other than victorin C (A.C.L. Churchill, S.W. Lu et al., unpubl.). These may represent intermediates in the victorin biosynthetic pathway and as such their chemical structures may lead to an understanding of the evolution of toxin production.

Since victorin is a cyclized peptide, its biosynthesis should depend on the action of a cyclic peptide synthetase. As described above for HC-toxin, cyclic peptide synthetases are very large and have conserved amino acid activation domains. This latter feature allows amplification of PCR products generated from primers that correspond to appropriate conserved regions (Turgay and Marahiel 1994). These products can then be used to identify which cyclic peptide synthetase is required for victorin biosynthesis.

Table 2. Segregation of toxin producing ability and host specificity in progeny of an interspecific cross between the corn pathogen *C. carbonum* and the oat pathogen *C. victoriae*

	Parents		Progeny (ratio)			
	C. carbonum	*C. victoriae*	1 :	1 :	1 :	1
TOX gene	*TOX2*	*TOX3*	*TOX2* *TOX3*	None	*TOX2*	*TOX3*
Toxin produced	HC-toxin	Victorin	HC-toxin victorin	None	HC-toxin	Victorin
Pathogenicity to	Corn	Oats	Corn, oats	None	Corn	Oats

C. Genetics of Host Resistance

The *Vb* allele confers dominant susceptibility of oats to *C. victoriae* and to victorin, and is genetically associated with *Pc-2*, a gene conferring dominant resistance to certain races of the crown rust fungus *Puccinia coronata* (Meehan and Murphy 1946; Litzenberger 1949; Finkner 1953). The genetic linkage among these three phenotypes has never been broken (Luke and Wheeler 1964; Luke et al. 1966; Rines and Luke 1985), even when rigorously tested with a large number of induced mutants selected for resistance to victorin; susceptibility to *P. coronata* occurred simultaneously (Wallace et al. 1967). Thus, cloning of the *Vb* gene may lead immediately to identification of the *Pc-2* gene for rust resistance. It has been suggested that *Vb* determines toxin sensitivity by coding for a toxin receptor (Scheffer and Livingston 1984). The particular structural features required for toxic activity, revealed by structure-activity studies (Wolpert et al. 1988), suggested that victorin may interact with its active site through a covalent association mediated by the aldehyde group of the glyoxylic acid residue. The toxin structure also indicated procedures for the production of biologically active derivatives of victorin that could be used to identify the toxin receptor. One of these derivatives, the ^{125}I-labeled Bolton-Hunter derivative of victorin C, was used to identify victorin-binding proteins (VBP) in oat leaf slices (Wolpert and Macko 1989). These results indicated that victorin binds in a covalent and ligand-specific manner to two proteins, a 100-kDa VBP and a 15-kDa VBP. The 15-kDa VBP binds victorin in vivo in both susceptible and resistant oat genotypes. However, the 100-kDa VBP binds victorin in vivo only in susceptible and not in resistant genotypes. The genotype-specific binding of the 100-kDa VBP suggested that this protein may be causally involved in the plant response to victorin. More recently, immunocytochemical localization of biotinylated victorin, conducted with goat antibiotin antibodies and gold-conjugated antigoat antibodies, revealed genotype-specific localization of victorin to susceptible and not resistant oat leaf mitochondria (H. Israel, T.J. Wolpert, and V. Macko, unpubl.). Thus, victorin localization, as well as victorin binding by the 100-kDa VBP, has indicated genotype specificity.

The 100-kDa VBP was purified and used for the production of antisera specific to the 100-kDa VBP (Wolpert and Macko 1991). Anti-100-kDa VBP antibody was used for high resolution immunocytochemical localization of the 100-kDa VBP to the mitochondria of susceptible and resistant oat leaf tissue (Israel et al. 1992). Antisera were also used to screen cDNA libraries and a full-length cDNA was isolated. The identity of the cDNA was confirmed by comparative peptide mapping of the oat 100-kDa VBP protein and the protein expressed from the cDNA in *E. coli*. Nucleotide sequence analysis, in vitro translations, subcellular fractionation and genotype-distribution studies all indicated that the 100-kDa VBP was the P-protein component of the multienzyme complex, glycine decarboxylase (Wolpert et al. 1994). Glycine decarboxylase is a nuclear-encoded, mitochondrial, multienzyme complex composed of four constituent enzymes: the pyridoxal phosphate-containing 100-kDa P-protein; the lipoamide-containing 15-kDa H-protein; the tetrahydrofolate-containing 45-kDa T-protein; and the L-protein, a 61-kDa lipoamide dehydrogenase (Oliver et al. 1990). Investigations of victorin binding to mitochondrial extracts identified a 15-kDa protein which binds victorin in vitro in a ligand-specific manner. Isolation, purification, and N- terminal amino acid sequence analysis of this protein revealed that it is the H-protein component of glycine decarboxylase (Navarre and Wolpert 1995). Thus, victorin binds in a ligand-specific manner to two components, the P- and the H-proteins of glycine decarboxylase. The current interpretation of these binding data is that the 15-kDa VBP observed in vivo is likely the same protein observed in vitro, the H-protein component of glycine decarboxylase (J.M. Lorang and T.J. Wolpert, unpubl.).

An independent approach for the detection of VBPs employed anti-victorin antibodies and resulted in the identification of 100-, 65-, and 45-kDa VBPs (Akimitsu et al. 1992). Although this study concluded that the 65- and 45-kDa VBPs were proteolytic products of the 100-kDa VBP, it is interesting to note that these molecular weights are very close to the molecular weights of the L- and T-proteins of the glycine decarboxylase complex.

In green tissues, glycine decarboxylase plays a critical role in the photorespiratory pathway. Mutations in the photorespiratory cycle are lethal to plants (Artus et al. 1986; Husic et al. 1987). Glycine decarboxylase is also present in nongreen tissues and etiolated tissues of plants (Walker and Oliver 1986) and in other organisms. In *E. coli* the serine-glycine pathway constitutes a major metabolic route that has been estimated to accommodate up to 15% of the carbon assimilated from glucose (Stauffer 1987). In humans, the lack of glycine decarboxylase leads to the fatal disease, hyperglycinemia (Kume et al. 1988). Thus, glycine decarboxylase activity appears to be an essential function in green tissues of plants and in nonplants and presumably in the nongreen tissues of plants. Based on the importance of the enzyme function, it is possible that if the sole effect of victorin is the inhibition of the enzyme complex, such inhibition would be sufficient to cause cell death. Assuming that the in vivo-labeled 15-kDa VBP is the H-protein, the H-protein binds victorin in vivo in both susceptible and resistant genotypes, whereas the P-protein component of glycine decarboxylase binds victorin in vivo only in the toxin-sensitive, susceptible genotypes. The P- and H-proteins interact to catalyze the decarboxylation of glycine and the H-protein acts as a carrier for the resultant methylamine moiety.

Although a number of interpretations of the victorin-binding data are being considered (Wolpert et al. 1995), current speculation is that the H-protein binds victorin and then transfers victorin to the P-protein in susceptible genotypes, leading to an accumulation of victorin at the P-protein and inhibition of the enzyme complex, resulting in cell death. This transfer apparently does not take place in the resistant genotypes, and cell viability is not affected. The hypothesis suggests that the biological consequences of victorin action arise from the cumulative binding of toxin to the P-protein and that specificity is determined by the interaction of the P- and H-proteins. Based on the above interpretation of the victorin-binding data, the product of the *Vb* gene is most likely either the P- or H-protein or a protein which effects the interaction of the P- and H-proteins. Because the H-protein interacts with all the component enzymes of the complex, the most likely candidates for proteins which affect the interaction of the P- and H-proteins are the remaining components of the complex, the T- and L-proteins. However, it is possible that another protein, as yet undetected, influences the interaction of the P- and H-proteins.

Oats susceptible to Victoria blight are allohexaploids and assumed to have arisen from the combination of three distinct progenitor diploids. Because of the essential role of glycine decarboxylase, it is likely that each progenitor had at least one copy of the gene for each enzyme component of the complex. Reconstruction and Southern analysis have indicated that there are from three to six copies of the gene for the P-protein in *Avena sativa* (D.A. Navarre and T.J. Wolpert, unpubl.). Undoubtedly, there are multiple copies of the genes for the other components of the complex as well. Because the *Vb* gene behaves as a dominant disomic (Rines and Luke 1985), it is reasonable to assume that the *Vb* gene is restricted to one chromosome set. Thus, if *Vb* is a gene for one of the components of glycine decarboxylase, it must exist in a background of other alleles encoding the same product. The effort to determine which, if any, of the components of glycine decarboxylase is the product of *Vb* requires a technique which will distinguish differences conceivably as minor as a single base pair from among a number of loci encoding the same gene product. Current efforts are utilizing single-stranded conformational polymorphism analyses (SSCP) (Orita et al. 1989; Hayashi 1991) of polymerase chain reaction products of discrete segments of the coding region of the gene in question. When a difference between susceptible and resistant genotypes is detected from among the genes encoding a specific component of the complex, the difference is analyzed for cosegregation with toxin sensitivity in segregating lines of oats. The approach necessitates nucleotide sequence analysis of a cDNA for each component of the complex. Such information is available for the P-protein (Wolpert et al. 1994) and recently, a cDNA clone for the H-protein has been isolated (D.L. Moore, M.K. Assaf, and T.J. Wolpert, unpubl.). Efforts are underway for the isolation of cDNA clones for the T- and L-protein components. Hopefully, this approach will lead to the identification of the *Vb* gene.

D. Role of Victorin in Pathogenesis

All isolates of *C. victoriae* collected from the field which produce victorin are pathogenic to oats with the *Vb* allele; all isolates which have lost ability to produce victorin are nonpathogenic to oats. Progeny of interspecific crosses between *C. victoriae* and *C. carbonum* which produce victorin are pathogenic to *Vb* oats; all other progeny are nonpathogenic. Although these genetic associations are impressive, an even more persuasive line of evidence comes from the analysis of the *Tox*$^-$ REMI mutants described above. The *Tox*$^-$ null mutant is nonpathogenic to oats, whereas the *Tox*$^-$ leaky mutants kill oat plants at a much slower rate than does wild type. Thus, victorin appears to be required by the fungus for its pathogenicity to oats carrying the *Vb* gene.

V. Bipolaris sacchari, Sugar Cane, and HS-Toxin

A. Need for Genetic Analysis

B. sacchari produces a glycosylated sesquiterpene (HS-toxin, Fig. 2) that has been implicated in the eyespot disease of sugarcane (Macko 1983; Macko et al. 1983; Scheffer and Livingston 1984). Although HS-toxin has been isolated and chemically characterized, lack of a sexual cycle for *B. sacchari* has suspended serious attempts to study the molecular mechanisms of its synthesis or its role in disease. Fewer than 50% of *Bipolaris* spp. have a known sexual stage. *Bipolaris* is a segregate out of *Helminthosporium* reserved for species with *Cochliobolus* teleomorphs, as well as for certain morphologically similar conidial fungi such as *B. sacchari*, which lack sexual stages (Sivanesan 1984; Alcorn 1988). The discovery of a sexual stage of *B. sacchari* would prove a major advance in the genetic analysis of HS-toxin biosynthesis. This would make available segregating progenies of naturally occurring isolates, and would facilitate molecular analyses. For example, *Tox*$^-$ mutants produced by the REMI procedure could be quickly screened to identify those that are tagged.

B. Chemistry of HS-Toxins

B. sacchari produces three isomeric $C_{39}H_{64}O_{22}$ HS-toxins, the composition of which corresponds to a fourfold galactosidation of a sesquiterpene aglycone, $C_{15}H_{24}O_2$ (Macko et al. 1981, 1983). The structures were confirmed and their absolute stereochemistries were established by the total synthesis of a biologically fully active HS-toxin A (C. Hildenbrand, T. Dabre, V. Macko, D. Arigoni, 1986, unpubl.; Hildenbrand 1986).

During the isolation of toxins from culture fluids of *B. sacchari*, the extracts were found to contain an additional 21 closely related but less polar and nontoxic compounds, the so-called lower homologues of HS-toxins (Duvick et al. 1984; Livingston and Scheffer 1983, 1984). The multiplicity and distribution of molecular weights of the compounds suggest that they result from a random sequential loss of galactofuranose units from the three isomeric HS-toxins A, B, and C. Detailed analysis of the NMR spectra of each of the 21 compounds allowed for the derivation of their structures (F. Weibel, C. Hildenbrand, V. Macko, D. Arigoni, 1981, unpubl.; Macko 1983; Hildenbrand 1986). HS-toxins A, B, and C can also be designated $A_{2,2}$, $B_{2,2}$, and $C_{2,2}$, respectively. In this nomenclature, the A, B, and C denote the structures of the aglycone moieties, and the appended suffixes indicate the number of the β-D galactofuranose units in the sugar extensions. A loss of galactofuranose units from either side of their molecules is indicated as $A_{2,1}$, $A_{1,2}$, $B_{2,1}$, etc. Three of these lower homologues, $A_{1,0}$, $A_{1,1}$, and $A_{2,0}$, were synthesized de novo in their optically active form (Hildenbrand 1986).

A direct correlation between the toxins and their lower homologues was provided by use of an enzyme, β-D-galactofuranosidase, partially purified from culture filtrates of *B. sacchari*. The enzyme is capable of hydrolyzing the β-1-5 linkage, as demonstrated by hydrolysis of a synthetic substrate, benzyl-5-0-(β-galactofuranosyl)-β-galactofuranoside. Treatment of HS-toxin C ($C_{2,2}$) with this enzyme preparation resulted in the appearance of $C_{2,1}$, $C_{1,2}$, $C_{1,1}$, $C_{0,2}$, $C_{2,0}$, $C_{1,0}$, and $C_{0,1}$ in the reaction mixture (Macko 1983; Hildenbrand 1986). Pretreatment of susceptible sugarcane tissue with solutions of the isolated lower homologues prevented the toxic action of HS-toxin (Duvick et al. 1984; Livingston and Scheffer 1984). A biologically inactive form of HS-toxin was isolated from the culture filtrate as well. It was identified as a bis-α-D-glucopyranosyl derivative of HS-toxin $C_{2,2}$ and is obviously a latent form of the toxin, as treatment with α-glucosidase led to the corresponding HS-toxin (C. Hildenbrand, V. Macko, D. Arigoni, 1983, unpubl.; Macko 1983).

The aglycones of HS-toxins A and B have been obtained by acid hydrolysis of the corresponding toxins, and the structure of aglycones A and C, including their absolute configuration, verified by total synthesis (C. Hildenbrand, D. Arigoni, 1983, unpubl.; Hildenbrand 1986; Tanaka et al. 1992). Aglycone C was also isolated from the mycelium of the fungus and also could be prepared by enzymatic hydrolysis of HS-toxin C with β-D-galactofuranosidase (Nakajima and Scheffer 1987). Oxidation of the aglycone to a keto-aldehyde and consecutive reduction of the latter with sodiumborotritiite and lithiumaluminumhydride yielded 13-[^{3}H]aglycone C (11.2 mCi/mmol). When this labeled aglycone was fed to *B. sacchari*, approximately 6.2% of the radioactivity was incorporated into HS-toxin C, suggesting that the aglycones are precursors of the corresponding HS-toxins (Nakajima and Scheffer 1987). The results suggest the metabolic route aglycone → metabolite Y → HS toxin → metabolite X for the biosynthesis of HS toxin where metabolites Y and X are lower molecular weight homologues of the toxin. Perhaps the aglycone is a biosynthetic precursor of the toxin, and the toxin is formed after galactofuranose is attached to the aglycone.

C. Role of HS-Toxin in Pathogenesis

The main evidence that this toxin has a role in disease derives from the fact that sugarcane clones susceptible to the fungus tend to be sensitive to the toxin, whereas resistant clones tend to be insensitive. One study found a perfect correlation between toxin sensitivity and pathogen susceptibility (Bournival et al. 1994), whereas others have noted exceptions to this correlation (Steiner and Byther 1971; Scheffer and Livingston 1980), raising a question about the role that HS-toxin plays in disease development. An additional cautionary note was signaled by the observation that toxin activity was detected in culture filtrates of some strains of the fungus, but not in others (Steiner and Byther 1976). Cloning and manipulation of genes involved in HS-toxin biosynthesis should help to clarify the role of HS-toxin in the eyespot disease.

VI. Perspectives and Conclusions

A comparative summary of toxin/plant interactions for the systems described here is presented in Table 3. All of the toxins have very high biological activity toward plants sensitive to them. Victorin and T-toxin are extremely specific for their respective sensitive genotypes, whereas only a 100-fold increase in the concentration of HC-toxin has the same effects on resistant and nonhost plants as lower concentrations have on susceptible ones (Kuo et al. 1970). This relatively low specificity is probably due to different bases for resistance: detoxification in the case of HC-toxin (Sect. III.C) vs. lack of a toxin-sensitive site for victorin (Sect. IV.C) and T-toxin (Sect. II.C).

The study of host-specific toxins has progressed to the point where it seems reasonable to speculate about their evolutionary origins. Each fungus that produces a host-specific toxin was unknown until it appeared suddenly on a crop plant genotype which was highly and uniquely sensitive to the toxin. This pattern of occurrence prompts questions about the origin of genes for toxin production on the part of the fungus and the genes for host plant sensitivity on the part of the plant. T-toxin, produced by *Cochliobolus heterostrophus*, and HC-toxin, produced by *C. carbonum*, represent rare examples in which something is known not only about the mechanism of attack on the part of the pathogen, but also about the mechanism of susceptibility and/or resistance on the part of the plant. In the case of T-toxin, it is clear that the key molecular interaction determining enhanced development of disease is the binding of the toxin molecule to the URF13-protein located

Table 3. Characteristics of toxin/plant interactions[a]

	C. heterostrophus	*C. carbonum*	*C. victoriae*
Toxin activity	Lethal	Fungistatic (stimulatory at certain concentrations)[b]	Lethal (stimulatory at certain concentrations)[b]
Toxin specificity			
EC_{50} Sus plants	5 ng ml^{-1}	0.2 µg ml^{-1}	0.1 ng ml^{-1}
EC_{50} Res plants	5 µg ml^{-1}	20.0 µg ml^{-1}	>0.1 mg ml^{-1}
Site of action	URF-13 poreforming protein	Histone deacetylase	Glycine decarboxylase
Resistance mechanism	No URF-13	Carbonyl reductase	Resistant glycine decarboxylase

[a] Sites of action and resistance mechanisms are described in the text. No reliable information is available for HS-toxin.
[b] Stimulatory effects include increased growth (Kuo et al. 1970), uptake of solutes (Yoder and Scheffer 1973a,b), negative electropotential across membranes (Gardner et al. 1974), respiration (Scheffer and Pringle 1963), and biosynthesis of ethylene (Shain and Wheeler 1975) and extracellular polysaccharide (Walton and Earle 1985).

in the inner mitochondrial membrane of T cytoplasm corn (Levings and Siedow 1992). Resistance (or lack of sensitivity) of corn (and presumably all other plants as well) is due to absence of the URF13-protein. The unique specificity of T-toxin to corn containing URF13, and its lack of effect on all other living things, suggest that this molecule should not be regarded as a "toxin", since it is biologically inert except in the presence of one unique gene product normally found in just one genotype of plant.

The case of HC-toxin offers a sharp contrast. Plant resistance is caused not by lack of a toxin binding site, but by presence of a plant-produced enzyme, carbonyl reductase, which inactivates the toxin (Meeley and Walton 1991; Johal and Briggs 1992; Meeley et al. 1992). Most, if not all, plant species, both hosts and nonhosts of *C. carbonum*, appear to have this enzyme. Only corn with a specific mutation in the *Hm1* gene, which causes loss of ability to detoxify HC-toxin, is susceptible to *C. carbonum* race 1, the producer of HC-toxin. This observation suggests that any plant with a mutation in its *Hm1* homologue would be susceptible to *C. carbonum* race 1, a hypothesis yet to be tested. Thus, resistance of plants to *C. carbonum* and to its toxin is active, whereas susceptibility is active in the case of *C. heterostrophus* and T-toxin (Fig. 4). The mechanism of plant susceptibility to T-toxin is clear, but for HC-toxin it is unknown, although inhibition of histone deacetylase is currently being investigated (Walton et al. 1994a, see above).

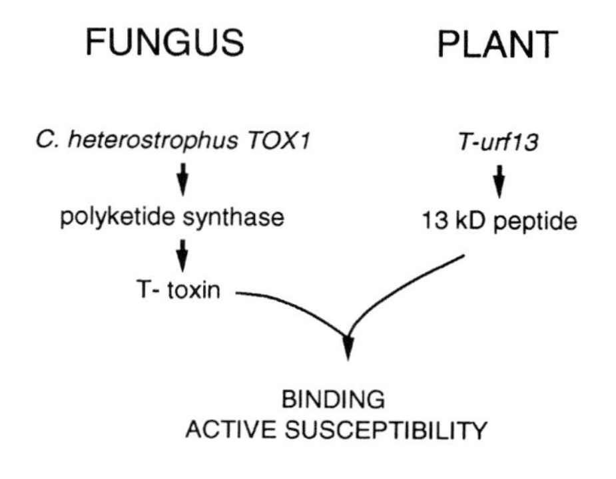

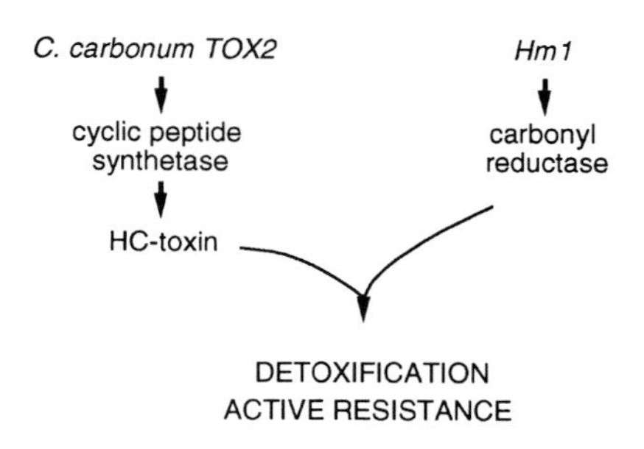

Fig. 4. Molecular mechanisms of host-parasite specificity are different in diseases caused by *C. heterostrophus* and *C. carbonum*. In both cases, specificity is determined by interaction of a fungal toxin with a plant protein. However, in the case of *C. heterostrophus*, the specific molecular interaction leads to susceptibility, whereas in the case of *C. carbonum* molecular specificity leads to resistance. The molecular mechanism of specificity of *C. victoriae* for oats probably resembles that of *C. heterostrophus* for corn, although the plant and fungal molecules involved are different. See text for details

How do genes for toxin production arise in fungi? Circumstantial evidence in the case of *C. heterostrophus* supports the idea that toxin production occurred following a mutation in a resident of the field population. In 1970, the first year in which the incidence of race T (the pathotype that produces T-toxin) was noticed to be high, virtually the entire field population of race T had the same mating type (*MAT-1*), whereas the two mating type genes were equally distributed in the field population of race O (the pathotype that does not produce T-toxin). By 1975, the *MAT* genes were equally distributed in the field populations of both race O and race T (Leonard 1971, 1973, 1977). These data are consistent with the hypothesis that race T is a mutant form of race O. Since race T was first observed on corn seed collected in 1968, even though the highly susceptible T cytoplasm corn was widely grown long before that, it seems likely that the mutation from race O to race T is a recent occurrence. Molecular analyses have confirmed that the ability to produce toxin is inseparable from the breakpoint of a reciprocal translocation (Tzeng et al. 1992). Moreover, race O is lacking DNA that is present in race T which is required for T-toxin production, suggesting that race T carries an insertion with respect to race O. How these two rearrangements of the race T genome affect production of T-toxin is currently under investigation.

The only other insight available on the genetic origin of a host-specific toxin comes from analysis of HC-toxin biosynthesis by *C. carbonum* race 1. Tox^{+} strains of this fungus have all or part of a chromosome that is missing in Tox^{-} strains and in other fungi (Walton et al. 1994a). On this chromosome are clustered at least two genes involved in toxin production (one encoding a cyclic peptide synthetase, the other an enzyme suspected of being involved in biosynthesis of one component of HC-toxin), and a third gene with homology to "antibiotic pump" encoding genes found in bacteria (Walton et al. 1994a). One interpretation of these observations is that *C. carbonum* acquired the ability to produce HC-toxin by importing DNA from another organism, perhaps a cyclic peptide-producing bacterium. Here, as with *C. heterostrophus*, toxin production is associated with a gross difference between the genomes of toxin-producing and nonproducing strains.

The two foregoing case studies are only a beginning. We can now determine how many plant diseases are mechanistically more like the T-toxin case or more like the HC-toxin case, or different from both. Another goal is to understand how and why fungi produce secondary metabolites such as host-specific toxins. The knowledge gained from these studies may have relevance beyond plant pathology. Several of the host-specific toxins are structurally similar to secondary metabolites that affect organisms other than plants. The fumonisins, which are similar to the host-specific AAL toxins produced by *Alternaria alternata lycopersici*, are toxic to mammals (Chen et al. 1992). This is also true for the HC-toxin-related cyclic peptides chlamydocin (Closse and Huguenin 1974), cyl-2 (Hirota et al. 1973), trapoxin (Itazaki et al. 1990), and WF-3161 (Umehara et al. 1990); a chlamydocin analog is also toxic to plants (Gupta et al. 1994). These observations suggest that, in nature, the host-specific toxins may have a broader, less specific, function than their selective toxicity to plants suggests.

References

Akimitsu K, Hart LP, Walton JD, Hollingsworth R (1992) Covalent binding sites of victorin in oat leaf tissues detected by anti-victorin polyclonal antibodies. Plant Physiol 98:121–126

Alcorn JL (1983) On the genera *Cochliobolus* and *Pseudocochliobolus*. Mycotaxon 16:353–379

Alcorn JL (1988) The taxonomy of "*Helminthosporium*" species. Annu Rev Phytopathol 26:37–56

Artus NN, Somerville SC, Somerville CR (1986) The biochemistry and cell biology of photorespiration. CRC Crit Rev Plant Sci 4:121–147

Bournival BL, Ginoza HS, Schenck S, Moore PH (1994) Characterization of sugarcane response to *Bipolaris sacchari*: inoculation and host-specific HS-toxin. Phytopathology 84:672–676

Braun CJ, Siedow JN, Levings CS (1990) Fungal toxins bind to the Urf13 protein in maize mitochondria and *Escherichia coli*. Plant Cell 2:153–161

Briggs SP, Johal GS (1994) Host-selective toxins and gene-for-gene interactions: mutually exclusive or faces of a coin? In: Kohmoto K, Yoder OC (eds) Host-specific toxin: biosynthesis, receptor and molecular biology. Faculty of Agriculture, Tottori University, Tottori, pp 219–226

Bronson CR (1988) Ascospore abortion in crosses of *Cochliobolus heterostrophus* heterozygous for the virulence locus *Tox1*. Genome 30:12–18

Bronson CR, Taga M, Yoder OC (1990) Genetic control and distorted segregation of T-toxin production in field isolates of *Cochliobolus heterostrophus*. Phytopathology 80:819–823

Ceballos H, Gracen VE (1988) A new source of resistance to *Bipolaris maydis* race T in maize. Maydica 33:233–246

Chen JP, Mirocha CJ, Xie WP, Hogge L, Olson D (1992) Production of the mycotoxin fumonisin b-1 by *Alternaria* alternata f. sp. *lycopersici*. Appl Environ Microbiol 58:3928–3931

Ciuffetti LM, Pope MR, Dunkle LD, Daly JM, Knoche HW (1983) Isolation and structure of an inactive product derived from the host-specific toxin produced by *Helminthosporium carbonum*. Biochemistry 22:3507–3510

Ciuffetti LM, Yoder OC, Turgeon BG (1992) A microbiological assay for host-specific fungal polyketide toxins. Fungal Genet Newsl 39:18–19

Closse A, Huguenin R (1974) Isolierung und Strukturaufklärung von Chlamydocin. Helv Chim Acta 57: 533–545

Comstock JC, Martinson CA, Gengenbach BG (1973) Host specificity of a toxin from *Phyllosticta maydis* for Texas cytoplasmically male sterile maize. Phytopathology 63:1357–1360

Craig J, Daniel-Kalio LA (1968) Chlorotic lesion resistance to *Helminthosporium maydis* in maize. Plant Dis Rep 52:134–136

Craig J, Fajemisin JM (1969) Inheritance of chlorotic lesion resistance to *Helminthosporium maydis* in maize. Plant Dis Rep 53:742–743

Danko SJ, Kono Y, Daly JM, Suzuki Y, Takeuchi S, McCrery DA (1984) Structural and biological activity of a host-specific toxin produced by the fungal corn pathogen *Phyllosticta maydis*. Biochemistry 23:759–766

Duvick JP, Daly JM, Kratky Z, Macko V, Acklin W, Arigoni D (1984) Biological activity of the isomeric forms of *Helminthosporium sacchari* toxin and of homologs produced in culture. Plant Physiol 74:117–122

Dynes JL, Clark AM, Shaulsky G, Kuspa A, Loomis WF, Firtel RA (1994) *LagC* is required for cell-cell interactions that are essential for cell-type differentiation in *Dictyostelium*. Genes Dev 8:948–958

Finkner VC (1953) Inheritance of susceptibility to *Helminthosporium victoriae* in crosses involving victoria and other crown rust resistant oat varieties. Agron J 45:404–406

Gardner JM, Scheffer RP, Higinbotham N (1974) Effects of host-specific toxins on electropotentials of plant cells. Plant Physiol 54:246–249

Gross ML, McCrery DA, Crow F, Tomer DB, Pope MR, Ciuffetti LM, Knoche HW, Daly JM, Dunkle LD (1982) The structure of the toxin from *Helminthosporium carbonum*. Tetrahedron Lett 23:5381–5384

Gupta S, Peiser G, Nakajima R, Hwang YS (1994) Characterization of a phytotoxic cyclotetrapeptide, a novel chlamydocin analogue, from *Verticillium coccosporum*. Tetrahedron Lett 35:6009–6012

Hayashi K (1991) A simple and sensitive method for detection of mutations in the genomic DNA. PCR Methods Appl 1:34–38

Hildenbrand C (1986) Neue Metabolite aus *Helminthosporium sacchari* und ein Beweis für die Struktur der HS-Toxine. Swiss Fed Inst Technol, Zürich, Thesis 8161, 241 pp

Hirota A, Suzuki A, Aizawa K, Tamura S (1973) Structure of Cyl-2, a novel cyclotetrapeptide from *Cylindrocladium scoparium*. Agric Biol Chem 37:955–956

Holley RN, Goodman MM (1989) New sources of resistance to Southern Corn Leaf Blight from tropical hybrid maize derivatives. Plant Dis 73:562–564

Hooker AL (1974) Cytoplasmic susceptibility in plant disease. Annu Rev Phytopathol 12:167–179

Hopwood DA, Khosla C (1992) Genes for polyketide secondary metabolic pathways in microorganisms and plants. In: Chadwick DJ, Whelan J (eds) Secondary metabolites: their function and evolution. John Wiley, Chichester, pp 88–112

Husic DW, Husic HD, Tolbert NE (1987) The oxidative photosynthetic carbon cycle or C2 cycle. CRC Crit Rev Plant Sci 5:45–100

Israel HW, Macko V, Wolpert TJ (1992) The victorin receptor site in oats. 1. Autoradiographical localization of a victorin C derivative. Phytopathology 82:244

Itazaki H, Nagashima K, Sugita K, Yoshida H, Kawamura Y, Yasuda Y, Matsumoto K, Ishi K, Uotani N, Et A (1990) Isolation and structural elucidation of new cyclotetrapeptides trapoxin a and trapoxin b having detransformation activities as antitumor agents. J Antibiot (Tokyo) 43:1524–1532

Johal GS, Briggs SP (1992) Reductase activity encoded by the *HM1* disease resistance gene in maize. Science 258:985–987

Kaspi CI, Siedow JN (1993) Cross-linking of the cms-T maize mitochondrial pore-forming protein URF13 by N,N′-dicyclohexylcarbodiimide and its effect on URF13 sensitivity to fungal toxins. J Biol Chem 268:5828–5833

Kawai M, Rich DH (1983) Total synthesis of the cyclic tetrapeptide, HC-toxin. Tetrahedron Lett 24:5309–5312

Kawai M, Rich DH, Walton JD (1983) The structure and conformation of HC-toxin. Biochem Biophys Res Commun 111:398–403

Kijima M, Yoshida M, Sugita K, Horinouchi S, Beppu T (1993) Trapoxin, an antitumor cyclic tetrapeptide, is an irreversible inhibitor of mammalian histone deacetylase. J Biol Chem 268:22429–22435

Kim SD, Knoche HW, Dunkle LD, McCrery DA, Tomer KB (1985) Structure of an amino acid analog of the host-specific toxin from *Helminthosporium carbonum*. Tetrahedron Lett 26:969–972

Kim SD, Knoche HW, Dunkle LD (1987) Essentiality of the ketone function for toxicity of the host-selective toxin produced by *Helminthosporium carbonum*. Physiol Mol Plant Pathol 30:433–440

Kinoshita T, Kono Y, Takeuchi S, Daly JM (1989) Structure of HV-toxin M, a host-specific toxin-related compound produced by *Helminthosporium victoriae*. Agric Biol Chem 53:1283–1290

Klittich CRJ, Bronson CR (1986) Reduced fitness associated with *Tox1* of *Cochliobolus heterostrophus*. Phytopathology 76:1294–1298

Kono Y, Daly JM (1979) Characterization of the host-specific pathotoxin produced by *Helminthosporium maydis* race T affecting corn with Texas male sterile cytoplasm. Bioorg Chem 8:391–397

Kono Y, Knoche HW, Daly JM (1981a) Structure: fungal host-specific. In: Durbin RD (ed) Toxins in plant disease. Academic Press, New York, pp 221–257

Kono Y, Takeuchi S, Kawarada A, Daly JM, Knoche HW (1981b) Studies on the host-specific pathotoxins produced in minor amounts by *Helminthosporium maydis* race T. Bioorg Chem 10:206–218

Kono Y, Danko SJ, Suzuki Y, Takeuchi S, Daly JM (1983) Structure of the host-specific pathotoxins produced by *Phyllosticta maydis*. Tetrahedron Lett 24:3803–3806

Korth KL, Kaspi CI, Siedow JN, Levings CS (1991) URF13, a maize mitochondrial pore-forming protein, is oligomeric and has a mixed orientation in *Escherichia coli* plasma membranes. Proc Natl Acad Sci USA 88:10865–10869

Kume A, Shigeo K, Keiya T, Koichi H (1988) The impaired expression of glycine decarboxylase in patients with hyperglycinemias. Biochem Biophys Res Commun 154: 292–297

Kuo MS, Yoder OC, Scheffer RP (1970) Comparative specificity of the toxins from *Helminthosporium carbonum* and *Helminthosporium victoriae*. Phytopathology 60:365–368

Kuspa A, Loomis WF (1992) Tagging developmental genes in *Dictyostelium* by restriction enzyme-mediated integration of plasmid DNA. Proc Natl Acad Sci USA 89:8803–8807

Leach J, Lang BR, Yoder OC (1982a) Methods for selection of mutants and in vitro culture of *Cochliobolus heterostrophus*. J Gen Microbiol 128:1719–1729

Leach J, Tegtmeier KJ, Daly JM, Yoder OC (1982b) Dominance at the *Tox1* locus controlling T-toxin production by *Cochliobolus heterostrophus*. Physiol Plant Pathol 21:327–333

Leonard KJ (1971) Association of virulence and mating type among *Helminthosporium maydis* isolates collected in 1970. Plant Dis Rep 55:759–760

Leonard KJ (1973) Association of mating type and virulence in *Helminthosporium maydis*, and observations on the origin of the race T population in the United States. Phytopathology 63:112–115

Leonard KJ (1977) Races of *Bipolaris maydis* in the Southeastern U.S. from 1974–1976. Plant Dis Rep 61:914–915

Levings CS, Siedow JN (1992) Molecular basis of disease susceptibility in the Texas cytoplasm of maize. Plant Mol Biol 19:135–147

Liesch JM, Sweeley CC, Staffeld GD, Anderson MS, Weber DJ, Scheffer RP (1982) Structure of HC-toxin, a cyclic tetrapeptide from *Helminthosporium carbonum*. Tetrahedron 38:45–48

Litzenberger SC (1949) Nature of susceptibility to *Helminthosporium victoriae* and resistance to *Puccinia coronata* in Victoria oats. Phytopathology 39:300–318

Livingston RS, Scheffer RP (1983) Conversion of *Helminthosporium sacchari* toxin to toxoids by β-galactofuranosidase from *Helminthosporium*. Plant Physiol 72:530–534

Livingston RS, Scheffer RP (1984) Selective toxins and analogs produced by *Helminthosporium sacchari*. Production, characterization and biological activity. Plant Physiol 76:96–102

Lu SW, Lyngholm L, Yang G, Bronson C, Yoder OC, Turgeon BG (1994) Tagged mutations at the *Tox1* locus of *Cochliobolus heterostrophus* using Restriction Enzyme Mediated Integration. Proc Natl Acad Sci USA 91:12649–12653

Luke HH, Wheeler H (1964) An intermediate reaction to victorin. Phytopathology 54:1492–1493

Luke HH, Murphy HC, Petr FC (1966) Inheritance of spontaneous mutations of the Victoria locus in oats. Phytopathology 56:210–212

Luttrell ES (1973) Loculoascomycetes. In: Ainsworth GC, Sparrow FK, Sussman AS (eds) The fungi. Academic Press, New York, pp 155–219

Luttrell ES, Rogerson CT (1959) Homothallism in an undescribed species of *Cochliobolus* and in *Cochliobolus kusanoi*. Mycologia 51:195–202

Macko V (1983) Structural aspects of toxins. In: Daly JM, Deverall BJ (eds) Toxins and plant pathogenesis. Academic Press, New York, pp 41–80

Macko V, Goodfriend K, Wachs T, Renwich JAA, Acklin W, Arigoni D (1981) Characterization of the host-specific toxins produced by *Helminthosporium sacchari*, the causal organism of eyespot disease of sugarcane. Experientia 37:923–924

Macko V, Acklin W, Hildenbrand C, Weibel F, Arigoni D (1983) Structure of three isomeric host-specific toxins from *Helminthosporium sacchari*. Experientia 39:343–347

Macko V, Wolpert TJ, Acklin W, Jaun B, Seibl J, Meili J, Arigoni D (1985) Characterization of victorin C, the major host-selective toxin from *Cochliobolus victoriae*: structure of degradation products. Experientia 41:1366–1370

Macko V, Wolpert TJ, Acklin W, Arigoni D (1989) Biological activities of structural variants of host-selective toxins from *Cochliobolus victoriae*. In: Graniti A, Durbin RD, Ballio A (eds) Phytotoxins and plant pathogenesis. Springer, Berlin Heidelberg New York, pp 31–42

Matthews DE, Gregory P, Gracen VE (1979) *Helminthosporium maydis* race T toxin induces leakage of NAD+ from T cytoplasm corn mitochondria. Plant Physiol 63:1149–1153

Meehan F, Murphy HC (1947) Differential phytotoxicity of metabolic by-products of *Helminthosporium victoriae*. Science 106:270–271

Meehan FL, Murphy HC (1946) A new *Helminthosporium* blight of oats. Science 104:413–414

Meeley RB, Walton JD (1991) Enzymatic detoxification of HC-toxin, the host-selective cyclic peptide from *Cochliobolus carbonum*. Plant Physiol 97:1080–1086

Meeley RB, Johal GS, Briggs SP, Walton JD (1992) A biochemical phenotype for a disease resistance gene of maize. Plant Cell 4:71–77

Mullin PG, Turgeon BG, Yoder OC (1993) Complementation of *Cochliobolus heterostrophus trp-* mutants produced by gene replacement. Fungal Genet Newsl 40:51–53

Nakajima H, Scheffer RP (1987) Interconversions of aglycone and host-selective toxin from *Helminthosporium sacchari*. Phytochemistry 26:1607–1611

Navarre DA, Wolpert TJ (1995) Inhibition of the glycine decarboxylase multienzyme complex by the host-selective toxin victorin. Plant Cell 7:463–471

Nelson OE, Ullstrup AJ (1964) Resistance to Leaf Spot in maize. Genetic control of resistance to race 1 of *Helminthosporium carbonum* Ull. J Hered 55:195–199

Nelson RR (1957) Heterothallism in *Helminthosporium maydis*. Phytopathology 47:191–192

Nelson RR (1959) *Cochliobolus carbonum*, the perfect stage of *Helminthosporium carbonum*. Phytopathology 49:807–810

Nelson RR (1960) *Cochliobolos victoriae*, the perfect stage of *Helminthosporium victoriae*. Phytopathology 50: 774–775

Oliver DJ, Neuburger M, Bourguignon J, Douce R (1990) Glycine metabolism by plant mitochondria. Physiol Plant 80:487–491

Orita M, Suzuki Y, Sekiya T, Hayashi K (1989) Rapid and sensitive detection of point mutations and DNA polymorphisms using the polymerase chain reaction. Genomics 5:874–879

Panaccione DG, Scott-Craig JS, Pocard JA, Walton JD (1992) A cyclic peptide synthetase gene required for pathogenicity of the fungus *Cochliobolus carbonum* on maize. Proc Natl Acad Sci USA 89:6590–6594

Pate JB, Harvey PH (1954) Studies on the inheritance of resistance in corn to *Helminthosporium maydis* leaf spot. Agron J 46:442–445

Payne GA, Yoder OC (1978) Effect of the nuclear genome of corn on sensitivity to *Helminthosporium maydis* race T toxin and on susceptibility to *H. maydis* race T. Phytopathology 68:331–337

Pope MR, Ciuffetti LM, Knoche HW, McCrery D, Daly JM, Dunkle LD (1983) Structure of the host-specific toxin produced by *Helminthosporium carbonum*. Biochemistry 22:3502–3506

Rasmussen JB, Scheffer RP (1988) Isolation and biological activities of four selective toxins from *Helminthosporium carbonum*. Plant Physiol 86:187–191

Rines HW, Luke HH (1985) Selection and regeneration of toxin-insensitive plants from tissue cultures of oats (*Avena sativa*) susceptible to *Helminthosporium victoriae*. Theor Appl Genet 71:16–21

Scheffer RP (1976) Host-specific toxins in relation to pathogenesis and disease resistance. In: Heitefuss R, Williams PH (eds) Encyclopedia of plant physiology, New Ser, vol 4, Physiological Plant Pathology. Springer, Berlin Heidelberg New York, pp 247–269

Scheffer RP (1989a) Ecological consequences of toxin production by *Cochliobolus* and related fungi. In: Graniti A, Durbin RD, Ballio A (eds) Phytotoxins and plant pathogenesis. Springer, Berlin Heidelberg New York, pp 285–300

Scheffer RP (1989b) Host-specific toxins in phytopathology: origins and evolution of the concept. In: Kohmoto K, Durbin RD (eds) Host-specific toxins: recognition and specificity factors in plant disease. Tottori University Press, Tottori, pp 1–17

Scheffer RP (1991) Role of toxins in evolution and ecology of plant pathogenic fungi. Experientia 47:804–811

Scheffer RP, Livingston RS (1980) Sensitivity of sugarcane clones to toxin from *Helminthosporium sacchari* as determined by electrolyte leakage. Phytopathology 70:400–404

Scheffer RP, Livingston RS (1984) Host-selective toxins and their role in plant diseases. Science 223:17–21

Scheffer RP, Pringle RB (1963) Respiratory effects of the selective toxin of *Helminthosporium victoriae*. Phytopathology 53:465–468

Scheffer RP, Nelson RR, Ullstrup AJ (1967) Inheritance of toxin production and pathogenicity in *Cochliobolus carbonum* and *Cochliobolus victoriae*. Phytopathology 57:1288–1291

Schiestl RH, Petes TD (1991) Integration of DNA fragments by illegitimate recombination in *Saccharomyces cerevisiae*. Proc Natl Acad Sci USA 88:7585–7589

Scott-Craig JS, Panaccione DG, Pocard JA, Walton JD (1992) The cyclic peptide synthetase catalyzing HC-toxin production in the filamentous fungus *Cochliobolus carbonum* is encoded by a 15.7-kilobase open reading frame. J Biol Chem 267:26044–26049

Scott-Craig JS, Pitkin JW, Ahn JH, Walton JD (1995) Molecular genetic analysis of the *TOX2* locus in *Cochliobolus carbonum*. Proc US-Japan Project on Molecular Biology of Host-Specific Toxins, Tottori University, Tottori Japan, Abstract

Shain L, Wheeler H (1975) Production of ethylene by oats resistant and susceptible to victorin. Phytopathology 65:88–89

Sharon A, Yamaguchi K, Christiansen S, Horwitz BA, Yoder OC, Turgeon BG (1996) An asexual fungus has the potential for sexual development. Mol Gen Genet 251:60–68

Sivanesan A (1984) The bitunicate Ascomycetes and their anamorphs. Strauss & Cramer, Hirschberg, 701 pp

Smith DR (1975) Expression of monogenic chlorotic-lesion resistance to *Helminthosporium maydis* in corn. Phytopathology 65:1160–1165

Smith DR, Hooker AL (1973) Monogenic chlorotic-lesion resistance in corn to *Helminthosporium maydis*. Crop Sci 13:330–331

Stauffer GV (1987) Biosynthesis of serine and glycine. In: Neidhardt FC, Ingraham JL, Low KB, Magasanik B, Schaechter M, Umbarger HE (eds) *Escherichia coli* and *Salmonella typhimurium* cellular and molecular biology. American Society for Microbiology, Washington, DC

Steiner GW, Byther RS (1971) Partial characterization and use of a host-specific toxin from *Helminthosporium sacchari* on sugarcane. Phytopathology 61:691–695

Steiner GW, Byther RS (1976) Comparison and characterization of toxin produced by *Helminthosporium sacchari* from Australia, Florida and Hawaii. Phytopathology 66:423–425

Suzuki Y, Knoche HW, Daly JM (1982a) Analogs of host-specific phytotoxin produced by *Helminthosporium maydis* race T. I. Synthesis. Bioorg Chem 11:300–312

Suzuki Y, Tegtmeier KJ, Daly JM, Knoche HW (1982b) Analogs of host-specific phytotoxin produced by *Helminthosporium maydis* race T. II. Biological activities. Bioorg Chem 11:313–321

Suzuki Y, Danko SJ, Daly JM, Kono Y, Knoche HW, Takeuchi S (1983) Comparison of activities of the host-specific toxin of *Helminthosporium maydis* race T and a synthetic C-41 analog. Plant Physiol 73:440–444

Tanaka A, Watanabe S, Yamashita K (1992) Synthesis of HS-toxin A aglycone. Biosci Biotechnol Biochem 56:104–107

Tanis SP, Horenstein BA, Scheffer RP, Rasmussen JB (1986) A new host specific toxin from *Helminthosporium carbonum*. Heterocycles 24:3423–3431

Tegtmeier KJ, Daly JM, Yoder OC (1982) T-toxin production by near-isogenic isolates of *Cochliobolus heterostrophus* races T and O. Phytopathology 72:1492–1495

Thompson DL, Bergquist RR (1984) Inheritance of mature plant resistance to *Helminthosporium maydis* race 0 in maize. Crop Sci 24:807–811

Turgay K, Marahiel MA (1994) A general approach for identifying and cloning peptide synthetase genes. Peptide Res 7:238–241

Turgeon BG, Garber RC, Yoder OC (1985) Transformation of the fungal maize pathogen *Cochliobolus heterostrophus* using the *Aspergillus nidulans amdS* gene. Mol Gen Genet 201:450–453

Turgeon BG, Garber RC, Yoder OC (1987) Development of a fungal transformation system based on selection of sequences with promoter activity. Mol Cell Biol 7:3297–3305

Turgeon BG, Bohlmann H, Ciuffetti LM, Christiansen SK, Yang G, Schafer W, Yoder OC (1993a) Cloning and analysis of the mating type genes from *Cochliobolus heterostrophus*. Mol Gen Genet 238:270–284

Turgeon BG, Christiansen SK, Yoder OC (1993b) Mating type genes in Ascomycetes and their Imperfect relatives. In: Reynolds DR, Taylor JW (eds) The fungal holomorph: mitotic, meiotic and pleomorphic speciation in fungal systematics. CAB International, Wallingford, pp 199–215

Turgeon BG, Kodama M, Yang G, Rose MS, Lu SW, Yoder OC (1995a) Function and chromosomal location of the *Cochliobolus heterostrophus Tox1* locus. Can J Bot 73:S1071–S1076

Turgeon BG, Sharon A, Wirsel S, Christiansen SK, Yamaguchi K, Yoder OC (1995b) Structure and function of mating type genes in *Cochliobolus* spp. and asexual fungi. Can J Bot 73:S778–S783

Tzeng TH, Lyngholm LK, Ford CF, Bronson CR (1992) A restriction fragment length polymorphism map and electrophoretic karyotype of the fungal maize pathogen *Cochliobolus heterostrophus*. Genetics 130:81–96

Umehara K, Nakahara K, Kiyoto S, Iwami M, Okamoto M, Tanaka H, Kohsaka M, Akoi H, Imanaka H (1990) Studies on WF-3161, a new antitumor antibiotic. J Antibiot 36:478–483

Waki T, Taga M, Tsuda M, Ueyama A (1979) Culture conditions for the ascocarp formation of *Cochliobolus kusanoi*. Trans Mycol Soc Jpn 20:73–81

Walker JL, Oliver DJ (1986) Light-induced increases in the glycine decarboxylase multienzyme complex from pea leaf mitochondria. Arch Biochem Biophys 248:626–638

Wallace AT, Singh RM, Browning RM (1967) Induced mutations at specific loci in higher plants III. Mutation response and spectrum of mutations at the *Vb* locus in *Avena byzantina* C. Kach. In: Gröber K, Scholz F, Zacharias M (eds) Induced mutations and their utilization. Erwin Baur Gedächtnisvorlesungen IV. Akademie-Verlag, Berlin, pp 47–57

Walton JD, Earle ED (1983) The epoxide in HC-toxin is required for activity against susceptible maize. Physiol Plant Pathol 22:371–376

Walton JD, Earle ED (1985) Stimulation of extracellular polysaccharide synthesis in oat protoplasts by the host-specific phytotoxin victorin. Planta 165:407–415

Walton JD, Holden FR (1988) Properties of two enzymes involved in the biosynthesis of the fungal pathogenicity factor HC-toxin. Mol Plant-Microbe Interact 1:128–134

Walton JD, Panaccione DC (1993) Host-selective toxins and disease specificity: perspectives and progress. Annu Rev Phytopathol 31:275–303

Walton JD, Earle ED, Gibson BW (1982) Purification and structure of the host-specific toxin from *Helminthosporium carbonum* race 1. Biochem Biophys Res Commun 107:785–794

Walton JD, Ahn JH, Akimitsu K, Pitkin JW, Ransom R (1994a) Leaf-spot disease of maize: chemistry, biochemistry and molecular biology of a host-selective cyclic peptide. In: Daniels MJ, Downie JA, Osbourn AE (eds) Advances in molecular genetics of plant-microbe interactions. Kluwer, Dordrecht, pp 231–237

Walton JD, Akimitsu K, Ahn JH, Pitkin JW (1994b) Towards an understanding of the *Tox2* gene of *Cochliobolus carbonum*. In: Kohmoto K, Yoder OC (eds) Host-specific toxin: biosynthesis, receptor and molecular biology. Faculty of Agriculture, Tottori University, Tottori, pp 227–237

Wirsel S, Turgeon BG, Yoder OC (1996) Deletion of the *cochliobolus heterostrophus* mating type (*MAT*) locus promotes function of *MAT* transgenes. Curr Genet 29:241–249

Wolpert TJ, Macko V (1989) Specific binding of victorin to a 100-kDa protein from oats. Proc Natl Acad Sci 86:4092–4096

Wolpert TJ, Macko V (1991) Immunological comparison of the in vitro and in vivo labeled victorin binding protein from susceptible oats. Plant Physiol 95:917–920

Wolpert TJ, Macko V, Acklin W, Jaun B, Seibl J, Meili J, Arigoni D (1985) Structure of victorin C, the major host-selective toxin from *Cochliobolus victoriae*. Experientia 41:1524–1529

Wolpert TJ, Macko V, Acklin W, Jaun B, Arigoni D (1986) Structure of the minor host-selective toxins from *Cochliobolus victoriae*. Experientia 42:1296–1299

Wolpert TJ, Macko V, Acklin W, Arigoni D (1988) Molecular features affecting the biological activity of the host-selective toxins from *Cochliobolus victoriae*. Plant Physiol 88:37–41

Wolpert TJ, Navarre DA, Moore DL, Macko V (1994) Identification of the 100-kD victorin binding protein from oats. Plant Cell 6:1145–1155

Wolpert TJ, Navarre DA, Lorang JM, Moore DL (1995) Molecular interactions of victorin and oats. Can J Bot 73:S475–S482

Yang G, Turgeon BG, Yoder OC (1994) Toxin-deficient mutants from a toxin-sensitive transformant of *Cochliobolus heterostrophus*. Genetics 137:751–757

Yoder OC (1973) A selective toxin produced by *Phyllosticta maydis*. Phytopathology 63:1361–1365

Yoder OC (1980) Toxins in pathogenesis. Annu Rev Phytopathol 18:103–129

Yoder OC (1981) Assay. In: Durbin RD (ed) Toxins in plant disease. Academic Press, New York, pp 45–78

Yoder OC, Gracen VE (1975) Segregation of pathogenicity types and host-specific toxin production in progenies of crosses between races T and O of *Helminthosporium maydis* (*Cochliobolus heterostrophus*). Phytopathology 65:273–276

Yoder OC, Scheffer RP (1973a) Effects of *Helminthosporium carbonum* toxin on absorption of solutes by corn roots. Plant Physiol 52:518–523

Yoder OC, Scheffer RP (1973b) Effects of *Helminthosporium carbonum* toxin on nitrate uptake and reduction by corn tissues. Plant Physiol 52:513–517

Yoder OC, Turgeon BG, Ciuffetti LM, Schäfer W (1989) Genetic analysis of toxin production by fungi. In: Graniti A, Durbin R, Ballio A (eds) Phytotoxins and plant pathogenesis. Springer, Berlin Heidelberg New York, pp 43–60

Yoder OC, Yang G, Adam G, DiazMinguez JM, Rose M, Turgeon BG (1993) Genetics of polyketide toxin biosynthesis by plant pathogenic fungi. In: Baltz RH, Hegeman GD, Skatrud PL (eds) Industrial microorganisms: basic and applied molecular genetics. American Society for Microbiology, Washington, DC, pp 217–225

Yoder OC, Yang G, Rose MS, Lu SW, Turgeon BG (1994) Complex genetic control of polyketide toxin production by *Cochliobolus heterostrophus*. In: Daniels MJ, Downie JA, Osbourn AE (eds) Advances in molecular genetics of plant-microbe interactions. Kluwer, Dordrecht, pp 223–230

Zaitlin D, Demars S, Ma Y (1993) Linkage of *rhm*, a recessive gene for resistance to southern corn leaf blight, to RFLP marker loci in maize (*Zea mays*) seedlings. Genome 36:555–564

9 Fungal Phytohormones in Pathogenic and Mutualistic Associations

B. TUDZYNSKI

CONTENTS

I. Introduction

Phytohormones are a naturally occurring group of organic substances that at low concentration control several stages of plant growth and development such as cell elongation, cell division, tissue differentiation, and apical dominance (Costacurta and Vanderleyden 1995) and occur in all major groups of flowering plants and gymnosperms. Five major groups of phytohormones regulate plant growth and development (Table 1), and all have been found in microorganisms. Both indole-3-acetic acid (Thimann 1936) and the gibberellins (Kurosawa 1926) were found very early in the study of plant growth regulators, long before their recovery in flowering plants. Abscisic acid is produced in large amounts by fungi of the genera *Cercospora* and *Botrytis* (Assante et al. 1977) and ethylene also by some fungal groups (Arshad and Frankenberger 1988). Cytokinins have been isolated from the bacterium *Pseudomonas syringae*, the causal organism of olive knot, and from some fungi as well (Mandahar and Suri 1983; Gulati and Mandahar 1986).

Fungi affect the growth and development of plants in many different ways. Mutualistic associations between fungi and plants (e.g., mycorrhizae) result in increased survival rates, more biomass production, better health, and better resistance to attacks by pathogens. On the other hand, infections by fungal pathogens can cause reduced growth rate and yields in agricultural crop plants, and are thus of economic importance.

Many factors influence the interaction between host plant and fungal pathogen, beginning when a spore or hyphae contacts the plant surface. The course of events following contact depends on the abilities of both organisms involved and on the environment in which they interact. Biotrophic fungi (rusts, powdery mildews) use the resources provided by the living host and cause minimal damage on invasion (Chaps. 4, 5, Vol. V, Part B). In contrast, necrotrophic fungi often kill plant cells in order to spread, often by secretion of toxins and cell wall-degrading enzymes (Chaps. 1–3, Vol. V, Part B). Many fungal pathogens cause distinctive symptoms in host plants such as rust, mildews, necrosis, blights, leaf spots, cankers, leaf curl, galls, and epinasty, among many others (Isaac 1992). Some of these symptoms arise from increased levels of growth regulators in plant tissues. Because of the involvement of plant growth regulators in the control of growth and differentiation of healthy plants (Table 1), any changes in the level of one or another hormone may result in an imbalance between different regulators and therefore in abnormal growth.

Plant growth regulators are often produced in plants in response to stress situations. However, phytopathogenic fungi can also produce and secrete plant hormones, and consequently, the source of increased levels of plant hormones during pathogenesis is often unclear. Phytopathogenic fungi in particular have been shown to synthesize compounds from all phytohormone

Lehrstuhl für Allgemeine Botanik/Mikrobiologie, Institut für Botanik, Westfälische Wilhelms-Universität Münster, Schlossgarten 3, D-48149 Münster, Germany

The Mycota V Part A
Plant Relationships
Carroll/Tudzynski (Eds.)

Table 1. Biological function of plant hormones

Plant hormone	Biological role
Auxins	Control of stem elongation, cell enlargement, stimulation of cell division and differentiation, induction of gene expression, root initiation
Gibberellins	Stimulation of internode extension, reversal of dwarfism, induction of flowering and seed germination, induction of enzymes
Cytokinins	Induction of DNA synthesis and cell enlargement, stimulation of cell division, promotion of differentiation, inhibition of senescence
Ethylene	Leaf abscission, senescence of plants, stimulation of enzyme formation, ripening of fruits
Abscisic acid	Growth inhibitor, leaf abscission, senescence, dormancy of seeds and buds, stomatal closure, induction of protein storage

classes, often via the same or similar biosynthetic pathways as in plants. The conservation of similar biosynthetic pathways between plant and fungus suggests that these compounds arose early in the history of life, that their synthesis has been maintained by selective pressures during the course of evolution, and that they must therefore be important for the survival and well-being of both fungi and plants.

Most fungi are able to produce growth regulators as secondary metabolites under laboratory conditions as well as in planta. The culture fluids containing these hormones often induce symptoms in plants similar to those caused by pathogens. The most dramatic of these changes involve superelongations (gibberellins) and tissue overgrowths, e.g., galls and cankers (cytokinins).

While plant hormones may be involved in plant-pathogen interactions, their significance in disease development has not been well understood until recently. The present chapter summarizes the most important results on biosynthesis of plant growth regulators by fungi in relation to disease. In addition to the five main classes of phytohormones, some new fungal compounds and their effects on plants are described.

II. Auxins

Auxins are growth regulators with various stimulatory effects. They are involved in the control of growth processes such as stem elongation, cell enlargement, and differentiation. Natural auxins are indole derivatives produced via the shikimic pathway as is the amino acid tryptophane. The major active auxin is β-indolyl acetic acid (IAA) (Fig. 1). In plants, auxins are found both free and bound to sugars, in which case they serve for both active transport and storage. Auxins are produced in meristem and leaves. In stems, the auxin conjugates are transported to the vegetation tips, where free auxin is released. IAA increases the plasticity of cell walls and therefore causes a wall-loosening effect in plant cells, allowing the cells to elongate very rapidly (Leopold and Kriedemann 1975). It has been supposed that auxins causes a rise in H^+ions in cell walls, resulting in destruction of bonds between the wall polymers. The other possibility – the induction of mRNA and cell wall-lysing enzymes – would take too much time (Hadwiger et al. 1970).

Auxins are known to be produced by a large number of fungal pathogens and endophytes (Table 2). Many fungal diseases of plants are associated with an increase in the auxin level in infected tissues. This results in tissue overgrowth, including epinasty (downward growth of shoots), hypertrophy (increase in the size of cells), and hyperplasia (increases in the number of cells by uncontrolled cell division) (Isaac 1992). Auxins are also known to play a significant part in the development of cankerous growth (Pegg 1984). The best-known example of a parasitic infection connected with increase in IAA levels is club root disease of *Brassica* spp. and crucifers caused by the phytopathogenic protist *Plasmodiophora brassicae*. Active zoospores of this organism invade the roots via root hairs and disrupt the normal regulatory control of IAA synthesis in plants, leading to root distortion in infected individuals and development of root galls. It has been suggest-

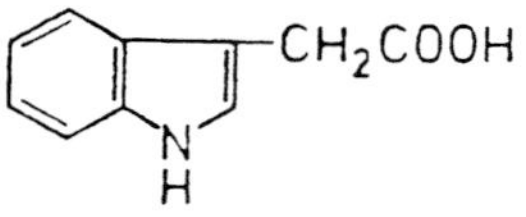

CH_2COOH N H

Fig. 1. Structure of indole-3-acetic acid

Table 2. Fungi producing auxins

Fungus	Effect on plants	Reference
Plasmodiophora brassicae	Club root disease of brassicas	Pegg (1984)
Phytophthora infestans	Root rot disease of potato plants	Pegg (1984)
Ustilago zeae	Corn smut disease	Wolf (1952)
Ustilago maydis	Tumor formation	Saville and Leong (1992)
Puccinia graminis f. sp. *tritici*	Infection of wheat	Isaac (1992)
Taphrina deformans	Production of IAA in culture	Isaac (1992)
Fusarium oxysporum	Production of IAA in culture	Isaac (1992)
Acremonium roseum	Acremoauxin A, plant growth inhibiting activity	Yoshida and Sassa (1990)
Aureobasidium pullulans	IAA-producing endophyte	Petrini (1991)
Epicoccum purpurascens	IAA-producing endophyte	Petrini (1991)

ed that the resting spores form IAA and induce root distortion (club roots). Another explanation for abnormal hormone physiology in infected plants is the breakdown of membranes which compartment myrosinase. This enzyme liberates the IAA precursor indolylacetonitrile (IAN) from 3-indolylmethyl-glucosinolate, the natural IAA form of brassicas (Pegg 1984). Searle et al. (1982) showed that ^{14}C IAN was found in *P. brassicae*-induced clubbed swede tissue fed with ^{14}C indolemethylglucosinolate. The availability of IAA to support the growth of clubbed tissue is under the control of nitrilase, which is present in higher levels in clubbed tissue than in controls.

A tenfold increase in IAA concentration has been measured in potato tuber tissues infected by the root-invading pathogen, *Phytophthora infestans*, and tumor tissue formed by *Ustilago maydis*, causative agent of corn smut. When grown in culture, *U. maydis* synthesizes and releases IAA to the media. Young, rapidly developing tumors of maize plants infected by this fungus contain 20 times more IAA than uninfected tissue, while older sporulating tumors maintain a level five times higher (Wolf 1952; Turian and Hamilton 1960). Although the source of the higher level of IAA has not been determined, these observations suggest that the production of IAA by *U. maydis* may play a role in tumor formation (Saville and Leong 1992).

Tumors or galls might provide refuges for the survival of plant pathogens and have been shown to act as sinks for plant assimilates. However, it is unclear why other nongall-producing bacteria and fungi can produce IAA as well as other auxin-like substances, or whether these microorganisms produce them in nature.

Recently, the isolation and structural identification of a novel fungal auxin derivative, acremoauxin A (Fig. 2), from a culture of *Acremonium roseum* has been reported (Yoshida and Sassa 1990). This compound is the first example of an arabitol-conjugated auxin-derivative from a fungus. Interestingly, this auxin shows high plant growth-inhibiting activity. 2-(3-indolyl) propionic acid (IPA), as well as alditol conjugates, are uncommon as microbial products. If *A. roseum* is fed different indolyl compounds and sugar moieties it can synthesize a spectrum of new auxin-like substances by conversion.

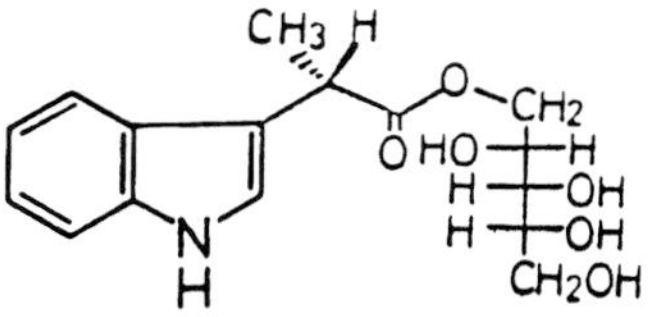

Fig. 2. Structure of acremoauxin A. (Yoshida and Sassa 1990)

The formation of phytohormones also seems to be an important factor in the development of highly specialized mutualistic interactions such as mycorrhizae. One example involves the formation of auxin and other phytohormones in symbiont interaction between *Dactylorhiza incarnata* (L.) and *Rhizoctonia* sp. This system was used to study the conditions for mycorrhizal or pathogenic interactions between the orchid and the fungus *Rhizoctonia*. Beyrle et al. (1991) found that increases in nitrogen concentration can lead to a shift from mutualism to pathogenesis. The effects of nitrogen on the interaction provided an experimental system for investigating hormone production during different stages of symbiosis using enzyme immunoassays. Phytohormones, especially IAA and cytokinins, are found in low levels under asymbiotic conditions. The presence of the fungus led to a large increase in phytohormone concentration, correlated with a growth stimulation in mycorrhizal protocorms. In pathogenically parasitized protocorms, auxins and cytokinins remained at higher concentration for longer than in mycorrhizal protocorms. The evidence that the fungi are involved in phytohormone synthesis comes from the finding that high auxin concentrations were observed in parasitized protocorms even after the death of orchid tissue (Beyrle et al. 1991).

Durand et al. (1992; Chap. 6, Vol. V, Part B) developed an interesting approach for clarifying the role of fungal IAA in ectomycorrhiza formation went. They isolated IAA-overproducing mutants of the ectomycorrhizal basidiomycete *Hebeloma cylindrosporum* Romagnesi by selecting for resistance to the indole analog, 5-fluoroindole. These mutants are tryptophane-overproducers, which in contrast to the wild type, are able to metabolize endogenous tryptophane to IAA. In order to compare these mutants with the wild parental dikaryons and to perform genetic analysis, the mutants were fused with a compatible wild monokaryon. The resulting dikaryons were able to form mycorrhizae with the normal host plant, *Pinus pinaster*, and to fruit under optimal culture conditions.

The ectomycorrhizal fungus *L. laccata*, common in coniferous forests of the Pacific Northwest, produces IAA as well as cytokinins (Ho 1987). A shift in the balance between the two phytohormones may strikingly alter plant morphological response and response to environmental stress. Isolates of *Laccaria laccata* differed significantly in production of both IAA and cytokinins. Therefore, different isolates of the fungus could effect significantly different root:shoot ratios in the host plant. Strains of the ectomycorrhizal fungi *Suillus bovinus*, *Rhizopogon luteolus*, and *Hebeloma hiemale* increase their IAA formation with tryptophane supplementation. IAA was also detected in culture filtrates obtained in the absence of exogenous tryptophane. This indicates that ectomycorrhizal fungi are able to metabolize endogenous tryptophane to IAA (Rudawska et al. 1992).

Different plant hormones were found also in endophytes, and epiphytes as well, which are reported to influence the growth of the host plant. Auxin-like hormones produced by these fungi may play a role in the plant senescence process (Jochmann and Fehrmann 1989). Production of indole-3-acetic acid (IAA) and indole-3-acetonitrile has been demonstrated in vitro for a number of epiphytes, including *Aureobasidium pullulans* (De Bary) Arnaud, *Epicoccum purpurascens*, and endophytic strains of *Balansia* spp. (Petrini 1991). Culture filtrates of these organisms have affected seed germination either positively or negatively. Bergamin-Strotz (1988) showed that the endophyte *Hypoxylon serpens* isolated from tobacco stimulated the flowering process of its host, while other strains of the same species induced wilting and inhibited the growth of tobacco seedlings. However, the latter effect could be due also to toxins.

Acremonium coenophialum, an endophytic fungus commonly found in tall fescue, has been observed to exhibit several adaptive morphological and physiological responses to drought stress compared with endophyte free cultivars (Joost 1995; Chap. 10, this Vol.). Drought-induced leaf rolling, leaf senescence, stomatal closure, and osmotic adjustment are more prevalent in endophyte-infected than in noninfected plants, and may be mediated through endophyte enhancement of the production of phytohormones. On the other hand, endophyte-infected tall fescue plants have been shown to be more productive and competitive (improvement of germination, tillering, biomass production per tiller), which may be due to fungal secretion of IAA (Joost 1995).

Although hyperauxiny is often associated with tissue overgrowths such as galls, neoplasms, and epinasty, there are several reasons for interpreting IAA-gall relations with caution (Pegg 1984):

1. Many diseases, e.g., cereal rusts, exhibit hyperauxiny in the absence of overgrowth symptoms.
2. Levels of auxins found in diseased tissue do not induce galls in healthy tissue.
3. A diseased plant showing tissue proliferation may show simultaneous increases in IAA, gibberellins, cytokinins, and ethylene.

Although many problems remain to be solved, it is true that most pathogens are able to synthesize IAA and related compounds, which may be variously involved in plant-fungus relationships. One way to clarify the role of fungal auxins in pathogenicity is the isolation of fungal genes, coding enzymes involved in phytohormone formation. Gene disruption experiments could show whether auxin formation plays a role as a phytopathogenic factor.

The role of auxins in the growth and development of phytohormone-producing fungi is not yet definitely elucidated. Gruen (1959) concluded that there is "no evidence for a growth-regulating role of auxins", particularly of IAA in fungi. On the other hand, Japanese workers demonstrated that auxin may act as a regulator of conidial germination (Nakamura et al. 1978), mycelial growth, and, in *Neurospora crassa*, fungal cell elongation (Tomita et al. 1984). The authors suggested that fungal auxins play a role in growth and differentiation of fungi similar to that in higher plants. Michniewicz and Rozej (1987) studied the influence of exogenous auxins (10^{-5}–10^{-9}M IAA) on growth and sporulation of *Fusarium culmorum*. The results showed that only at inadequate pH of medium and lowered temperature, both of which inhibit growth and development of the fungus, did exogenous IAA stimulate these processes. Therefore, IAA is not a growth regulator in fungi or at least not in *Fusarium culmorum*. The differences between these results and those for *Neurospora* suggest that the roles of auxin in growth and development may differ in different species of fungi.

III. Gibberellins

Gibberellins are present in very low concentrations in tissues of all higher plants, where they function as an important group of phytohormones regulating various growth processes. The gibberellin story begins in 1898 with a report by Hori about the bakanae disease of the rice plant (Hori 1898). This superelongation of rice seedlings is one of the first rice diseases to be described in the Orient, especially in Japan and China. Diseased plants are slender, pale yellowish in color, and taller than their undiseased counterparts. Farmers have called such plants foolish seedlings. After growing very rapidly, infected plants die. Hori (1898) detected the causative agent and identified it the soil-borne hyphomycete *Fusarium moniliforme* (Sheldon). The perfect sexual stage of the fungus is *Gibberella fujikuroi* (Saw.) Wr. In 1926, Kurosawa demonstrated the presence of an active principle in the culture filtrates of the bakanae fungus. Sterile, mycelium-free culture fluid induced the typical bakanae symptoms in rice seedlings. In 1934, Yabuta and coworkers reported the isolation of a pure crystalline substance that was be called gibberellin after the producer *G. fujikuroi* (Fig. 3).

After World War II, a new phase of gibberellin research began. The Japanese findings initiated

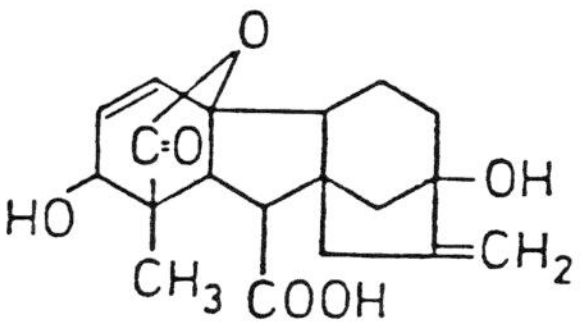

Fig. 3. Structure of the ent-gibberellane skeleton and gibberellic acid (GA_3)

interest and research in this field, mainly in Great Britain (Imperial Chemical Industries, ICI) and the USA (Abbott Laboratories). Interestingly, not until 1954 did Brian et al. report the elongation of a pea dwarf cultivar by exogenous gibberellin. In 1956, West and Phinney presented the first evidence for the presence of gibberellic acid in higher plants. These unidentified compounds with plant growth-inducing biological effects were called gibberellin-like substances. In 1958, Mac Millan and Suter succeeded in identifying a gibberellin from a higher plant, the gibberellin A_1. In the following years, the universal occurrence of gibberellins among plants was established. Approximately 80 gibberellins have been discovered since that time, from both plants and fungi.

In a healthy plant, the action of gibberellins is responsible for internode extension, reversal of dwarfism, induction of flowering, maintenance of cell division, and the induction of enzymes, particularly those involved with cell wall synthesis. Gibberellic acid also influences the development of the embryo as well as protein and RNA synthesis. These effects are very similar to those of auxins and include increased auxin synthesis.

In 1972, Lozano at the Centro Internacionale de Agricultura Tropical (CIAT) in Cali (Colombia) reported a superelongation disease of cassava plants. Rademacher and Graebe (1979), as well as Zeigler et al. (1980), established the occurrence of GA_4 in the fungal pathogen *Sphaceloma manihoticola*. This fungus induces necrotic leaf spots and stem and petiolar canker as primary symptoms. The secondary symptoms include a striking elongation of the internodes, which become brittle and spindly, with misshapen laminar growth around the leaf spots. When 30 μg authentic GA_4 or 3×10 μg GA_4 was applied over 6 days to young cassava plants, symptoms identical to natural infection were obtained. It remains to detect increased levels of GA_4 in infected tissue and to show that GA_4 is not a native gibberellin of cassava (Pegg 1984).

Table 3. Fungi producing gibberellins

Fungus	Gibberellin	Effect on plants	Reference
Gibberella fujikuroi	GA_3, GA_4, GA_7, GA_9, GA_{13}, GA_1	Bakanae disease of rice	Bearder (1983)
Sphaceloma manihoticola	GA_4	Superelongation of cassava, necrotic leaf spots stem canker	Zeigler et al. (1980) Rademacher and Graebe (1979)
Cercospora rosicola	GA_3, GA_4, GA_7, GA_{13}	Black spot disease of roses	Coolbaugh et al. (1985)
Botryodiplodia theobromae	GA_3	–	Prema et al. (1988)
Puccinia punctiformis	GA_3	Rust infection of creeping thistle	Bailiss and Wilson (1967)
Phaeosphaeria sp.	GA_1	–	Kawaide and Sassa (1993)
Glomus mosseae	GA3 and other GA-like substances	VA mycorrhizal fungus	Barea and Azcon-Aguilar (1982)
Sorosporium reilianum	GA_1, GA_3, GA_{19}, GA_{20}, GA_{53}	Head smut of *Soghum bicolor*	Matheussen et al. (1991)
Ustilago maydis	GA_3	Smut of maize	Saville and Leong (1992)

In the past 10 years a few other fungi have been found to produce gibberellins (Table 3); but to date, bakanae disease of rice and superelongation of cassava are the only described plant diseases caused by gibberellin formation in fungal pathogens.

Takenaka et al. (1992) isolated mutants of *G. fujikuroi* by selection for resistance against lethal concentrations of the sterol demethylation inhibitor, pefurazoate. The mutants isolated from agar plates, containing pefurazoate, were tested by bioassay of gibberellins and assay of GA_3 using HPLC. Mutants were also tested for their virulence to rice plants. Decreased ability to induce bakanae symptoms due to decreased production of gibberellins was observed in the mutants. It is interesting that the ability to produce high amounts of GA_3 is restricted to the members of *Gibberella fujikuroi*, mating group C (B. Brückner, unpubl.). All strains identified as mating group C by RAPD technique or by sexual crossings were isolated from rice and are able to induce superelongation symptoms in a rice biotest system. Conversely, none of the members of the other five mating groups A, B, D, E, and F (Leslie et al. 1992), isolated from host plants other than rice, produced significant amounts of gibberellins. These results indicate that the *Gibberella fujiukuroi* strains inducing bakanae symptoms are likely to have undergone a close coevolution with their host. In fact, these strains show sexual incompatibility with isolates from other hosts and probably represent a distinct species.

In 1991, Matheussen and coworkers purified and assayed extracts of head smut fungus *Sorosporium reilianum* by GC-MS-SIM and dwarf rice bioassay, and detected specific gibberellins. This is the first report on the formation of gibberellins by a heterobasidiomycete and on their possible involvement in head smut disease of sorghum. Reduction of sorghum plant height across a wide range of sorghum varieties and hybrids is the characteristic symptom of head smut. The greatest inhibition of elongation occurred in the last internode (predungle) and the reduction in internode length was progressively less in internodes farther away from the panicle. The question is, why does a fungus that produces GA_3 in culture reduce rather than increase plant heights? Quantitative analysis revealed that panicles of healthy control plants contained 60–100% more GA_3 than panicles of smutted plants. It was proposed that the decrease in plant height is due to fungal interference with host plant biosynthesis of GA_3 and, further, that this may be accomplished through the diversion of plant GA precursors to other uses by the fungus (Matheussen et al. 1991).

Another example of fungal gibberellin production occurs in *U. maydis*, the causative agent of maize smut. The fungus produces and releases IAA, cytokinins, and gibberellins (Saville and Leong 1992). The content of gibberellins is three times greater in diploid than in haploid cells. Interestingly, haploid cells are incapable of inducing gall formation upon infection. On the other hand, pathogenic haploid transformants, which have an additional different *U. maydis* b allele, display mycelial growth and are weakly pathogenic on maize (Kronstad and Leong 1989; Schulz et al. 1990). Gillissen et al. (1992) hypothesized that the formation of a bW-bE gene-heterodimer is instrumental in the switch from growth by budding to filamentous, pathogenic growth. Since the b gene complex has been shown to be essential for pathogenicity on maize, it would be interesting to clarify whether the GA_3 formation is under the control of this gene complex. *U. maydis* mutants defective in phytohormone formation would be useful in determining the role of these compounds in tumor production.

In a culture of the fungus *Phaeosphaeria* sp. *L487*, GA_9, GA_1, and GA_4 were isolated, and the metabolic conversion of GA_9 via GA_4 to GA_1 was demonstrated by feeding experiments with deuterium-labeled GA_9 (Kawaide and Sassa 1993). The 3β hydroxylation is a very important step releasing physiologically active GA_4 and GA_1. This pathway, however, has not been found in *G. fujikuroi* and *S. manihoticola*, although it is found in various higher plants (Graebe 1987). The fungus *Phaeosphaeria* produces GA_1 as the end product of its GA biosynthesis in concentrations of ca. 50mg of GA_1 in 1l of maltose-yeast extract medium. GA_3 could not be detected in the culture filtrate by HPLC analysis. The authors suggest that this fungus could be used for production of pure GA_1 for agricultural applications.

Glomus mosseae, A VA mycorrhizal fungus (Chap. 7, Vol. V, Part B) has been investigated for phytohormone production. Sterile spores were germinated axenically in a liquid medium, which was then assayed for the presence of phytohormones. These assays revealed the presence of substances with the properties of gibberellins and cytokinins (Barea and Azcón-Aguilar 1982). The authors suggested that the phytohormones produced by this mutualistic fungus might be involved with morphological and physiological changes in host plants which occur upon infection with VA mycorrhizal fungi. Microbial phytohormones may favor the establishment of the VA mycorrhizal symbiosis, resulting in higher rates of nutrient absorption by the plant. It is well known that gibberellins increase leaf area and induce the growth of lateral roots.

Recently, it was shown that the ability to produce gibberellins is widespread among the genus *Fusarium* (Voigt et al. 1995) and other Ascomycotina such as *Aspergillus*, *Penicillium*, and *Cladosporium* (Hasan 1994a,b). Although these compounds are widespread among fungi, the role of gibberellins in plant disease, other than their involvement in superelongation effects, requires further study.

Some authors suggest that gibberellins, as well as the other hormones, may play an intracellular role for the producing fungus. In *N. crassa*, gibberellins, although produced in small amounts and not secreted into the environment, enhance the rate of conidial germination and hyphal elongation (Tomita et al. 1984). On the other hand, experiments with *F. culmorum* in which the effects of the growth retardant AMO 1618, an inhibitor of gibberellin biosynthesis, on spore germination were examined indicate that GA_3 does not significantly affect this process (Michniewicz and Rozej 1988a).

IV. Cytokinins

Cytokinins are adenyl derivatives with an isoprenoid side chain. In many plant families, three to eight different cytokinins and their respective ribotides, ribosides, and glucosides can be found, the activity of which depends on the properties of the side chains. More than 40 cytokinins have been identified to date (Nieto and Frankenberger 1990), although the most important are zeatin and kinetin (Fig. 4). Cytokinins are produced in active root cells and they are transported to the distant

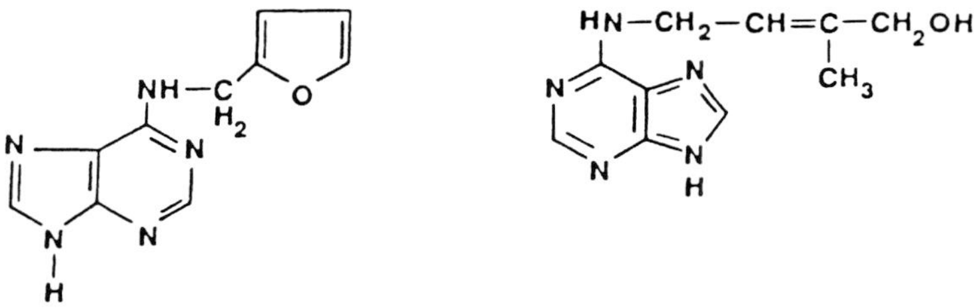

Fig. 4. Structure of two cytokinins: zeatin (*left*) and kinetin (*right*). (Isaac 1992)

plant organs via the xylem. They are known to stimulate cell division in dormant buds, to promote differentiation, and to encourage growth of plant cell cultures. Furthermore, they inhibit senescence by stimulating protein synthesis in tissues and by increasing stress-resistance.

The ratio of cytokinins to auxin is important for the development of plants. In callus cultures it is known that a high ratio induces the differentiation of shoot and bud tissue, whereas a low ratio promotes the differentiation of roots from callus material (Isaac 1992). Cytokinins are found also in bacteria and fungi (Greene 1980). Investigations on microbial production of cytokinins have been focused primarily on plant pathogenic fungi that cause overgrowths, galls, tumors, and rusts (Table 4). Exogenous application of cytokinins causes similar abnormalities in plants, which suggests a possible involvement of these hormones in the regulation of organ differentiation. Very often, cytokinin levels are higher in plants infected by fungal pathogens than in healthy plants. This situation has been found in maize infected by *U. maydis*, where the auxin as well as cytokinin content are increased drastically. Infection of 6-week-old maize apical meristems with *U. maydis* teliospores resulted in increased size correlated with increased cytokinin activity (Mills and Van Staden 1978). The authors suggested that the fungus produces and secretes cytokinins which accumulate in infected plant tissues. The accumulation of cytokinins is associated with gall formation and abnormal development. The less polar cytokinins, which cochromatographed with zeatin and zeatin riboside, were present in both healthy and infected plants, while the polar compounds, which are produced only by the fungus, accumulated only in the infected plants.

Considerable work has been published on *Taphrina* spp., which cause leaf hyperplasias and abnormal tissue growth. In peach leaves infected by *T. deformans*, leaves become curled due to greater growth of cells in the upper layers (palisade mesophyll) than in the lower (spongy mesophyll) layers. Correspondingly elevated levels of cytokinin have been found in infected leaves. Three cytokinins detected by chromatography were the same as those in healthy tissues; one additional form was not detected in healthy plants at all (Sziraki et al. 1975). Johnston and Trione (1974) also detected cytokinin-like compounds in the culture fluid of this fungus. Exogenous application of cytokinins or culture filtrates of *T. deformans* to peach buds and leaves did not produce disease symptoms. It was suggested that the hormones may not have been able to penetrate the

Table 4. Fungi producing cytokinins

Fungus	Cytokinin	Effect on plants	Reference
Taphrina cementorum	Zeatin riboside Zeatin Isopentenyladenin	Leaf hyperplasia in orchard trees	Kern and Naef-Roth (1975)
Taphrina betulina	Zeatin Zeatin riboside	Witches' broom in orchard trees	Kern and Naef-Roth (1975)
Taphrina deformans	Zeatin Zeatin riboside	Peach leaf curl	Johnston and Trione (1974)
Helminthosporium sp.	Different cytokinins	Green islands on barley	Mandahar and Angra (1987)
Ustilago maydis	Different cytokinins	Maize smut disease	Mills and Van Staden (1978)
Uromyces sp.	Different cytokinins	Rust disease of bean	Greene (1980)
Rhizopogon roseolus	Zeatin Zeatin riboside	Ectomycorrhiza, induction of root branching	Miller (1967)
Botrytis cinerea	Zeatin Zeatin riboside Kinetin	Pathogenic to many host plants	Talieva and Filimonova (1992)
Erysiphe cichoracearum	Zeatin Zeatin riboside Isopentenyladenine Kinetin	Phlox mildew	Talieva et al. (1991)
Alternaria brassicae	Kinetin	Green islands in mustard leaves	Mandahar and Suri (1983)

cuticle or bud scales, whereas fungal hyphae have the ability to penetrate leaf tissues and supply a direct source of cytokinins. Cytokinin was also identified in culture filtrates of *T. cerasi* and *T. betulina*, two fungal pathogens known to cause witches broom in orchard trees (Kern and Naef-Roth 1975).

The phenomenon of green island formation at infection sites of rusts, downy mildews, and powdery mildews seems to be caused by cytokinins (Poszar and Kiraly 1964). *Helminthosporium sativum*, a facultative pathogen of many crops, *H. oryzae*, a rice pathogen and *H. turcicum*, a maize pathogen, induce green islands on barley, wheat, maize and rice leaves, respectively, 20–32h after inoculation with spores. Detection of cytokinins quite early at the infection sites of facultative parasites indicates their possible role in the initial stages of the pathogen attack (Gulati and Mandahar 1984, 1986). Cytokinins are known to regulate the transport of nutrients (Leopold and Kriedemann 1975), and, indeed, green islands shown much higher sugar and starch contents in addition to higher cytokinin contents than comparable healthy tissues. Analysis of cell-free culture filtrates demonstrated secretion of cytokinins by *H. sativum* in vitro. Paper chromatography of the isolated fraction of purified extracts revealed the presence of two biological active cytokinin-like compounds which, when placed on healthy leaves of barley, induced the green islands typical for fungal infection (Angra and Mandahar 1985). *H. turcicum* causes the green island disease in maize leaves within 48h. Increases of 178, 166, 75, and 86% in chlorophyll, sugars, starch, and total carbohydrate contents, respectively, were recorded in green islands compared with comparable patches of surrounding aging tissue. Bioassays with culture fluid of *Helminthosporium turcicum zeae* on radish showed retardation of cotyledon expansion and root elongation compared to controls using water-treated leaves (Gulati and Mandahar 1986). Application of culture filtrate extracts from 30-day-old cultures of this fungus to excised maize leaves and their incubation in the dark for 72h also induces green islands. The presence of higher quantities of sugars in these areas could possibly be explained on the basis of mobilization of carbohydrates to this region from surrounding tissue. Nutrient mobilization by fungal cytokinins could, therefore, serve the pathogen to great advantage during infection and establishment within the host tissue.

Green islands were formed also in mustard leaves infected by *Alternaria brassicae*, which produces kinetin in vitro. Carbohydrates and protein contents were higher in the green sports formed beneath the water drops containing conidia of this fungus. Since the leaves were kept in darkness, only one explanation was possible, i.e., mobilization and retention of nutrients at the infection sites of these hemibiotrophs. The germinating conidia of the fungus secrete cytokinins inducing translocatory sinks early on. Nutrients moved from the surrounding areas to these sinks, suggesting that the sinks supplied food and energy to the pathogen for penetration and subsequent development of the lesion (Mandahar and Suri 1983).

Conidia and mycelium of the phlox mildew causative agent, *Erysiphe cichoracearum*, contain significant amounts of cytokinins, including zeatine, zeatine riboside, isopentenyladenine, and even traces of abscisic acid (Talieva et al. 1991). The presence of isopentenyladenine, a characteristic fungal cytokinin, was detected in infected plant tissue only.

Numerous other fungi have been tested for cytokinin formation using the soybean callus tissue assay. Some of them (e.g., *Suillus cothurnatus*, *S. punctipes*, and *Rhizopogon ochraceorubens*) induced callus growth and seem to secrete cytokinin-like substances (Crafts and Miller 1974). Hormonal substances with cytokinin activity were also demonstrated in culture filtrate and in the mycelium of pathogenic *Botrytis* spp. using GC-MS and HPLC (Talieva and Filimonova 1992).

From all these data, it is evident that cytokinins play an important role in fungal pathogenesis, especially in association with fungal diseases like rusts, powdery mildews, cankers, and leaf spots (Mandahar and Angra 1987).

V. Ethylene

Ethylene (Fig. 5) was the first phytohormone to be implicated in the induction of disease-like symptoms in plants, e.g., tissue swelling, chlorosis, and epinasty (Isaac 1992). It is produced by all higher plants and some microorganisms as a growth regulator effective at very low concentrations (Fukuda et al. 1993). In healthy plants, ethylene stimulates leaf abscission and leaf fall, probably as a result of its effect on cell wall-degrading enzymes and senescence. Furthermore, ethylene

Table 5. Fungi producing ethylene

Fungus	Effect an plants	Reference
Acremonium falciforme	–	Arshad and Frankenberger (1991)
Verticillium daliliae	Defoliation in hop plants	Tzeng and Devay (1984)
Botrytis cinerea	Leaf drop, epinasty	Williamson (1950)
Fusarium oxysporum	–	Tzeng and Devay (1984)
Fusarium culmorum	–	Michniewicz and Rozej (1987)
Verticillium albo-atrum	Leaf drop of tomato, senescence	Pegg (1976)
Pythium ultimum	Root infection	El-Sharouny (1984)

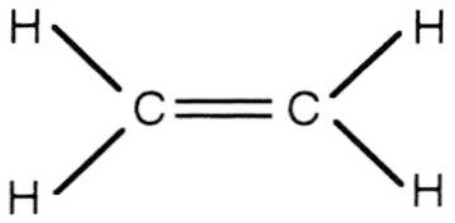

H H C=C H H

Fig. 5. Structure of ethylene

stimulates the synthesis of enzymes directly affecting the integrity of cell walls, e.g., cellulases, β-(1,3)-glucanases, and enzymes, which influence the levels of hydroxyproline-rich proteins. Methionine is the sole precursor for C_2H_4 biosynthesis in plants and microorganisms (Adams and Yang 1979). Arshad and Frankenberger (1991) observed that corn rhizosphere contained a number of bacteria and fungi capable of deriving C_2H_2 from methionine. The fungi reported as ethylene producers are listed in Table 5. Among the soil fungal isolates, *Acremonium falciforme* was the most active producer of this hormone (Arshad and Frankenberger 1989).

Ethylene is produced in plants in response to wounding and mechanical stress. Therefore, the use of detached leaves provides material which will obviously be subject to increased levels of ethylene as a response to wounding. Plant tissues infected by fungi show increased levels of ethylene as a consequence of wounding caused by penetration and fungal invasion. On the other hand, ethylene triggers germination of conidia, branching of hyphae, and multiple appressoria formation in *Colletotrichum*, thus allowing fungi to time their infection to coincide with ripening of the host (Kolattukudy et al. 1995). Therefore, ethylene seems to play an important role in signal transduction of plant-pathogen interaction.

Seedling roots of different varieties of soybean infected by the soybean pathogen *Phytophthora megasperma* f. sp. *glycinea* displayed an increase in ethylene biosynthesis (Reinhardt et al. 1991); the rate started to increase ca. 6h after infection and reached ca. 10–15-fold maximum 11–12h after inoculation. In incompatible interactions between soybean and *Phytophthora* races, however, ethylene production was stimulated 5–10-fold 3h after infection and reached the 50-fold level only 6h after inoculation.

Other phytohormones, e.g., absicsic acid, also interact with ethylene, and for this reason, the interpretation of plant disease symptoms in relation to ethylene production proves obscure. Nevertheless, a number of disease symptoms might plausibly be attributed to ethylene production triggered by fungal infection.

Several senescence symptoms of plant disease such as foliar chlorosis, petiolar epinasty, foliar abscission, tissue distortion, and necrosis can be induced in response to fungal infections (Pegg 1984), but to date it is unclear whether such symptoms are induced by ethylene of fungal or host origin.

Verticillium wilt of hops is accompanied by leaf drop, epinasty, and premature senescence, symptoms attributed to increased levels of ethylene in plants (Isaac 1992). The stimulation of aging and senescence of plants by ethylene may increase the general susceptibility of plants to fungal invasions. On the other hand, the response of plants to ethylene may increase levels of disease resistance. Thus, the increase in ethylene levels after *Verticillium* infection may induce higher disease resistance in infected plants by acting on other metabolic pathways. Pegg (1976) described the alleviation of symptoms of *V. albo-atrum* in tomato caused by ethylene. Similarly, the effects of *Sclerotinia* root rot (Isaac 1991) and bitter rot in apples (Lockhart et al. 1968) are reduced and the synthesis of the phytoalexin pisatin is increased in pea (Chalutz and Strahmann 1969). For some other ethylene-producing fungi, e.g., *Penicillium notatum* and *Septoria musiva*, no action on plants has been studied.

Again, one must question whether a plant hormone serves any biological function in the fungus that produces it. Michniewicz and Rozey (1987) studied the role of ethrel, an ethylene-releasing compound, in the growth and development of the ethylene-producing fungus, *Fusarium culmorum*. The authors found no correlation between growth rate, sporulation, or germination of spores and ethylene formation. The results show clearly that ethylene is not an endogenous factor controlling growth and development. The isolation of mutants blocked at ethylene biosynthesis would help to elucidate the role of ethylene for the producing fungus as well as in the plant-pathogen interaction.

VI. Abscisic Acid

Abscisic acid (ABA) has been implicated in the inhibition of growth and RNA synthesis, the promotion of senescence, abscission, and in the adaptation of plants to stress conditions (Norman et al. 1983). Furthermore, ABA induces dormancy in plants, reduces meristematic activity, and inhibits bud break (Isaac 1992). It also plays an important role in stomatal closure.

ABA is an isoprenoid (Fig. 6) synthesized via mevalonic acid, isopentenylpyrophosphate, geranylpyrophosphate, and farnesylpyrophosphate. The natural ABA consists entirely of (+)-enantiomer while the synthetic (−)-enantiomer shows almost the same physiological effects, though at lower magnitude. Although the chemical structure is similar to gibberellic acid, these two plant hormones induce opposite effects on plant tissues.

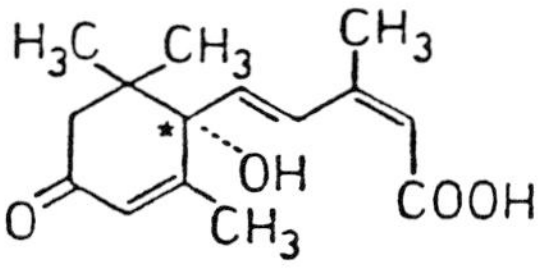

H_3C CH_3 CH_3 OH COOH O CH_3

Fig. 6. Structure of abscisic acid

The discovery of ABA production by a plant pathogenic fungus has been reported rather recently (Assante et al. 1977), and thus, the role of ABA in the higher plant-parasitic fungus system is doubtless the least explored. Since all ABA-producing fungi described so far are phytopathogens, fungal ABA may well have a function in phytopathogenesis. ABA has been found in *Cercospora rosicola*, *C. cruenta*, *Botrytis cinerea*, *Fusarium culmorum*, *Alternaria brassicae*, and some other fungal pathogens (Table 6). Physiological studies on ABA-producing fungi have shown that plant extracts from *Narcissus*, carnation, rose, stock, and *Chrysanthemum* enhanced phytohormone formation in pure cultures of the fungus (Takayama et al. 1983). *Narcissus* extract is most effective, increasing the production of ABA by 36 times.

The role of ABA in disease expression is not yet clear, since ABA, like most plant growth regulators, interacts with other hormones in the plant. Further, it is difficult to determine whether increased production of ABA is a response of the infected plant or the result of fungal hormone synthesis during the infection process. Kettner and Doerffling (1995) studied the metabolism of ABA in tomato leaves infected with two virulent strains of *Botrytis cinerea*, one producing ABA, the other without ability to form ABA. The hormone level in tomato leaves infected with the ABA-producing isolate rose greatly but not after infection with the nonproducing *Botrytis* strain. It is concluded that at least four processes control the level of ABA in wild-type tomato leaves infected with *Botrytis cinerea*: stimulation of fungal ABA biosynthesis by the host; release of ABA or its pre-

Table 6. Phytopathogenic fungi producing absicisic acid

Fungus	Effect an plants	Reference
Cercospora rosicola	Black spot disease of roses	Assante et al. (1977)
Cercospora cruenta	–	Oritani et al. (1982)
Botrytis cinerea	Gray mold disease	Marumo et al. (1982)
Fusarium culmorum	–	Michniewicz et al. (1987)
Ceratocystis coerulescens	Pine pathogen	Doerffling and Petersen (1984)
Rhizoctonia solani	Tomato pathogen	Doerffling and Petersen (1984)
Agrocybe praecox	–	Crocoll et al. (1991)
Alternaria brassicae	Black spot disease of canola	Dahiya and Tewari (1991)
Macrophoma castaneicola	Black root rot disease of chestnut trees	Arai et al. (1989)
Penicillium italicum	Infection of oranges	Rudnicki et al. (1969)

cursor by the fungus; stimulation of biosynthesis of plant ABA by the fungus; and inhibition of its metabolism by the fungus. Application of ABA together with fungal spores to tomato leaves caused a faster development of necrotic leaf area than spore inoculation only (Kettner and Doerffling 1995).

Recently, a mutant strain of *C. cruenta* deficient in the synthesis of ABA was isolated after transformation with a foreign gene via integration into genomic DNA (Kitagawa et al. 1995). Mutants defective in plant hormone synthesis will be useful for functional analysis of plant hormones as putative pathogenicity factors.

Certain plant metabolic responses to fungal infection have been shown to be suppressed by exogenously supplied plant growth regulators (Li and Heath 1990). Thus, ABA strongly inhibited accumulation of the phytoalexins rishitin and lubimin in potato tubers in response to fungal infection; the tubers subsequently proved susceptible to *Phytophthora infestans* and even to *Cladosporium cucumerinum*, which is nonpathogenic on potato (Henfling et al. 1979). Exogenous ABA also increased the susceptibility of a resistant cultivar of barley to *Erysiphe graminis* (Edwards 1983) and of tobacco to *Peronospora tabacina* (Salt et al. 1986), whereas kinetin increased resistance. Furthermore, ABA and gibberellins increased susceptibility of nonhost bean leaves to the cowpea or corn rust fungi that are normally nonpathogenic on beans (Li and Heath 1990).

These results are important for understanding the nature of plant-fungal interactions, since several phytopathogenic fungi have been shown to produce phytohormones, including ABA. ABA and cytokinins obviously play an important role in the control of phytoalexin synthesis in both plants and cell suspension cultures. Ward et al. (1989) suggested that ABA may affect the transcription of plant defense genes.

In addition to repressing the biosynthesis of phytoalexins, ABA inhibits the uptake of K^+ ions, and thus prevents any increase in guard cell turgor; as a result, it maintains stomata in the closed state, preventing water loss through stomatal pores in wilted plants. Pegg (1981) suggested that loss of turgor in a plant due to wilting (e.g., in vascular wilt disease) may lead to increased levels of ABA in infected tissues. The development of stunting in these tissues may therefore be due to the ABA produced by the plant in response to water deficit. There is no general evidence to show that ABA causes stunting in diseased plants. Where wilting is a feature of disease, enhanced ABA will automatically be produced; however, it usually accumulates at levels below those required for inhibition of growth (Pegg 1984).

Until now, there are no studies on the role of glycosylated ABA or the enzymic release of free ABA during pathogenesis. Sembdner et al. (1980) and Milborrow (1978) found that the level of ABA bound as glucosyl esters and glucosides may exceed free ABA in the leaf. The sequestering of ABA by chloroplasts is another reason why difficulties arise in interpreting its true role in infected plants.

VII. Other Compounds with Growth Regulatory Properties

In the past years, a number of additional fungal metabolites with properties of plant growth regulators have been isolated and characterized. These include the following.

A. Jasmonic Acid and Related Compounds

One new group of native plant growth regulators includes jasmonic acid and related compounds. Methyljasmonate, first isolated from jasmine oil (Demole et al. 1962), shows ABA-like, senescence-promoting activity. Their major representatives-(−)-jasmonic acid and (+)-7-isojasmonic acid (Fig. 7) are distributed widely in plants. Free jasmonic acid has been found as a plant growth inhibitor in culture filtrates of *Lasiodiplodia theobromae* (Aldridge et al. 1971) as well as in leaves and the galls of chestnut (Takagi et al. 1975).

It is suggested that leaf senescence promoted by jasmonate is characterized not only by the degradation of chloroplast constituents, but also by the de novo synthesis of new cytoplasmic polypeptides. These may be the cause rather than the consequence of the senescence syndrome (Weidhase et al. 1987). Incorporation experiments with

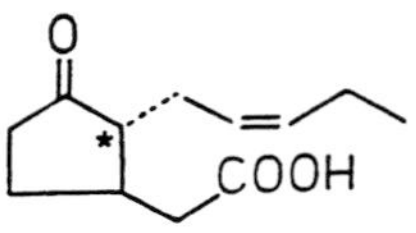

Fig. 7. Structure of jasmonic acid

labeled amino acids indicate that barley leaf segments treated for 48h with methyl-jasmonate incorporated radioactive amino acids almost exclusively into the most abundant proteins. Furthermore, addition of methyl jasmonate initiates de novo transcription of genes, such as that for phenylalanine ammonia lyase, which is known to be involved in the chemical defense of plants. These data demonstrate the integral role of jasmonic acid and its derivatives in the intracellular signal cascade that begins with interaction of an fungal elicitor molecule with the plant cell surface (Grundlach et al. 1992).

Miersch et al. (1993) screened the occurrence of jasmonates in culture filtrates of 46 fungal strains belonging to 23 different genera. In addition to *Botryodiplodia theobromae* (Miersch et al. 1991) and *G. fujikuroi* (Miersch et al. 1992), species of the genera *Collybia*, *Coprinus*, *Mycena*, *Phellinus*, and *Trametes* secrete jasmonic acid and 7-iso-jasmonic acid.

The physiological function of jasmonates in fungi is unknown, but they may serve to foster pathogenesis. Induction of senescence in ornamental roses caused by *B. theobromae* leading to the death of infected plant organs could be stimulated by fungal jasmonates (Miersch et al. 1993).

B. Fusicoccin

Fusicoccin (FC) is a carbotricyclic diterpene glucoside (Fig. 8) which is produced by the wilt and canker pathogen of almond and peach, *Fusicoccum amygdali*. The fungus infects twigs and leaves causing first necrotic spots, even in leaves located well above the infection wound and not yet invaded by the fungus. Later, the infection results in formation of canker around buds and nodes as a result of a defense mechanisms of the plant (Ballio, 1978). It was shown that leaf necrotic spots, similar to those caused by the fungus, appear after uptake of small amounts of *Fusicoccum amygdali* culture filtrate.

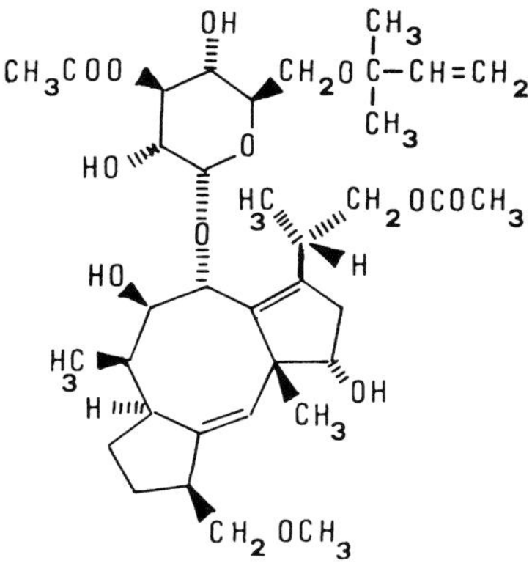

Fig. 8. Structure of Fusicoccin. (Ballio 1978)

Fusicoccin was first described as a phytotoxin causing excessive transpiration and leading to irreversible wilt and death of infected trees (Graniti 1964). A range of dramatic physiological processes are induced in higher plants by fusicoccin. These include:

- Stimulation of stomatal opening, reverse of seed dormancy and stem, root, and leaf cell enlargement (Bottalico et al. 1978).
- Stimulation of H^+ extrusion from plant cells with resulting acidification of cell walls leading to increased growth (Schröder et al. 1990).
- Stimulation of cation, anion, sugar, and amino acid uptake, stimulated respiration, and increased membrane permeability to water (Pegg 1984).

Proteins that bind to FC have now been found in plasma membranes from several plants (Scott 1992). The finding suggests that FC might induce changes in gene expression involved in the metabolic adjustment to some biochemical effects of FC, perhaps as a stress response to osmotic, drought, or temperature stresses (Skriver and Mundy 1990). To date, FC has not been found in uninfected plants, and therefore it can be used as a model system for studying the molecular basis of the interactions between plants and microbial pathogens. In addition to the phytotoxic effects in almond and peach trees, FC and some of its metabolites induce important physiological responses similar to IAA, cytokinins, and gibberellins, and opposite to those of ABA, e.g., stimulation of H^+ extrusion associated with a hyperpolarization of the transmembrane electrical potential, the promotion of ion and solute transport, the stimulation of cell enlargement, increasing fresh weight of tissue, and the breaking of seed dormancy (Radice et al. 1981). Unlike IAA, however, FC promotes root growth.

A number of compounds chemically related to FC have been isolated from culture fluid of the fungus. Most of them differ from FC only in acetylation pattern (O-acetyl compounds). Other compounds have no oxygen at C-19 (deoxyfusicoccins). One of the 19-deoxyderivatives carries an extra oxygen at C-3 and is related to cotylenins, another family of fungal plant growth regulators. Interestingly, some nontoxic analogs still exert an effect in several physiological processes and might open perspectives for applications to agriculture, e.g., as antistress factors.

C. Other New Fungal Compounds with Plant Hormone-Like Effects

In the course of the search for new plant growth regulators among fungal metabolites, Kimura et al. (1992b) succeeded in isolating new metabolites promoting the growth of roots and stems. One of the interesting new metabolites produced by *Penicillium* sp. is 7-hydroxy-2-hydroxymethyl-5-methylchromone (Fig. 9). At the concentration of $300\,mg\,l^{-1}$, this substance promotes root growth by more than 40% (Kimura et al. 1992a).

The fungus *Alternaria* sp. produces two growth regulators of the chromone type, the altechromones 5-hydroxy-2,7-dimethyl chromone and 2-acetonyl-5-hydroxy-7-methyl chromone. Both metabolites promote root growth of lettuce seedlings by about 50% at a concentration of $3\,mg\,l^{-1}$ and $100\,mg\,l^{-1}$, respectively (Kimura 1992b). Kimura et al. (1990) also isolated a new root promoting substance from the culture fluid of the fungus *Sesquicillium candelabrum*, which was named sescandelin (Fig. 10). Sescandelin was suggested to be an isocoumarin analog. The promoting effect on *Azukia* cuttings at $30\,mg\,l^{-1}$ was about 40% over the water control. The promoting activity was most effective when IAA ($35\,mg\,l^{-1}$) was applied to *Azukia* cuttings on the 2nd day, after an application of sescandelin on the 1st day (Kimura et al. 1990).

Phytoactive substances were present in a culture broth of the fungus *Drechslera gigantea*, a pathogen of several grasses (Sugawara et al. 1993). Twelve eremophilane sesqiterpenes were

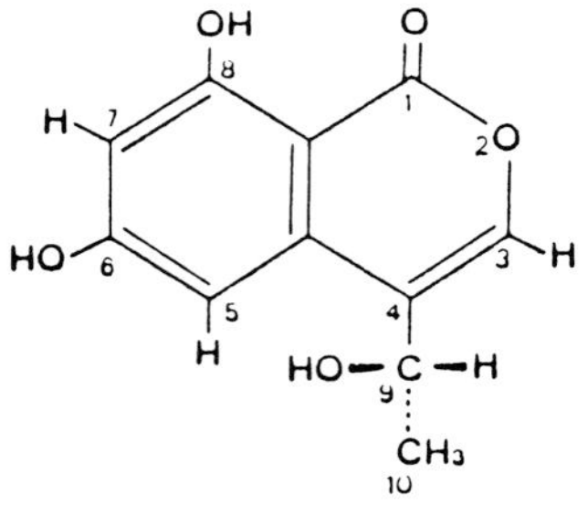

CH3 O H H 6 5 4a 4 3 7 8 8a 2 HO O CH2OH H

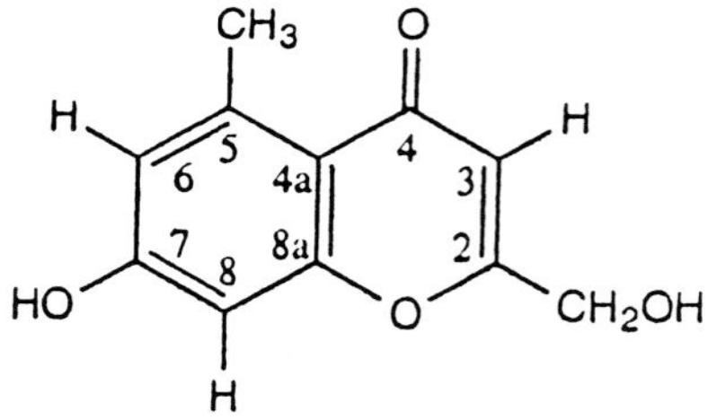

Fig. 9. Structure of a 7-hydroxy-chromone analogue. (Kimura et al. 1992a)

OH O 8 1 H 7 2O HO 6 3 H 5 4 H HO C H 9 CH3 10

Fig. 10. Structure of sescandelin. (Kimura et al. 1990)

isolated and structurally characterized. Most of these terpenoids were phytotoxic; however, two compounds, gigantenone and petasol, caused chlorophyll retention, an activity associated with phytohormones. Petasol is a known eremophilane from higher plants.

VIII. Conclusions

Phytohormones are important substances, present in healthy plants, which are involved in the organization of all growth and differentiation processes. Any changes in the levels of these compounds and the relationship between them may result in abnormal growth. The changed relationship may be the result of increased levels of one or more plant hormone present in plant tissues, generated as a response of the plant to fungal invasion. On the other hand, some of the symptoms of disease develop as the result of the action of phytohormones of fungal rather than plant origin. It may often prove difficult to determine whether the observed effects are due to growth regulators produced and liberated by the invading fungus or to those induced in the plant by direct stimulatory effects of the pathogen.

New analytical methods with higher specificity and sensitivity, together with genetic and molecular genetic methods, should provide more knowledge on the role and physiological activities of growth regulators in fungal-plant interactions, pathogenic as well as mutualistic.

Natural plant growth regulators are becoming an industrially important group of chemicals in agriculture and horticulture. Auxins are compounds that cause enlargement of plant cells; gibberellins and cytokinins are used to stimulate cell division in plants, gibberellins increase size of grapes and other fruits, and induce flowering and growth in sugarcane and other crop plants. Ethylene influences a wide variety of plant processes even in extremely small amounts, and many other natural and synthetic plant growth regulators seem to exert their influence through their effect on ethylene production. Induction of flowering of pineapple and ripening of pears are some examples of ethylene action (Nickell 1982).

To date, GA_3 and some other gibberellins are the only plant hormones which are produced by fermentation processes on a large scale (Lonsane and Kumar 1992). The costs are lower than those for cytokinins and abscisic acid produced by chemical synthesis. Exploratory studies on the production of cytokinins by fungi have largely dealt with bioprospecting; only a few reports with industrially relevant information on medium optimization and fermentation processes have appeared (Miller 1967; Crafts and Miller 1974).

The auxins IAA and IBA are produced even more cheaply than GA_3 by large-scale chemical synthesis. As a result, the use of IAA and IBA has become routine throughout the world (Thimann 1974).

Abscisic acid (ABA) is the most expensive phytohormone. Because of low yields from all producing fungi, (maximum titers of $52\,mg\,l^{-1}$ in 7 days from *Botrytis cinerea*; Marumo et al. 1982), no technology for producing the hormone by fermentation at present exists.

Ethylene, a simple organic compound, is the cheapest of all plant growth regulators; because it is a gas, in situ production is required for its action, and production of the product per se by industrial fermentation is economically unfeasible.

In addition to the well-known phytohormones discussed above, the search for new plant growth regulators among fungal metabolites continues.

References

Adams DO, Yang SF (1979) Ethylene biosynthesis: identification of 1-aminocyclopropane 1-carboxylic acid as an intermediate in the conversion of methionine to ethylene. Proc Natl Acad Sci USA 176:170–174

Aldridge DC, Galt S, Giles D, Turner WB (1971) Metabolites of *Lasiodiplodia theobromae*. J Chem Soc (C) 1623–1627

Angra R, Mandahar WB (1985) Pathogenesis of maize leaves by *Helminthosporium* spp.: Production and possible significance of "green islands". Res Bull Panjab Univ 36:239–243

Arai K, Shimizu S, Miyajima H, Yamamoto Y (1989) Castaneiolide, abscisic acid and monorden, phytotoxic compounds isolated from fungi (*Macrophoma castaneicola* and *Didymosporium radicicola*) cause black root rot disease in chestnut trees. Chem Pharm Bull 37(10):2870–2872

Arshad M, Frankenberger WT Jr (1988) Influence of ethylene produced by soil microorganisms on etiolated pea seedlings. Appl Environ Microbiol 54:2728–2732

Arshad M, Frankenberger WT Jr (1989) Biosynthesis of ethylene by *Acremonium falciforme*. Soil Biol Biochem 21:633–638

Arshad M, Frankenberger WT Jr (1991) Microbial production of plant hormones. In: Kleister DL, Cregan PB (eds) The rhizosphere and plant growth. Kluwer, Dordrecht

Assante WB, Merlini L, Nahini G (1977) (+) Abscisic acid, a metabolite of the fungus *Cercospora rosicola*. Experientia 33:1556–1557

Bailiss KW, Wilson IM (1967) Growth hormones and the creeping thistle rust. Ann Bot 31:195–211

Ballio A (1978) Fusicoccin the vivotoxin of *Fusicoccum amygdali* Del., chemical properties and biological activity. Ann Phytopathol 10(2):145–156

Barea JM, Azcon-Aguilar C (1982) Production of plant growth-regualting substances by the vesicular-arbuscular mycorrhizal fungus *Glomus mosseae*. Appl Environ Microbiol 43:810–814

Barthe P (1971) Recherche de substances de type cytokinine synthetisées par. *Nectria galigena* Bres. var. major Wr., en culture pure. C R Acad Sci Ser D 272:2881–2883

Bearder JR (1983) In vivo diterpenoid biosynthesis in *Gibberella fujikuroi*: the pathway after ent-kaurene. In: Crozier A (ed) The biochemistry and physiology of gibberellins, vol 1. Praeger, New York, p 251

Bergamin-Strotz LM (1988) Floristische und ökophysiologische Aspekte im Zusammenleben von Tabakpflanzen und endophytischen Pilzen. Dissertation. ETH Nr 8520, Swiss Federal Institute of Technology, Zürich

Beyrle H, Penningsfeld F, Hock B (1991) The role of nitrogen concentration in determining the outcome of the interaction between *Dactylorhiza incarnata* (L.) Soo and *Rhizoctonia*. New Phytol 117:665–672

Bottalico A, Graniti A, Lerario P (1978) Further investigation on the biological activity of some fusicoccins and cotylenins. Phytopathol Mediterr 17(2):127–134

Brian PW, Elson GW, Hemming HG, Radley M (1954) The plant growth promoting properties of gibberellic acid, a metabolic product of the fungus *Gibberella fujikuroi*. J Sci Food Agric 5:602–609

Chalutz E, Strahmann MA (1969) Induction of pisatin by ethylene. Paytopathology 59:1972

Coolbaugh RC, Al-Nimri LF, Nester JE (1985) Evidence for gibberellins in culture filtrates of *Cercospora rosicola*. 12th Int conf Plant Growth Substances, Heidelberg, Abstr 12

Costacurta A, Vanderleyden J (1995) Synthesis of phytohormones by plant associated bacteria. Crit Rev Microbiol 21:1–18

Crafts CB, Miller CO (1974) Detection and identification of cytokinins produced by mycorrhizal fungi. Plant Physiol 54:586–588

Crocoll C, Kettner J, Doerffling K (1991) Abscisic acid in saprophytic and parasitic species of fungi. Phytochemistry 30:1059–1060

Dahiya JS, Tewari JP (1991) Plant growth factors produced by the fungus *Alternaria brassicae*. Phytochemistry 30:2825–2828

Demole E, Ledrer E, Mercier D (1962) Isolement et détermination de la structure du jasmonate de méthyle, constituant odorant charactéristique de l'essence de jasmin. Helv Chim Acta 45:675–685

Doerffling K, Petersen W (1984) Abscisic acid in phytopathogenic fungi of the genera *Botrytis*, *Ceratocystis*, *Fusarium* and *Rhizoctonia*. Z Naturforsch 39c:683–684

Durand N, Debaud JC, Casselton LA, Gay G (1992) Isolation and preliminary characterization of 5-fluoroindole-resistant and IAA-overproducer mutants of the ectomycorrhizal fungus *Hebeloma cylindrosporum Romagnesi*. New Phytol 121:545–553

Edwards HH (1983) Effect of kinetin, abscisic acid and cations on host-parasite relations of barley inoculated with *Erysiphe graminis* f. sp. *hordei*. Phytopathol Z 107:22–30

El-Sharouny HM (1984) Screening of ethylene-producing fungi in Egyptian soil. Mycopathologia 85:13–15

Fukuda H, Ogawa T, Tanase S (1993) Ethylene production by microorganisms. Adv Microb Physiol 35:275–306

Gillissen B, Bergemann J, Sandmann C, Schroeer B, Bölker M, Kahmann R (1992) A two-component regulatory system for self non-self recognition in *Ustilago maydis*. Cell 86:647–657

Graebe JE (1987) Gibberellin biosynthesis and control. Annu Rev Physiol 38:419–465

Graniti A (1964) Qualche dato di tossicita della fusicoccina A, una tossina prodotta in vitro da *Fusicoccum amygdali* Del. Phytopathol Mediterr 3:125–128

Greene EM (1980) Cytokinin production in microorganisms. Bot Rev 46:25–74

Gruen H (1959) Auxin and fungi. Annu Rev Plant Physiol 10:405–440

Grundlach H, Müller MJ, Kutchan TM, Zenk MH (1992) Jasmonic acid is a signal transducer in elicitor-induced plant cell cultures. Proc Natl Acad Sci USA 89:2389–2393

Gulati A, Mandahar CL (1984) Pathogenesis of rice leaves by *Helminthosporium oryzae*: secretion of cytokinins in vitro by the fungus. Res Bull Panjab Univ 35:115–121

Gulati A, Mandahar CL (1986) Host-parasite relationship in *Helminithosporium turcicumzea mays* complex: involvement of cytokinin-like substances in pathogenesis. Ind J Exp Biol 24:309–314

Hadwiger LA, Hess SL, Von Broembsen S (1970) Stimulation of phenylalanine ammonia lyase activity and phytoalexin production. Phytopathology 60:332–336

Hasan HA (1994a) Action of carbamate biocides on sterols, gibberellin and aflatoxin formation. J Basic Microbiol 34:225–230

Hasan HA (1994b) Production of hormones by fungi. Acta Microbiol Pol 43:327–333

Henfling JWDM, Bostock R, Kuc J (1979) Effect of abscisic acid on rishitin and lubimin accumulation and resistance to *Phytophthora infestans* and *Cladosporium cucumerinum* in potato tuber tissue slices. Phytopathology 70:1074–1078

Ho I (1987) Enzyme activity and phytohormone production of a mycorrhizal fungus, *Laccaria laccata*. Can J For Res 17:855–858

Hori S (1898) Some observations on "bakanae" disease of the rice plant. Man Agric Res Sta Tokyo 12:110

Isaac S (1992) Fungal-plant interactions. Chapman & Hall, London, pp 252–265

Jochmann H-Th, Fehrmann H (1989) Effects of phyllosphere microorganisms on the senescence of wheat leaves. Z Pflanzenkr Pflanzenschutz 96:124–133

Johnston JC, Trione EJ (1974) Cytokinin production by the fungi *Taphrina cerasi* and *Taphrina deformans*. Can J Bot 52:1583–1589

Joost RE (1995) *Acremonium* in fescue and ryegrass: boon or bane: a review. J Anim Sci 73:881–888

Kawaide H, Sassa T (1993) Accumulation of gibberellin A_1 and the metabolism of gibberellin A_9 to gibberellin A_1 in a *Phaeosphaeria* sp. L487 culture. Biosci Biotech Biochem 57:1403–1405

Kawanak Y, Yamane H, Murayama T, Takahashi N, Teruka N (1983) Identification of Gibberellin A_3 in mycelia of *Neurospora crassa*. Agric Biol Chem 47:1693–1698

Kern H, Naef-Roth (1975) Zur Bildung von Auxinen und Cytokininen durch *Taphrina*-Arten. Phytopathol Z 83:193–222

Kettner J, Doerffling K (1995) Biosynthesis and metabolism of abscisic acid in tomato leaves infected with *Botrytis cinerea*. Planta 196:627–634

Kimura Y, Nakjima H, Hamasaki T (1990) Sescandelin, a new root promoting substance produced by the fungus, *Sesquicillium candelabrum*. Agric Biol Chem 54:2477–2479

Kimura Y, Mizuno T, Nakajima H, Hamasaki T (1992a) Altechromones A and B, new plant growth regulators produced by the fungus *Alternaria* sp. Biosci Biotech Biochem 56:1664–1665

Kimura Y, Shiojima K, Nakajima H, Hamasaki T (1992b) Structure and biological activity of plant growth regulators produced by *Penicillium* sp. No 31. Biosci Biotech Biochem 56(7):1138–1139

Kitagawa Y, Yamamoto H, Oritani T (1995) Biosynthesis of abscisic acid in the fungus *Cercospora cruenta*; stimulation of biosynthesis by water stress and isolation of a transgenic mutant with reduced biosynthetic capacity. Plant Cell Physiol 36:557–564

Kolattukudy PE, Rogers LM, Hwang CS, Flaishman MA (1995) Surface signaling in pathogenesis. Proc Natl Sci USA 92:4080–4087

Kronstad JW, Leong SA (1989) Isolation of two alleles of the b locus of *Ustilago maydis*. Proc Natl Acad Sci USA 86:978–982

Kurosawa E (1926) Experimental studies on the nature of the substance excreted by the bakanae fungus. Trans Nat Hist Soc Formosa 16:213–216

Leopold AC, Kriedemann PE (1975) Plant growth and development. McGraw-Hill, New York

Leslie JF, Plattner RD, Desjardins AE, Klittich CJR (1992) Fumonisin B1 production by strains from different mating populations of *Gibberella fujikuroi* (*Fusarium* section *Liseola*). Phytopathology 82:341–345

Li A, Heath MC (1990) Effect of plant growth regulators on the interactions between bean plants and rust fungi non-pathogenic on beans. Physiol Mol Plant Pathol 37:245–254

Lockhart CL, Foryth FR, Eaves CA (1968) Effect of ethylene on development of *Gloeosporium album* in apple and on growth of the fungus culture. Can J Plant Sci 48:557–559

Lonsane BK, Kumar PKR (1992) Fungal plant growth regulators. In: Arora DK, Elander RP, Mukerji KG (eds) Handbook of applied mycology, vol 4. Fungal biotechnology. Marcel Dekker, New York, pp 565–602

MacMillan J, Suter PJ (1958) The occurence of gibberellin A_1 in higher plants: isolation from the seed of runner bean (*Phaseolus multiflorus*). Naturwissenschaften 45: 46–51

Mandahar CL, Angra R (1987) Involvement of cytokinins in fungal pathogenesis. Res Bull Panjab Univ 38:35–49

Mandahar CL, Suri RA (1983) Secretion of cytokinins in vivo and in vitro by *Alternaria brassicae* and their role in pathogenesis. Trop Plant Sci Res 1(4):285–288

Marumo S, Katayama M, Komori E, Ozaki Y, Natsume M, Kondo S (1982) Microbial production of abscisic acid by *Botrytis cinerea*. Agric Biol Chem 46:1967–1968

Matheussen A-M, Morgan PW, Frederiksen RA (1991) Implication of gibberellins in head smut (*Sorosporium reilianum*) of *Sorghum bicolor*. Plant Physiol 96:537–544

Michniewicz M, Rozej B (1987) Further studies on the role of ethylene in the growth and development of *Fusarium culmorum* (W.G.Sm.) Sacc Bull Pol Acad Sci Biol Sci 35:10–12, 323–328

Michniewicz M, Rozej B (1988a) Is gibberellin the limiting factor for the growth and development of *Fusarium culmorum* (W.G.Sm.) Sacc?. Acta Physiol Plant 10:227–236

Michniewicz M, Rozej B (1988b) Further studies on the role of auxin on the growth and development of *Fusarium culmorum* (W.G.Sm.) Sacc. Acta Physiol Plant 9(4):219–227

Michniewicz M, Michalski L, Rozej B (1987) Control of growth and development of *Fusarium culmorum* (W.G.Sm.) Sacc. by abscisic acid under unfavorable pH values of the media and temperature. Bull Pol Acad Sci Biol Sci 35:4–6, 143–151

Miersch O, Schneider G, Sembdner G (1991) Hydroxylated jasmonic acid and related compounds from *Botryodiplodia theobromae*. Phytochemistry 30:4049–4051

Miersch O, Brückner B, Schneider G, Sembdner G (1992) Cyclopentane fatty acids from *Gibberella fujikuroi*. Phytochemistry 31:3875–3877

Miersch O, Günther Th, Fritsche W, Sembdner G (1993) Jasmonates from different fungal species. Nat Prod Lett 2:293–299

Milborrow BV (1978) Abscisic acid. In: Letham DS, Goodwin PB, Higgins TJV (eds) Phytohormones and related compounds: a comprehensive treatise, vol 1, Elsevier/North Holland Biomedical Press, Amsterdam, pp 295–347

Miller CO (1967) Zeatin and zeatin-riboside from a mycorrhizal fungus. Science 157:1055–1056

Mills LJ, Van Staden J (1978) Cytokinins from soils. Physiol Plant Pathol 13:73–80

Nakamura T, Kawanabe Y, Takiyama E, Takahaschi N, Murayama T (1978) Effects of auxin and gibberellin on conidial germination in *Neurospora crassa*. Plant Cell Physiol 19(4):705–709

Nickell LG (1982) Plant growth regulators. Agricultural Uses. Springer, Berlin Heidelberg New York

Nieto KF, Frankenberger WT Jr (1990) Microbial production of cytokinins. In: Bollag J-M, Stotzky G (eds) Soil Biochemistry, vol 6. Marcel Dekker, New York

Norman S, Poling SM, Maier VP, Orme ED (1983) Inhibition of abscisic acid biosynthesis in *Cercospora rosicola* by inhibitors of gibberellin biosynthesis and plant growth retardants. Plant Physiol 71:15–18

Oritani T, Ichimura M, Yamashita K (1982) The metabolism of analogs of abscisic acid in *Cercospora cruenta*. Agric Biol Chem 46:1959–1960

Pegg GF (1976) The response of ethylene-treated tomato plants to infection by *Verticillium albo-atrum*. Physiol Plant Pathol 9:215–226

Pegg GF (1981) The involvement of growth regulators in the diseased plant. In: Ayres PG (ed) Effects of disease on the physiology of the growing plant. Cambridge University Press, Cambridge, pp 149–177

Pegg GF (1984) The role of growth regulators in plant disease. In: Wood RKS, Jellis GJ (eds) Plant diseases: infection, damage and loss. Blackwell, Oxford, pp 29–48

Petrini O (1991) Physiology of endophytes, In: Andrews JH, Hirano SS (eds) Microbial ecology of leaves. Springer, Berlin Heidelberg New York

Poszar BI, Kirali Z (9164) Cytokinin-like effect of rust infection in the regulation of phloem transport and senescence. Symp on host-parasite relations in plant pathology, Budapest, 1964, 199 pp

Prema P, Thakur MS, Prapulla SG, Ramakrishna SV, Lonsane BK (1988) Gibberellin preparation using *Gibberella fujikuroi*, *Fusarium moniliforme* or *Botryodiplodia theobromae*. Indian J Microbiol 28:78–86

Redemacher W, Graebe JE (1979) Gibberellin A_4 produced by *Sphaceloma manihoticola*, the cause of superelongation disease of cassava (*Manihot esculenta*). Biochem Biophys Res Commun 91:35–42

Radice M, Sacchi A, Pesci P, Beffagna N, Marré MT (1981) Comparative analysis of the effects of fusicoccin (FC) and of its derivative dideacetylfusicoccin (DAF) on maize leaves and roots. Physiol Plant 51:215–221

Reihhardt D, Wiemaken A, Booller Th (1991) Induction of ethylene biosynthesis in compatible and incompatible interactions of soybean roots with *Phytophthora megasperma* f. sp. *glycinea* and its relation to phytoalexin accumulation. J Plant Physiol 138:394–399

Rudawska M, Bernillon J, Gay G (1992) Indole compounds released by the ectendomycorrhizal fungal strain Mrg X isolated from a pine nursery. Mycorrhiza 2:17–23

Rudnicki R, Borecka H, Pieniazek J (1969) Abscisic acid in *Penicillium italicum*. Planta 86:195–196

Salt SD, Tuzun S, Kuc J (1986) Effect of b-ionone and abscisic acid on the growth of tobacco and resistance of blue mold. Mimicry of effects of stem infection by *Peronospora tabacina* Adam. Physiol Mol Plant Pathol 28:287–297

Sassa T, Suzuki K, Haruki E (1989) Isolation and identification of gibberellins A_4 and A_9 from a fungus *Phaeosphaeria* sp. Agric Biol Chem 53:303–304

Saville BJ, Leong SA (1992) Phytohormone production by *Ustilago maydis*. In: Setlow JK (ed) Genetic engineering principles and methods, vol 14. Plenum Press, New York

Schröder M, Schulz S, Weiles EW (1990) The growth promoting fungal toxin fusicoccin does not act through an ester-hydrolysis mechanism in plants. Naturwissenschaften 77:82–83

Schulz B, Banuett F, Dahl M, Schlessinger R, Schafer W, Martin T, Herskowitz I, Kahmann R (1990) The b alleles of *Ustilago maydis*, whose combination program pathogenic development, code for polypeptides containing a homeodomain-related motif. Cell 60:295–306

Scott IM (1992) Fusicoccin-induced changes in the translatable RNAs of etiolated pea stem tissue. J Exp Bot 43:255, 1361–1365

Searle LM, Chamberlain K, Rausch T, Butcher DN (1982) Conversion of 3-indolemethylglucosinolate to 3-indolylacetonitrile by myrosinase, and its relevance to the clubroot disease of the Cruciferae. J Exp Bot 33:935–842

Sembdner G, Dathe W, Kefeli VI, Kutacek M (1980) Ab-

scisic acid and other naturally occurring plant growth inhibitors. In: Skoog F (ed) Plant growth substances. Springer, Berlin Heidelberg New York, pp 254–261
Striver K, Mundy J (1990) Gene expression in response to abscisic acid and osmotic stress. Plant Cell 2:503–512
Sugawara F, Hallock YF, Bunkers GD, Kenfield DS, Strobel G, Yoshida Sh (1993) Phytoactive eremophilanes produced by the weed pathogen *Drechslera gigantea*. Biosci Biotech Biochem 57:236–239
Sziraki I, Balazs E, Kiraly Z (1975) Increased levels of cytokinin and indolacetic acid in peach leaves infected with *Taphrina deformans*. Physiol Plant Pathol 5:45–50
Takagi M, Okabayashi M, Yokota T, Takahashi N, Shimura I, Umeya K (1975) Biochemical studies of chestnut gall tissue. National Meet of the Agric Chem Society of Japan, Sapporo, 1975, Abstr of papers, 349 pp
Takayama T, Yoshida H, Araki K, Nakayama K (1983) Microbial production of abscisic acid with *Cercospora rosicola*. I. Stimulation of abscisic acid accumulation by plant extracts. Biotechnol Lett 5:55–58
Takenaka M, Hayashi K, Ogawa T, Kimura S, Tanaka T (1992) Lowered virulence to rice plants and decreased biosynthesis of gibberellins in mutants of *Gibberella fujikuroi* selected with perfurazoate. J Pestic Sci 213–220
Talieva MN, Filimonova MV (1992) On parasitic specialization of the *Botrytis* species in the light of new experimental data. J Gen Biol (Moscow) 53:225–231
Talieva MN, Filimonova MV, Andreev LN (1991) Compounds with cytokinine activity in *Erysiphe cichoracearum* phlox mildew causative agent. Izv Akad Nauk SSSR Ser Biol 2:194–200
Thimann KV (1936) On the physiology of the formation of module of legume roots. Proc Natl Acad Sci USA 22: 511–514
Thimann KV (1974) Fifty years of plant hormone research. Plant Physiol 54:450–453
Tomita K, Murayama T, Nakamura T (1984) Effects of auxin and gibberellin on elongation of young hyphae in *Neurospora crassa*. Plant Cell Physiol 25(2):355–358
Turian G, Hamilton RH (1960) Chemical detection of 3-indolyacetic acid in *Ustilago zeae* tumors. Biochem Biophys Acta 148–150
Tzeng DD, Devay JE (1984) Ethylene production and toxigenicity of methionine and its derivatives with riboflavin in cultures of *Verticillium*, *Fusarium* and *Colletotrichum* spp. exposed to light. Physiol Plant 62: 545–552
Voigt K, Schleier S, Brückner B (1995) Genetic variability in *Gibberella fujikuroi* and some related species of the genus *Fusarium* based on random amplification of polymorphic DNA (RAPD). Curr Genet 27:528–535
Ward EWB, Cahill DM, Bhattacharyya MK (1989) Abscisic acid suppression of phenylalanine ammonia-lyase activity and m-RNA, and resistance of soybeans to *Phytophthora megasperma* f. sp. *glycinea*. Plant Physiol 91:23–27
Weidhase RA, Lehmann J, Kramell H, Sembdner G, Parthier B (1987) Degradation of ribulose-1,5-biphosphate carboxylase and chlorophyll in senescing barley ester, and counteraction by cytokinin. Physiol Plant 69:161–166
West CA, Phinney BO (1956) Properties of gibberellin-like factors from extracts of higher plants. Plant Physiol 31 Suppl:20–26
Williamson CE (1950) Ethylene, a metabolic product of diseased or injured plants. Phytopathology 40:205–208
Wolf FT (1952) The production of indoleacetic acid by *Ustilago zeae* and its possible significance in tumor formation. Proc Natl Acad Sci USA 5:38, 106
Yoshida N, Sassa T (1990) Synthesis of acremoauxin A, a new plant growth regulator produced by *Acremonium roseum* 14267. Agric Biol Chem 54:2681–2687
Zeigler RS, Powell LE, Thurston HD (1980) Gibberellin A_4 production by *Sphaceloma manihoticola*, causal agent of cassava superelongation disease. Phytopthology 70:589–596

10 Toxin Production in Grass/Endophyte Associations

M.R. Siegel and L.P. Bush

CONTENTS

Departments of Plant Pathology and Agronomy, University of Kentucky, Lexington, Kentucky 40546, USA

I. Introduction

While the discovery that fungal endophytes were present in some pooid grass species was made at the turn of the 20th century (White et al. 1993), knowledge of their widespread distribution (Leuchtmann 1992; Chap. 12, Vol. V, Part B) and the ecological and evolutionary consequences for host and fungi (Clay 1994; Leuchtmann 1994; Schardl et al. 1994; Tsai et al. 1994; Chap. 14, Vol. V, Part B), as well as their impact on livestock production (Prestidge 1993; Strickland et al. 1993), is of more recent origin. The association of grasses with endophytes of the genus *Epichloë* has been shown to affect the growth and/or reproduction not only of mammalian herbivores, but also of insect herbivores (Latch 1993; Popay and Rowan 1994) and, to lesser extents, nematodes, (Latch 1993), plants (Petroski et al. 1990), and fungi (Guo et al. 1992; Gwinn and Gavin 1992). In many cases, the toxicity has been associated with specific secondary metabolites produced by grass/endophyte associations (symbiota). Additionally, endophyte compatibility and toxin production is under the influence of both host and fungal genomes (Leuchtmann 1992; Christensen et al. 1993; Leuchtmann and Clay 1993; Agee and Hill 1994; Roylance et al. 1994; Welty et al. 1994; Christensen 1995) as well as the environment (Agee and Hill 1994; Arechavaleta et al. 1992; Bacon 1993). These constraints upon toxin biosynthesis result in a wide spectrum of biological activity for symbiota and offer the opportunity for manipulation of the endophytes for improvement of the grasses when used in forage, turf, and soil conservation (Schardl 1994; Siegel and Bush 1994).

The Mycota V Part A
Plant Relationships
Carroll/Tudzynski (Eds.)

A. Fungi and Symbiosis

The mycosymbionts belong to the family Clavicipitaceae (Ascomycota), tribe Balansieae. The Balansieae are endophytic (growing between host cells) or epiphytic (between epidermal layers), but do not penetrate host cells. They typically maintain perennial, biotrophic associations with grasses (Poaceae) or sedges (Cyperaceae) and are triggered to produce sporogenous stromata or sclerotia only at certain developmental stages of their hosts. For many grasses infected with Balansieae, the production of stromata and/or sclerotia occurs on immature flowering tillers, and maturation of the florets is halted. Where known, infection occurs by ascospores (horizontal transmission) via the flowers and ovule. The latent period, from infection to the appearance of external symptoms of the fungus, is long (up to 10 months), and in the holomorphic species of genus *Epichloë* (Pers.:Fr.) Tul (*E. typhina* sensu lato) certain anamorphs (*Acremonium* spp. Link sect. *Albo-lanosa* Morgan-Jones et Gams = *Neotyphodium*, Glenn, Bacon et Hanlin¡ Glenn et al. 1996) remain totally endophytic, nonreproductive, maternally (vertically) transmitted seed disseminated entities.

The taxonomic relationships of the *Epichloë* spp. are not firmly established (White 1993), but they are a diverse group of fungi (Christensen et al. 1993; Leuchtmann 1994; Schardl et al. 1994; Tsai et al. 1994; Schardl 1996; Chap. 12, Vol. V, Part B) that grow mostly in C_3 grasses (subfamily Pooideae). Their ectophytic plectenchymatous stromata prevent emergence of the host inflorescence (choke disease). The relative importance of ectophytic sporulation versus endophytic seed dissemination forms the basis for classifying the types of symbiotic associations of *Epichloë* and anamorphs (White 1993; Schardl 1996). The symptomatic expression of the ectophytic stromata is agonistic or detrimental to the fitness of the plant because it prevents flowering/sexual reproduction and reduces the genetic diversity of the plant genome. Conversely, if the endophyte is asymptomatic, normal host sexual reproduction can occur. The relationship becomes mutualistic because reproduction of the fungus is now associated with the host, via seed dissemination. The infected host receives the benefits of toxin production for protection against herbivores and other pests, as well as other factors which effect persistence and yield (Hill et al. 1991a; Bacon 1993; Bouton et al. 1993; Belesky and Fedders 1995). However, how the benefits and detriments are valued is not as clearly defined. Three factors complicate the valuation. Ectophytic expression ranges from total sterilization of the flower panicles to production of stromata on only a few panicles, and specific anamorphs are totally endophytic (White 1993; Schardl 1996; Chap. 14, Vol. V, Part B); toxin production is independent of fungal expression (Siegel et al. 1990; Christensen et al. 1993); and toxin-induced resistance to grazing animals, while an ecological fitness enhancement for the host, is a major detriment for livestock production (Hoveland 1993; Prestidge 1993; Strickland et al. 1993).

B. Toxin Production – Types of Compounds

A number of different biologically active classes of chemicals are associated with symbiota. Many are alkaloids, and have individual and/or multiple activity against different classes of organisms. Four classes of symbiota-specific alkaloids have received the most intensive study (Porter 1994; Siegel and Bush 1996). These are the pyrrolizidines (lolines), ergots (clavines, lysergic acids, and derivative alkaloids), indolediterpenoids (lolitrems), and pyrrolopyrazine (peramine) alkaloids. No more than three alkaloid classes have been observed in any one symbiotum, and many symbiota contain combinations of three or two classes, some only one class, and a few none (Table 1). Additionally, Table 1 also illustrates that grass species can be hosts for more than one *Epichloë/Neotyphodium* sp. or biotype, and that symbiota, naturally infected or derived from artificially introduced isolates, can produce either the same or different alkaloid profiles. While these data support the hypothesis that the host and/or fungus genomes may both exert a strong influence on alkaloid synthesis, the alkaloids (except for the lolines) are of known fungal origin (Bush et al. 1993a; Rowan 1993; Porter 1994).

The variation in alkaloid profiles of symbiota can, in certain instances, be correlated with the phylogenetic relationships of the endophytes, as indicated by cluster analysis of allozyme profiles (Leuchtmann 1992, 1994) and molecular analysis of DNA sequences (Schardl et al. 1994; Tsai et al. 1994; Schardl 1996). These observations have important ramifications for endophyte diversity, as related to alkaloid profiles and other host-beneficial characteristics in the grass species

Table 1. Toxic alkaloids produced by Balansieal-infected grasses

Grass	Endophyte[b]	Alkaloid class[a]				Reference[c]
		Pyr.	Er.	PyrPy	Idt	
Agrostis hiemalis	*Epichloë amarillans*	–	+	+	–	3
Achnathercum inebrians	*Neotyphodium* sp.	–	+	–	Tr[d]	5
Andropogon virginicus	*Balansia henningsiana*	ND	+	ND	ND	2
Bromus anomalus	*Neotyphodium starri*	–	+	+	–	3
Cenchrus echinatus	*B. obtecta, B. cyperi*	ND	+	ND	ND	2
Dactylis glomerata	*Epichloe typhina*	–	–	–	–	[e]
Elymus canadensis	*Epichloë* sp.	–	–	+	–	3
Festuca arundinaceae	*N. coenophialum*	+	+	+	–	1, 3, 4
	N. coenophialum	+	–	+	–	1
	Neotyphodium sp. (FaTG-2)[f]	–	+	–	–	1
	Neotyphodium sp. (FaTG-2)	–	+	+	–	1
	Neotyphodium sp. (FaTG-2)	–	+	+	+	1
	Neotyphodium sp. (FaTG-3)	+	–	+	–	1
F. arizonica	*N. huerfanum*	–	–	–	–	3
F. gigantea	*Neotyphodium* sp.	+	+	–	–	3
F. glauca	*Epichloë* sp.	–	+	+	–	3
F. longifolia	*Epichloë festucae*	–	+	+	+	3
	Epichloë sp. (AI[g]/*F. rubra communtata*)	–	+	–	–	3
F. obtusa	*N. starrii*	–	–	–	–	3
F. versuta	*Neotyphodium* sp.	–	–	–	+	3
F. paradoxa	*Neotyphodium* sp.	–	–	+	–	3
F. pratensis	*N. uncinatum*	+	–	–	–	1
F. rubra commutata	*Epichloë festucae*	–	+	+	–	3
	Epichloë festucae	–	+	–	–	3
F. rubra litoralis	*Epichloë festucae* (AI/*F. rubra commutata*)	–	+	–	–	3
F. rubra rubra	*Epichloë festucae*	–	+	–	–	3
Lolium multiflorum	*Neotyphodium* sp.	+	–	ND	ND[h]	4
L. persicum	*Neotyphodium* sp.	+	–	ND	ND	4
L. perenne	*N. lolii*	–	+	+	+	3
	Epichloë typhina	–	–	+	–	3
	N. coenophialum (AI/*F. arundinaceae*)	+	+	+	–	3
	Epichloë festucae (AI/*F. longifolia*)	–	+	+	–	3
	N. lolii	–	–	+	+	1
	N. lolii	–	–	–	–	1
	N. lolii	–	+	+	–	1
	Neotyphodium sp. (LpTG-2)[f]	–	+	+	–	1
L. rigidum	*Neotyphodium* sp.	+	–	ND	ND	4
L. temulentum	*Neotyphodium* sp.	+	–	ND	ND	4
Hordeum bogdanii	*Neotyphodium* sp.	–/Tr[i]	–/Tr	ND	ND	4
H. brevisubulatum	*Neotyphodium* sp.	–	–	ND	ND	4
	Neotyphodium sp.	Tr	+	ND	ND	4
Poa alsodes	*Neotyphodium* sp.	+	+	ND	ND	4
	Neotyphodium sp.	–	+	ND	ND	4
P. ampla	*Neotyphodium* sp.	–	–	+	–	3
P. autumnalis	*Neotyphodium* sp.	+	–	+	–	3
Sporobolus poirettii	*B. epichloë*	ND	+	ND	ND	2
Sitanion longifolium	*Epichloë* sp.	–	+	+	–	3
Stipa robusta	*Neotyphodium chisosum*	+	+	ND	ND	4

[a] Pyr, pyrrolizidines (lolines); Er, ergots; PyrPy, pyrrolopyrazine (perarmine): Idt, indolediterpenoids (lolitrems).
[b] Names reflect present understanding of taxonomy (Glenn et al. 1996).
[c] 1, Christensen et al. (1993); 2, Powell and Petroski (1992); 3, Siegel et al. (1990); 4, TePaske and Powell (1993); 5, Miles et al. (1996).
[d] Tr, trace amounts; ND, not determined.
[e] M.R. Siegel and L.P. Bush (unpubl. data).
[f] FaTG, *F. arundinacea* endophyte taxonomic grouping; LpTG, *Lolium perenne* endophyte taxonomic grouping.
[g] AI, artificially introduced isolate from a different grass species.
[h] Lolitrems not present except when *A. lolii* was artificially introduced (Prestidge 1991).
[i] –/Tr, none to a trace when multiple symbiota analyzed.

evolved, given the extent to which grass endophytes propagate clonally by seed infection.

C. Toxin Production – Altered Host Metabolism

Little information exists on the nature of altered host metabolism as it relates to alkaloid toxin synthesis, but parasitism of plants by nonphotosynthetic organisms must, by definition, involve the utilization of products that the host produces from sunlight or other energy sources. This utilization could affect myco toxin synthesis by the direct requirement of specific host metabolites, as intermediates, or indirectly through the sequestration of simple carbon and nitrogen sources required for fungal growth and metabolism. Uptake and utilization of simple carbon (White et al. 1991; Richardson et al. 1992) and nitrogen metabolites (Lyons et al. 1990; Ferguson et al. 1993) by endophytes in culture and in planta have been reported. Additionally, fungal invertase (Lam et al. 1995) and proteinase (Lindstrom and Belanger 1994) activity is expressed in *Neotyphodium/Epichloë*-infected grasses. These enzymes may serve to enhance breakdown and utilization of host carbohydrates and proteins (via establishment of metabolite gradients between source and sink regions) and/or reduce host structural barriers to the intercellular hyphal ramification in host tissue. The utilization of host metabolites by endophytes did not result in major changes in forage quality (crude protein, acid detergent fiber, neutral detergent fiber, and in vitro dry matter disappearance) and agronomic performance as indicated by a comparison of *N. coenophialum* Glenn, Bacon et Hanlin (Morgan-Jones et Gams)-infected and noninfected *Festuca arundinacea* Schreb. (tall fescue; Bush and Burrus 1988; Fritz and Collins 1991).

If carbon and nitrogen compounds of the plant are mobilized at the source to sink regions for fungal growth, the alkaloid toxins may remain or be mobilized away from the site of synthesis, and this varies for the individual alkaloid types. Some of the alkaloids are recovered in high concentrations in the upper vegetative portions of the plant (leaf blade) where the endophyte is not present or present at low levels. The endophytes are usually not present in the roots and few or no alkaloids have been recovered from the root tissue. Tissue distribution and seasonal accumulation patterns of alkaloids are important factors, as they indicate what portions of seedlings or mature plants are likely to contain alkaloids in sufficient quantity to be toxic to insect and mammalian herbivores.

D. Scope of the Chapter

The primary focus of this chapter will be the toxic alkaloids associated with *Epichloë*- and *Neotyphodium*-infected hosts, although alkaloids of other Balansieae will be included. Where applicable, each of the toxin classes will be subdivided into a discussion of biological considerations – biological activity, distribution patterns, and seasonal accumulation as related to plant growth, effect of host, and fungal genomes on synthesis; and chemical considerations – structures, biosynthetic and chemical synthesis.

II. Ergot Alkaloids

Ergotism, caused by ingestion of sclerotia of species of *Claviceps* Tul, has a long history of afflicting humans and livestock. The pathological, physiological, and pharmacological activities involved in ergotism have been associated with clavines, lysergic acid, and derivative alkaloids (Berde and Schilid 1978). The discovery that ergot alkaloids were present in *N. coenophialum*-infected tall fescue suggested that toxicosis associated with this grass was also due to these compounds (Lyons et al. 1986). Ergot toxicity was also reported in an *N. lolii* Glenn, Bacon et Hanlin (Latch, Christensen et Samuels)-infected *Lolium perenne* L. (perennial ryegrass) association that was modified to contain an isolate of the endophyte that did not produce lolitrem (Tapper 1993). Ergot alkaloids associated with the Balansieae, their chemistry, and aspects of biological activity have been reviewed (Garner et al. 1993; Strickland et al. 1993; Porter 1994; Siegel and Bush 1996).

A. Biological Considerations

While aspects of the chemistry and biosynthesis of ergot alkaloids will be discussed in Section II.B, the types and nomenclature of the compounds involved in biological considerations are elucidated here. Ergot alkaloids include numerous com-

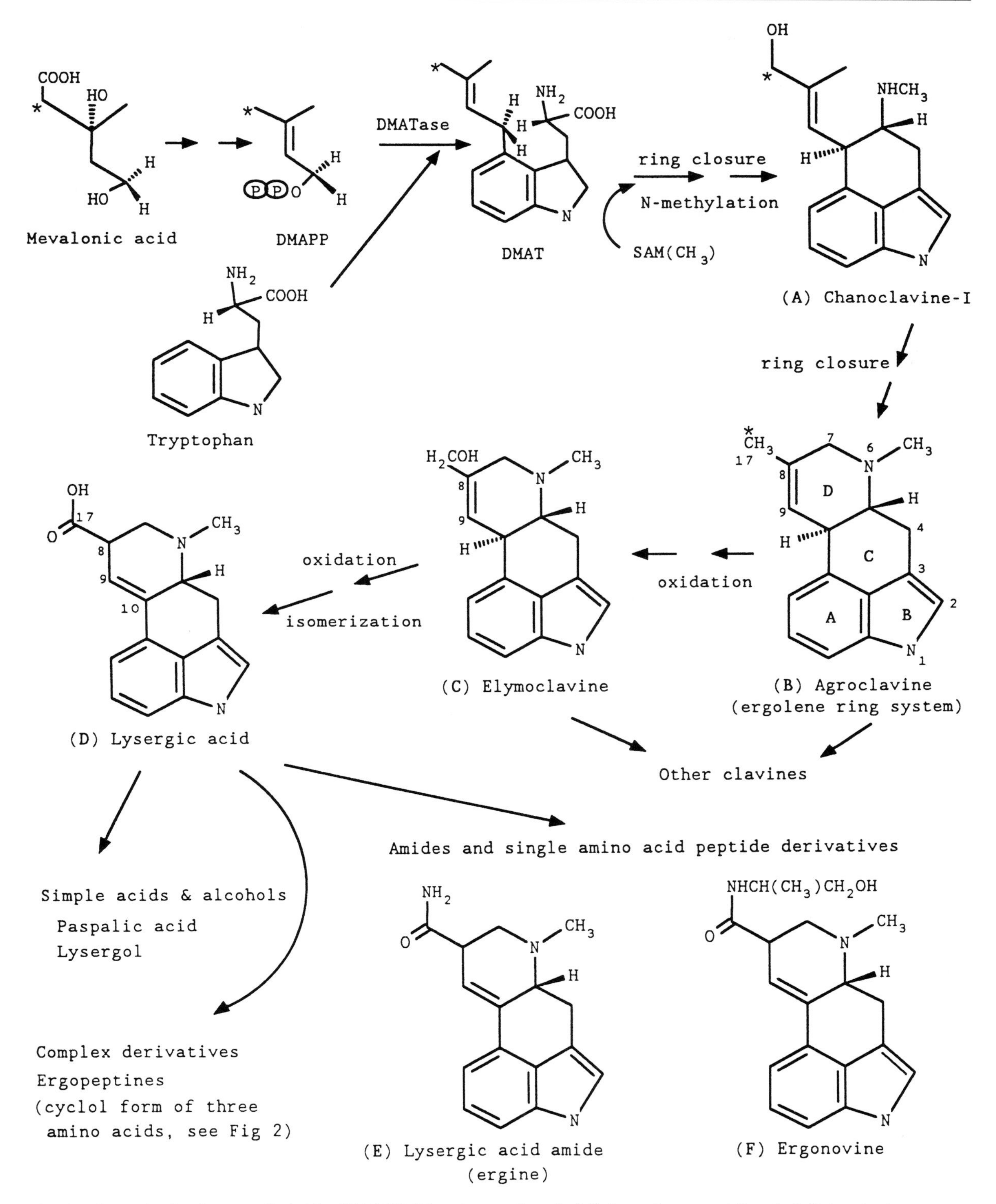

Fig. 1. Biosynthesis of ergot alkaloids. *DMAPP* Dimethylallyl pryrophosphate; *DMAT* 4-(γ,γ-dimethylallyl)tryptophan; *DMATase* dimethylallyl diphosphate:L-tryptophan dimethylallyltransferase (dimethylallytryptophan synthase); * represents carbon 2 from mevalonic acid into the ergoline ring

pounds under the general headings of clavines, ergolene acids and alcohols, simple lysergic acid derivatives, and complex lysergic acid derivatives (ergopeptines). Figure 1 illustrates the structures of these compounds, as well as the biosynthetic pathway. The clavines (Fig. 1A–C) include chanoclavine – open D ring – and agroclavine, elymoclavine, and other closed D ring 8-ergolene class of alkaloids with Δ8,9 double bonds. Clavines are converted to an ergolene acid, lysergic acid, by oxidation of the C17 hydroxyl to an acid moiety and a shift in the double bond from Δ8,9 to Δ9,10 (Fig. 1C). Simple acids and alcohols (paspalic acid and lysergol) are derived from lysergic acid by epimerization, double bond shift or decarboxylation at the C8 position. Other simple derivatives include lysergic acid amide or ergine (Fig. 1E), lysergic acid diethyl amide and ergonovine (Fig. 1F) and complex derivatives (ergopeptines) which are synthesized from lysergic acid via amide formation at the C17 position. With ergonovine there is an amino acid at this position. Ergopeptines are produced with an amino group derived from one of the three amino acids (at position I) of the cyclol moiety, with the proline ring moiety the only amino acid in position III (Fig. 2). The ergopeptines are further divided into three major groups based on the amino acid substituents at the position I [R_1] group.

1. Presence in the Balansieae

The different types of ergot alkaloids described previously have been isolated from various species of the genus *Balansia* Speg. and *Neotyphodium*/*Epichloë*-infected grasses or in culture as major and minor components of the total ergot alkaloid fraction. The *Balansia* spp. studied [*B. epichloë* (Weese) Diehl, *B. henningsiana* (Moell.) Diehl, *B. strangulans* (Mont.) Diehl, *B. claviceps* Speg.] contained primarily clavines and the simple lysergic acid derivative, ergonovine, as the major alkaloid components (Porter 1994). While no ergopeptines with proline were isolated, ergobalansine, a cyclol derivative without proline, was detected in *B. obtecta* Diehl and *B. cyperi* Edr. as the major component (Powell and Petroski 1992). Endophyte-infected tall fescue has received the most attention because of the associated toxicoses. Ergovaline is the major component, 60–80% of the ergopeptine fraction, recovered from tall fescue (Arechavaleta et al. 1992). In addition, ergine has also been reported to be an important component (Powell and Petroski 1992). The minor ergot alkaloid components include clavines, ergonovine, ergosine, and ergotamine (Arechavaleta et al. 1992; Powell and Petroski 1992). Detection of ergovaline and its epimer ergovalinine alone have been used as indicators of ergot alkaloid synthesis

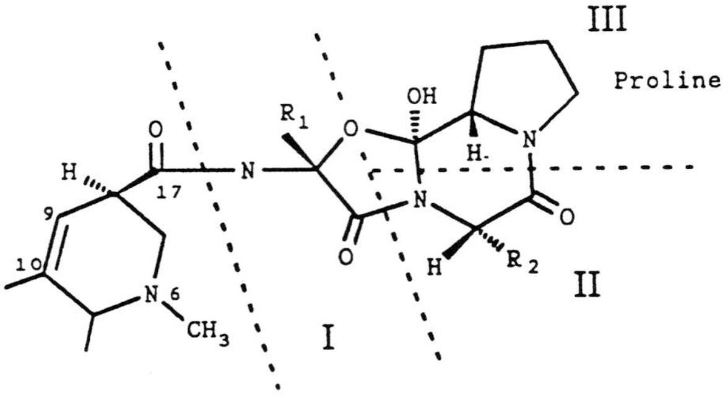

Alkaloid	Position I [R1]	Position II [R_2]
Ergotamine[1]	Alanine [-CH_3]	Phenylalanine [-$CH_2C_6H_6$]
Ergovaline[1]	Alanine [-CH_3]	Valine [-$CH(CH_3)_2$]
Ergosine[1]	Alanine [-CH_3]	Leucine [-$CH_2CH(CH_3)_2$]
Ergonine[2]	Amino Butyric Acid	Valine [-$CH(CH_3)_2$]
Ergocrystine[3]	Valine [-$CH(CH_3)_2$]	Phenylalanine [-$CH_2C_6H_6$]
Ergocornine[3]	Valine [-$CH(CH_3)_2$]	Valine [-$CH(CH_3)_2$]

[1] Ergotamine group.
[2] Ergoxine group.
[3] Ergotoxine group.
[1–3] Identified in culture (except ergocrystine) and in endophyte infected tall fescue (Garner 1993; Porter 1994).

Fig. 2. Amino acids of the cyclol moiety of ergopeptine alkaloids. (Garner et al. 1993)

in *Epichloë/Neotyphodium*-infected grasses, with 56% of the symbiota listed in Table 1 producing this alkaloid. However, it is possible that the lack of ergovaline in symbiota, other than tall fescue, where the ergot alkaloid profile is generally known, may not be an indication that other ergot alkaloids are present. The detection of mainly other types of ergot alkaloids in *Balansia* species (Porter 1994) and high levels of lysergic acid derivatives in *Neotyphodium*-infected *Stipa robusta* Schribn. (sleepy grass; Petroski et al. 1992) and *Achnatherum inebrians* (Hance) Keng (drunken horse grass; Miles et al. 1996) indicates the need for testing symbiota for other types of ergot alkaloids. As with *Claviceps* spp., the list of detected ergot alkaloids is expected to grow as more symbiota are subjected to detailed analysis and as endophyte biotypes are isolated and grown in culture.

2. Biological Activity

Certain types of ergot alkaloids have been reported to have some insecticidal activity (antibiosis and antifeedent effects) and, along with the other classes of alkaloids, may contribute to the spectrum of activity against insects in endophyte-infected grasses. Activity was dependent upon insect species, type of alkaloid, and its concentration (Clay and Cheplick 1989; Siegel et al. 1989; Yates et al. 1989; Prestidge and Ball 1993). In many studies, much of the activity occurred at higher concentrations than found in infected grasses. However, the ergot alkaloids used in various bioassays did not include ergovaline due to unavailability. Consequently, the lack of activity reported for specific ergot alkaloids may not be indicative of the potential activity of ergovaline and ergine, the prominent types found in *Neotyphodium*-infected tall fescue and other grasses.

While the pyrrolizidines alkaloids were detected first and initially thought to be responsible for fescue toxicosis (see Sect. III), many of the physiological and pathological effects occurring with large and small herbivores mimic those known for ergot alkaloids (Siegel and Bush 1996; Strickland et al. 1993). Because ergovaline was unavailable for testing, a direct relationship could not be established between toxicosis and ergot alkaloids until recently. Animal toxicity based on feeding endophyte-infected seed or forage has been subdivided into effects on growth (reduced weight gains), temperature (elevated), blood flow (restricted-vasoconstriction-fescue foot syndrome), reproduction (reduced), and milk production (reduced and agalactia-decreased serum prolactin) (Strickland et al. 1993). The manifestations of many of these symptoms may be dependent on vertebrate species or type/concentration of alkaloids. Other factors that affect the physiological activity of ergot alkaloids include the type of forage, its chemical and physical composition, rumen microflora, location of alkaloid absorption sites, and availability and solubility of the alkaloids. Pathological expression is obviously dependent upon maintenance of sufficient blood alkaloid levels and availability of the sites of action in the animal.

The biological activity of ergot alkaloids, primarily ergopeptines and lysergic acid and its derivatives, has been measured in numerous in vitro systems, and many of the physiological effects noted help explain the pathology of toxicosis in whole animals (Abney et al. 1993; Dyer 1993; Moubarak et al. 1993; Oliver et al. 1993; Strickland et al. 1994; Larson et al. 1995). From these studies, much of the physiological activity of ergot alkaloids appears to be based on their interaction with biogenic amine receptors (e.g., dopamine, norepinephrine, serotonin) which causes changes in normal homeostatic mechanisms resulting in the pathology of toxicosis (Strickland et al. 1993).

3. Distribution And Seasonal Accumulation

Like lolitrem and peramine, ergot alkaloids, measured primarily as ergovaline, are generally recovered at levels (ca. 0.5–5.0$\mu g\,g^{-1}$) 1000 times lower than those of loline alkaloids. Ergovaline was detected in infected plants alone or in combination with the other classes of toxic alkaloids (Table 1). In the important forage and turf grasses, ergovaline was detected in tall fescue along with peramine and loline alkaloids, while in perennial ryegrass, ergovaline was recovered along with peramine and lolitrem. However, ergovaline has not been detected in *N. uncinatum* Glenn, Bacon et Hanlin (Gams Petrini et Schmidt)-infected *F. pratensis* Huds (meadow fescue). The levels of ergot alkaloids in grasses and their seasonal distribution are dependent on the type of plant tissue, environment (fertilization and water availability), and host-fungal genome interactions.

a) Distribution Within the Plant

Ergovaline was recovered in highest concentrations from pseudostems and to a lesser extent in leaves of vegetative perennial ryegrass which contain little or no endophyte (Table 2). The pseudostem and senescent leaf sheath contained 78% of the total ergovaline, and these tissue types also contain the highest concentration of endophyte. In tall fescue, the order of concentration, from highest to lowest, was crown, stems, and leaves with levels in seed equal to or greater than those in the crown tissue (Rottinghaus et al. 1991; Azevedo et al. 1993). Translocation of ergovaline to the leaves of tall fescue and perennial ryegrass occurred readily, as indicated by the presence of little or no endophyte in this tissue. The low levels of ergovaline in the senescent leaf sheaths (Table 2) indicate either remobilization, concurrent with loss of endophyte viability, or metabolism. Ergovaline has also been recovered from tall fescue in low amounts in the roots (Azevedo et al. 1993). Roots of the tall fescue seedlings have been reported to be infected with endophyte (Azevedo and Welty 1995).

b) Effect of Environment

Two environmental factors, nitrogen fertilization and soil water deficit, affected ergovaline concentrations in plants. These factors may act directly or indirectly on host and/or endophyte metabolism. High nitrogen greatly increased ergovaline levels (2–20 times, depending on tissue type) in all portions of tall fescue plants, including roots (Azevedo et al. 1993; Rottinghaus et al. 1991). Three cycles of water deficit increased ergovaline concentrations by 117% in endophyte-infected (isolate 187BB) ecotype perennial ryegrass in comparison to the watered control (Barker et al. 1993). In this study, peramine and lolitrem concentrations were unaffected by water deficit. Long-term (40 days) water stress also increased ergovaline concentrations, particularly with high nitrogen (Arechavaleta et al. 1992). While ergot alkaloids increased with deficiencies in water, they should not be viewed as stress metabolites because they were present in sufficient quantities in nonstressed plants to act as biologically active defensive compounds. Well-fertilized plants store excess nitrogen in proteins which, under N deficit, can be metabolized to low molecular weight compounds (amino acids) and remobilized to other portions of the plant to maintain growth (Azevedo et al. 1993). In infected plants, the reduction in available N compounds should reduce growth and/or ergot alkaloid synthesis. Conversely, excess nitrogen not only stimulates plant and fungal growth, but nitrogen and its metabolites could be used directly or be converted by the mycosymbiont to precursors used in ergot alkaloid biosynthesis (Lyons et al. 1990). The relationship between water deficit and increased ergot alkaloid synthesis has been postulated to involve the accumulation of specific carbohydrates (Arechavaleta et al. 1992). Infected plants grown under drought stress contain higher concentrations of nutrients than nonstressed plants. In particular, sugar alcohols and other soluble carbohydrates have been identified (Richardson et al. 1992). While these compounds may contribute to osmotic potential and regrowth capacity of infected plants, it has also been suggested that they serve as precursors for ergot alkaloid synthesis by the mycosymbiont (Arechavaleta et al. 1992).

c) Seasonal Accumulation

Comprehensive multi-year field studies involving interactions among the major alkaloids (ergots, lolines, peramine, or lolitrems) in the presence of biotic and/or abiotic factors (grazing, N fertilization, temperature, and precipitation) are lacking.

Table 2. Percent distribution of lolitrem B, paxilline, ergovaline, and peramine in *Neotyphodium lolii*-infected perennial ryegrass (cv. Grasslands Marsden) grown in New Zealand. (After Davies et al. 1993)

Plant part[a]	Lolitrem B	Paxilline	Ergovaline	Peramine
Leaves 1–4	17	52	22	56
Upper pseudostem	15	20	16	17
Lower pseudostem	22	13	44	20
Senescent leaf sheath	46	15	18	7

[a] Harvested (fall).

In an 8-month study of *N. lolii*-infected cultivars of perennial ryegrass in Australia, ergovaline concentrations were highest in late summer-fall and lowest in midwinter (Woodburn et al. 1993). In a 9-month study of *N. coenophialum*-infected tall fescue in Georgia, US, ergovaline levels were highest in late fall-early winter and lowest in midsummer (Belesky et al. 1988). However, Agee reported and Hill (1994) that ergovaline concentrations, for a number of infected plant genotypes (low to high alkaloid-producing) grown at three sites in Georgia, were low during the entire year except in the spring, when levels reached a maximum. In the 2-year study, the differences in ergovaline levels were determined to be significantly affected by location, genotype, and harvest date.

4. Host/Endophyte Genome Interactions

Ergovaline synthesis appears to be under control of both endophyte and/or host genomes. Wild-type *N. lolii*-infected Nui perennial ryegrass contained 3.5 times less alkaloid than when the grass was infected with isolate 196. Nui infected with isolate 187BB and 196 had the same levels of ergovaline (Davies et al. 1993). A host/endophyte interaction effect is indicated by the fourfold decrease in ergovaline when 187BB isolate was introduced into cv. Greenstone and the level compared to that in infected Nui. Control of the genetic potential of the tall fescue endophyte by the host, as it pertains to ergot alkaloid synthesis, has been reported in tall fescue as well. The introduction of an isolate from a high ergovaline-producing symbiotum into a noninfected genotype, that previously produced low ergovaline levels when infected, resulted in a low alkaloid-producing symbiota (Hill et al. 1991b). Reciprocal crosses between high- and low-producing tall fescue genotypes produced F_1 progeny with a frequency distribution of low to high ergovaline levels. The data suggest that it will be possible to select for low ergovaline-producing populations that will remain low across environments (Agee and Hill 1994). When low ergovaline progeny were analyzed for peramine, it was determined that the alkaloids were independently regulated (Roylance et al. 1994). These data indicate that the low ergovline-producing plant population will maintain sufficient levels of peramine to control insects, an important aspect in situations where reduced animal toxicosis is desired.

B. Chemical Considerations

1. Biosynthesis

The ergopeptide alkaloids found in the endophyte and in the endophyte-host system are indole derivatives of tryptophan and mevalonic acid (Fig. 1). As indicated, much of the ergot alkaloid biosynthesis pathway is known in *Claviceps* species (Shibuya et al. 1990; Garner et al. 1993), and a similar pathway is expected in the endophytes, given their relationship to *Claviceps* spp. (Schardl et al. 1994). The prenylation of tryptophan, formation of the ergoline ring system, and assembly of the peptide portion are significant steps in synthesis. The first intermediate in the committed pathway is 4-(γ, γ-dimethylallyl)tryptophan (DMAT), derived from dimethylallyl pryrophosphate (DMAPP) and L-tryptophan. The enzyme that catalyzes its formation is dimethylallylpyrophosphate: L-tryptophan dimethylallyltransferase, also known as dimethylallyltryptophan (DMAT) synthase. The enzyme has been purified to homogeneity and characterized (Shibuya et al. 1990; Gebler and Poulter 1992). Labeling experiments indicate that C17 in the ergoline ring is derived from the C2 of mevalonic acid during the first sequence reactions. Since DMAT synthase catalyzes the first committed step in ergot alkaloid synthesis, it can serve as a model for regulation of alkaloid synthesis genes and the role of the ergopeptines in fitness (abiotic and biotic stress tolerances) of symbiota. This step is clearly an appropriate target for gene manipulation, with the practical objective of eliminating a class of alkaloids associated with toxic effects on livestock (Schardl 1994; Tsai et al. 1995).

Many of the biologically active compounds isolated from symbiota are derived from lysergic acid. This key intermediate can be converted into simple or complex derivatives by amidation. The tricyclic peptide portion of the complex derivatives is assembled linearly from proline and two other variable amino acids (Fig. 2) possibly by a multienzyme complex (Porter 1994). However, these enzymes, as well as most others required for synthesis of the different types of ergot alkaloids, have not been characterized. The relationship between tryptophan, ergot alkaloids, lolitrem, and indoleacetic acid biosynthesis has been noted for symbiota (Porter 1994). The key intermediate, tryptophan, can be converted to paxilline and lolitrems or indolegylcerols and indole auxins.

Consequently, any shifts in host and/or endophyte metabolism induced by changes in the genomes or in environment could result in changes in tryptophan utilization involving ergot and lolitrem biosynthesis and auxin-induced growth. When low ergot or lolitrem-producing *Neotyphodium* isolates are introduced into new hosts to control livestock toxicoses, the potential for metabolic shifts involving these compounds may be profound, with the examples given for the lolitrem minus infected perennial ryegrass cultivars which had increased levels of ergovaline.

2. Detection

The methods for extraction and measurement of the various types of ergot alkaloids have been reviewed (Garner et al. 1993). Because of the importance of ergovaline in symbiota, this compound and its measurement have received much attention. Like many other ergot alkaloids, it exists as an epimer. The epimers ergovaline and ergovalinine are easily extracted and separated by the specific methods described by Garner et al. (1993). The results are usually reported as total ergovaline, but the distribution of the epimers in the symbiota may not be reflected in what is recovered by the analytical analysis due to artifacts in the extraction and detection methods.

Extracting infected tissue or mycelium and detecting the ergot alkaloids present a number of problems associated with their chemistry (e.g., solubility, stability, and isomerization) and the condition of the material to be extracted. The methods of extraction that result in maximum recovery for one group or individual compounds may serve less well for others.

Two methods of detection offer the optimum quantative recovery of ergopeptide and other ergot alkaloids. These are tandem mass spectrometry (MS/MS) and high-pressure liquid chromatography (HPLC). MS/MS detects pg amounts of the various types of ergot alkaloids, but equipment requirements make the detection costly. HPLC with fluorescent detection is ideally suitable for rapid and inexpensive detection of ergopeptine alkaloids, specifically ergovaline and its epimer. A modification of the HPLC method resulted in a highly reproducible increase in recovery of ergovaline, from 82–93%, and analysis time, permitting the analysis of 60–80 samples per day (Hill et al. 1993). Enzyme-linked immunosorbent assay offers another potential rapid and relatively sensitive approach for detection of individual and total ergot alkaloids in infected grasses (Shelby and Kelley 1992; Hill and Agee 1994).

III. Pyrrolizidine Alkaloids

Most of the well-known pyrrolizidine alkaloids are esters of hydroxylated 1-methylpyrrolizidine and have one double bond in the pyrrolizidine ring. This unsaturation between carbons 1 and 2 provides the hepatotoxic and carcinogenic characteristic of these alkaloids. However, the saturated amino pyrrolizidines, loline and derivatives, are found in abundance in grasses infected with certain endophytic fungi. The unique characteristics of the amino pyrrolizidine alkaloids are that they are saturated in the pyrrolizidine ring and they contain an oxygen bridge between carbons 2 and 7 of the pyrrolizidine base (Fig. 3). Loline was first isolated in 1955 from seed of *Lolium cuneatum* Nevski (Bush et al. 1993a; Powell and Petroski 1993). Early pharmacological studies of lolines were done in the 1960s, and structural confirmation was accomplished in the mid-1960s. Isolation, identification, and research into the pharmacological action of these alkaloids were begun in the absence of the knowledge that they were present only in symbiota. Notwithstanding, the lolines are

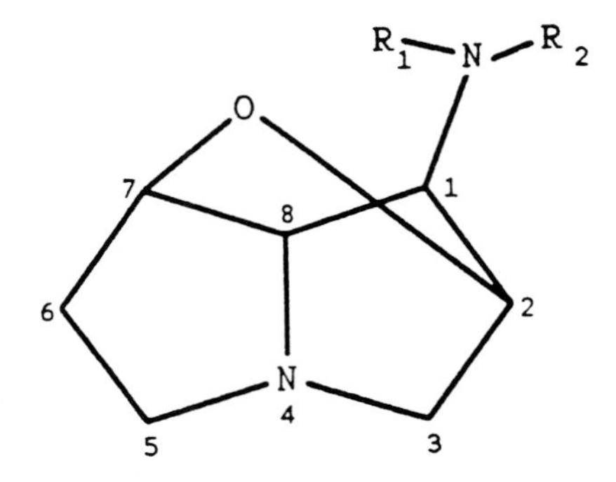

	R_1	R_2
Loline	H	CH_3
N-formylloline	CHO	CH_3
N-acetylloline	CH_3CO	CH_3
N-methylloline	CH_3	CH_3
Norloline	H	H
N-acetylnorloline	CH_3CO	H
N-formylnorloline	CHO	H

Fig. 3. Saturated amino pyrrolizidine (loline) alkaloids

present in the greatest concentration of any of the alkaloids identified to date, with levels in forage that may exceed 10 g kg^{-1}. Consequently, they cannot be ignored in the total toxic syndrome of grazing herbivores.

A. Biological Considerations

Although loline is found in many symbiota, most alkaloids in this class are derivatives of loline (Fig. 3). Loline alkaloids have not been detected in endophytes grown in culture or in the plant without the presence of a fungal endophyte. Therefore, the source of these alkaloids is not known. Loline alkaloids have been found in 36% of the symbiota evaluated in Table 1. This percentage represents a population much smaller than for the ergot alkaloids and peramine, but greater than for the lolitrems.

1. Biological Activity

Broad-spectrum insecticidal activity of N-formylloline (NFL) and N-acetylloline (NAL) as a contact substance has been reported (Siegel et al. 1989; Rowan et al. 1990; Dahlman et al. 1991; Popay et al. 1993; Prestidge and Ball 1993; Schmidt 1993). Longer acyl chain N′-loline derivatives significantly deterred feeding in choice tests by selected insects (Riedell et al. 1991). Japanese beetle grub were significantly deterred (34%) from feeding (Riedell et al. 1991) by addition of NFL and NAL to the diet in an amount found in roots of *N. coenophialum*-infected tall fescue (Siegel et al. 1989). When administered to vertebrates, loline alkaloids have caused many different animal responses (Yates and Tookey 1965; Robbins et al. 1972; Solomons et al. 1989; Abney et al. 1993; Powell and Petroski 1993; Strickland et al. 1994). However, it is difficult to reach a general conclusion on the effect of these alkaloids from the in vitro and in vivo data available. Lolines also have allelopathic properties as inhibition of rate and amount of seed germination has been reported (Petroski et al. 1990; Bush et al. 1993a).

2. Distribution Within the Plant

NFL, NAL, loline, N-acetyl norloline (NANL), and N-methylloline (NML) have been detected in *Neotyphodium*-infected tall fescue (Yates et al. 1990). NFL, NAL, and NANL are found in greatest abundance, with NFL frequently present in concentrations three times greater than NAL and NANL. Loline alkaloids are found in the leaves, stem, and roots of *Neotyphodium*-infected tall fescue (Siegel et al. 1989). The amount present in the root was dependent upon the root growth medium, as none was detected in roots grown in solution culture (Bush et al. 1993a). The lolines are found in all the shoot tissues, with the greatest accumulation most often occurring in the seed and the least accumulation in the leaf blade. In stem, rachis, and blade of flowering *N. coenophialum*-infected tall fescue, the amount of NAL + NFL in each tissue had a positive and significant linear correlation coefficient with the amount of *N. coenophialum* detected. Translocation of loline alkaloids is indicated by its detection in roots and leaf blades, where the endophyte is not present. In vegetative plants the pseudostems contained a greater accumulation than the leaf blade (Belesky et al. 1989; Bush et al. 1993b).

3. Seasonal Distribution

Seasonal accumulation of NAL + NFL in tall fescue is less responsive to growth conditions than perloline, an alkaloid known to be under genetic control of the host (Bush et al. 1993a). Increased accumulation of NAL + NFL occurred during late summer in southeastern USA and may have been the consequence of reduced plant dry weight accumulation and normal alkaloid production, resulting in an apparent increased alkaloid accumulation (Putnam et al. 1991). More recently, Bush and Schmidt (1994) found greatest NAL + NFL accumulation in *F. pratensis*-infected *N. uncinatum* to occur in August and September in both Switzerland and Kentucky, USA. Generally, in Kentucky, the NAL + NFL concentration in forage was low, 200–300 $\mu g\, g^{-1}$, during winter, increased during early spring growth, but decreased on a dry weight basis during the rapid growth of late spring, and had the maximum accumulation in late summer.

Environmental and management factors also influence accumulation of NAL and NFL. An optimum growth temperature increased NAL + NFL accumulation compared to higher or lower growth temperature regimes. Water stress also significantly increased forage tissue accumulation of NAL and NFL; however, these alkaloids, in contrast to the ergot alkaloids, did not increase with added N fertilizer. Increased plant age and clipping the vegetative tissue did stimulate accumulation of NAL and NFL (Bush et al. 1993a,b). NFL was present in the greatest amount in green leaf tissue, and as herbage tissue matured and senescence occurred NFL concentration decreased (Eichenseer et al. 1991).

4. Host/Endophyte Genome Interactions

Greatest accumulation of NAL and NFL occurs in *N. uncinatum*-infected *F. pratensis* (Bush and Schmidt 1994). A survey of many symbiota for the presence of loline alkaloids generally found high amounts of the loline alkaloids in *F. arundinacea* infected with *N. coenophialum* (Table 1). Seed of *F. versuta* Beal contained high amounts of NFL and NANL but the *Neotyphodium* endophyte was not identified. One entry each of *Hordeum bogdanii* Wil. and *H. brevisubulatum* (Trin.) Link contained trace amounts of NFL. Within the *Lolium* species tested *L. rigidum* Gaudin, *L. temulentum* L., and *L. persicum* Boiss. & Hohen. contained NFL and NAL and *L. temulentum* contained loline and NML. *L. multiflorum* contained small amounts of NFL, NAL, and NANL. Forage of *Stipa robusta* (Vasey) Scribn. contained NFL and seed of *Poa alsodes* A. Gray contained NFL and NAL.

Siegel et al. (1990) reported the presence of loline alkaloids in naturally infected symbiota of *F. arundinacea*/*N. coenophialum*; *F. gigantea* L./*Neotyphodium* spp., *Poa autumnalis* Muhl./*N. coenophialum* and in the artificial infection of *L. perenne*/*N. coenophialum* symbiotum. The naturally infected *L. perenne*/*N. lolii* symbiotum usually does not accumulate loline alkaloids. However, Huizing et al. (1991) reported the presence of NFL and NAL in *L. perenne*/*N. lolii* symbiotum grown at 30 °C but not when grown at 23 °C. The presence of loline alkaloids in the artificial *L. perenne*/*N. coenophialum* association and the temperature response of *L. perenne*/*N. lolii* indicate an environmental-host-endophyte interaction. Certainly in this instance, the endophyte, *N. coenophialum*, is a significant parameter in loline alkaloid accumulation. Loline alkaloid accumulation in *F. pratensis*/*N. uncinatum* and *F. arundinacea*/*N. coenophialum* was only slightly different between Changins, Switzerland, and Kentucky, USA, indicating a small environmental by symbiotum interaction (Bush and Schmidt 1994).

B. Chemistry

1. Separation and Detection

Extraction of loline alkaloids from tissue has been achieved by standard procedures for alkaloids (Bush et al. 1993a; Powell and Petroski 1993). The only caution to be taken in extraction and purification is the use of HCL, as derivatives of loline are not stable to treatment with HCL and will result in formation of loline hydrochloride. This characteristic has been exploited by Petroski et al. (1989) in semisynthesis of several derivatives of loline. Separation of the individual loline alkaloids has been achieved to greater or lesser degree by paper, thin-layer, and gas chromatography. Perhaps the best procedure for separation is use of a 0.53-mm id capillary column and detection with flame ionization, ion-specific or a mass spectrometer detector (Takeda et al. 1991; Powell and Petroski 1993).

2. Synthesis

Total synthesis of loline has been achieved (Tufariello et al. 1986), but the reactions in the multistep synthesis do not appear to be readily amenable to large-scale synthesis of loline and loline derivatives. Synthesis of the compounds accumulated in several symbiota has been done by isolating loline from seed of tall fescue and forming the individual alkaloids by varying the substitutions at the N′-atom (Petroski et al. 1989). NFL and NAL are easily prepared by these procedures and they have been used to prepare standards and do bioassays.

3. Biosynthesis

Total biosynthesis of the lolines has not been elucidated, but biosynthesis of other pyrrolizidine bases has been investigated (Böttcher et al. 1994). Based upon these studies, we have postulated that ornithine is the initial precursor for the pyrrolizidine structure and ornithine is decarboxylated to putrescine by ornithine decarboxylase (Fig. 4; Bush et al. 1993a). Putrescine and decarboxylated S-adenosylmethionine are substrate for spermidine synthetase to form spermidine. Spermidine may be oxidized by a polyamine oxidase to allow cyclization and with the addition of the oxygen bridge between carbon atoms 2 and 7 to form norloline. Also, spermidine could be methylated via spermidine methyltransferase to methylsper-

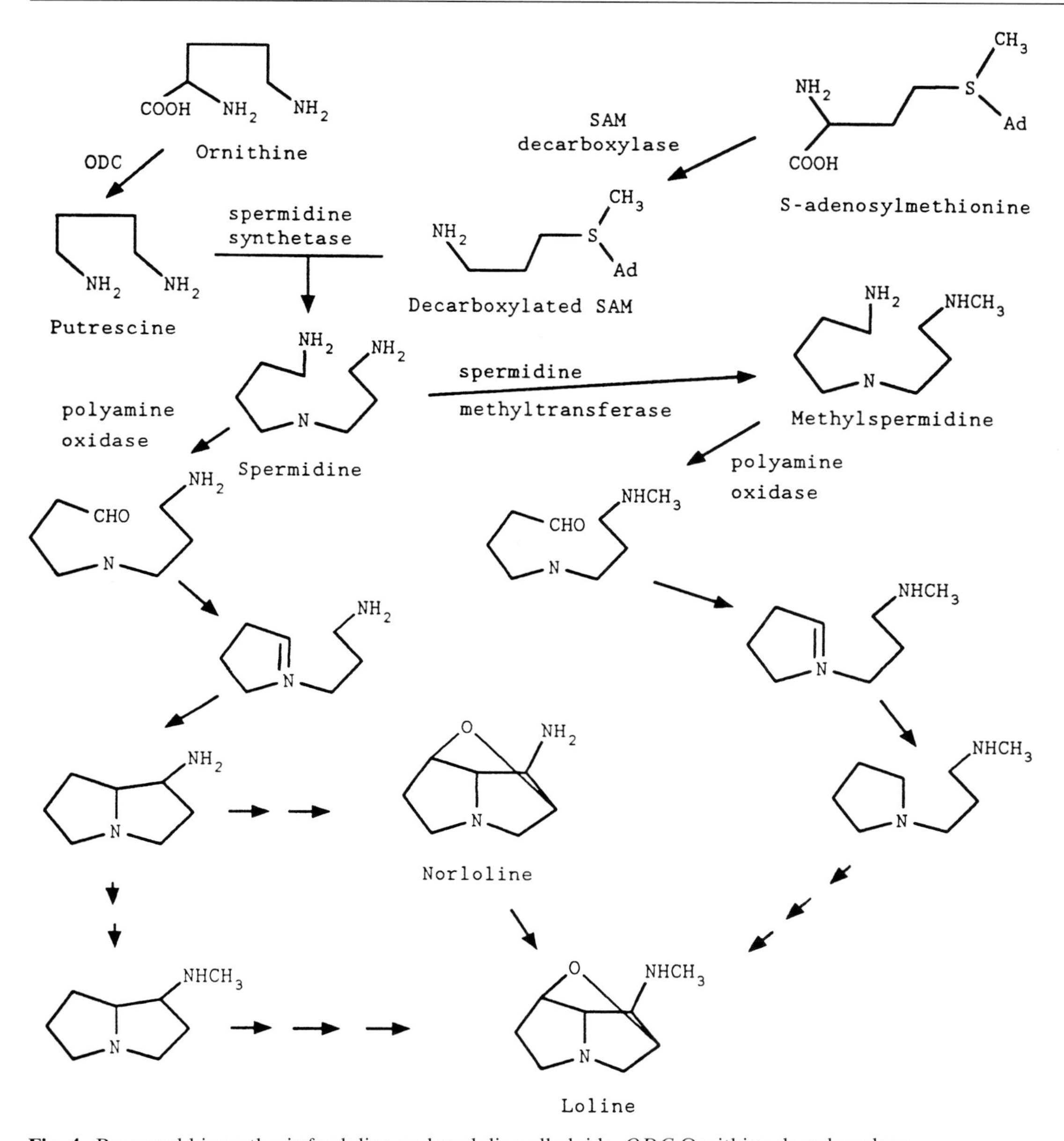

Fig. 4. Proposed biosynthesis for loline and norloline alkaloids. *ODC* Ornithine decarboxylase

midine, oxidized to allow ring closure and then addition of the oxygen bridge for loline formation. Studies on loline alkaloid biosynthesis are made difficult because it is not known whether the alkaloids are formed by the plant, the endophyte, or by a synergistic interaction of the symbionts.

4. Structure-Activity Relationship

Of the naturally occurring loline alkaloids, only NFL demonstrated significant inhibition of *L. multiflorum* seed germination (Petroski et al. 1990) with 50% inhibition at 7.7×10^{-4}mmol per seed. Also, as the acyl chain length decreased from 8 or increased from 14 carbons, germination inhibition activity decreased. Activity was slightly greater, but consistently so at acyl carbon lengths of 4–16, against *L. multiflorum* than *Medicago sativa* L. Cyclic derivatives of loline were relatively nontoxic. A response similar to that in the phytotoxicity tests to acyl chain length was observed in choice and no-choice feeding experiments with fall armyworm (Riedell et al. 1991). NFL was as active as any of the derivatives tested. However, loline and the 2, 3, and 4 acyl chain length derivatives were not different from the control. Straight acyl chain lengths from 6–16 carbons were equal to NFL. Greater selectivity was observed with European corn borer, as acyl chain length of 10 and 16 carbons were most effective, and only NAL of the shorter chain derivatives was different from the controls in no choice tests. The short-chain, 1–3 carbon, derivatives were more toxic to greenbug aphids than the longer-chain derivatives. In general, the naturally occurring NFL has greatest pharmacological activity and, among the acyl synthetic lolines, the intermediate chain length is most active.

IV. Lolitrems

A. Biological Considerations

The lolitrems are tremorgenic neurotoxins produced mainly in *L. perenne*/*N. lolii* symbiotum and are generally referred to as the principal causative agent of ryegrass staggers. The lolitrems are exemplified best by the principal lolitrem, lolitrem B, and most of this discussion on lolitrems will be directly related to lolitrem B. Ryegrass staggers has been reported in horses, cattle, sheep, and deer, and affected animals show tetanic muscle spasms and a hypersensitivity to external stimuli (Rowan 1993).

1. Biological Activity

Lolitrem B (Fig. 5) and its precursor, paxilline, were shown to be a feeding deterrent to Argentine stem weevil larvae. Reported tremorgenic activity of the lolitrems is based primarily on a mouse bioassay (Gallagher and Hawkes 1985). Of the many lolitrems isolated from *L. perenne*/*N. lolli* associations, lolitrem B is present in greatest abundance. Lolitrems A, B, and C are equally tremorgenic (Miles et al. 1995a,b). Paxilline and lolitrem F also have tremorgenic activity in the mouse bioassay, whereas the precursors α-paxitriol and lolitriol do not have tremorgenic activity (Miles et al. 1992).

2. Distribution Within the Plant

Lolitrem B concentration increased from $0.6\,\mu g\,g^{-1}$–$5.7\,\mu g\,g^{-1}$ in the youngest emerging leaf to the oldest live leaf on a vegetative tiller of *L. perenne* infected with *N. lolii*. The basal pseudostem contained nearly twice the amount of lolitrem B as the upper pseudostem and over ten times the amount present in the leaf blade (Table 2; Keogh and Tapper 1993). Also, in these older leaf blades, lolitrem B concentration increased from tip to ligule. Within a young infected seedling, lolitrem B increased in the shoot for the first 24 days after planting, decreased to a minimum 48 days after planting, and then increased (Ball et al. 1993). The initial increase of lolitrem B in the seedlings may be translocation from the seed into the new seedling as concentration in the seed and seed residue decreased during this time. Dry weight increase was rapid during days 24–48, and a dilution or metabolism probably occurred before a more steady-state growth and alkaloid production occurred after day 48. Very small amounts of lolitrem B were detected in roots of these seedlings. In flowering plants, lolitrem B concentration was greatest in basal leaf sheath and, in descending order, lesser concentrations were measured in upper leaf sheath, seed, stem, and leaf blade (DiMenna et al. 1992).

3. Seasonal Distribution

Lolitrem B levels in *L. perenne*/*N. lolii* symbiotum in New Zealand were very low ($<1.0\,\mu g\,g^{-1}$) during winter, increased during spring and summer, and reached a maximum ($>5\,\mu g\,g^{-1}$) in autumn (Rowan 1993). Ball et al. (1995) reported a similar pattern of accumulation in New Zealand with minimum accumulation in September and maximum in January. Lolitrem B and peramine were assayed in this experiment and mean levels in each sample were positively correlated ($r^2 \cong 0.50$), suggesting that the biosynthesis and accumulation of these two alkaloids could be linked by physiological or environmental factors. DiMenna et al. (1992) have shown a similar pattern of lolitrem B accumulation in flowering herbage of infected *L. perenne*. However, as lolitrem B decreased between March and April (autumn) harvests, the concentration of paxilline, a precursor for lolitrem B biosynthesis, increased (Davies et al. 1993).

4. Host/Endophyte Genome Interaction

Lolitrem A and B were isolated initially from *N. lolii*-infected *L. perenne*, and most lolitrem studies have been done with this naturally occurring sym-

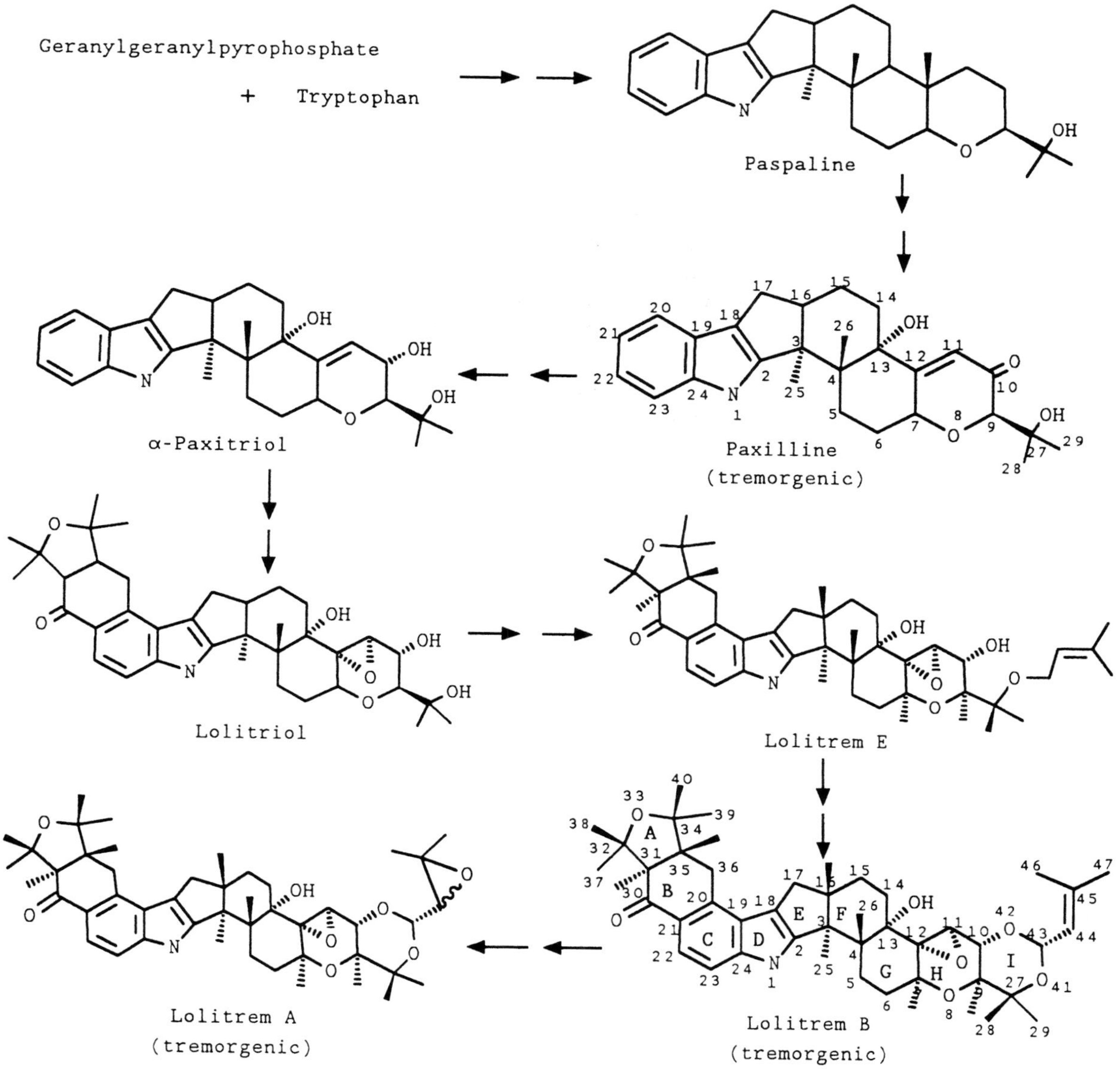

Geranylgeranylpyrophosphate + Tryptophan → → Paspaline → → Paxilline (tremorgenic) ← ← α-Paxitriol → Lolitriol → → Lolitrem E → Lolitrem B (tremorgenic) ← ← Lolitrem A (tremorgenic)

Fig. 5. Biosynthesis of lolitrem B

biotum. However, Siegel et al. (1990) found lolitrem B in tall fescue artificially infected with *N. lolii* but not in perennial ryegrass artificially infected with *N. coenophialum*, suggesting the primary significance of the endophyte in alkaloid accumulation. *L. multiflorum* naturally infected with an *Neotyphodium*-like endophyte contained low concentration (0.01–0.09 $\mu g g^{-1}$) of lolitrem B; however, with *N. lolii* artificially introduced into *L. multiflorum*, lolitrem B concentration increased to 0.2–4.5 $\mu g g^{-1}$ (Prestidge 1991). Lolitrem B has also been reported in *F. longifolia* Thuill./*E. typhina* symbiotum, but when *E. typhina* was artificially introduced into *L. perenne* and other hosts, lolitrem B was not detected (Siegel et al. 1990). In addition, Penn et al. (1993) have found paxilline and lolitrem B in agar culture of *N. coenophialum*, the common endophyte of tall fescue, but the symbiotum did not accumulate lolitrem in herbage. Lolitrem B has been detected in seed of tall fescue infected with *N. coenophialum* (C.O. Miles, pers. comm.).

The New Zealand researchers have manipulated *L. perenne*/*N. lolii* genotypes to study accumulation of lolitrem B and other alkaloids in the symbiotum. In one *L. perenne* genotype infected with three different *N. lolii* genotypes, three significantly different levels of lolitrem accumulation occurred (Davies et al. 1993). Also, when one specific endophyte genotype was introduced into three different host genotypes there were two different levels of lolitrem B accumulation. In this study, paxilline accumulation in the same host but with the different endophytes was not different, whereas when the same endophyte was used in three different host genotypes, three different levels of paxilline accumulation occurred. Based on all the information available, both the host and the endophyte genotype significantly alter accumulation of lolitrem B and most likely precursors of lolitrem B in symbiota. Consequently, results of manipulation of one symbiont cannot be accurately predicted in the absence of the other. This indicates that evaluation of genetically altered host or mycosymbiont must be made at the symbiotum level.

B. Chemistry

1. Separation and Detection

The lolitrems are lipophilic substituted indoles of approximately 460–700 molecular weight and most readily extracted, separated, and detected by liquid chromatographic techniques. Lolitrem B may be separated and measured with high sensitivity by HPLC with fluorescence detection (Gallagher et al. 1985). A procedure for the immunological detection of the lolitrems has been described (Garthwaite et al. 1993). Antibodies to protein conjugates of lolitrem B precursors, paxilline and lolitriol, and lolitrem B had minimum cross-reactivity. The minimum detection in methanolic herbage extracts was $0.2 \mu g g^{-1}$ dry weight. Only a simple dichloromethane:methanol extraction procedure was required before assay, and this procedure may be essential for studies on lolitrem accumulation that include genetic manipulation of the endophyte or host.

In vivo animal studies with lolitrems had not been possible until development of a large-scale isolation of lolitrem B (Miles et al. 1994). Lolitrems were extracted from ground seed by Soxhlet extraction, partially purified by solvent partitioning and flash chromatography, and crystallized. With this procedure they were able to isolate a few hundred mg of lolitrem B and lesser amounts of lolitrem E and lolitrem F, plus many additional lolitrems not previously identified. The amount of lolitrem B isolated was sufficient for in vivo sheep studies on pharmacological properties of lolitrem B.

2. Biosynthesis

Complete biosynthesis of the lolitrems including paxilline has not been elucidated; however, much is known from analogous tremorgens. As with the ergot alkaloids, tryptophan was the starting intermediary metabolite of the indole moiety for paxilline biosynthesis (Laws and Mantle 1989) and, consequently, should be for the indole of lolitrem B (Fig. 5). The diterpene moiety of paxilline is from 12 acetate units which form four mevaloate-derived isoprene units which yield a geranylgeranylpyrophosphate (deJesus et al. 1983). Every atom in paxilline is present in an equivalent position in lolitriol and lolitrem B (Miles et al. 1992), suggesting that paxilline is a precursor of lolitrem B (Fig. 5). Miles et al. (1992) proposed the paxilline $\rightarrow$ α-paxitriol $\rightarrow$ lolitriol $\rightarrow$ lolitrem B biosynthetic pathway in Fig. 5. Miles et al. (1995a) suggested that the ability to synthesize paspaline is a crucial step in the tremorgen biosynthesis. Lolitriol, lolitrem B, and paxilline have been found together in *L. perenne* infected with *N. lolii* and in cultures of *N. lolii*, further indicating their biosynthetic relationship (Miles et al. 1992). If lolitriol is to be a near precursor of lolitrem B, as Miles et al. (1992) proposed, the epoxidation of the olefinic double bond at C11 of paxilline and the diprenylation to form the bicyclic terpenoid on the indole moiety to yield lolitriol must occur prior to the addition of the isoprene moiety at the oxygen atoms at C10 and C27 to yield lolitrem B. They proposed lolitrem E as an intermediate with the open ring (I) after addition of the last isoprene unit between lolitriol and lolitrem B. The sequence of the modifications at C44–C47 is not known chemically or physiologically.

3. Structure-Activity Relationship

Selala et al. (1989) proposed that the tremorgens containing a single nitrogen atom had activity based on conformation of a pharmacophore comparable to the conformational form of GABA (γ-amino butyric acid). This would include the atoms 1, 2, 3, 4, 5, 12, 13, 14, 16, 18, 25, and 26 of lolitrem

B. This hypothesis met the proposed GABA-mediated effects of the lolitrems (Eldefrawi et al. 1990) but the proposed active portion of lolitrem B is constant in all the compounds in the biosynthetic pathway from paxilline onward. As noted earlier, paxilline, lolitrem A, and lolitrem B are tremorgenic, but Miles et al. (1992) found paxilline to be only 20% as tremorgenic as lolitrem B in the mouse bioassay. However, α-paxitriol, lolitriol, and lolitrem E were not tremorgenic (Miles et al. 1992, 1994). Structure-activity relationship is apparently very subtle as the ring structure A to H of lolitrem B, lolitriol, and lolitrem E are identical. The sterochemistry of the A/B ring fusion and the presence of a side chain at C43 are important for tremorgenic activity (Ede et al. 1994; Miles et al. 1995b).

V. Pyrrolopyrazine Alkaloids

A. Biological Considerations

Peramine (Fig. 6) is the only known member of its class isolated from *Epichloë* and *Neotyphodium* spp.-infected grasses. When analyses of symbiota included all the classes of alkaloids, peramine was detected most often (67% of the associations listed in Table 1) in various combinations with other alkaloids or alone. Levels in endophyte-infected plants range from 0.5–132 $\mu g\,g^{-1}$ (Siegel et al. 1990; Popay and Rowan 1994; Ball et al. 1995). Peramine is a fungal product as it is recovered from mycelia and culture filtrates (Rowan 1993).

1. Biological Activity

Peramine has not been reported to have activity against mammalian herbivores or be phytotoxic; its primary activity is insecticidal (Rowan 1993). Feeding deterrence to the Argentine stem weevil, cutworm *Graphaina mutans* Walker, and beetle *Phaedon cochleariae* F. was demonstrated in bioassays using diets of peramine containing extract from *A. lolii*-infected perennial ryegrass or pure peramine. Deterrence occurred at concentrations found in the grass. Endophyte-infected perennial ryegrass generally contains lolitrems and ergot alkaloids, in addition to peramine, which are also active in bioassays, against the weevil. The Argentine stem weevil is a major pest of perennial ryegrass in New Zealand and endophyte-free grass does not survive the combination of overgrazing and insect predation. The spectrum of insects affected by infected tall fescue and perennial ryegrass has been reviewed (Latch 1993; Popay and Rowan 1994), but because these grasses contain the other classes of alkaloids, it is not possible, except for the greenbug *Schizaphis graminum* Rondani, to ascertain whether peramine alone is responsible for deterrence and/or toxicity (Siegel et al. 1990).

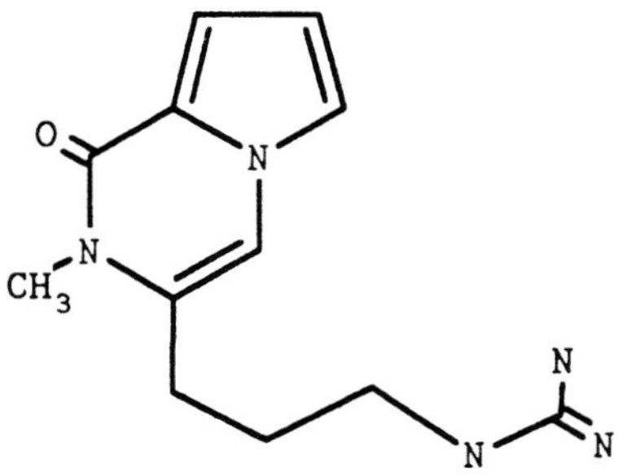

Fig. 6. Structure of peramine

2. Distribution and Seasonal Accumulation

In vegetative plants, the distribution of peramine in perennial ryegrass follows the distribution of the endophyte in the plant. Leaf sheaths contained greater concentrations of peramine and endophyte than leaf blades. The oldest green leaf sheath contained the highest concentrations and the youngest emerging leaf blade the very least (Keogh and Tapper 1993). This suggests that peramine is translocated from the sheaths, the site of endophyte synthesis, to the leaf blade, and from older to younger leaves (sheaths and blades). While lolitrem and peramine moved into germinating seedling tissue, prior to substantial endophyte growth, a higher proportion of peramine was more rapidly translocated than lolitrem (Ball et al. 1993). Peramine was not recovered from roots of mature plants, but was found initially in roots of seedlings, possibly offering transient protection against soil-inhabiting insects. In flowering plants, the highest concentration of endophyte and peramine were in youngest tissue, culms, and panicles.

A distribution analysis of the three classes of alkaloids, plus the lolitrem intermediate paxilline, detected in vegetative perennial ryegrass indicated that the patterns of accumulation in the different tissue types differed for the groups (Table 2).

Lolitrem and ergovaline did not accumulate to high concentrations in the leaf blade, while they did accumulate in the pseudostems. Of particular interest is the accumulation pattern in the senescent leaf sheath. Both peramine and ergovaline had low levels (7–18%) in this tissue while lolitrem had high concentrations (46%) of the total alkaloids. This suggests that peramine and ergovaline, in contrast to lolitrem, were subjected to either plant metabolism or remobilization from the senescent tissue.

The distribution pattern of peramine in flowering plants of tall fescue genotypes was similar to that in perennial ryegrass (Roylance et al. 1994). Highest concentrations were present in locations of the endophyte, culm and panicle. However, about a third of the peramine was present in the leaf blade, where no endophyte was present.

Seasonal variation occur in the accumulation of peramine in tall fescue or perennial ryegrass. This is in comparison to lolines, lolitrem, and ergovaline levels which show consistent seasonal variations. In New Zealand, peramine was lowest in the winter in both basal and regrowth tissue and highest in the spring through fall (Ball et al. 1995). Yearly mean concentrations of peramine in individual plants were closely related to, and probably determined by, yearly mean concentrations of endophyte. Levels of peramine produced by three different endophyte-infected perennial ryegrass associations were unaffected by three cycles of water deficit (Barker et al. 1993). Peramine levels in tall fescue grown in the greenhouse (Roylance et al. 1994) and field (Bush and Schmidt 1994) also did not show significant seasonal variations. However, when the same *N. coenophialum*-infected isoline of tall fescue (cv. KY 31) was grown in Chargins, Switzerland, and Lexington, Kentucky, USA, peramine could be recovered only from the plants at the USA location, even though lolines and ergovaline were recovered at both locations. The data suggest subtle differences in environment affecting a specific alkaloid, either by changes in host and/or endophyte responses.

3. Host/Endophyte Genome Interactions

A wide range (2–132 $\mu g g^{-1}$) of peramine concentrations has been determined in symbiota (Siegel et al. 1990), indicating a high degree of natural variation. However, in *Neotyphodium*-infected perennial ryegrass peramine levels appear to be under regulatory control of the host genome (Davies et al. 1993). Peramine levels were not significantly different between isolates of *N. lolii* in the same cultivar, indicating no endophyte effect. However, concentrations decreased ca. 50%, from that of 187BB-infected cv. Ruanui, when this endophyte strain was introduced into cv. Greenstone. The lack of lolitrem in both cultivars, but with the presence of peramine, also suggests that accumulation of the two alkaloids is independently regulated. Independent regulation of ergovaline and peramine accumulation in *N. coenophialum*-infected tall fescue has also been demonstrated (Roylance et al. 1994).

The relatively even distribution of peramine in the above-ground portions of plants, its lack of a pronounced seasonal distribution, and the independent regulation of synthesis/accumulation from the other known mammalian toxic alkaloids are important biological features in endophyte-infected perennial ryegrass and tall fescue. Location of toxin and seasonal distribution patterns are characteristics that will affect feeding habits, and hence control, of susceptible insects. Independent regulation of synthesis is important in those grass cultivars that have been manipulated to have greatly reduced levels of lolitrems or ergot alkaloids to alleviate ryegrass staggers or fescue toxicosis, respectively, but still maintain high levels of peramine for insect control.

B. Chemical Considerations

Chemical synthesis and measurement of peramine and related products have been reviewed by Rowan (1993). The peramine molecule consists of a lipophilic pyrrolopyrazine heterocyclic ring with a hydrophilic quanidinyl-terminated side chain. Peramine is relatively less lipophilic than the other classes of alkaloids, which may explain its distribution patterns in plants. It has been suggested that peramine is derived from proline and arginine, and that the biosynthesis occurs via cyclization and condensation reactions, with the final N-methylation via an N-methyltransferase and S-adenosylmethionine (Porter 1994). Measurement is by reverse-phase or normal phase HPLC with UV detection at 280 nm or by reverse-phase thin layer chromatography with colorimetric detection at 600 nm after spraying with Van Urk's reagent.

Peramine has been synthesized and various chemical intermediates and analogs have been used in structure-activity relationship studies (Rowan 1993). In agar bioassay feeding deterrent studies the most active compounds were peramine and homoperamine, a homolog with the addition of a fourth methylene carbon into the aliphatic chain. Activity was substantially reduced by removal of the quanidinium moiety, suggesting the importance of the side chain for full activity. However, substitution or addition in the heterocyclic ring system resulted in a complete loss of activity. The biochemical mode(s) of action are unknown for peramine, but it interferes with microsomal cytochrome P450, causing the carbamate insecticide carbaryl to be twice as toxic to *Spodoptera eridania* (Cramer), an insect not directly affected by peramine (Dubis et al. 1992).

VI. Miscellaneous Toxins

A. Sterols and Steroid Metabolites

Sterols and related metabolites have been extracted from cultures of *N. coenophialum* and infected tall fescue (Davis et al. 1986) and cultures of *B. epichloë* and infected *Sphenophlis poiretti* (Roem et Schmidt) Hitchc. (smut grass) (Porter et al. 1975). The compounds include ergosterol, ergosterol peroxide, and ergostatetraeneone. In bioassays only ergosterol peroxide, a light-induced metabolite, was toxic to brine shrimp and chick embryos (Davis et al. 1986). It is not possible to evaluate the importance of the sterols in toxicity to herbivores, but presence of these compounds does indicate fungal growth in infected plant tissue and potential for synthesis of toxic alkaloids previously discussed.

B. Antifungal Compounds

Agar bioassay studies have indicated that cultures of *Neotyphodium* spp. from numerous hosts exhibit antifungal activity against specific grass pathogens (Christensen and Latch 1991; Christensen et al. 1991; Siegel and Latch 1991). However, resistance of endophyte-infected grasses has been reported for only crown rust (*Puccinia coronata* Cda), seedling disease caused by *Rhizoctonia zeae* Vorrhees (Gwinn and Gavin 1992), and the leaf spot pathogen *Cladosporium phlei* (Gregory) de Vries (Shimanuki 1987). Additionally, infected tall fescue affected colonization and reproduction of mycorrhizal Glomus species (Guo et al. 1992). Peramine, lolines, and several ergot alkaloids were not active in the agar bioassays against various turf pathogens, indicating that antifungal activity is not apparently associated with most of the compounds toxic to herbivores (Siegel and Latch 1991).

Numerous antifungal compounds have been isolated from stromata of *E. typhina* on timothy (*Phleum pratensis* L.). The groups of compounds include sesquiterpenoids, chokols A-G (types differ only in the side chain); oxygenated fatty acids; and phenolic compounds (Koshino et al. 1992). These compounds may be produced directly as fungitoxic metabolites either by the fungus or plant (induced as phytoalexins) or as nonantifungal metabolites of either fungal or plant origin converted by either symbiont into toxins. Resveratrol, a stilbene reported to be a phytoalexin, has been isolated from endophyte-infected seed of *F. versuta* Beal and other grasses (Powell et al. 1994). The induction of the phytoalexin requires not only the presence of endophyte, but some unknown abiotic stress factors, as the compound was not recovered from all lots of infected seed. Consequently, the exact relationship between resveratrol presence and endophytes is yet to be determined.

VII. Conclusion

The symbiotic relationship between fungal species of the genus *Epichloë* and their associated grass hosts involves both agonism and mutualism. While the ecological and evolutionary consequence of this symbiotic duality is highly unique, the impact on agricultural production is also of significance. The production of mycotoxins is a relatively common phenomenon, but as a product of endophyte mediation in symbiota, its importance in defense against herbivores is one of the essential aspect of mutualism. We have discussed in this chapter primarily biological and chemical significances of four classes of alkaloids produced by endophytes and symbiota. The unequal and widespread distribution of alkaloids among grass species, as well as within the grass tissue, was noted. Also of profound importance was the impact of host and fungal genomes and environment on regulation of alkaloid accumulation. Since the mycosymbionts must receive nutrition from the host, the various parameters that regulate alkaloid synthesis will greatly affect host mediated metabolism in symbiota. Because these endophytes are seed-disseminated, required in many instances for host survival, and because the alkaloid profiles can be manipulated, they are subjects of intense research. The deletion, reduction, or addition of specific alkaloids in symbiota primarily responsible for herbivore activity will require careful evaluation of all interactive factors affecting host/fungus associations, i.e., management practices as they pertain to grass productivity, animal utilization, and insect control.

Acknowledgment. This is publication number 94-11-190 of the Kentucky Agricultural Experiment Station, published with the approval of the director.

References

Abney LK, Oliver JW, Reinemeyer CR (1993) Vasoconstrictive effects of tall fescue alkaloids on equine vasculature. J Equine Vet Sci 13:340

Agee CS, Hill NS (1994) Ergovaline variability in *Acremonium*-infected tall fescue due to environment and plant genotype. Crop Sci 34:221–226

Arechavaleta M, Bacon CW, Plattner RD, Hoveland CS, Radcliffe DE (1992) Accumulation of ergopeptide alkaloids in symbiotic tall fescue grown under deficits of soil water and nitrogen fertilizer. Appl Environ Microbiol 58:857–861

Azevedo MD, Welty RE (1995) A study of the fungal endophyte *Acremonium coenophialum* in the roots of tall fescue seedlings. Mycologia 87:289–297

Azevedo MD, Welty RE, Craig AM, Bartlett J (1993) Ergovaline distribution, total nitrogen and phosphorus content of two endophyte-infected clones. In: Hume DE, Latch GCM, Easton HS (eds) Proc 2nd Int Symp on *Acremonium*/grass interactions. AgResearch Grasslands, 4–7 Feb, Palmerston North, New Zealand, pp 59–62

Bacon CW (1993) Abiotic stress tolerances (moisture, nutrients) and photosynthesis in endophyte-infected tall fescue. Agric Ecosyst Environ 44:123–141

Ball OP-J, Prestidge RA, Sprosen JM (1993) Effect of plant age and endophyte viability on peramine and lolitrem B concentrations in perennial ryegrass seedlings. In: Hume DE, Latch GCM, Easton HS (eds) Proc 2nd Int Symp on *Acremonium*/grass interactions. AgResearch Grasslands, 4–7 Feb, Palmerston North, New Zealand, pp 63–66

Ball OP-J, Prestidge RA, Sprosen JM (1995) Interrelationships between *Acremonium lolli*, peramine, and lolitrem B in perennial ryegrass. Appl Environ Microbiol 61:1527–1533

Barker DJ, Davies E, Lane GA, Latch GCM, Nott HM, Tapper BA (1993) Effect of water deficit on alkaloid concentrations in perennial ryegrass endophyte associations. In: Hume DE, Latch GCM, Easton HS (eds) Proc 2nd Int Symp on *Acremonium*/grass interactions. AgResearch Grasslands, 4–7 Feb, Palmerston North, New Zealand, pp 67–71

Belesky DP, Fedders JM (1995) Tall fescue development in response to *Acremonium coenophialum* and soil acidity. Crop Sci 35:529–533

Belesky DP, Stuedemann JA, Plattner RD, Wilkinson SR (1988) Ergopeptine alkaloids in grazed tall fescue. Agron J 80:209–212

Belesky DP, Stringer WC, Plattner RD (1989) Influence of endophyte and water regime upon tall fescue accessions. II. Pyrrolizidine and ergopeptine alkaloids. Ann Bot 64:343–349

Berde B, Schild HO (eds) (1978) Handbook of experimental pharmacology, vol 49. Ergot alkaloids and related compounds. Springer, Berlin Heidelberg New York, p 1003

Böttcher F, Ober D, Hartmann T (1994) Biosynthesis of pyrrolizidine alkaloids: putrescine and spermidine are essential substrates of enzymatic homospermidine formation. Can J Chem 72:80–85

Bouton JH, Gates RN, Belesky DP, Owsley M (1993) Yield and persistence of tall fescue in the southeastern coastal plain after removal of its endophyte. Agron J 85:52–55

Bush LP, Burrus PB (1988) Tall fescue forage quality and agronomic performance as affected by the endophyte. J Prod Agric 1:55–60

Bush LP, Schmidt D (1994) Alkaloid content of meadow fescue and tall fescue grown with natural endophytes. In: Krohn K, Paul VH, Thomas J (eds) International conf on harmful and beneficial microorganisms in grassland, pasture and turf, vol 17. Int Organization Biological and Integrated Control of Noxious Animals and Plants; Gent, pp 259–265

Bush LP, Fannin FF, Siegel MR, Dahlman DL, Burton HR (1993a) Chemistry, occurrence and biological effects of saturated pyrrolizidine alkaloids associated with endophyte-grass interactions. Agric Ecosyst Environ 44:81–102

Bush LP, Gay S, Burhan W (1993b) Accumulation of pyrrolizidine alkaloids during growth of tall fescue. Proc 17th Int Grassland Congr, Palmerston North, New Zealand, pp 1378–1379

Christensen MJ (1995) Variation in the ability of *Acremonium* endophytes of perennial rye-grass (*Lolium perenne*), tall fescue (*Festuca arundinacea*) and meadow fescue (*Festuca pratensis*) to form compatible associations in the three grasses. Mycol Res 99:466–470

Christensen MJ, Latch GCM (1991) Variation among isolates of *Acremonium* endophytes (*A. coenophialum* and possibly *A. typhinum*) from tall fescue (*Festuca arundinacea*). Mycol Res 95:1123–1126

Christensen MJ, Latch GCM, Tapper BA (1991) Variation within isolates of *Acremonium* endophytes from perennial ryegrass. Mycol Res 95:918–923

Christensen MJ, Leuchtmann DD, Rowan DD, Tapper BA (1993) Taxonomy of *Acremonium* endophytes of tall fescue (*Festuca arundinacea*), meadow fescue (*F. pratensis*), and perennial ryegrass (*Lolium perenne*). Mycol Res 97:1083–1092

Clay K (1994) The potential role of endophytes in ecosystems. In: Bacon CW, White JF (eds) Biotechnology of endophytic fungi of grasses. CRC Press, Boca Raton, pp 63–86

Clay K, Cheplick GP (1989) Effect of ergot alkaloids from fungal endophyte-infected grasses on fall armyworm (*Spodoptera frugiperda*). J Chem Ecol 15:169–182

Dahlman DL, Eichenseer H, Siegel MR (1991) Chemical perspectives of endophyte-grass interactions and their implications to insect herbivory. In: Barbosa P, Krischik VA, Jones CG (eds) Microbial mediation of plant-herbivore interactions. John Wiley, New York, pp 227–252

Davies E, Lane GA, Latch GCM, Tapper BA, Garthwaite I, Towers NR, Fletcher LR, Pownall DB (1993) Alkaloid concentrations in field-grown synthetic perennial ryegrass endophyte associations. In: Hume DE, Latch GCM, Easton HS (eds) Proc 2nd Int Symp on *Acremonium*/grass interactions. AgResearch Grasslands, 4–7 Feb, Palmerston North, New Zealand, pp 72–76

Davis ND, Cole RJ, Dorner JW, Weete JD, Backman PA, Clark EM, King CC, Schmidt SP, Diener UL (1986) Steroid metabolites of *Acremonium coenophialum*, an endophyte of tall fescue. J Agric Food Chem 34:105–108

de Jesus AE, Gorst-Allman CP, Steyn PS, van Heerden FR, Vleggaar R, Wessels PL, Hull WE (1983) Tremorgenic mycotoxins from *Penicillium crustosum*. biosynthesis of penitrem A. J Chem Soc Perkin Trans I (8):1863–1868

Di Menna ME, Mortimer PH, Prestidge RA, Hawkes AD, Sprosen (1992) Lolitrem B concentrations, counts of *Acremonium lolii* hyphae, and the incidence of ryegrass staggers in lambs on plots of *A. lolii*-infected perennial ryegrass. NZ J Agric Res 35:211–217

Dubis NN, Brattsten LB, Dungan LB (1992) Effects of the endophyte-associated alkaloid peramine on southern armyworm microsomal cytochrome P450. In: Mullin CA, Scott GS (eds) Molecular mechanisms of insecticide resistance. ACS Symp Ser 505:125–136

Dyer DC (1993) Evidence that ergovaline acts on serotonin receptors. Life Sci 53:223–228

Ede RM, Miler CO, Meagher LP, Munday SC, Wilkins AL (1994) Relative sterochemistry of the A/B range of the tremorgenic mycotoxin lolitrem B. J Agric Food Chem 42:231–233

Eichenseer H, Dahlman DL, Bush LP (1991) Influence of endophyte infection, plant age and harvest interval on *Rhopalosiphum padi* L. survival and its relation to quantity of N-formyl and N-acetyl loline in tall fescue. Entomol Exp Appl 60:29–38

Eldefrawi ME, Gant DB, Eldefrawi AT (1990) The GABA receptor and the action of tremorgenic mycotoxins. In: Pohland AE, Dowell VR, Richard JL (eds) Microbial toxins in foods and feeds. Plenum Press, New York, pp 291–295

Ferguson NH, Rice JS, Allgood NG (1993) Variation in nitrogen utilization in *Acremonium coenophialum* isolates. Appl Environ Microbiol 59:3602–3604

Fritz JO, Collins M (1991) Yield, digestibility, and chemical composition of endophyte free and infected tall fescue. Agron J 83:537–541

Gallagher R, Hawkes AD (1985) High-performance liquid chromatography with stop-flow ultraviolet spectral characterization of lolitrem neurotoxins from perennial ryegrass. J Chromatogr 322:159–167

Gallagher R, Hawkes AD, Stewart JM (1985) Rapid determination of the neurotoxin lolitrem B in perennial ryegrass by high-performance liquid chromatography with fluorescence detection. J Chromatogr 321:217–226

Garner GB, Rottinghaus GE, Cornell CN, Testereci H (1993) Chemistry of compounds associated with endophyte grass interaction: ergovaline-related and ergopeptine-related alkaloids. Agric Ecosyst Environ 44:65–80

Garthwaite I, Miles CO, Towers NR (1993) Immunological detection of the indole diterpenoid tremorgenic mycotoxins. In: Hume DE, Latch GCM, Easton HS (eds) Proc 2nd Int Symp on *Acremonium*/grass interactions. AgResearch Grasslands, 4–7 Feb, Palmerston North, New Zealand, pp 77–80

Gebler JC, Poulter CD (1992) Purification and characterization of dimethylallyltryptophan synthase from *Claviceps purpurea*. Arch Biochem Biophys 296:308–313

Glenn AE, Bacon CW, Price R, Hanlin RT (1996) Molecular phylogeny of *Acremonium* and its taxonomic implications. Mycologia 88:369–383

Guo BZ, Hendrix JW, An ZQ, Ferriss RS (1992) Role of *Acremonium* endophyte of fescue on inhibition of colonization and reproduction of mycorrhizal fungi. Mycologia 84:882–885

Gwinn KD, Gavin AM (1992) Relationship between endophyte infestation level of tall fescue seed lots and *Rhizoctonia zeae* seedling disease. Plant Dis 76:911–914

Hill NS, Agee CS (1994) Detection of ergoline alkaloids in endophyte-infected tall fescue by immunoassay. Crop Sci 34:530–534

Hill NS, Belesky DP, Stringer WC (1991a) Competitiveness of tall fescue as influenced by *Acremonium coenophialum*. Crop Sci 31:185–195

Hill NS, Parrott WA, Pope DD (1991b) Ergopeptine alkaloid production by endophytes in a common tall fescue genotype. Crop Sci 31:1545–1547

Hill NS, Rottinghaus GE, Agee CS, Schultz LM (1993) Simplified sample preparation for HPLC analysis of ergovaline in tall fescue. Crop Sci 33:331–333

Hoveland CS (1993) Importance and economic significance of the *Acremonium* endophytes to performance of animals and grass plant. Agric Ecosyst Environ 44:3–12

Huizing HJ, van der Molen W, Kloek W, den Nijs APM (1991) Detection of lolines in endophyte-containing meadow fescue in The Netherlands and the effect of elevated temperature on induction of lolines in endophyte-infected perennial ryegrass. Grass Forage Sci 46:441–445

Keogh RG, Tapper BA (1993) *Acremonium lolii*, Lolitrem B, and peramine concentrations within vegetative tillers of perennial ryegrass. In: Hume DE, Latch GCM, Easton HS (eds) Proc 2nd Int Symp on *Acremonium*/grass interactions. AgResearch Grasslands, 4–7 Feb, Palmerston North, New Zealand, pp 81–84

Koshino H, Yoshihara T, Ichihara A, Tajimi A, Shimanuki T (1992) Two shingoid derivatives from stromata of *Epichloë typhina* on *Phleum pratense*. Phytochemistry 31:3757–3759

Lam CK, Belanger FC, White JF, Daie J (1995) Invertase activity in *Epichloë*/*Acremonium* fungal endophytes and its possible role in choke disease. Mycol Res 99:867–873

Larson BT, Samford MD, Camden JM, Piper EL, Kerley MS, Paterson JA, Turner JT (1995) Ergovaline binding and activation of d2 dopamine receptors in GH(4)ZR(7) cells. J Anim Sci 73:1396–1400

Latch GCM (1993) Physiological interactions of endophytic fungi and their hosts – biotic stress tolerance imparted to grasses by endophytes. Agric Ecosyst Environ 44:143–156

Laws I, Mantle PG (1989) Experimental constraints in the study of the biosynthesis of indole alkaloids in fungi. J Gen Microbiol 135:2679–2692

Leuchtmann A (1992) Systematics, distribution, and host specificity of grass endophytes. Nat Toxins 1:150–162

Leuchtmann A (1994) Isozyme relationships of *Acremonium* endophytes from 12 *Festuca* Species. Mycol Res 98:25–33

Leuchtmann A, Clay K (1993) Nonreciprocal compatibility between *Epichloë typhina* and 4 host grasses. Mycologia 85:157–163

Lindstrom JT, Belanger FC (1994) Purification and characterization of an endophytic fungal proteinase that is abundantly expressed in the infected host grass. Plant Physiol 106:7–16

Lyons CL, Plattner RC, Bacon CW (1986) Occurrence of peptide and clavine ergot alkaloids in tall fescue. Science 232:487–489

Lyons CP, Evans JJ, Bacon CW (1990) Effects of the fungal endophyte *Acremonium coenophialum* on nitrogen accumulation and metabolism in tall fescue. Plant Physiol 92:726–732

Miles CO, Wilkins AL, Gallagher RT, Hawkes AD, Munday SC, Towers NR (1992) Synthesis and tremorgenicity of paxitriols and lolitriol: possible biosynthetic precursors of lolitrem B. J Agric Food Chem 40:234–238

Miles CO, Munday SC, Wilkins AL, Ede PM, Towers NR (1994) Large scale isolation of lolitrem B and structure determination of lolitrem E. J Agric Food Chem 42:1488–1492

Miles CO, Munday-Finch SC, Meagher LP, Wilkins AL (1995a) Lolitrem structure-bioactivity relationships. In: Garthwaite I (ed) Toxinology and food safety research report 1992–1995. Ag Research, Hamilton, N2, pp 15–17

Miles CO, Munday-Finch SC, Wilkins AL (1995b) Improved lolittrem B isolation and the isolation and identification of new lolitrems. In: Garthwaite I (ed) Toxinology and food safety research report 1992–1995. Ag Research, Hamilton, N2, pp 12–14

Miles CO, Lane GA, di Menna ME, Garthwaite I, Piper EL, Ball OJ-P, Latch GCM, Allen JM, Hunt MB, Bush LP, Min FK, Fletcher I, Harris PS (1996) High levels of ergonovine and lysergic acid amide in toxic *Achnatherum inebrians* accompany infections by an *Acremonium*-like endophytic fungus. J Agric Food Chem 44:1285–1290

Moubarak AS, Piper EL, West CP, Johnson ZB (1993) Interaction of purified ergovaline from endophyte-infected tall fescue with synaptosomal ATPase enzyme system. J Agric Food Chem 41:407–409

Oliver JW, Abney LK, Strickland JR, Linnabary RD (1993) Vasoconstriction in bovine vasculature induced by the tall fescue alkaloid lysergamide. J Anim Sci 71:2708–2713

Penn J, Garthwaite I, Christensen MJ, Johnson CM, Towers NR (1993) The importance of paxilline in screening for potentially tremorgenic *Acremonium* isolates. In: Hume DE, Latch GCM, Easton HS (eds) Proc 2nd Int Symp on *Acremonium*/grass interactions. AgResearch Grasslands, 4–7 Feb, Palmerston North, New Zealand, pp 88–92

Petroski RJ, Yates SG, Weisleder D, Powell RG (1989) Isolation, semi-synthesis, and NMR spectral studies of loline alkaloids. J Nat Prod 52:810–817

Petroski RJ, Dornbos DL, Powell RG (1990) Germination and growth inhibition of annual ryegrass (*Lolium multiflorum* L.) and alfalfa (*Medicago sativa* L.) by loline alkaloids and synthetic N-acylloline derivatives. J Agric Food Chem 38:1716–1718

Petroski RJ, Powell RG, Clay K (1992) Alkaloids of *Stipa robusta* (sleepy grass) infected with an *Acremonium* endophyte. Nat Toxins 1:84–88

Popay AJ, Rowan DD (1994) Endophytic fungi as mediators of plant insect interactions. In: Bernays EA (eds) Insect-plant interactions, vol V. CRC Press, Boca Raton, pp 83–103

Popay AJ, Mainland RA, Saunders CJ (1993) The effect of endophytes in fescue grass on growth and survival of third instar grass grub larvae. In: Hume DE, Latch GCM, Easton HS (eds) Proc 2nd Int Symp on *Acremonium*/grass interactions. AgResearch Grasslands, 4–7 Feb, Palmerston North, New Zealand, pp 174–176

Porter JK (1994) Chemical constituents of grass endophytes. In: Bacon CW, White JF (eds) Biotechnology of endophytic fungi of grasses. CRC Press, Boca Raton, pp 103–123

Porter JK, Bacon CW, Robbins JD, Higman HC (1975) A field indicator in plants associated with ergot-type toxicities in cattle. J Agric Food Chem 23:771–775

Powell RG, Petroski RJ (1992) Alkaloid toxins in endophyte-infected grasses. Nat Toxins 1:163–170

Powell RG, Petroski RJ (1993) The loline group of pyrrolizidine alkaloids. In: Pelletier SW (ed) The alkaloids: chemical and biological perspectives, vol 8. Springer, Berlin Heidelberg New York, pp 320–338

Powell RG, TePaske MR, Plattner RD, White JF, Clement SL (1994) Isolation of resveratrol from *Festuca versuta* and evidence for the widespread occurrence of this stilbene in the Poaceae. Phytochemistry 35:335–338

Prestidge RA (1991) Susceptibility of Italian ryegrass (*Lolium multiflorum* Lam) to Argentine stem weevil (*Listronotus bonariensis* Kuschel) feeding and oviposition. NZ J Agric Res 34:119–125

Prestidge RA (1993) Causes and control of perennial ryegrass staggers in New Zealand. Agric Ecosyst Environ 44:283–300

Prestidge RA, Ball OJ-P (1993) The role of endophytes in alleviating plant biotic stress in New Zealand. In: Hume DE, Latch GCM, Easton HS (eds) Proc 2nd Int Symp on *Acremonium*/grass interactions: plenary papers. AgResearch Grasslands, 4–7 Feb, Palmerston North, New Zealand, pp 141–152

Putnam MR, Bransby DI, Schumacher J, Boosinger TR, Bush LP, Shelby RA, Vaughan JT, Ball D (1991) The effects of the fungal endophyte *Acremonium coenophialum* in fescue on pregnant mares and foal viability. Am J Vet Res 52:2071–2074

Richardson MD, Chapman GW, Hoveland CS, Bacon CW (1992) Sugar alcohols in endophyte-infected tall fescue under drought. Crop Sci 32:1060–1061

Riedell WE, Kieckhefer RE, Petroski RJ, Powell RG (1991) Naturally-occurring and synthetic loline alkaloid derivatives: insect feeding behavior modification and toxicity. J Entomol Sci 26:122–129

Robbins JD, Sweeny JG, Wilkinson SR, Burdick D (1972) Volatile alkaloids of Kentucky 31 tall fescue seed (*Festuca arundinacea* Schreb.). J Agric Food Chem 29:653–657

Rottinghaus GE, Garner GB, Cornell CN, Ellis JL (1991) HPLC method for quantitating ergovaline in endophyte-infected tall fescue: variation of ergovaline levels in stems with leaf sheaths, leaf blades, and seed heads. J Agric Food Chem 39:112–115

Rowan DD (1993) Lolitrems, peramine and paxilline – mycotoxins of the ryegrass-endophyte interaction. Agric Ecosyst Environ 44:103–122

Rowan DD, Dymock JJ, Brimble MA (1990) Effect of the fungal metabolite peramine and analogs on feeding and development of Argentine stem weevil (*Listronotus bonariensis*). J Chem Ecol 16:1683–1695

Roylance JT, Hill NS, Agree CS (1994) Ergovaline and peramine content in endophyte-infected tall fescue. J Chem Ecol 20:2171–2183

Schardl CL (1994) Molecular and genetic methodologies and transformation of grass endophytes. In: Bacon CW, White JF (eds) Biotechnology of endophytic fungi of grasses. CRC Press, Boca Raton, pp 151–165

Schardl CL (1996) *Epichloë* species: fungal symbionts of grasses. Annu Rev Phytopathol 34:109–130

Schardl CL, Leuchtmann A, Tsai H-F, Collett MA, Watt DM, Scott DB (1994) Origin of a fungal symbiont of perennial ryegrass by interspecific hybridization of a mutualist with the ryegrass choke pathogen, *Epichloë typhina*. Genetics 136:1307–1317

Schmidt D (1993) Effects of *Acremonium uncinatum* and a *Phialophora*-like endophyte on vigour, insect and disease resistance of meadow fescue. In: Hume DE, Latch GCM, Easton HS (eds) Proc 2nd Int Symp on *Acremonium*/grass interactions. AgResearch Grasslands, 4–7 Feb, Palmerston North, New Zealand, pp 185–188

Selala MI, Daelemans F, Schepens PJC (1989) Fungal tremorgens: the mechanism of action of single nitrogen containing toxins – a hypothesis. Drug Chem Toxicol 12:237–257

Shelby RA, Kelley VC (1992) Detection of ergot alkaloids from *Claviceps* species in agricultural products by competitive ELISA using a monoclonal antibody. J Agric Food Chem 40:1090–1092

Shibuya M, Chou H-M, Fountoulakis M, Hassam M, Kim S-U, Kobayashi K, Otsuka H, Rogalska E, Cassady JM, Floss HG (1990) Sterochemistry of the isoprenylation of tryptophan catalyzed by 4-(γ,γ-dimethlyally) tryptophan synthase from *Claviceps*, 1st pathway-specific enzyme in ergot alkaloid biosynthesis. J Am Chem Soc 112:297–304

Shimanuki T (1987) Studies on the mechanisms of the infection of timothy with purple spot disease caused by *Cladosporium phlei* (Gregory) de Vries. Res Bull Hokkaido Natl Agric Exp Stn 148:1–56

Siegel MR, Bush LP (1994) Importance of endophytes in forage grasses, a statement of problems and selection of endophytes. In: Bacon CW, White JF (eds) Biotechnology of endophytic fungi of grasses. CRC Press, Boca Raton, pp 135–150

Siegel MR, Bush LP (1996) Defensive chemicals in grass-fungal endophyte associations. Recent Adv Phytochem 30:81–118

Siegel MR, Latch GCM (1991) Expression of antifungal activity in agar culture by isolates of grass endophytes. Mycologia 83:525–537

Siegel MR, Dahlman DL, Bush LP (1989) The role of endophytic fungi in grasses: new approaches to biological control of pests. In: Leslie AR, Metcalf RL (eds) Integrated pest management for turfgrass and ornamentals. US Environmental Protection Agency, Washington, DC, pp 169–186

Siegel MR, Latch GCM, Bush LP, Fannin FF, Rowan DD, Tapper BA, Bacon CW, Johnson MC (1990) Fungal endophyte-infected grasses: alkaloid accumulation and aphid response. J Chem Ecol 16:3301–3315

Solomons RN, Oliver JW, Linnabary RD (1989) Reactivity of dorsal pedal vein of cattle to selected alkaloids associated with *Acremonium coenophialum*-infected fescue grass. Am J Vet Res 50:235–238

Strickland JR, Oliver JW, Cross DL (1993) Fescue toxicosis and its impact on animal agriculture. Vet Hum Toxicol 35:454–464

Strickland JR, Cross DL, Birrenkott GP, Grimes LW (1994) Effect of ergovaline, loline, and dopamine antagonists on rat pituitary cell prolactin release in vitro. Am J Vet Res 55:716–721

Takeda A, Suzuki E, Kamei K, Nakata H (1991) Detection and identification of loline and its analogues in house urine. Chem Pharm Bull 39:964–968

Tapper BA (1993) A new Zealand perspective on endophyte metabolites. In: Hume DE, Latch GCM, Easton HS (eds) Proc 2nd Int Symp on *Acremonium*/grass interactions: plenary papers. AgResearch Grasslands, 4–7 Feb, Palmerston North, New Zealand, pp 89–93

TePaske MR, Powell RG (1993) Analyses of selected endophyte-infected grasses for the presence of loline-type and ergot-type alkaloids. J Agric Food Chem 41: 2299–2303

Tsai H-F, Liu J-S, Staben CS, Christensen MJ, Latch GCM, Siegel M, Schardl CL (1994) Evolutionary diversification of fungal endophytes of tall fescue grass by hybridization with *Epichloë* species. Proc Natl Acad Sci USA 91:2542–2546

Tsai H-F, Wang H, Gebler JC, Poulter DC, Schardl CL (1995) The *Claviceps purpurea* gene encoding dimethylyalltryptophan synthase, the commited step for ergot alkaloid biosynthesis. Biochem Biophys Res Commun 216:119–125

Tufariello JJ, Meckler H, Winzenberg K (1986) Synthesis of the *Lolium* alkaloids. J Org Chem 51:3356–3357

Welty RE, Craig AM, Azevedo MD (1994) Variability of ergovaline in seeds and straw and endophyte infection in seeds among endophyte-infected genotypes of tall fescu. Plant Dis 78:845–849

White JF (1993) Endophyte-host associations in grasses. XIX. A systematic study of some sympatric species of *Epichloë* in England. Mycologia 85:444–455

White JF Jr, Morrow AC, Morgan-Jones G, Chambless DA (1991) Endophyte-host associations in forage grasses. XIV. Primary stromata formation and seed transmission in *Epichloë typhina*: developmental and regulatory aspects. Mycologia 83:72–81

White JF, Morgan-Jones G, Morrow AC (1993) Taxonomy, life cycle, reproduction and detection of *Acremonium* endophytes. Agric Ecosyst Environ 44:13–37

Woodburn OJ, Walsh JR, Foot JZ, Heazlewood PG (1993) Seasonal ergovaline concentrations in perennial ryegrass cultivars of differing endophyte status. In: Hume DE, Latch GCM, Easton HS (eds) Proc 2nd Int Symp on *Acremonium*/grass interactions. AgResearch, Grasslands, 4–7 Feb, Palmerston North, New Zealand, pp 100–104

Yates SG, Tookey HL (1965) Festucine, an alkaloid from tall fescue (*Festuca arundinacea* Schreb.): chemistry of the functional groups. Aust J Chem 18:53–60

Yates SG, Fenster JC, Bartelt RJ (1989) Assay of tall fescue seed extracts, fractions, and alkaloids using the large milkweed bug. J Agric Food Chem 37:354–357

Yates SG, Petroski RJ, Powell RG (1990) Analysis of loline alkaloids in endophyte-infected tall fescue by capillary gas chromatography. J Agric Food Chem 38:182–185

11 Metabolic Interactions at the Mycobiont-Photobiont Interface in Lichens

R. Honegger

CONTENTS

I. Introduction

Lichen-forming fungi are, like plant pathogens or mycorrhizal fungi, a polyphyletic, taxonomically heterogeneous group of nutritional specialists (Table 1; Gargas et al. 1995). As ecologically obligate biotrophs, they acquire fixed carbon from a population of minute, extracellularly located algal and/or cyanobacterial cells. In contrast to pathogenic interactions of fungi and fungus-like organisms with unicellular algae (e.g., van Donk and Bruning 1992), the photobiont cells of lichen thalli are not severely damaged by the fungal partner. *Geosiphon pyriforme*, the only known fungal (zygomycetous) symbiosis with an intracellularly located cyanobacterium (*Nostoc* sp.; Mollenhauer 1992), is not normally considered as a lichen.

The species names of lichens refer to the fungal partner. Lichen taxonomy is based on the criteria of general fungal systematics and on morphotypic and chemotypic features of the symbiotic phenotype. There is no evidence for any fundamental difference between lichen-forming and nonlichenized fungi. Concentric bodies, proteinaceous cell organelles of unknown origin and function which were formerly supposed to be a peculiarity of ascomycetous lichen mycobionts, have been found in long-lived, desiccation-tolerant cells of a range of saprotrophic and plant pathogenic Ascomycetes (Honegger 1993). It is by their manifold adaptations to the cohabitation with a population of minute photobiont cells that lichen mycobionts differ from nonlichenized fungi.

Institute of Plant Biology, University of Zürich, Zollikerstrasse 107, CH-8008 Zürich, Switzerland

II. Peculiarities of the Lichen Symbiosis

Approximately 85% of lichen-forming fungi are symbiotic with unicellular or filamentous green algae, 10% associate with cyanobacteria, and an estimated 3–4%, the so-called cephalodiate species, simultaneously with both green algae and cyanobacteria (Tschermak-Woess 1988). The majority of lichen-forming fungi produce morphologically simple thalli by either overgrowing or ensheathing compatible photobiont cells which grow either on or within the substratum (Table 1). Particularly interesting are the symbiotic phenotypes of morphologically advanced taxa which enter the third dimension by forming either foliose or fruticose thalli with a distinct internal stratification due to a remarkable hyphal polymorphism, the main building blocks being conglutinate, hydrophilic pseudoparenchyma and gas-filled zones formed by systems of aerial hyphae with hydrophobic wall surfaces (Figs. 1–4, 7). The photobiont cells of such thalli are maintained, carried along by growth processes, and controlled by the quantitatively predominant mycobiont (Hill 1985, 1989; Honegger 1993). It is the fungal partner of morphologically complex lichens which mimics plant-like structures by forming leaf- or shrub-like thalli, with the photobiont layer arranged similarly to the palisade parenchyma of plants, competes for

The Mycota V Part A
Plant Relationships
Carroll/Tudzynski (Eds.)

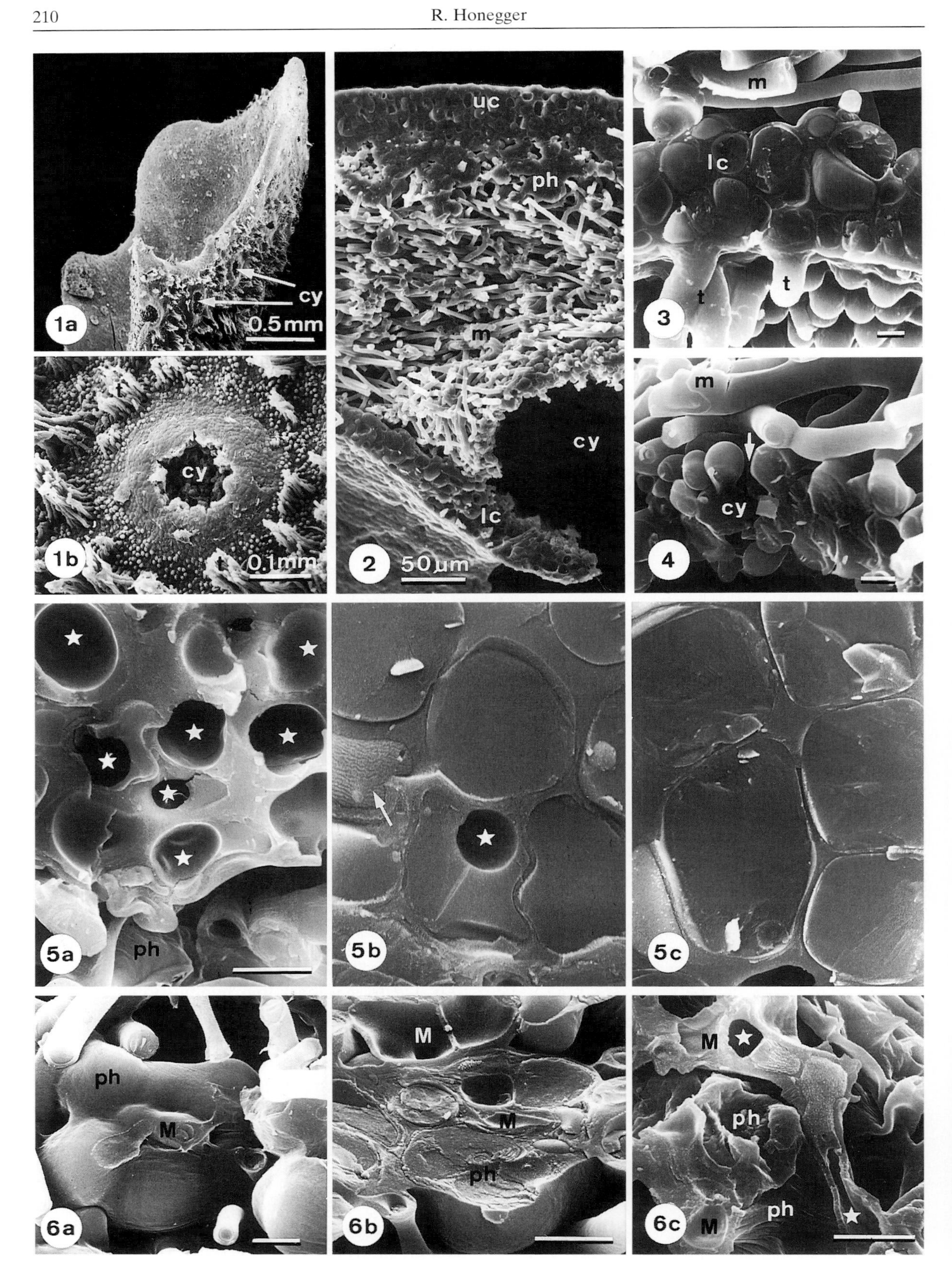
cy
1a
0.5mm
t
cy
1b
t
0.1mm
uc
ph
m
cy
lc
2
50µm
m
lc
t
t
3
m
cy
4
5a
ph
5b
5c
ph
M
6a
M
M
ph
6b
M
ph
ph
6c
M

space above ground, and facilitates gas exchange (partly by differentiating very peculiar aeration pores; Figs. 1a–b, 2, 4).

All of these peculiarities of morphologically complex lichens are unique features within the fungal kingdom. The photobiont cell population of morphologically complex lichens is enclosed by fungal structures and has no direct access to water and dissolved mineral nutrients from the thalline exterior except via passive translocation in the fungal apoplast (Fig. 7).

III. Lichen Photobionts

So far, about 100 species of lichen photobionts have been identified (Tschermak-Woess 1988; Honegger 1990; Büdel 1992; Gärtner 1992). The majority are unicellular or filamentous, aerophilic Chlorophyta. As in other algal symbioses with heterotrophic partners (e.g., with marine and freshwater protoctista and invertebrates), only comparatively few taxa out of the large numbers of potentially available (aerophilic) algae are acceptable partners of lichen-forming fungi. Some lichen photobionts are widespread and common in the nonsymbiotic state (e.g., representatives of the Trentepohliales, some of which are considered as pests in tropical agriculture; Hawksworth 1988b), others are very rarely found outside lichen thalli (Tschermak-Woess 1988; Honegger 1990). This is especially true of *Trebouxia* spp. (Chlorophyta; Friedl 1995; approx. 20 spp.), which have been identified as photobionts of more than 50% of lichen species.

The compatible photobiont triggers, in a yet unknown manner, the phenotypic expression of the fungal genotype. Particularly interesting situations occur in some cephalodiate species and so-called morphotype pairs, where a morphologically different thallus (morphotype) is formed by probably the same fungal partner in association with either a green algal or a cyanobacterial photobiont (James and Henssen 1976; Hawksworth and Hill 1984; Jahns 1988). So far, it has not been possible to resynthesize heteromorphic morphotype pairs from cultured mycobiont and photobiont isolates under axenic conditions.

IV. Recognition and Specificity

In less than 2% of all lichen species has the photobiont ever been identified at the species level, and very little is known about the range of compatible photobiont species per fungal taxon. Germ tubes

Fig. 1a–6c. Thallus anatomy and hyphal polymorphism and multifunctionality in the foliose macrolichen *Sticta sylvatica* (Peltigerales; cyanobacterial photobiont: *Nostoc* sp.) as observed in conventionally prepared (**1a**, **1b**) and in cryofixed, frozen-hydrated specimens with low-temperature scanning electron microscopy (**2–6c**). **1a** Lateral view of a thallus lobe with smooth upper and tomentose lower surface with numerous cyphellae (*cy*; specialized aeration pores). **1b** Laminal view of a cyphella (*cy*) and of the tomentum (*t*). **2** Vertical cross section (fracture) of the fully hydrated thallus with conglutinate upper cortex (*uc*), photobiont layer (*ph*), gas-filled medullary layer (*m*), and conglutinate lower cortex (*lc*). Part of a cyphella (*cy*) is fractured. Note: free water within the thalline interior is confined to the symplast and to the apoplast. The medullary and photobiont layers are gas-filled even at water saturation. **3** Detail of the lower cortex and adjacent medullary layer and tomentum, revealing multiple transitions in fungal growth: filamentous medullary hyphae (*m*; aerial hyphae of the thalline interior) with hydrophobic cell wall surface layer switch to apolar growth and secrete hydrophilic β-glucans when participating in the formation of the conglutinate lower cortex (*lc*). Globose cortical cells may switch back to filamentous growth and form the hair-like hyphae of the tomentum (*t*) with a hydrophilic wall surface (water uptake). **4** Detail of a vertically fractured cyphella (*cy*): medullary hyphae with hydrophobic cell wall surface layer switch from filamentous to globose/apolar growth but retain the hydrophobic cell wall surface layer. The *arrow* points to pores between the globose cells of the cyphella which facilitate the gas exchange. **5a–c** Cross-fractures of the upper cortex at different levels of hydration: **5a** in the desiccated state (<20% water dw^{-1}); the *asterisks* refer to a cytoplasmic gas bubble within the strongly condensed protoplast of each cell. Most cells reveal irregular outlines. **5b** during the rehydration phase, 17 s after the addition of a water droplet to the thallus surface. The top cells are already ovoid in shape, their cytoplasmic gas bubble has disappeared, and their plasmamembrane appears smooth. The lower cells still have irregular outlines, their cytoplasmic gas bubble is vanishing within the rehydrating protoplast. The *arrow* points to the plasmamembrane of an adjacent, not yet fully rehydrated, cell with drought-stress-induced, fine foldings. **5c** Fully hydrated cortical cells. Note the dramatic changes in cell volume between **5a** and **5c**. **6a–c** The mycobiont-photobiont interface in the fully hydrated, (**6a**, **6b**) and in the desiccated state (**6c**). The mycobiont (*M*) forms intragelatinous protrusions within the gelatinous sheath of the cyanobacterial photobiont (*ph*). *Asterisks* refer to cytoplasmic gas bubbles in desiccated, fractured fungal cells. *Bars* 5 μm unless otherwise stated. Same magnification in **5a–c**

Table 1. Orders of Ascomycotina and Basidiomycotina, and classes of Deuteromycotina, which include lichenized taxa. (Hawksworth 1988a; Honegger 1992; Hawksworth and Honegger 1994)

Order Class	Nutritional strategies	Thallus anatomy in lichenized taxa
Ascomycotina[a,b,c]		
Arthoniales	**l**, nl (*lp*, *sap*)	**ns**
Caliciales	**l**, nl (*f*, *lp*, *sap*)	**ns**, s
Dothideales	**nl** (*sap*, *pp*, *lp*), l	**ns**
Graphidales	**l**, nl (*lp*)	**ns**
Gyalectales	**l**	**ns**
Lecanorales	**l**, nl (*sap*, *lp*)	**ns**, **s**
Leotiales	**nl** (*sap*, *pp*, *lp*), l	**ns**, s
Lichinales	**l**	**ns**, s
Opegraphales	**l**, nl (*lp*, *sap*)	**ns**, s
Ostropales	**nl** (*sap*, *pp*, *lp*), l	**ns**
Patellariales	**nl** (*sap*, *lp*), l	**ns**
Peltigerales	**l**	**s**, ns
Pertusariales	**l**	**ns**
Pyrenulales	**l**, nl (*sap*)	**ns**
Teloschistales	**l**, nl (lp)	**ns**, **s**
Verrucariales	**l**, nl (*sap*, *lp*)	**ns**, s
Agaricales	**nl** (*sap*, *myc*, *pp*, *lp*), l	**ns**, **s**
Aphyllophorales	**nl** (*sap*, *myc*, *pp*, *f*, *lp*), l	**ns**, s
Basidiomycotina[d,e]		
Deuteromycotina[f,g]		
Coelomycetes	**nl** (*sap*, *pp*, *f*, *lp*), l	**ns**
Hyphomycetes	**nl** (*sap*, *pp*, *f*, *lp*), l	**ns**
Sterile taxa (with no known reproductive structures)[h]		**ns**, **s**

Approx. 55% of lichen-forming fungi form nonstratified (crustose, microfilamentous etc.) thalli, 20% form either squamulose or placodioid thalli, and 22% form either foliose, or fruticose, internally stratified thalli.

Abbreviations

f, fungicolous, l, lichenized, lp, lichenicolous (lichen parasites), myc, mycorrhiza, nl, nonlichenized, ns, nonstratified, pp, plant pathogens, s, internally stratified, sap, saprotrophic.

[a] Approx. 98% of lichen-forming fungi are Ascomycotina.
[b] 16 out of 46 orders of Ascomycotina include lichenized taxa.
[c] ~46% of Ascomycotina are lichenized (approx. 13250 spp.).
[d] Approx. 0.4% of lichen-forming fungi are Basidiomycotina.
[e] Only ~0.3% of Basidiomycotina are lichenized (approx. 50 spp.).
[f] Approx. 1.6% of lichen-forming fungi are Deuteromycotina.
[g] ~0.3% of Deuteromycotina are lichenized (~200 spp.).
[h] Approx. 75 spp. (11 genera).

or other free hyphae of lichen-forming fungi may temporarily associate with ultimately incompatible algal cells, but morphologically complex symbiotic phenotypes are expressed exclusively in compatible associations (Ahmadjian and Jacobs 1981; Ott 1987; Honegger 1993). Sexually reproducing or macroconidia-bearing lichen mycobionts must reestablish the symbiotic state after each reproductive cycle, but large numbers of species disperse very efficiently by means of either thallus fragments (e.g., reindeer lichens of arctic tundras) or by symbiotic propagules (Fig. 8). These enable a species to invade large areas regardless of whether compatible photobiont cells are present in the environment as free-living populations.

The majority of lichen mycobionts so far investigated are physiologically facultative biotrophs, i.e., can be cultured under sterile conditions, their aposymbiotic phenotype being completely different from the symbiotic one. Unfortunately, the symbiotic phenotype cannot be routinely resynthesized by combining axenic, compatible fungal and algal isolates. It remains unknown whether this failure is due to microclimatic factors (e.g., continuously moist culturing conditions instead of

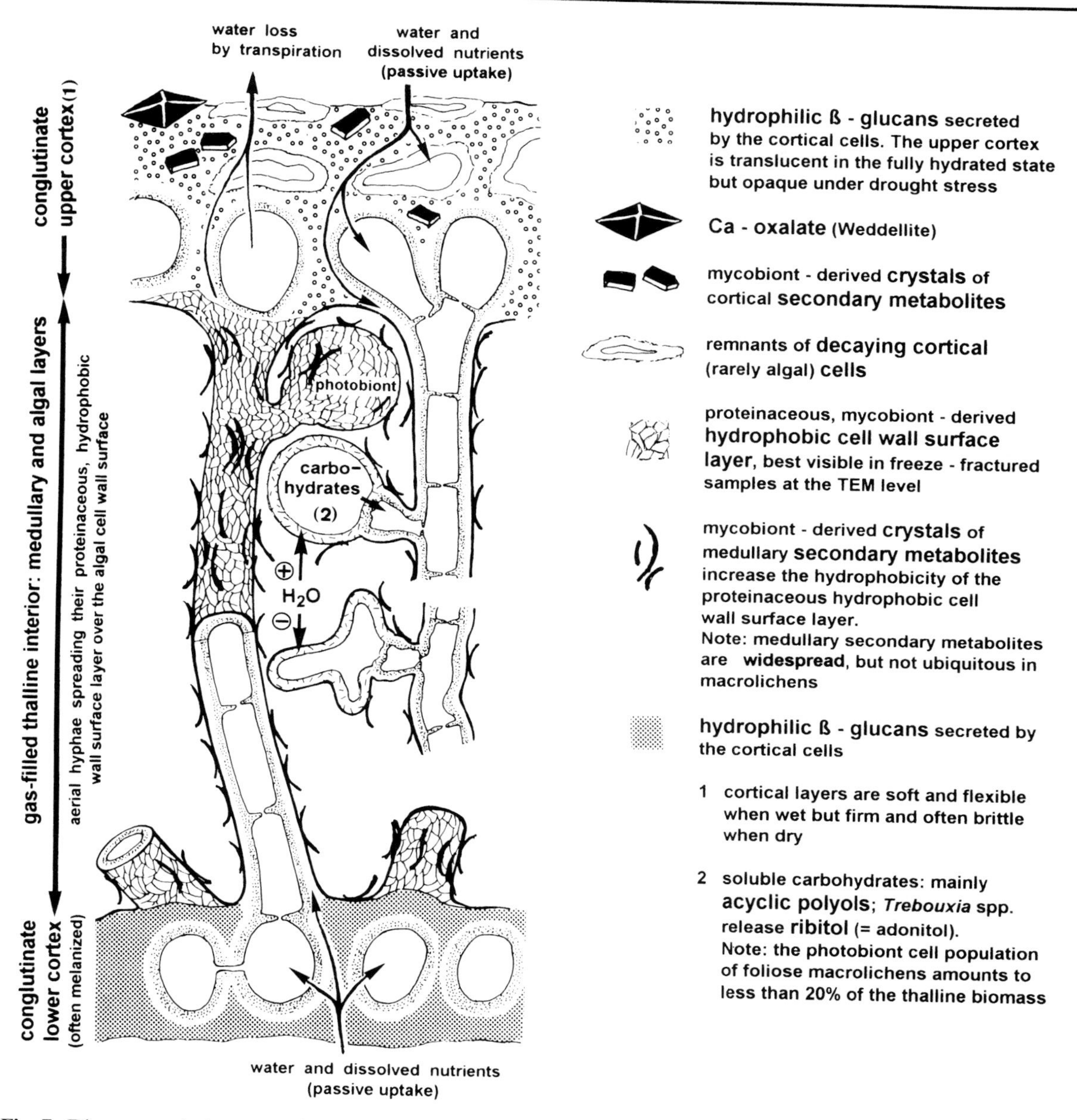

Fig. 7. Diagrammatic interpretation of the functional morphology of foliose macrolichens

periodic harsh wetting and drying cycles), or to the absence of other, potentially beneficial microorganisms. Due to the near impossibility of routinely resynthesizing lichens from axenic fungal and algal isolates, our present, very limited, knowledge on the range of compatible photobiont species per mycobiont taxon has been achieved by isolation of algae in pure culture (e.g., Friedl 1989; Ihda et al. 1993). These studies show that mycobionts of morphologically complex lichens are either moderately specific (accepting several species of the same algal genus as compatible partners) or specific (associating with only one photobiont species) with regard to their photobiont, specificity being defined by the degree of taxonomic relatedness of compatible partners (Smith and Douglas 1987).

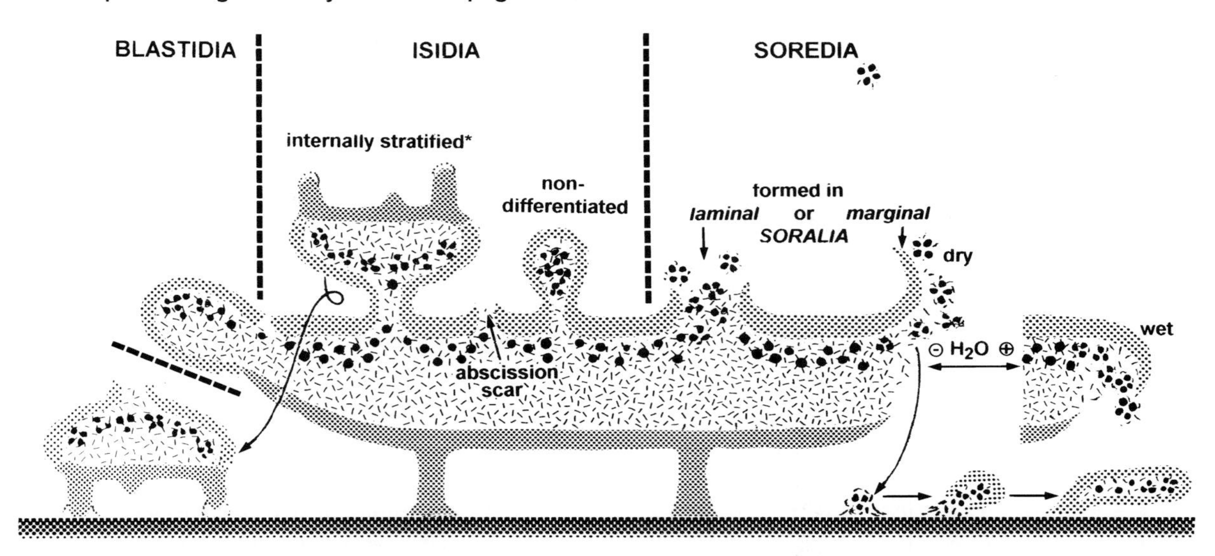

Fig. 8. Diagram illustrating examples of vegetative symbiotic propagules in lichens. Further examples in Hawksworth and Hill (1984) and Jahns (1988). *Asterisk* refers to an uncommon type of isidium (as observed in the lecanoralean *Parmelia pastillifera*; Honegger 1987) but an example of a remarkable level of fungal differentiation: juvenile thalli with strongly melanized lower cortex develop in an inverted position on the nongrowing part of the mother thallus

According to this definition, cephalodiate lichen mycobionts which associate simultaneously with particular green algal and cyanobacterial taxa, i.e., with representatives of different kingdoms, reveal a very low specificity, but they are highly selective with regard to photobiont acquisition (Bubrick et al. 1985). So far, it has not proved possible to identify key features by which compatible photobiont taxa differ from incompatible ones or from nonsymbiotic algae.

V. The Mycobiont-Photobiont Interface in Different Types of Lichens

Different types of mycobiont-photobiont interactions occur in lichens, depending on the taxonomic affiliation of the partners involved, on the fine structure and composition of the cell wall of the photobiont, and on the degree of morphological complexity of the symbiosis (Tschermak 1941; Plessl 1963; Honegger 1984, 1985, 1986a, 1991, 1992). Transparietal ("intracellular") haustoria which penetrate the cellulosic algal cell wall or the cyanobacterial murein sacculus are typically found in lichens with nonstratified thallus (Honegger 1992). Mycobionts that express morphologically complex symbiotic phenotypes, i.e., foliose or fruticose thalli with internal stratification, do not pierce the cell wall of the photoautotrophic partner. They form intragelatinous protrusions within the sheaths of cyanobacterial partners (Figs. 4, 5) or, in association with green algal photobionts, either wall-to-wall appositions or very peculiar intraparietal haustoria which are the result of coordinated growth and development of both partners (Fig. 7). Intraparietal haustoria are formed by thousands of fungal species within the Lecanorales and Teloschistales in association with unicellular green algae of the genus *Trebouxia* (Microthamniales, Chlorophyta).

An inconspicuous, but functionally important element of the mycobiont-photobiont interface in lichens is a thin, proteinaceous, and highly hydrophobic cell wall surface layer of mycobiont origin which, at the very first contact of the growing hypha with the juvenile cell of the photoautotrophic partner, spreads over the wall surface of the algal cell (Honegger 1984, 1991). This hydrophobic cell wall surface layer reveals a distinct rodlet pattern in freeze-etch preparations (Honegger 1984, 1986b, 1991). The rodlets may become obscured by mycobiont-derived secondary metabolites which crystallize in and on the hydrophobic protein layer of the wall surface of either partner, thus enhancing the hydrophobicity of the wall surface. The hydrophobic coat of the wall surfaces in the thalline interior is essential for the functioning of the symbiosis. It has the properties of a cuticle equivalent and seals the apoplastic continuum between the partners, thus directing the fluxes of solutes from the thalline exterior to the interior and vice versa during the harsh wetting and drying cycles (Fig. 7). At the same time, it keeps the wall surfaces in the thalline interior free from water even in the water-saturated state (Fig. 6a; Brown et al. 1987; Honegger and Peter 1994; Scheidegger 1994), a prerequisite for efficient gas exchange.

Proteinaceous, hydrophobic cell wall surface layers with a distinct rodlet pattern are very widespread in aerial structures of lichenized and nonlichenized fungi, and are likely to play an important functional role not only in the lichen symbiosis, but also in a wide range of fungal interactions with plant cells (Honegger 1984; Wessels 1993; deVries et al. 1993; Templeton et al. 1994). A group of small, cystein-rich hydrophobic proteins, the hydrophobins, have been shown to self-assemble at the liquid-air interface into a membrane-like rodlet layer (Wösten et al. 1994). It remains to be seen whether the hydrophobic cell wall surface proteins of lichen mycobionts are homologous or at least comparable with the well-characterized hydrophobins of nonlichenized taxa.

The most astonishing feature of mycobiont-photobiont interfaces in morphologically complex lichens is their structural simplicity as compared with the elaborate haustorial complexes of highly evolved, biotrophic plant pathogens such as rusts (see Chaps. 2, 3, this Vol.; 5, Vol. V, Part B) or powdery mildews (see Chap. 4, Vol. V, Part B) or with the arbuscules of zygomycetous mycorrhizae (see Chap. 7, Vol. V, Part B). Mycobionts of morphologically advanced lichens do not attempt to create a large exchange surface by forming transparietal, branched, or even arbuscular haustoria in close contact with the plasma membrane of the photoautotroph, a rather irritating situation since the photobiont cell population of such associations amounts to less than 20% of the thalline biomass (Fig. 2). Lichens differ from the abovementioned fungal symbioses by their poikilohydric water relations. The main fluxes of solutes are not mobilized between metabolically fully active cells but during the regularly occurring transitions between water saturation and desiccation. The plasma membrane of either partner of the lichen symbiosis becomes leaky during stress events, and considerable amounts of soluble compounds are passively released into the apoplastic continuum, whence they can be recovered and analyzed (MacFarlane and Kershaw 1985; Dudley and Lechowicz 1987; Melick and Seppelt 1994).

VI. Mobile Carbohydrates

Lichens were the first symbiosis of fungi and photoautotrophs in which the mobile photosynthates which move from the autotrophic to the heterotrophic partner were qualitatively determined. By introducing the so-called inhibition technique (or isotope trapping technique) Smith and coworkers identified acyclic polyols as mobile carbohydrates in green algal photobionts and glucose in cyanobacterial taxa (Table 2; reviews: Richardson 1973; Smith and Douglas 1987). The inhibition technique is based on the assumptions that (1) the mycobiont-photobiont interface is easily accessible when horizontally dissected lichen fragments are kept on an incubation medium; (2) no tight connections exist between the partners, and (3) the nutritional requirements of the fungal partner can be satisfied by incubation on a carbohydrate-containing medium, thus inhibiting further uptake (Fig. 9). The very low recovery of mobile carbohydrates in lichens with green algal photobionts (Table 2) indicated that probably only part of the mobile fraction can be detected with this technique.

During the major part of the year, lichen photobionts of extreme climates can be photosynthetically active for only very short periods of time per day, yet a positive carbon balance is gained (reviews: Kappen 1988, 1993). If less than 5% of total fixed carbon could be translocated within a few hours in *Trebouxia*-containing lichens (those which predominate in extreme climates), as concluded from inhibition technique experiments, the symbiosis would be unlikely to function and persist. ^{13}C NMR studies in the foliose macrolichen *Xanthoria calcicola* (photobiont: *Trebouxia arboricola*) confirmed the qualitative pattern of mobile photosynthates, but indicated a more efficient translocation (Lines et al. 1989). Ultrastructural studies of the mycobiont-photobiont interface show that (1) the mycobiont-photobiont interface is not so easily accessible as previously assumed

Table 2. Carbohydrate movement between symbionts of lichens
Data based on experiments using the inhibition technique according to Smith and coworkers. Reviews: Richardson (1973), Smith and Douglas (1987)

Photobiont genus	Mobile carbohydrate	Rate of movement of fixed $^{14}CO_2$ between symbionts
Lichens with green algal photobionts:		
Verrucariales, Peltigerales (n = 3) with *Myrmecia* or *Dictyochloropsis*	Ribitol	Slow (~2–4% in 3 h, 20–40% in 24 h)
Lecanorales, Teloschistales, Caliciales (n = 8) with *Trebouxia*	Ribitol	Slow
Gyalectales, Opegraphales (n = 5) with *Trentepohlia*	Erythritol	Very slow (~1–2% in 3 h, 5–12% in 24 h)
Pyrenulales (n = 1) with *Phycopeltis*	Erythritol	Slow
Verrucariales (n = 3) with *Heterococcus* or *Hyalococcus*	Sorbitol	Intermediate (~10–15% in 3 h)
Verrucariales (n = 1) with *Trochiscia*	Sorbitol	Slow
Lichens with cyanobacterial photobionts:		
Lichinales (n = 1) with *Calothrix*	Glucose/glucan	Slow
Lecanorales, Peltigerales (n = 8) with *Nostoc*	Glucose	Fast (~20–40% in 3 h)
Lecanorales (n = 1) with *Scytonema*	Glucose	Slow
Lichens with green algal and cyanobacterial photobionts (cephalodiate spp.):		
Peltigerales (n = 2) with		
Myrmecia and *Nostoc*	Ribitol and glucose	Slow/fast
Coccomyxa and *Nostoc*	Ribitol and glucose	Intermediate/fast

due to the mycobiont-derived hydrophobic wall surface layer on either biont, and (2) the fungal and algal partners are tightly interlinked with each other via particular appressorial or haustorial structures. Only those soluble carbohydrates were recovered in inhibition technique experiments which were leaching out of cut or otherwise damaged cells (Fig. 9; reviews: Honegger 1991, 1992). A negative correlation between the degree of wall surface hydrophobicity in the thalline interior and the quantitative pattern of photosynthate release in inhibition technique experiments is obvious (Honegger 1991).

VII. Poikilohydric Water Relations

As poikilohydric microorganisms, lichen mycobionts and photobionts tolerate dramatic fluctuations in cellular water contents between saturation (>150% water dw^{-1}) and desiccation (<20% water dw^{-1}). During drought stress events, both fungal and algal cells shrivel dramatically (Fig. 6; Brown et al. 1987; Honegger and Peter 1994; Scheidegger 1994). Both partners lose water, first from the apoplast, later from the symplast. Soluble compounds are passively released into the apoplastic space due to (fully reversible) leaching from the cell membrane under extreme drought stress (MacFarlane and Kershaw 1985; Dudley and Lechowicz 1987; Melick and Seppelt 1994). The fungal and algal cytoplasm of desiccated lichens is strongly condensed, but the cellular membrane systems are well preserved (Honegger and Peter 1994; Honegger 1995; Honegger et al. 1996), a prerequisite for rapid recovery of the metabolic activity after stress events. Fungal and algal polyols are likely to act as compatible solutes and to protect the cellular membrane systems during severe drought stress events (Farrar 1988).

During such drought stress events, most cells of desiccating lichen-forming fungi cavitate since they cannot deform their wall sufficiently to keep it in close contact with the plasma membrane during the dramatic volume changes within the desic-

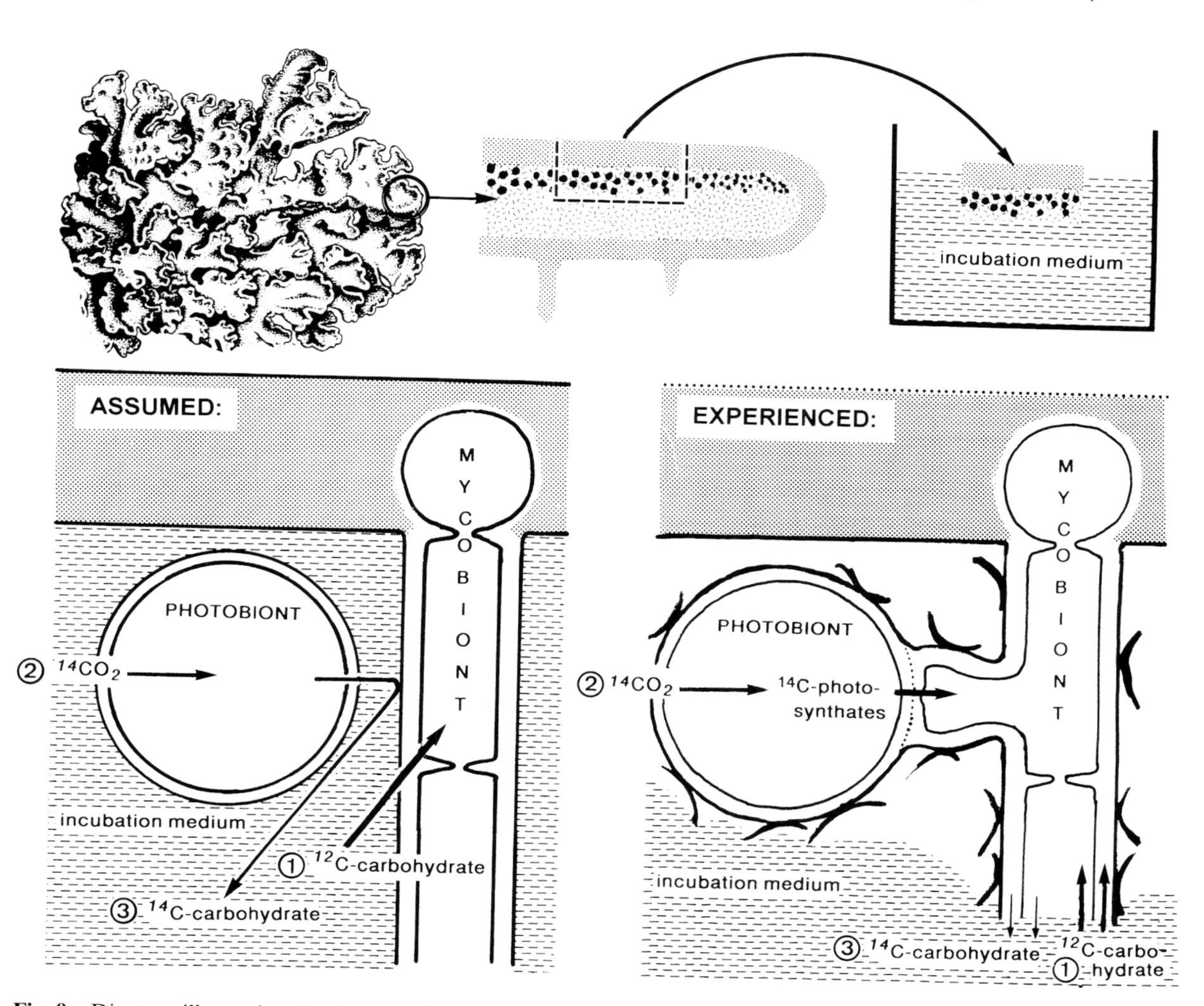

Fig. 9. Diagram illustrating the design and outcome of experiments using the inhibition technique. Results are summarized in Table 2. See text for detailed explanations

cating protoplast. A rapidly expanding cytoplasmic gas bubble of yet unknown contents (Figs. 5a,b, 6c) acts as an "internal air bag" and presses the shrinking protoplast against the wall (Honegger 1995; Scheidegger et al. 1995). Such cytoplasmatic gas bubbles are not bound by a biomembrane, as concluded from transmission electron microscopy studies of cryofixed, freeze-substituted specimens (Honegger 1995). They disappear rapidly upon rehydration, and the cells stay fully viable. The cytologically remarkable phenomenon of cytoplasmic cavitation was first observed in the ascospores of a range of nonlichenized fungi with rigid, often melanized walls (deBary 1866; Ingold 1956; Milburn 1970).

Desiccated lichen thalli can partially rehydrate by passive uptake of water vapor, but the mycobiont and photobiont reach full turgescence only when liquid water is available. However, the photoautotrophic partner of most lichen species reestablishes photosynthesis at surprisingly low water potentials, long before full turgescence is attained (Kappen 1988, 1993; Scheidegger et al. 1995).

Desiccated cells of lichen mycobionts and photobionts survive extreme temperature stress periods unharmed in a state of dormancy and thus are able to colonize climatically extreme habitats (arctic/antarctic, alpine, or desert) where higher plants are at their physiological limits (Kappen 1988, 1993). Accordingly, an estimated 8% of terrestrial ecosystems are lichen-dominated.

VIII. Secondary Metabolites

More than 600 mycobiont-derived secondary metabolites have been described in lichens (Fahselt 1994a; Huneck and Yoshimura 1996). Some of these compounds belong to classes with a wide distribution among organisms (e.g., anthraquinones: lichenized and nonlichenized fungi and higher plants) but the majority (depsides, depsidones, usnic acids, etc.) are restricted to lichen mycobionts. In the symbiotic state, lichen-forming fungi release their secondary compounds in an as yet unknown soluble form into the apoplast whence they are passively carried along by the main fluxes of solutes until they reach their site of crystallization: either the gelatinous matrix of conglutinate pseudoparenchyma (e.g., the upper cortex) or the proteinaceous, hydrophobic cell wall surface layer in the thalline interior (Fig. 7). By passive translocation, the unknown soluble forms reach also the photobiont cells and crystallize on their surfaces within the mycobiont-derived hydrophobic protein layer (Honegger 1986b). Secondary lichen metabolites are almost insoluble in aqueous systems at low or neutral pH ranges once they are crystallized. Large amounts of secondary compounds (up to 10% of thallus dw^{-1}) may thus accumulate within thalli. A considerable number of axenically cultured lichen mycobionts have been shown to produce and release secondary metabolites under appropriate culturing conditions in the absence of compatible photobiont cells (Fahselt 1994a).

Various biological functions have been proposed and partly demonstrated for lichen products, ranging from light screening, allelopathic, antimicrobial, and antifeeding effects to rock mineralization and the formation of insoluble complexes with metals (Lawrey 1986; Fahselt 1994a). For the lichen symbiosis proper, the increase in hydrophobicity of particular thallus layers (e.g., the upper cortex) and of the hydrophobic wall surface layers in the gas-filled thalline interior are most important (Honegger 1991). Unequal wettability of the lower and upper cortical layers may have a directional impact on water uptake and translocation in the thallus, and thus on ecological niche preference in the natural habitat (Larson 1984). The significance of the water repellence of wall surfaces in the thalline interior for the functioning of the symbiotic system is discussed above. The as yet unknown soluble forms of phenolic, mycobiont-derived secondary metabolites were speculated to act as potential inhibitors or regulators of the algal cell turnover in thalline areas where the fungal partner has terminated laminar growth (Honegger 1993). The majority of photobiont cells in these thalline areas are oversized, i.e., have exceeded the size required for autospore formation without undergoing cell division (Hill 1985, 1989; Fiechter 1990; Honegger 1993).

IX. Mineral Nutrition and Acquisition of Fixed Nitrogen

Relatively little is known about the mineral requirements and nutrition in lichens (Crittenden et al. 1994). This is primarily due to the near impossibility of growing lichens for prolonged periods of time under controlled laboratory conditions. Best investigated are N concentrations and availability in lichen taxa with N_2-fixing cyanobacterial photobionts (Rai 1988), either as the sole photoautotrophic partner (in approx. 10% of lichen spp.), or as predominantly N_2-fixing specialists beside the green algal partner in the triple symbiosis of cephalodiate taxa (approx. 3–4% of lichen spp.). In most cephalodiate lichens the fungal partner creates microaerobic conditions and thus facilitates N_2 fixation by tightly ensheathing the cyanobacterial colonies in gall-like, corticate structures. Cephalodia may be located on the thallus surface or in the interior (Hawksworth and Hill 1984; Jahns 1988; Honegger 1993). Increasing heterocyst numbers and nitrogenase activity were recorded with increasing age of the cephalodium (Englund 1977).

A large number of loose associative symbioses of green algal lichens with free-living cyanobacteria have been described (Poelt and Mayrhofer 1988). Although the biology of these associations has not been studied experimentally, they can be considered as potentially beneficial, since products containing fixed N_2 which leach from free-living prokaryotes may become available for the lichen. The ubiquitous bacterial epibionts of lichen thalli have not yet been considered as potential donors of fixed N_2. This situation deserves attention especially in oligotrophic boreal-arctic ecosystems where most of the mat-forming, dominant species (*Cladonia* spp., the "reindeer lichens") are green algal lichens. Evidence has accumulated that these non-N_2-fixing, mat-forming lichens are N-limited (Crittenden et al. 1994).

X. Accumulation of Metals and Radionuclides

There is a large body of literature on ion uptake and accumulation in lichens (Hawksworth and Hill 1984; Lawrey 1984, 1993; Brown 1991). Selective accumulation of high amounts of metals (up to 5% thallus dw^{-1}) has been reported in some lichen taxa which colonize either metalliferous substrata or sites which are regularly subjected to severe heavy metal pollution (e.g., the wooden supporting structures in vineyards which are sprayed up to 12 times a year with copper-containing pesticides; Poelt and Huneck 1968). Metals are accumulated on and/or within the thallus in either particulate form or bound to cell wall carbohydrates or to secondary metabolites (Purvis et al. 1990). Lichens have often been used as biomonitors of heavy metal pollution or radioactive fallout (Hawksworth and Hill 1984; Lawrey 1984, 1993; Feige et al. 1990).

XI. Conclusions and Perspectives

The majority of biologists consider lichens as a rather aberrant group of microorganisms. Many lichenologists, on the other hand, are so bewitched by the morphologically and anatomically complex symbiotic phenotypes of lichen mycobionts that they refer to them, even at the end of the 20th century and long after the discovery of their symbiotic life-style (Schwendener 1867), as plants. However, lichenization is just one among other nutritional specializations of biotrophic fungi in association with photoautotrophic hosts (Smith 1978; Honegger 1991, 1993; Gargas et al. 1995). The taxonomic diversity, numeric significance and ecological success of lichen-forming fungi convincingly demonstrate that lichenization is indeed a successful nutritional strategy.

Many data on symbiotic interactions of lichen mycobionts with their photoautotrophic partners are relevant to plant pathogenic or mycorrhizal fungi. Recently, molecular biologists reported on the enhanced expression of developmentally regulated genes encoding hydrophobins during the establishment of a functional interface of plant pathogenic and ectomycorrhizal fungi with their hosts (Talbot et al. 1993, 1996; Templeton et al. 1994; Tagu et al. 1996). Lichens, on the other hand, were the first fungal symbionts with photautotrophic partners in which a hydrophobic, proteinaceous fungal wall surface layer was shown to spread over the cell wall of the photautotrophic host, thus sealing the apoplastic continuum and directing the fluxes of apoplastic solutes along source-sink gradients between the partners (Honegger 1985). It will be a fascinating task to characterize these hydrophobic cell wall surface proteins of the rodlet layer of lichen-forming fungi and to compare them with the hydrophobins of nonlichenized taxa. Another field of biological interest, with potentially interesting applications, is the biochemistry and molecular genetics of the impressive desiccation tolerance of lichen-forming fungi and their photoautotrophic partners in comparison with other poikilohydric organisms. The notoriously slow-growing lichen-forming fungi are unlikely to become important model systems in developmental biology or biotechnology, but promising results have been achieved in "tissue" culture experiments aiming at the search for new biologically active compounds (Yamamoto et al. 1993). Improvement of culture techniques in lichens is desirable not only for experimental purposes but also with regard to species preservation. In the near future, many lichen species are likely to be extinct before their biology, their role in natural ecosystems, or their potential uses are known.

References

Ahmadjian V, Jacobs JB (1981) Relationship between fungus and alga in the lichen *Cladonia cristatella* Tuck. Nature 389:169–172

Brown DH (1991) Lichen mineral studies – currently clarified or confused. Symbiosis 11:207–223

Brown DH, Rapsch S, Beckett A, Ascaso C (1987) The effect of desiccation on cell shape in the lichen *Parmelia sulcata* Taylor. New Phytol 105:295–299
Bubrick P, Frensdorff A, Galun M (1985) Selectivity in the lichen symbiosis. In: Brown DH (ed) Lichen physiology and cell biology. Plenum, New York, pp 319–334
Büdel B (1992) Taxonomy of lichenized procaryotic blue-green algae. In: Reisser W (ed) Algae and symbioses. Plants, animals, fungi, viruses, interactions explored. Biopress, Bristol, pp 301–324
Crittenden PD, Katucka I, Oliver E (1994) Does nitrogen supply limit the growth of lichens? Cryptogam Bot 4:143–155
deBary A (1866) Morphologie und Physiologie der Pilze, Flechten und Myxomyceten. W Engelmann, Leipzig
deVries OMH, Fekkes MP, Wösten HAB, Wessels JGH (1993) Insoluble hydrophobin complexes in the walls of *Schizophyllum commune* and other filamentous fungi. Arch Microbiol 159:330–335
Dudley SA, Lechowicz MJ (1987) Losses of polyol through leaching in subarctic lichens. Plant Physiol 83:813–815
Englund B (1977) The physiology of the lichen *Peltigera aphthosa*, with special reference to the blue-green phycobiont (*Nostoc* sp.). Physiol Plant 41:298–304
Fahselt D (1994a) Secondary biochemistry of lichens. Symbiosis 16:117–165
Fahselt D (1994b) Carbon metabolism in lichens. Symbiosis 17:127–182
Farrar JF (1988) Physiological buffering. In: Galun M (ed) CRC Handbook of lichenology, vol 2. CRC Press, Boca Raton, pp 101–105
Feige GB, Niemann L, Jahnke S (1990) Lichens and mosses – silent chronists of the Chernobyl accident. Bibl Lichenol 38:63–77
Fiechter E (1990) Thallusdifferenzierung und intrathalline Sekundärstoffverteilung bei Parmeliaceae (Lecanorales, lichenisierte Ascomyceten). Inaugural-Dissertation, Philosophische Fakultät II, Universität Zürich, Zürich
Friedl T (1989) Systematik und Biologie von *Trebouxia* (Microthamniales, Chlorophyta) als Phycobiont der Parmeliaceae (lichenisierte Ascomyceten). Inaugural-Dissertation, Fak Biol, Universität Bayreuth, Bayreuth
Friedl T (1995) Inferring taxonomic positions and testing genus level assignments in coccoid lichen algae: a phylogenetic analysis of 18s ribosomal RNA sequences from *Dictyochloropsis reticulata* and from members of the genus *Myrmecia* (Chlorophyta, Trebouxiophyceae cl. nov.). J Phycol 31:632–639
Friedl T, Zeltner C (1994) Assessing the relationships of lichen algae and the Microthamniales (Chlorophyta) with 18S rRNA gene sequence comparisons. J Phycol 30:500–506
Gargas A, Depriest PT, Grube M, Tehler A (1995) Multiple origins of lichen symbioses in fungi suggested by SSU rDNA phylogeny. Science 268:1492–1495
Gärtner G (1992) Taxonomy of symbiotic eucaryotic algae. In: Reisser W (ed) Algae and symbioses. Plants, animals, fungi, viruses, interactions explored. Biopress, Bristol, pp 325–338
Hawksworth DL (1988a) The fungal partner. In: Galun M (ed) CRC Handbook of lichenology, vol 1. CRC Press, Boca Raton, pp 35–38
Hawksworth DL (1988b) Effects of algae and lichen-forming fungi on tropical crops. In: Agnihotry VP, Sarbhoy KA, Kumar D (eds) Perspectives of mycopathology. Malhotra Publishing House, New Delhi, pp 76–83
Hawksworth DL, Hill DJ (1984) The lichen-forming fungi. Blackie, Glasgow
Hawksworth DL, Honegger R (1994) The lichen thallus: a symbiotic phenotype of nutritionally specialized fungi and its response to gall producers. In: Williams MAJ (ed) Plant galls: organisms, interactions, populations. Clarendon Press, Oxford, pp 77–98
Hill DJ (1985) Changes in photobiont dimensions and numbers during co-development of lichen symbionts. In: Brown DH (ed) Lichen physiology and cell biology. Plenum, New York, pp 303–317
Hill DJ (1989) The control of the cell cycle in microbial symbionts. New Phytol 112:175–184
Honegger R (1984) Cytological aspects of the mycobiont-phycobiont relationship in lichens. Haustorial types, phycobiont cell wall types, and the ultrastructure of the cell wall surface layers in some cultured and symbiotic myco- and phycobionts. Lichenologist 16:111–127
Honegger R (1985) Fine structure of different types of symbiotic relationships in lichens. In: Brown DH (ed) Lichen physiology and cell biology. Plenum Press, New York, pp 287–302
Honegger R (1986a) Ultrastructural studies in lichens. I. Haustorial types and their frequencies in a range of lichens with trebouxioid phycobionts. New Phytol 103:785–795
Honegger R (1986b) Ultrastructural studies in lichens. II. Mycobiont and photobiont cell wall surface layers and adhering crystalline lichen products in four Parmeliaceae. New Phytol 103:797–808
Honegger R (1987) Isidium formation and the development of juvenile thalli in *Parmelia pastillifera* (Lecanorales, lichenized Ascomycetes). Bot Helv 97:147–152
Honegger R (1990) Surface interactions in lichens. In: Wiessner W, Robinson DG, Starr RC (eds) Experimental phycology 1. Cell walls and surfaces, reproduction, photosynthesis. Springer, Berlin Heidelberg New York, pp 40–54
Honegger R (1991) Functional aspects of the lichen symbiosis. Annu Rev Plant Physiol Plant Mol Biol 42:553–578
Honegger R (1992) Lichens: mycobiont-photobiont relationships. In: Reisser W (ed) Algae and symbioses. Plants, animals, fungi, viruses, interactions explored. Biopress, Bristol, pp 255–275
Honegger R (1993) Developmental biology of lichens. New Phytol 125:659–677
Honegger R (1995) Experimental studies with foliose macrolichens: fungal responses to spatial disturbance at the organismic level and to spatial problems at the cellular level during drought stress events. Can J Bot 73 (Suppl 1):569–578
Honegger R, Peter M (1994) Routes of solute translocation and the location of water in heteromerous lichens visualized with cryotechniques in light and electron microscopy. Symbiosis 16:167–186
Honegger R, Peter M, Scherrer S (1996) Drought-induced structural alterations at the mycobiont-photobiont interface in a range of foliose macrolichens. Protoplasma 190:221–232
Huneck S, Yoshimura I (1996) Identification of lichen substances. Springer, Berlin
Ihda TA, Nakano T, Yoshimura I, Iwatsuki Z (1993) Phycobionts isolated from Japanese species of *Anzia* (lichenes). Arch Protistenkd 143:163–172
Ingold CT, (1956) A gas phase in viable fungal spores. Nature 177:1242–1243

Jahns HM (1988) The lichen thallus. In: Galun M (ed) CRC Handbook of lichenology, vol 1. CRC Press, Boca Raton, pp 95–143
James PW, Henssen A (1976) The morphological and taxonomic significance of cephalodia. In: Brown DH, Hawksworth DL, Bailey RH (eds) Lichenology: progress and problems. Academic Press, London, pp 27–77
Kappen L (1988) Ecophysiological relationships in different climatic regions. In: Galun M (ed) CRC Handbook of lichenology, vol 2. CRC Press, Boca Raton, pp 37–100
Kappen L (1993) Lichens in the antarctic region. In: Friedman EI (ed) Antarctic microbiology. Wiley-Liss, New York, pp 433–490
Larson DW (1984) Habitat overlap/niche segregation in two *Umbilicaria* lichens: a possible mechanism. Oecologia 62:118–125
Lawrey JD (1984) Biology of lichenized fungi. Praeger, New York
Lawrey JD (1986) Biological role of lichen substances. Bryologist 89:111–122
Lawrey JD (1993) Lichens as monitors of pollutant elements at permanent sites in Maryland and Virginia. Bryologist 96:339–341
Lines CEM, Ratcliffe RG, Rees TAV, Southon TE (1989) A ^{13}C NMR study of photosynthate transport and metabolism in the lichen *Xanthoria calcicola* Oxner. New Phytol 111:447–456
MacFarlane JD, Kershaw KA (1985) Some aspects of carbohydrate metabolism in lichens. In: Brown DH (ed) Lichen physiology and cell biology. Plenum Press, New York, pp 1–8
Melick DR, Seppelt RD (1994) The effect of hydration on carbohydrate levels, pigment content and freezing point in *Umbilicaria decussata* at a continental Antarctic localitiy. Cryptogam Bot 4:212–217
Milburn JA (1970) Cavitation and osmotic potentials of *Sordaria* ascospores. New Phytol 69:133–141
Mollenhauer D (1992) *Geosiphon pyriforme*. In: Reisser W (ed) Algae and symbioses. Plants, animals, fungi, viruses, interactions explored. Biopress, Bristol, pp 339–351
Ott S (1987) Reproductive strategies in lichens. Bibl Lichenol 25:81–93
Plessl A (1963) Über die Beziehungen von Pilz und Alge im Flechtenthallus. Oesterr Bot Z 110:194–269
Poelt J, Huneck S (1968) *Lecanora vinetorum* nova spec., ihre Vergesellschaftung, ihre Oekologie und ihre Chemie. Oesterr Bot Z 115:411–422
Poelt J, Mayrhofer H (1988) Über Cyanotrophie bei Flechten. Plant Syst Evol 158:265–281
Rai AN (1988) Nitrogen metabolism. In: Galun M (ed) CRC Handbook of lichenology, vol 1. CRC Press, Boca Raton, pp 201–237
Richardson DHS (1973) Photosynthesis and carbohydrate movement. In: Ahmadjian V, Hale ME (eds) The lichens. Academic Press, New York, pp 249–288
Scheidegger C (1994) Low-temperature scanning electron microscopy: the localization of free and perturbed water and its role in the morphology of the lichen symbionts. Cryptogam Bot 4:290–299
Scheidegger C, Schroeter B, Frey B (1995) Structural and functional processes during water vapour uptake and desiccation in selected lichens with green algal photobionts. Planta 197:399–409
Schwendener S (1867) Ueber die wahre Natur der Flechten. Verh Schweiz Naturforsch Ges 51:88–90
Smith DC (1978) What can lichens tell us about real fungi? Mycologia 70:915–935
Smith DC, Douglas A (1987) The biology of symbiosis. Edward Arnold, London
Tagu D, Python M, Cretin C, Martin F (1996) Cloning symbiosis-related cDNAs from eucalypt ectomycorrhiza by PCR-assisted differential screening. New Phytol 125:339–343
Talbot NJ, Ebbole DJ, Hamer JE (1993) Identification and characterization of MPG1, a gene involved in pathogenicity from the rice blast fungus, *Magnaporthe grisea*. Plant Cell 5:1575–1590
Talbot NJ, Kershaw MJ, Wakley GE, deVries OMH, Wessels JGH, Hamer JE (1996) MPG1 encodes a fungal hydrophobin involved in surface interactions during infection-related development of *Magnaporthe grisea*. Plant Cell 8:985–999
Templeton MD, Rikkerink EHA, Beever RE (1994) Small, cysteine-rich proteins and recognition in fungal-plant interactions. Mol Plant-Microbe Interact 3:320–325
Tschermak E (1941) Untersuchungen über die Beziehungen von Pilz und Alge im Flechtenthallus. Oesterr Bot Z 90:233–307
Tschermak-Woess E (1988) The algal partner. In: Galun M (ed) CRC Handbook of lichenology, vol 1. CRC Press, Boca Raton, pp 39–92
van Donk E, Bruning K (1992) Ecology of aquatic fungi in and on algae. In: Reisser W (ed) Algae and symbioses. Plants, animals, fungi, viruses, interactions explored. Biopress, Bristol, pp 567–592
Wessels JGH (1993) Wall growth, protein excretion, and morphogenesis in fungi. New Phytol 123:397–413
Wösten HAB, Asgeirsdottir SA, Krook JH, Drenth JHH, Wessels JGH (1994) The fungal hydrophobin Sc3P self-assembles at the surface of aerial hyphae as a protein membrane constituting the hydrophobic rodlet layer. Eur J Cell Biol 63:122–129
Yamamoto Y, Miura Y, Higuchi M, Kinoshita Y, Yoshimura I (1993) Using lichen tissue cultures in modern biology. Bryologist 96:384–393

12 Altered Gene Expression During Ectomycorrhizal Development

F. Martin, F. Lapeyrie, and D. Tagu

CONTENTS

I. Introduction

Mycorrhizas are widespread symbiotic associations involving soil fungi and the roots of most land plants. They have been of primary importance in the evolution of land plants (Harley and Smith 1983; Simon et al. 1993) and are a key component of plant and fungal communities (Allen 1991; Chap. 11, Vol. V, Part B). Mycorrhizal symbioses have a beneficial impact on plant growth in natural and agroforestry ecosystems and allow the completion of the fungal life cycle (Harley and Smith 1983; Read 1991, 1992). The formation of the symbiosis requires several days and includes major changes in cellular and tissue morphology as well as in the biochemistry and physiology of the partners. Numerous recent reviews have covered the physiological and biochemical aspects of endo- and ectomycorrhiza in detail (Smith and Gianinazzi-Pearson 1988; Jakobsen 1991; Söderström 1992; Martin and Botton 1993; Botton and Chalot 1995; Hampp and Schaeffer 1995). Alterations of the host plant and the fungal cell surface have been reviewed recently (Bonfante-Fasolo 1988; Bonfante and Perotto 1992; Bonfante 1994). Because mycorrhiza development involves the formation of specific multicellular structures, it incorporates the main features of developmental processes common in all organisms, i.e., temporal and spatial changes in cellular and tissular differentiation.

Although ectomycorrhizal development has been studied for many years, investigations have focused mainly on the morphological and cytological changes (Kottke and Oberwinkler 1986, 1987; Massicotte et al. 1987, 1989; Horan et al. 1988; Moore et al. 1989; Dexheimer and Pargney 1991; Bonfante and Perotto 1992; Scheidegger and Brunner 1995). These have stressed the drastic alterations in cell growth and tissue organization experienced by the symbionts. Exciting advances are starting to reveal the molecular mechanisms underlying these symbiosis-related alterations (Martin and Hilbert 1991; Bonfante and Perotto 1992; Martin et al. 1995a,b; Tagu and Martin 1996), but several developmental processes remained unexplored. Mycorrhiza development poses several questions: (1) What are the nature and the biochemical roles of signaling molecules that the symbionts exchange?, (2) How do the specialized symbiotic fungal structures (i.e., mantle and intercellular Hartig net) develop from the initial hyphae?, (3) How do the hyphae influence the root meristematic activity and cell shape?;

Equipe de Microbiologie Forestière, Institut National de la Recherche Agronomique, Centre de Recherches de Nancy, 54280 Champenoux, France

The Mycota V Part A
Plant Relationships
Carroll/Tudzynski (Eds.)

(4) How do the plant and fungal cells coordinate their symbiotic development and metabolic activity? To date, none of these questions have been properly answered, and the available data are too fragmented to put forward hypotheses for experimental examination.

In this chapter, we outline the major morphological, biochemical, and molecular events that occur during ectomycorrhizal development, using our studies of the interaction between *Eucalyptus* spp. and the ectomycorrhizal gasteromycete *Pisolithus tinctorius* as a focal point. We will then discuss what is known about the functions of the proteins and genes that are developmentally regulated, and will attempt to point out promising avenues of research for the future.

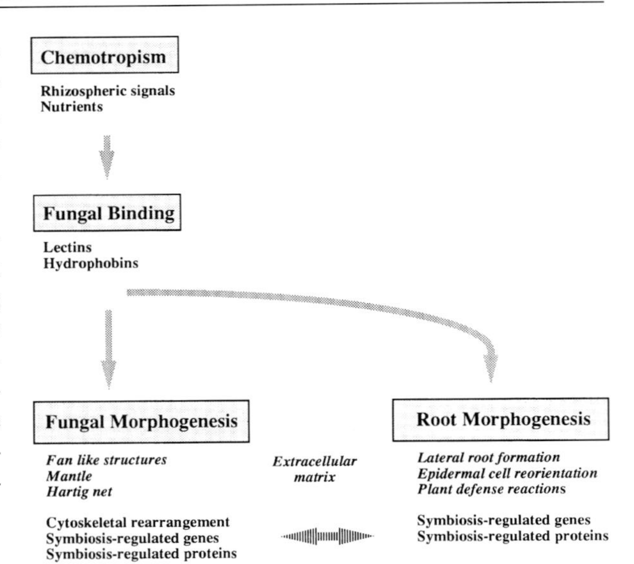

Fig. 1. Morphological events during ectomycorrhizal formation and possible modes of interaction

II. Overview of Ectomycorrhizal Development

We are fortunate that the morphological and anatomical changes that accompany ectomycorrhizal development have been studied and described in great detail in various associations (e.g., *Picea abies-Amanita muscaria*, Kottke and Oberwinkler 1986, 1987; *P. abies-Hebeloma crustuliniforme*, Scheidegger and Brunner 1995; *Eucalyptus* spp.–*Pisolithus tinctorius*, Horan et al. 1988, Massicotte et al. 1987; *Alnus rubra-Alpova diplophloeus*, Massicotte et al. 1989). The process of root colonization by the fungus, the differentiation of symbiotic structures, and the establishment of an active association involve a complex sequence of interactions between the fungal hyphae and the plant root cells (Fig. 1). The mature organization of ectomycorrhizae varies with the host and fungal species. However, although some of the details vary, early stages of ectomycorrhizal development have well-characterized similar morphological transitions as follows: (1) contact of the fungus with the surface of a young apical portion of a root; (2) growth on the exposed surface; (3) adhesion to roots; (4) penetration between epidermal cells and cortical cells; (5) formation of the mantle and Hartig net with concomitant coordinated alteration in the root structure. Developmental variants of the ectomycorrhizal basidiomycete *Laccaria bicolor* have been identified and characterized (Wong et al. 1989). Some of these variants are blocked at discrete points in the normal morphological continuum, and molecular study of these mutants should provide information about intermediate steps in the morphogenetic process for which the action of a specific protein is required. Although the morphological stages accompanying the ontogenesis of ectomycorrhizae have been described in detail many times, it is useful to summarize them briefly, as a reference point for subsequent discussions on the role of the developmentally regulated genes and proteins.

A. Rhizospheric Signals in the Early Interactions

1. A Role for Plant Flavonoids?

Early in the interactions between plants and microbes, signals are produced that elicit discrete responses in the respective partners, the first step in the cascade of biochemical events leading to contact at the host surface and the subsequent outcome of the symbiosis or pathogenesis (Lamb et al. 1989; Dixon and Lamb 1990). An interplay of signals coordinates and organizes the responses of the partner cells and modulates their respective differentiation. In contrast to some other plant-microbe interactions (Hirsch 1992), the nature of the signaling molecules and the molecular basis of signal perception and transduction in mycorrhiza are unknown or ill-defined. Identifying the processes that regulate the information flow between mycorrhizal fungi and host root is currently an active research area. Fungal spore germination, chemoattraction of the hyphae by the root cells, adhesion to root surface, root penetration, and development of fungal multicellular structures on/in the root are probably dependent on precisely tuned host-derived signaling molecules, and their study has general implications for the understanding of plant-fungus interactions.

Plant-bacteria interactions (e.g., rhizobia-legume symbiosis, *Agrobacterium* infection) involve various flavonoid derivatives (Lynn and Chang 1990; Hirsch 1992). Recently, attempts have been made to characterize root flavonoids that may act as signals for the initiation and development of endomycorrhizal and ectomycorrhizal symbioses. Products of the phenylpropanoid pathway apparently function as signaling molecules in endomycorrhizal interactions. The role of flavonoids is not yet clear, but a variety of flavanones, flavones, and isoflavones enhances spore germination and hyphal growth of arbuscular mycorrhizal (AM) fungi in vitro (Gianinazzi-Pearson et al. 1989; Nair et al. 1991; Tsai and Phillips 1991; Chabot et al. 1992) and promote AM fungal colonization. However, in an elevated CO_2 atmosphere, it was shown that among the flavonoids tested only the flavonols stimulated the hyphal growth, while the other derivatives were all inhibitory (Bécard et al. 1992). The establishment of the mycorrhizal symbiosis is accompanied by both qualitative and quantitative changes in many of the secondary metabolites (coumestrol, daidzein, formononetin, glyceollin) of the host root (Morandi et al. 1984; Harrison and Dixon 1993; Volpin et al. 1994). The flavonoid increase in root tissues is induced by the presence of the fungus in the rhizosphere, and invasion of the root tissues is not required (Volpin et al. 1994). Formononetin produced in the presence of endomycorrhizal hyphae may play a role in competition between AM species, favoring colonization of the roots by the species for which formononetin promotes spore germination and stimulates mycelial growth (Volpin et al. 1994). Recently, it has been argued that flavonoids are not necessary plant signal compounds in AM symbioses (Bécard et al. 1995).

The effects of plant flavonoids on spore germination and hyphal growth of ectomycorrhizal fungi are unknown. However, several metabolites released by the plant roots trigger events leading to fungal infection of the host root (Horan and Chilvers 1990; Gogala 1991). In the saprophytic phase, spores of several ectomycorrhizal fungi respond to the stimulation of the diterpene resin acid, abietic acid, in root exudates (Fries et al. 1987). It is not known how commonly such germination factors may occur in other ectomycorrhizal associations, but many species present a loose specificity, since they respond to several types of germination activators produced by other microorganisms.

2. A Role for Fungal Auxins?

A number of plant hormones (e.g., cytokinins, gibberellins, ethylene) are produced by mycorrhizal fungi, but auxins such as indolacetic acid (IAA) have been studied most intensively. Their role in mycorrhizal development is controversial (Gogala 1991; Beyrle 1995). Based on comparative root morphology and root ramification following either IAA application or fungal colonization, it has been suggested that fungal auxins play a key role in formation of the symbiosis (Gogala 1991; Beyrle 1995). Recent studies using fluoroindole-resistant mutant of *Hebeloma cylindrosporum* indicated that IAA controls major anatomical features of pine ectomycorrhiza. Overproduction of this fungal hormone induced an abnormal proliferation of the intercellular network of hyphae (Gay et al. 1994; Gea et al. 1994; Chap. 6, Vol. V, Part B). Therefore, ethylene and auxins appear to provide chemical signals modulating fungal ingress within roots and to control several anatomical features of ectomycorrhizae. It has been suggested that the intraspecific variations in symbiotic structures of *Laccaria bicolor–Pinus banksiana* associations (Wong et al. 1989) are related to the differences in IAA-producing activity among various fungal isolates.

However, the role of auxins in ectomycorrhizal development has been challenged by Horan (1991). High exogenous levels of IAA were applied to non-inoculated *Eucalyptus* seedlings to simulate the fungal secretion. Treatments using appropriate concentrations of hormones induced the formation of slow-growing, swollen, and highly branched roots that were superficially similar to ectomycorrhizae within the same short period of time. However, these roots exhibited anatomical alterations which are not observed in ectomycorrhiza, including prolific root hair production, increased number of cortical cell layers, and increased number of vascular bundles in the stele. Conversely, auxins failed to generate some features typical of ectomycorrhizae (e.g., radially elongated epidermal cells). It is, nevertheless, too early to rule out any role for auxins. Massive secretion of fungal auxins toward the roots is unlikely. Auxin concentration in the different root tissues is presumably regulated by the fungal partner during the various stages of the mycorrhizal ontogeny. This regulation may imply targeted secretion of auxins and auxin conjugates, and localized regulation of root endogenous auxin metabolism (i.e., through fungal auxin oxidase inhibitor). The role of other fungal indolic compounds should also be considered. For example, the main indolic compound accumulated and excreted by the ectomycorrhizal *Pisolithus tinctorius* is hypaphorine (Béguiristain et al. 1995), the tryptophan betaine. Hypaphorine presents an auxin-like action on lentil roots (Hofinger et al. 1976) and its concentration in *Pisolithus* mycelium is 1000-fold higher than IAA. The major observed morphogenetic effect of hypaphorine on the development of *Eucalyptus* roots is a strong reduction of root hair elongation (Béguiristain and Lapeyrie 1997), suggesting that fungal molecules, such as hypaphorine, could prevent root hair development as the lateral roots emerge from colonized roots.

Auxins are also known to modulate cell wall plasticity, and this role would allow the Hartig net development between cortical cells without fungal secretion of lytic enzymes. In addition, auxins inhibit chitinase induction in various plant-fungus interactions (Shinshi et al. 1987), including *Eucalyptus* roots treated with elicitors from ectomycorrhizal fungi (T. Béguiristain and F. Lapeyrie unpubl.). It is clear that fungal auxins and derivatives may modulate root morphogenesis during symbiosis development, but their precise role in the control of the cascade of molecular events leading to the mature ectomycorrhiza is still uncertain. To analyze the role of fungal auxins in ectomycorrhizae development, the effects of exogenous hormones on patterns of protein and gene expression should be considered. It will be interesting to see whether any of these IAA-induced genes show homology with symbiosis-related (SR) genes and whether they are also expressed during mycorrhiza development (see Sect. III.C.2). Fungal mutants unable to secrete auxins/hypaphorine would help in understanding the role of these hormones.

B. The Early Morphogenetic Events

1. Chemodifferentiation

The key steps in developing a compatible plant-mycorrhizal fungus interaction occur within the

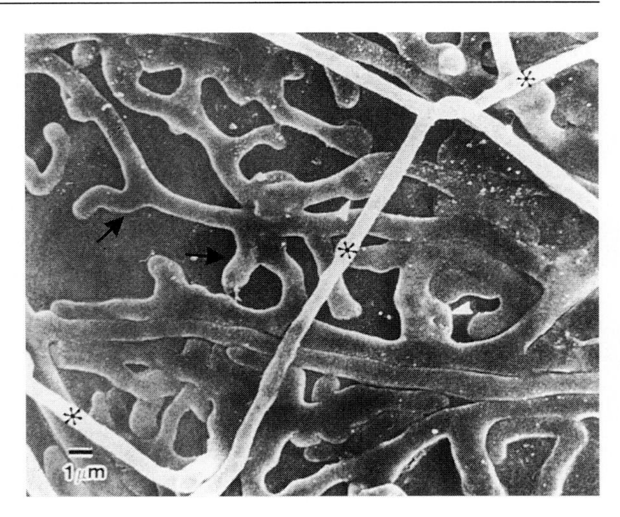

Fig. 2. Hyphae of the ectomycorrhizal *Pisolithus tinctorius* growing in the vicinity of the root surface rapidly switch from their original branching pattern and apical dominance (*) to differentiate in a new irregular, septate branching pattern with reduced interhyphal spacing and multiple apices, the so-called palmettes (*arrows*). (Photo courtesy of Prof. R. L. Peterson, University of Guelph)

first few hours after contact between the hyphae and the root. Root cells elicit a local change in the hyphal morphology of ectomycorrhizal (Jacobs et al. 1989; Fig. 2) and AM fungi (Giovannetti et al. 1993) prior to the differentiation of specialized symbiotic structures. Hyphae growing in the vicinity of the root surface rapidly switch from their original branching pattern and apical dominance to differentiate in a new irregular, septate branching pattern with reduced interhyphal spacing and multiple apices. This new mode of hyphal growth allows an intimate contact with the root surface and appears to be a prerequisite for subsequent contact and penetration of the root and further differentiation of the specialized symbiotic structures. It may allow the mycorrhizal fungus facing osmotic and nutrient stresses on the root surface to increase the likelihood of locating favorable environmental conditions. Under stress, many nonmycorrhizal filamentous fungi also produce fan-like structures as the result of multiple growing hyphal tips (Markham 1992).

In addition to trophic factors such as nutrient deprivation, signaling molecules presumably trigger hyphal elongation and the elicitation of a different branching pattern. It is tempting to speculate the occurrence of a concentration gradient of morphogens diffusing from the root surface through the rhizosphere. The signaling molecule(s) would induce the morphogenetic switch of the mycelium only at the high concentrations occurring at the root surface. Studies on the *E. globulus-P. tinctorius* association implicated a small, difusible, secreted plant compound as a morphogenetic chemoattractant (Horan and Chilvers 1990) and effector of early developmental processes. Giovannetti et al. (1993) provided evidence for a similar chemotaxis/differentiation signal in endomycorrhizal interactions.

Hyphal branching in filamentous fungi is controlled by several secreted metabolites, such as choline and betaine (Markham et al. 1993). The concentration of the tryptophane betaine, hypaphorine, is enhanced in *P. tinctorius* when hyphae grow onto the surface of host roots. This increased accumulation of hypaphorine is controlled by specific diffusible molecules released by eucalypt roots which are lacking in nonhost plants (Béguiristain and Lapeyrie 1997). As hypaphorine stimulates net K^+ uptake and H^+ extrusion in hyphae (Huang et al. 1997), it might contribute to (1) the regulation of ion fluxes through the plasma membrane during elongation and branching of hyphal tips (Levina et al. 1995) and (2) nutrient exchanges between hyphae and root cortical cells (Smith and Smith 1990).

2. Fungal Binding

Root surface recognition, the binding of the hyphal tips, and the conveyance of information to promote infection-related morphogenetic events are essential features of mycorrhizal interactions (Bonfante-Fasolo and Perotto 1992). It is likely that mycorrhizal fungi, like most biotrophic fungi, have developed a wide variety of mechanisms by which they might mediate binding to the host. Evidence is accumulating that they secrete specific binding molecules (adhesins) able to recognize and bind to specific proteins or polysaccharides present on the root surfaces (Piché et al. 1988; Martin and Tagu 1995; Tagu and Martin 1996). When these putative mycorrhizal adhesins interact with their respective ligand, they are capable of mediating a specific, high-affinity interaction between the host and the symbiont. Without adhesion to the root, mycorrhizal hyphae may not be able to persist at the initial infection site, and thus be unable to colonize and penetrate root tissues.

In the different types of mycorrhizas, adhesion appears to be mediated by a mucilaginous matrix covering the hyphae and the root surface (Bonfante-Fasolo 1988; Piché et al. 1988; Lapeyrie et al. 1989). Microscopic observations showed that orientated fimbriae containing Concanavalin A-recognized glycoproteins are involved in the attachment (Lei et al. 1991). The secretion of these extracellular glycoproteins appears to be crucial to the subsequent colonization of the root surface by hyphae in compatible associations. A layer of extracellular fibrillar polymers is present in the extracellular matrix of the free-living mycelium of *L. bicolor* (Lei et al. 1991) and *P. tinctorius* (Lei et al. 1990) even before the interaction with the root. At the contact sites between hyphae and root cells, an increased secretion of these extracellular fibrillar polymers takes place. It is likely that a concomitant reorganization of the extracellular fibrillar polymers occurs, observed on microscopic sections as an accumulation and orientation of the extracellular polymeric fimbriae towards the host cell (Fig. 3). In contrast, in the noncompatible in-

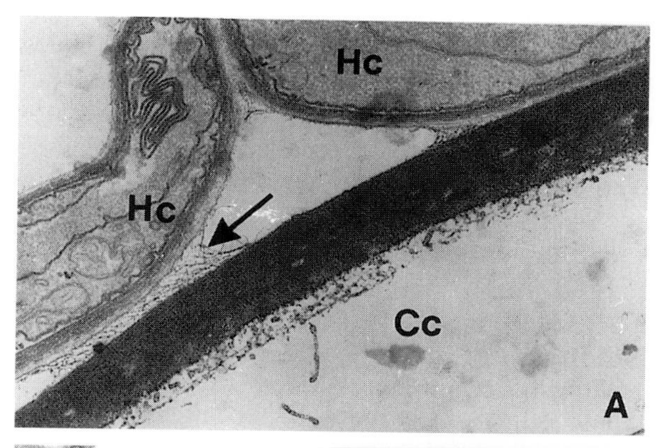

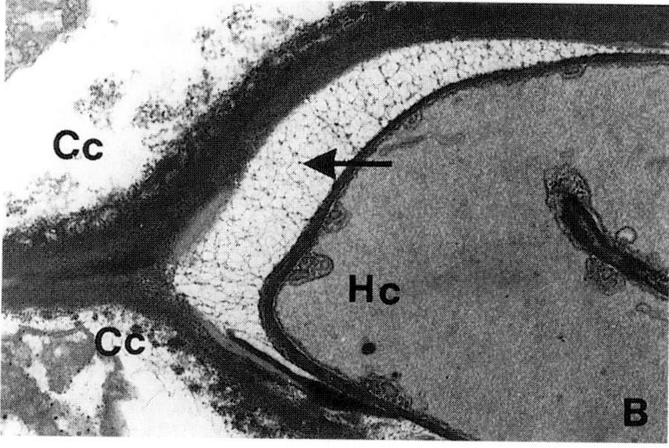

Fig. 3A,B. Secretion of glycoprotein fibrillae (*arrows*) by a compatible isolate of *Pisolithus tinctorius* in contact with *Eucalyptus urophylla* cortical cells (**A**) and during Hartig net formation (**B**). Polysaccharides of cell walls have been stained using the PATAG test. *Cc* Cortical cells; *Hc* hyphal cells. (Lei et al. 1990)

teractions, no fibrillar material is secreted and mycorrhiza development is delayed. Apart from the secretion of extracellular fibrillar polymers, evident alterations in polysaccharide composition of the host and fungal cell walls have been described at the interface formed during the early stages of contact (Dexheimer and Pargney 1991). Fungal lectins presumably involved in the adhesion have been described in the ectomycorrhizal species *Lactarius* spp. (Giollant et al. 1993), but the biochemical interactions between symbionts involving these lectins are not understood. However, affinity of the fungal walls to lectins is not affected during early steps of development of the eucalypt ectomycorrhiza (Lapeyrie and Mendgen 1993).

C. Development of Symbiotic Structures

After the attachment of the mycobiont and the initial colonization of the root surface, the development of ectomycorrhizae involves the differentiation of structurally specialized fungal tissues and an apoplastic interface between the symbionts (Kottke and Oberwinkler 1986, 1987; Massicotte et al. 1989; Dexheimer and Pargney 1991; Bonfante and Perotto 1992; Bonfante 1994).

Fungal hyphae grow to form initial aggregates and eventually bind to one another to form a specialized tissue, the mantle. This sheath then encloses the epidermal and the root cap cells. As with many fungal multicellular structures, mantle differentiation is certainly mediated by adhesion proteins that specify the cell-to-cell interactions. It has been recently realized how important the diverse macromolecules of the extracellular matrix (ECM) and adhesion proteins are in the establishment and maintenance of fungal complex organs (Wessels 1992, 1993, 1996; Mendgen and Deising 1993). The potential role of these morphogenetic proteins in ectomycorrhizal morphogenesis will be discussed in Section III.C.1.

During mantle formation, the intercellular hyphal network, the Hartig net, develops between the cortical cells. Uniseriate hyphae progressing within the root undergo rapid nuclear divisions without cytokinesis, resulting in multiple enlarged lobed nuclei. Increased numbers of cell organelles and endoplasmic reticulum networks are observed, and the cells are metabolically highly active (Kottke and Oberwinkler 1986, 1987; Dexheimer and Pargney 1991). A characteristic feature of the Hartig net hyphae is the elaborate network of membranous cell ingrowths, typical of transfer cells (Kottke and Oberwinkler 1987; Massicotte et al. 1989; Scheidegger and Brunner 1995). These cell wall ingrowths increase the surface area of the associated membrane and facilitate the import/export of metabolites between the symbionts.

Evidence from morphological observations and cytochemical assays has shown that the walls of the infection hyphae and the adjacent plant cells are modified to give rise to this specialized interfacial matrix. It is of host origin and rich in pectic material (Bonfante 1994). Changes in the composition of the protein/melanin fraction of the fungal cell walls during the eucalypt ectomycorrhizal development have been demonstrated by

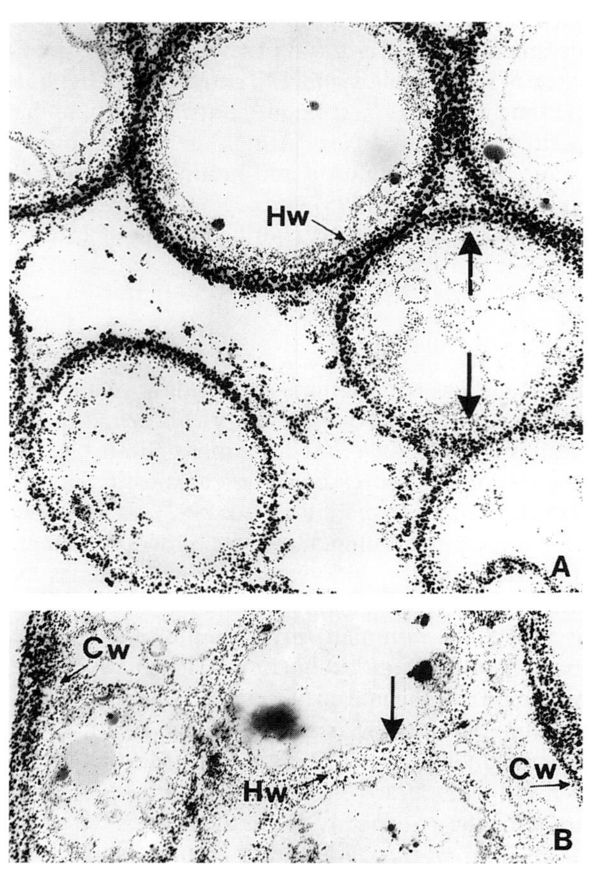

Fig. 4A,B. Changes in the fungal cell wall composition of *Cenococcum geophilum* hyphae in the mantle (**A**) and in the Hartig net (**B**) of *Eucalyptus globulus* mycorrhiza. *Thick arrows* show the centripetal decrease of protein labeling (Gomori-Swift test) in the bottom layers of the mantle and the weak staining in the Hartig net. *Hw* Hyphal wall; *Cw* cortical cell walls. (Paris et al. 1993)

microscopic and histochemical observations (Paris et al. 1993; Fig. 4). According to Bonfante (1994), the presence of the mycorrhizal hyphae can affect the assembly, but not the de novo synthesis of compounds which usually occur at the host cell surface. In many biotrophic fungi, a thick layer reinforced with melanin and hydrophobin-like proteins reduces the porosity of the appressorium wall to establish high turgor pressure therein (Mendgen and Deising 1993; Chap. 3, this Vol.). Similarly, high pressure is thought to be used by mycorrhizal fungi to grow between root cells, thus avoiding the secretion of large amounts of lytic enzymes. These fungus-induced alterations in plant cell walls are necessarily subtle, since the mycelium is able to avoid strong plant defense reactions (see Sect. IV).

Data indicate the key role played by the cell wall interface in mycorrhiza development. Biochemical characterization of the cell wall polysaccharides and proteins should provide insights into the early steps of symbiosis differentiation.

III. Differential Gene Expression in the Symbionts

Despite the wide morphological diversity of mature ectomycorrhizas, ectomycorrhizal fungi induce symbiotic organs with the same basic organization in several hundred host species (Harley and Smith 1983; see also Sect. II.C). This strongly suggests that the mycobiont interferes with master switches at the level of host genetic program and that the direct interaction with the root tissues turns on specific fungal morphogenetic programs leading to sophisticated multicellular structures. It is very likely that signaling molecules secreted by both the fungal and root cells interact reciprocally with their respective genomes to orchestrate the expression of sets of symbiosis-related (SR) genetic programs. The nutritional function of the symbiosis also implies a large reorganization of cell metabolism (Martin et al. 1987; Martin and Hilbert 1991; Botton and Chalot 1995; Hampp and Schaeffer 1995) and gene expression. Data gathered in recent years substantiate this contention, and show that protein biosynthesis and transcript populations are altered during the development of various types of mycorrhizas (Martin and Tagu 1995; Martin et al. 1995a,b).

A. Symbiosis-Related Proteins

1. Characterization and Functions

A detailed analysis of developmental changes in protein biosynthesis has been made in eucalypt ectomycorrhizae with an experimental system using aseptic cultures and controlled infection. Malajczuk et al. (1990) managed to synchronize to some extent the development of *Eucalyptus* spp.–*P. tinctorius* associations, and were able to isolate large numbers of ectomycorrhizas from defined stages in the infection process. This technique was used with both a compatible and an incompatible interaction between *E. urophylla* and *P. tinctorius* to characterize developmental stages in the symbiosis by microscopic analysis (Lei et al. 1990). This system was then used to study the changes in protein biosynthesis during early stages of development in *E. globulus-P. tinctorius* (Hilbert et al. 1991) and *E. grandis-P. tinctorius* (Burgess et al. 1995) associations. Inoculated seedlings were sampled over several days after contact, and used for labeling proteins in vivo with [^{35}S]methionine followed by two-dimensional polyacrylamide gel electrophoresis (2-D PAGE) analysis. Induction of a dozen of symbiosis-specific polypeptides, referred to as ectomycorrhizins (Hilbert and Martin 1988), was demonstrated. Ectomycorrhizins were specifically synthesized in inoculated roots and could not be detected in free-living hyphae and noninoculated roots or, more importantly, in roots that were challenged with nonmycorrhizal fungal isolates. In addition to the specific expression of ectomycorrhizins, these investigations revealed rapid and massive up- and downregulation in both plant and fungal protein biosynthesis as a result of the early interactions with compatible *Pisolithus*.

Most of the information currently available on symbiosis-regulated (SR) polypeptides with enhanced synthesis concerns a family of fungal acidic polypeptides. Based on 2-D PAGE analysis and peptide mapping, the early ectomycorrhizins (E_{32a}, E_{32b}), initially characterized in *E. globulus* mycorrhizas (Hilbert et al. 1991) and then in *E. grandis* mycorrhizas (Burgess et al. 1995), belong to a family of acidic alanine-rich polypeptides synthesized in the free-living hyphae. These symbiosis-regulated acidic polypeptides of 30 to 37 kDa, so-called SRAPs, are prominent polypeptides of the soluble fraction and fungal cell walls, and some isoforms are abundantly secreted by the hyphae (De Carvalho 1994; Burgess 1995; Martin et al. 1995a,b; Tagu and Martin 1996). Within the first hours following contact of the symbionts, upregulated and symbiosis-specific isoforms of these cell wall acidic polypeptides are rapidly synthesized and become the most abundant synthesized polypeptides of the inoculated roots. The massive synthesis of these SR-proteins coincides with the initial colonization of the root tissues and the secretion of glycoproteins forming the secreted fungal fimbriae (Lei et al. 1990). This suggests their role in fungal adhesion and assembly of multicellular symbiotic structures (i.e., the mantle). Confirmation of this assumption awaits further immunochemical analysis.

Besides upregulation, downregulation of synthesis of specific polypeptides will also affect the development and metabolism of the symbiotic tissues (Hilbert et al. 1991; Burgess et al. 1995). The biosynthesis of about one third of the polypeptides synthesized in root tissues of *E. globulus* and *E. grandis* was arrested within 2 days after contact. Several major plant polypeptides were not synthesized in developing symbiotic tissues, strongly supporting the hypothesis of an important alteration of the root developmental programmes (Martin and Hilbert 1991). The massive downregulation of plant protein synthesis may illustrate a possible drawback to such an approach; the massive infection rates that are obtained within a short period of time using the petri dish technique may have a disproportionally destructive effect on root growth. However, it is believed that this massive downregulation is not artifactual, and results from the arrest of the meristematic activity, disappearance of root hairs, and changes in plant metabolism (Martin and Hilbert 1991; Martin and Tagu 1995). The mycobiont presumably optimizes the functioning of its nursing structure by downregulating the expression of some or all genes not essential to the modified root tissues. This may allow a reorganization of the protein synthesis machinery to the development of the ectomycorhizal structures and the assimilation of the increased nutrient supply. Infection of plant roots by nematodes (Sijmons 1993) and pathogenic fungi (Zhang et al. 1993) similarly induced a strong downregulation of root gene expression. Downregulation of fungal proteins also occurs during the formation of the symbiosis. The most abundant protein in the cell wall of *P. tinctorius*, the acidic mannoprotein gp95, is only present at low concentration in the fungal tissues of ectomycorrhizal roots (Fig. 5; Laurent 1995; Martin et al. 1995a; Tagu and Martin 1996).

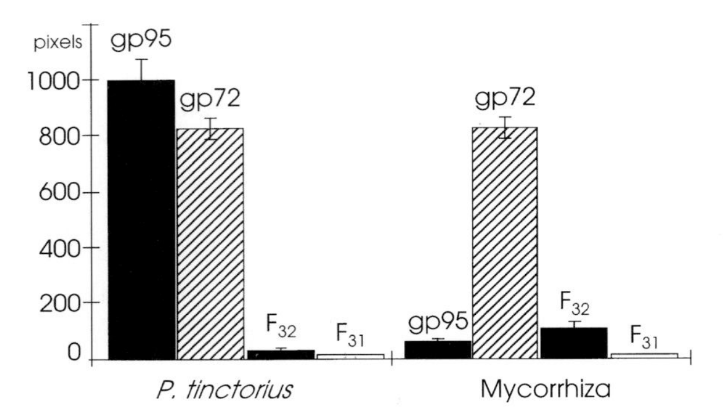

Fig. 5. Changes in the biosynthesis of major cell wall proteins (CWP) of *Pisolithus tinctorius* upon the development of the eucalypt ectomycorrhiza. The synthesis rate of the CWP gp95, gp72, and the acidic polypeptides, F_{32}, a SRAP, and F_{31}, a nonregulated CWP was estimated in cell walls of free-living mycelium of *P. tinctorius* 441 and in 4-day-old ectomycorrhiza of *P. tinctorius-Eucalyptus globulus*. The incorporation of ^{35}S in proteins was estimated by image analysis of 2-D fluorograms. Data are the mean of three replicates (±SD). Loadings of fungal proteins on 2-D-PAGE were normalized using ergosterol and chitin to take into account the dilution effect resulting from the presence of plant proteins. (Laurent 1995)

The upregulation of SRAPs biosynthesis and the downregulation of gp95 provide further evidence for biochemical differentiation of the fungal cell wall and extracellular matrix during the early stages of the ectomycorrhizal interaction. Alteration of the concentration of secreted proteins is most likely a way to regulate the symbiosis morphogenesis by changing the chemical structure of cell wall polymers. As a result, the mechanical properties of cell walls and extracellular matrix involved in the symbiotic interface may be strongly modified.

2. SR-Proteins: a Common Feature of Mycorrhizas

Ectomycorrhizins and up- and downregulated polypeptides were initially characterized during the early stages of development of eucalypt ectomycorrhizas (Hilbert and Martin 1988; Hilbert et al. 1991; Burgess et al. 1995). Alterations in protein biosynthesis were then detected in other ectomycorrhizal associations: *Pinus resinosa-Paxillus involutus* (Duchesne 1989), *Picea abies-A. muscaria* (Guttenberger and Hampp 1992), and *Betula pendula-Paxillus involutus* (Simoneau et al. 1993). The appearance of endomycorrhizins and alteration in protein biosynthesis following the infection by AM fungus infection is also known to occur (Dumas et al. 1989; Pacovski 1989; Wyss et al. 1990; Schellenbaum et al. 1992; Arines et al. 1993; Simoneau et al. 1994b; Gianinazzi-Pearson and Gianinazzi 1995), but their functions are not yet determined. Protein analysis carried out on the *Glomus*-soybean endomycorrhiza (Wyss et al. 1990) revealed an immunochemical cross-reactivity between endomycorrhizins and peribacteroid membrane nodulins of nodulated soybeans, suggesting that root cells react to fungal and bacterial associates by turning on a common set of genes. One of these endomycorrhizins may be related to nodulin-26, a protein homologous to the transmembrane intrinsic protein (Yamamoto et al. 1990), and probably involved in water and small metabolite transport. The fact that Nod$^-$ pea mutants produced by chemical mutagenesis also were Myc$^-$ (Duc et al. 1989) also suggests that common early infection events are required in both types of symbiosis. Upregulation of *ENOD* genes in endomycorrhizal roots of *Pea* and *Medicago* (C. Albrecht et al.; V. Gianinazzi et al., pers. comm.) supports this contention.

B. Symbiosis-Related Genes: Characterization and Functions

A fundamental tenet of developmental biology is that cell differentiation and tissue morphogenesis are governed in part by the expression of sets of genes in temporally ordered sequences. It is widely accepted that morphogenesis in fungal and plant systems is orchestrated by well-defined programs of gene expression that are played out over the course of several hours (Chasan and Walbot 1993; Timberlake 1993). The discovery of distinctive patterns of protein synthesis during mycorrhizal development suggests that soon after contact between the mycorrhizal symbionts, many genes are turned on/off in fungal and plant cells (Gianinazzi-Pearson and Gianinazzi 1989; Martin and Hilbert 1991; Martin and Tagu 1995; Martin et al. 1995a,b).

Alterations in gene expression during development of ectomycorrhizae and endomycorrhiza have been evaluated by isolating genes differentially expressed in symbiotic tissues and subsequently identifying the function of those genes. The basic experimental strategy relied on a comparison between mRNA populations expressed in symbiotic tissues versus free-living mycelium and noninoculated roots. A number of genes have been identified in barley and alfalfa endomycorrhiza (Harrison and Van Buuren 1995; Murphy et al. 1995) and eucalypt ectomycorrhizal (Tagu et al. 1993; Nehls and Martin 1995) that are enhanced or downregulated in expression in developing symbioses.

Plant and fungal cDNAs have been characterized that correspond to genes that are either up- or downregulated around the time of ectomycorrhizal mantle formation in the *E. globulus-P. tinctorius* mycorrhiza (Fig. 6; Tagu et al. 1993; Nehls and Martin 1995). The ectomycorrhiza-regulated transcripts represent approximately one third of the cDNA population initially screened, confirming early speculations, based on protein analysis, that early stages of ectomycorrhizal development initiate major changes in the gene expression. According to protein analysis, it appears that ectomycorrhizal cells are still expressing many of the genes that are expressed in free-living partners, but at a different level. This is confirmed by the analysis of mRNA populations, and the decreased expression of plant genes is especially noteworthy. Nucleotide sequencing of these cDNAs and search of homology in databases already provided worthwhile information on the putative role of these mycorrhiza-regulated genes (Table 1). As suggested by protein analysis, mycorrhiza development affects all aspects of the cellular activity. Among the cDNAs cloned from eucalypt ectomycorrhizas are genes encoding (1) components of the transduction pathways such as a GTP-binding *ras* protein and calmodulin, (2) cytoskeleton proteins (α-tubulins, kinesin), (3) proteins involved in protein turnover (ubiquitin-conjugating enzyme, proteasome, DnaJ chaperone), and (4) enzymes of the primary metabolism (acid phosphatase, alternate oxidase, cytochrome c1, homoserine kinase). High levels of transcripts known to be involved in other plant-microbe interactions, such as the pathogenesis-related protein, PR-1, thionins, and catalase, suggest that early mycorrhiza development and pathogen infection initiate similar cascades of molecular events.

Amongst the cloned genes, fungal hydrophobins, α-tubulins, and the auxin-regulated *EgPar* have been studied in detail, since their regulation is highly relevant to the understanding of the key morphogenetic events leading to the formation of the symbiosis.

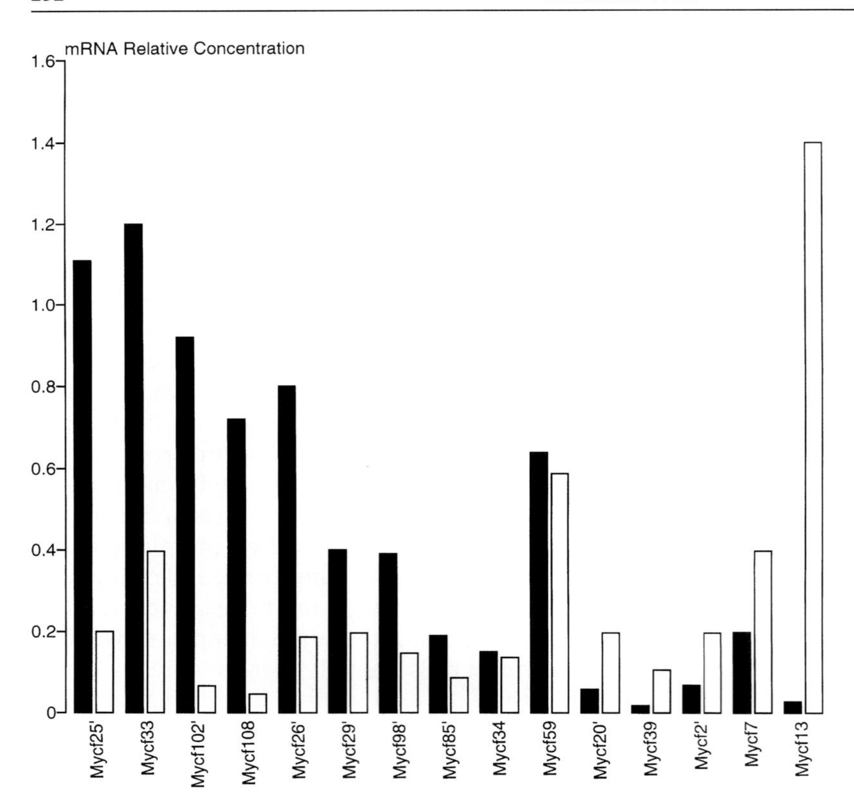

Fig. 6. Relative concentration of nonaffected, up- and downregulated fungal transcripts in free-living mycelium (■) and in ectomycorrhizas (□) of *Eucalyptus globulus-Pisolithus tinctorius*. (After Tagu et al. 1993)

Table 1. Sequence similarities of cDNAs from a eucalypt ectomycorrhiza library to known sequences (Nehls and Martin 1995; Tagu and Martin 1995; C. Voiblet and F. Martin, unpubl.)[a]

Accession no.	Clone	Protein	Species	Similarity (%)
U80615	*EgPar*	Auxin-regulated protein	*Nicotiana tabacum*	88
X97063	*EgIDH*	Isocitrate dehydrogenase	*N. tabacum*	94
L38736	EST 45	Metalloprotease	*Aspergillus fumigatus*	70
L38740	EST 114	S-adenosyl-L-methionine decarboxylase	*Catharanthus roseus*	70
L38747	EST 141	Hydrophobin *HydPt-2*	*Schizophyllum commune*	54
L38748	EST 144	Methylcrotonyl-CoA carboxylase	*Arabidopsis thaliana*	69
L38751	EST 149	Proteinase inhibitor	*Glycine max*	90
L38755	EST 155	Enoyl-acyl carrier protein reductase	*Brassica napus*	89
L38756	EST 158	Ubiquitin-conjugating enzyme E2	*A. thaliana*	63
L38757	EST 164	Thiol-specific antioxidant enzyme	*Onchocerca volvulus*	66
L38758	EST 167	PR1 protein STH-21	*S. tuberosum*	80
L38759	EST 173	Shikimate dehydrogenase	*N. tabacum*	76
L38762	EST 31	Elongation factor EF-1γ	*A. thaliana*	65
L38763	EST 32	Hydrophobin *HydPt-1*	*Sc. commune*	84
L38764	EST 35	Poly(A) RNA-binding protein methyltransferase	*Saccaromyces cerevisiae*	78
L38767	EST 44	Homoserine kinase	*S. cerevisiae*	74
L38768	EST 46	Ubiquitin-conjugating enzyme E2	*S. cerevisiae*	55
L38769	EST 49	Sodium channel α-subunit	*Equus caballus*	68
L38770	EST 54	Glyoxylate pathway regulator	*Yarrowia lipolytica*	64
L38771	EST 56	*copia*-like transposable element Tal-3	*A. thaliana*	67

Table 1. *Continued*

Accession no.	Clone	Protein	Species	Similarity (%)
L38772	EST 57	Sphingomyelinase	*Clostridium perfringens*	78
L38773	EST 60	Proteasome α-subunit	*S. cerevisiae*	74
L38774	EST 63	Cysteine-rich granulin	*Homo sapiens*	74
L38775	EST 7	Saposin A sphingolipid activator protein	*H. sapiens*	57
L38778	EST 73	SMY2 protein	*S. cerevisiae*	79
L38780	EST 74	SPO14 gene product	*S. cerevisiae*	71
L38782	EST 78	Aconitase	*S. cerevisiae*	54
L38785	EST 82	Hemoglobinase	*S. cerevisiae*	86
L38786	EST 84	Alternative oxidase	*Hansenula anomala*	66
L38789	EST 91	Transposase	*Lactococcus lactis*	70
L38790	EST 94	Cylicin	*B. taurus*	44
L38792	EST 98	Hypothetical protein HRB574	*S. cerevisiae*	94
L41693	ESTun16	Translation factor GOS2	*Oryza sativa*	87
L41694	ESTun32	Peroxisomal catalase A	*G. max*	91
L41695	ESTun38	Ubiquinol cytochrome c reductase	*Solanum tuberosum*	96
L41696	ESTun42	DnaJ protein	*Allium porrum*	91
L41697	ESTun86	α-Tubulin (*EgTubA2*)	*Zea mays*	90
L41698	ESTun517	Calmodulin-1	*A. thaliana*	100
L41699	ESTun585	NADP-malate dehydrogenase	*Sorghum bicolor*	66
L41702	ESTun360	IS1070 putative transposase	*Leuconostoc lactis*	81
L41703	ESTun144	PDC2 regulatory protein	*S. cerevisiae*	63
L41704	ESTun277	Auxin downregulated ADR11-2 protein	*G. max*	71
L41710	ESTun431	Protein disulfide-isomerase	*Medicago sativa*	64
L41713	ESTun052	Kininogen precursor	*Rattus norvegicus*	62
L41718	ESTud283	Periodic tryptophan protein/transducin	*S. cerevisiae*	63
L41722	ESTun178	Cyclophilin-related protein	*Bos taurus*	51
L41724	ESTun332	Cytochrome-c oxidase	*Crithidia fasciculata*	58
U37794	*EgTubA1*	α-Tubulin (*EgTubA1*)	*Z. mays*	95

[a] Similarities with sequences in international nucleotide and protein data banks were detected using the BLAST alignment program (Altschul et al. 1990). Only similarities considered to be biologically significant are shown. Columns show, respectively, the GenBank accession number, the clone name, the name of the protein for which a significant similarity has been detected, the organism for which the similarity was found, and the percent of similarity. These sequences are available from the WWW site MycorWeb (http://mycor.nancy.inra.fr).

1. Hydrophobin-Like Proteins in Ectomycorrhiza

As stressed above, a recurring theme of recent findings in developmental biology is that the same set of genes is used to execute the same type of operation in multiple developmental processes across a wide range of organisms. This is clearly illustrated by the expression of hydrophobin genes in eucalypt ectomycorrhiza. Two mRNA species are highly expressed during the early stages of the root infection (12 h after inoculation) when only a few percent by weight of fungal material is detected in the roots. These transcripts, corresponding to the cDNAs *hydPt-1* and *hydPt-2*, were classified as being of *Pisolithus tinctorius* origin by DNA and RNA gel blot analysis (Tagu et al. 1996). These transcripts are produced constitutively in free-living mycelium and in planta. Normalization of the Northern hybridization signals demonstrated that these cDNAs represent mRNA species severalfold more abundant in planta than in free-living mycelium.

The predicted amino acid sequences of *hydPt-1* and *hydPt-2* have characteristics of a class of proteins known as the hydrophobins (Fig. 7; Tagu et al. 1996). The amino acid sequences of the hydrophobins have two main features: the conservation of cysteine residues and the conservation of hydrophobic domains within the polypeptide sequence. Hydrophobins are morphogenetic proteins that allow – or cause – hyphae to emerge off the substrate and to adhere to each other during development of aerial multicellular reproductive

A

```
            5          15         25         35         45
HydPt-1  MKFAAVVVLA AAAAAVSAET NAQRMARGLP PKAPIRRHGT PADTEKRSHP
HydPt-2  EIVSLSLS.L .VVPL.VLVI AGPLNV..-- --------.. .SQ-------

            55         65         75         85         95
HydPt-1  SSTGGGQCNT GPIQCCNTVA TSGSQSGVDE LLTLLGLSVP VGTQVGASCS
HydPt-2  ------.... .TP...QQ.Q QTSDLQQFRS SFG.VDALAG ASAL...N.N

            105        115        125        135
HydPt-1  PISAVGTGSG AQCSGQTVCC EQNEWNGLVN IGCMPINLNA
HydPt-2  .V.VL...N. ...NT.P... TS.QML.A.. M....L.V..
```

B

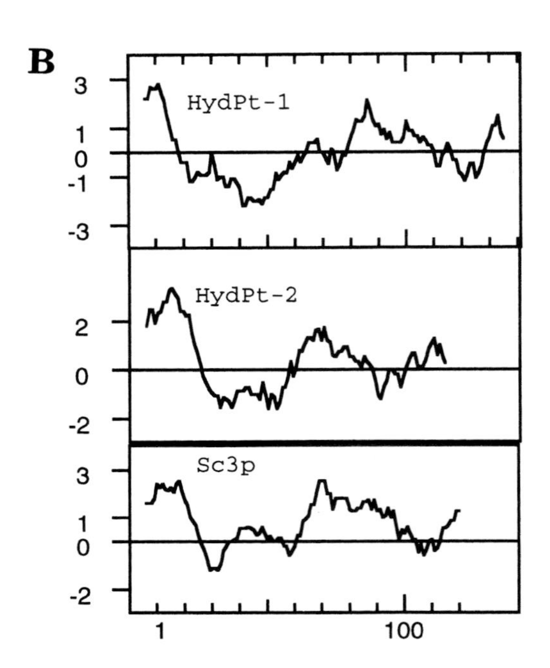

Fig. 7A,B. Conceptual sequence of the *Pisolithus* hydrophobins HydPt-1 and HydPt-2 (**A**) and their hydropathy plots (**B**). The hydropathy plots of *Pisolithus* hydrophobins were aligned with the hydropathy plot of the hydrophobin Sc3p of *Schizophyllum commune* (Wessels et al. 1991). Protein sequences were aligned based on the conserved cysteine residues (shown in *bold*)

structures (e.g., conidiophores, basidiocarps) in filamentous fungi (Wessels et al. 1991; Stringer et al. 1991). They presumably play this role through intermolecular hydrophobic interactions and/or formation of disulfide bridges between hyphal surfaces (Wessels 1992, 1993, 1996). Insoluble hydrophobins occur in the cell walls of several filamentous fungi (De Vries et al. 1993). We speculate that hydrophobin-like proteins synthesized during the mantle formation in eucalypt mycorrhizas are involved in the aggregation of fungal cells (see Sect. II.B). Vegetative and sexual multicellular structures involve similar developmental requirements in the fungal mycelium, and it would not be surprising for gene products involved in the aggregation of asexual and sexual multicellular structures to be similar to those used for the ectomycorrhizal mantle formation. Recent demonstrations that cerato-ulmin, a toxin involved in Dutch elm disease, is a fungal hydrophobin (Stringer and Timberlake 1993) and that the *MPG1* gene required for infection of rice by *Magnaporthe grisea* encodes a hydrophobin (Talbot et al. 1993; Chap. 3, Vol. V, Part B) strongly support an additional role for this class of proteins in plant-microbe interactions. In plant-fungus interactions, hydrophobins would be secreted by infecting fungi onto the plant surface, and presumably play a role in fungal attachment (Talbot et al. 1993; Wessels 1996).

2. *Pisolithus* Hypaphorine Stimulates the Expression of an Auxin-Regulated Gene in Eucalypt Roots

Amongst the identified SR-genes, a cDNA clone, called *EgPar*, from *E. globulus* shows a high homology with members of a hitherto unidentified new type of plant auxin-responsive glutathione S-transferases (GST) (Nehls and Martin 1995). GSTs are found in plants, insects, and mammals, and are involved in the protection of tissue against oxidative damages, detoxification of cytotoxic products, metabolism of lipidic signals such as leukotrienes and arachidonic acid, and transport of hydrophobic metabolites. Members of this auxin-responsive gene family are expressed during protoplast division (Takahashi and Nagata 1992), initiation of lateral root meristems (Vera et al. 1994), and fungal infection (Hahn and Strittmatter 1994). Furthermore, photoaffinity labeling demonstrated the binding of IAA to GSTs, suggesting that they are involved in the metabolism of this phytohormone.

Treatment of *E. globulus* seedlings with either the synthetic auxin 2,4-dichlorophenoxyacetic

A

```
          1                                                    50
EgPar     M.AEEVILLD FWPSPFGMRA KIALREKGVH FDLREEELLS NKSPLLLQMN
GTXA_TOB  MESNNVVLLD FWPSSFGMRL RIALALKGIK YEAKEENL.S DKSPLLLEMN
Consensus M.aeeViLLD FWPSpFGMRa kIALaeKGih fdakEEeL.S dKSPLLLeMN

          51                                                  100
EgPar     PVHKKIPVLI HNGKPVCESH IIVQYIDETW GPESPLLPSE PHERARARFW
GTXA_TOB  PVHKKIPILI HNSKAICESL NILEYIDEVW HDKCPLLPSD PYERSQARFW
Consensus PVHKKIPiLI HNgKaiCESh iIleYIDEtW gdecPLLPSd PhERaqARFW

          101                                                 150
EgPar     ADYVDKKIFP AGRAAWRSTG EAQEAAKKEY IEGLKMLEGE LGDTPYFGGE
GTXA_TOB  ADYIDKKIYS TGRRVWSGKG EDQEEAKKEF IEILKTLEGE LGNKTYFGGD
Consensus ADYiDKKIfp aGRaaWrgkG EaQEaAKKEf IEgLKmLEGE LGdkpYFGGd

          151                                                 200
EgPar     RFGFLDVSLI PFYSWFYAVE TLTGCSFEEE CPKLVGWAKR CMQRESVARS
GTXA_TOB  NLGFVDVALV PFTSWFYSYE TCANFSIEAE CPKLVVWAKT CMESESVSKS
Consensus nfGFlDVaLi PFtSWFYavE TcagcSfEaE CPKLVgWAKr CMerESVakS

          201                   221
EgPar     LPDQHKVYDF SRRSGRRFRR K
GTXA_TOB  LPHPHKIYGF VLELKHKLGL A
Consensus LPdpHKiYdF slelghkfgl a
```

B

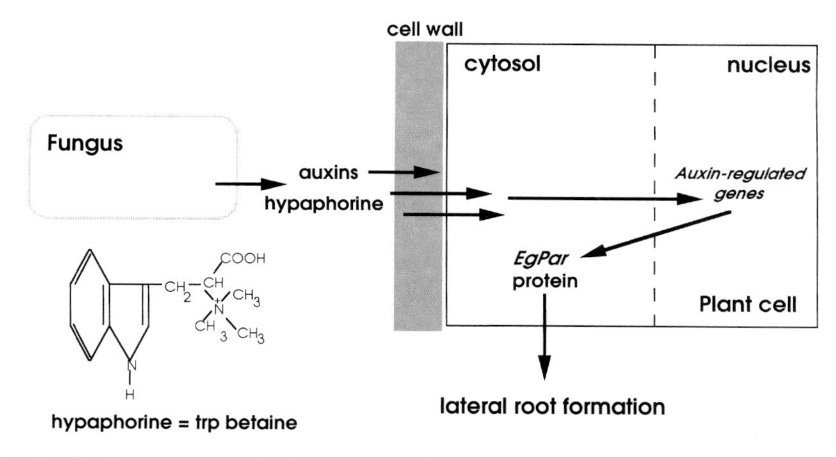

Fig. 8A,B. Comparison of the polypeptide sequences predicted by the symbiosis-regulated gene *EgPar* of *Eucalyptus globulus* with the auxin-responsive glutathione S-transferase of tobacco (GTXA_TOB) (**A**) and potential interactions between *Pisolithus*-secreted hypaphorine and expression of auxin-regulated genes in colonized eucalypt roots (**B**)

acid (2,4-D) or the natural auxin IAA clearly led to an increase in *EgPar* transcript level, as expected from a member of a family of auxin-responsive genes. The discovery of a strongly enhanced expression of *EgPar* in eucalypt ectomycorrhizae and in seedlings incubated in the presence of either *Pisolithus* acellular extracts or the *Pisolithus*-secreted hypaphorine (U. Nehls, T. Béguiristain, F. Lapeyrie, F. Martin unpubl. results) is fascinating. This is the first report of an alteration of the host plant gene expression by a diffusible signal from an ectomycorrhizal fungus. It can be envisaged that *EgPar* expression is triggered by auxin-related compounds released by ectomycorrhizal mycelium and involved in the initiation of rhizogenesis accompanying ectomycorrhizal formation (Fig. 8). The fungal symbiont appears to hijack parts of the genetic machinery used in normal plant development, e.g., induction of lateral roots. The auxin/hypaphorine-responsive elements present in *EgPar* promoters can serve to isolate binding proteins and subsequently other components of the signal transduction chain of plant auxins.

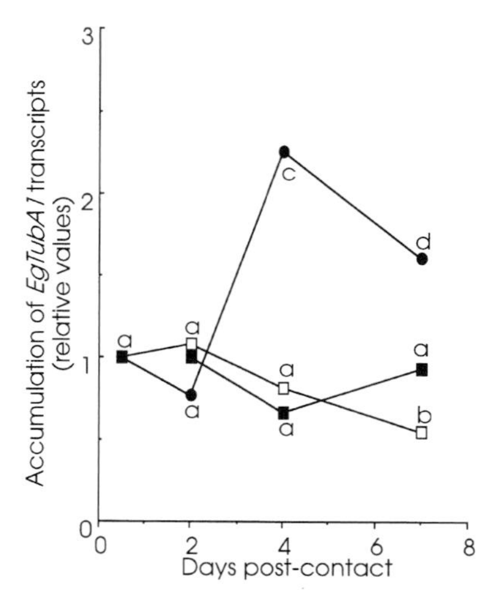

Fig. 9. Changes in the level of eucalypt α-tubulin transcripts during the development of *E. globulus-P. tinctorius* ectomycorrhiza. Steady-state levels of transcripts of the α-tubulin gene (*EgTubA1*) were assessed in non-inoculated roots (□), and roots of seedlings inoculated with the aggressive isolate 441 (●) or the poorly mycorrhizal isolate 270 (■) of *P. tinctorius* at different key time points following their contact. The α-tubulin signal intensity was standardized against the plant 5.8S rRNA. *Letters* indicate significantly different ($P < 0.05$) values based on the parametric Scheffé test from the ANOVA procedure. (Carnero Diaz et al. 1996)

3. Increased Level of α-Tubulin Transcripts in Ectomycorrhiza

Because symbionts are experiencing major morphological changes during ectomycorrhizal development, the expression of genes encoding cytoskeletal proteins is most likely altered. To test this contention, we cloned and characterized an α-tubulin cDNA (*EgTubA1*) from *E. globulus* (Carnero Diaz et al. 1996). The poorly aggressive isolate 270 of *P. tinctorius* causes no changes in root transcript levels of *EgTubA1*, whereas a drastic upregulation in its expression is observed at between 3 to 4 days after contact with the aggressive isolate 441 (Fig. 9; Carnero Diaz et al. 1996). This enhanced α-tubulin expression coincides with the increased lateral root formation induced by fungal colonization. The changes in α-tubulin expression support a role for cytoskeleton components in ectomycorrhizal development.

In addition, the molecular inventory of ectomycorrhizal genes identified several undescribed or novel sequences that are enhanced or downregulated in symbiotic tissues compared to free-living partners. The temporal and spatial expression of each of these genes during ectomycorrhizal development is currently being evaluated through in situ hybridization and RT-PCR. These cDNAs may prove to be useful molecular markers for specific cell and tissue types within the developing symbiosis.

IV. Host Defense Reactions Against Mycorrhizal Fungi

A. Induction of Pathogenesis-Related Proteins

Plant cells have a sensitive perception system for chemical signals (elicitors) derived from fungi. They react to elicitor stimulation with a concerted biochemical defense response (Dixon and Lamb 1990), including changes in membrane properties, enhanced production of the stress hormone ethylene, and transcriptional activation of the genes encoding enzymes involved in the phenylpropanoid synthesis. In addition, several new proteins are produced (Van Loon 1985). Some of these proteins, e.g., β-1,3-glucanases and chitinases, have the potential to hydrolyze polymers of the fungal cell walls (Mauch et al. 1988). Although mycorrhizal fungi are not considered to induce typical defense responses in host plants (Codignola et al. 1989), chitinase (Spanu et al. 1989; Dumas-Gaudot et al. 1992; Albrecht et al. 1994a,b; Volpin et al. 1994) and peroxidase (Spanu and Bonfante-Fasolo 1988; Albrecht et al. 1994b) activities were induced in roots during the early stages of root colonization by endomycorrhizal and ectomycorrhizal fungi (Fig. 10). An enhanced population of transcripts in *Glomus*-infected barley encodes for a chitinase (Murphy et al. 1995). Expression of thionins, catalase, and other defense proteins (Table 1) in ectomycorrhizae strongly supports the induction of plant defense reactions during the symbiosis formation.

The changes in the secondary metabolite profiles are reflected by the patterns of transcripts and activities corresponding to the enzymes involved in the phenylpropanoid, flavonoid, and phytoalexin biosynthesis. The observed increases in the activites of phenylalanine ammonia-lyase (PAL) (Simoneau et al. 1994a; Volpin et al. 1994), chalcone isomerase (Volpin et al. 1994, 1995) and transcripts encoding PAL, chalcone synthase, and isoflavone reductase (Harrison and Dixon 1994) in endo- and ectomycorrhizal roots are consistent with overall increased flavonoid biosynthesis. In endomycorrhiza, the distribution of the different transcripts (PAL, chalcone isomerase) is restricted to root cells containing arbuscules, indicating that the plant defense reaction is certainly reduced and tightly regulated.

It is generally assumed that the observed response is part of a general, nonspecific induction of phytoalexin and hydrolytic enzyme biosynthesis, as has been demonstrated in bean in response to inoculation with pathogenic and saprophytic bacteria (Jakobek and Lindgren 1993). How can the fungal growth in planta occur without succumbing to plant defenses, as would often occur with pathogenic fungi? Recent studies have shown that, after reaching maximum levels, the activity of these PR-proteins in inoculated roots rapidly declines to levels lower than that in noninoculated roots. Similarly, the transcription of endochitinase, β-1,3-endoglucanase, and chalcone isomerase was supressed in bean roots following AM mycorrhizal inoculation (Lambais and Mehdy 1993). This suggests that the initial nonspecific defense response is actively and specifically suppressed to allow the differentiation of symbiotic structures. The host plant reaction to mycorrhizal colonization is somewhat reminiscent of the early stages of biotrophic or hemibiotrophic compatible fungal plant-pathogen interactions, where the pathogen successfully invades large areas of the host and initially avoids eliciting the host defense reactions (Lamb et al. 1989). The fact that a mutation at a single genetic locus can change a fungal pathogen to a nonpathogenic, endophytic mutualist warrants reassessment of certain hypotheses that concern plant-fungus interactions (Freeman and Rodriguez 1993).

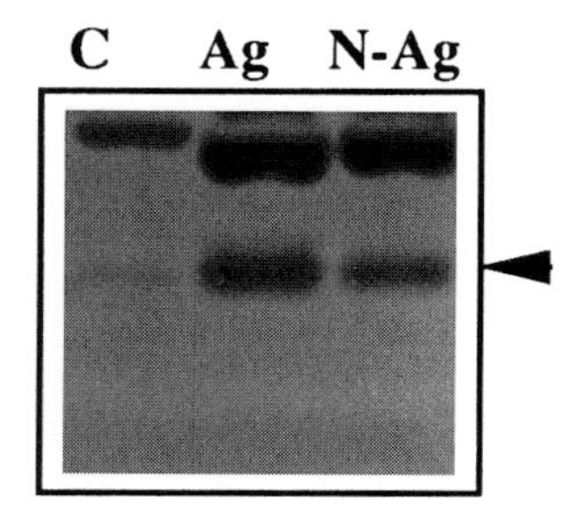

Fig. 10. Stimulation of chitinase activities of *Eucalyptus* roots treated with *Pisolithus tinctorius* cell-free extract indicating the occurrence of fungal elicitors. In treated roots, the activity of a mycorrhiza-regulated isoform (*arrowhead*) was drastically increased by comparison to control roots (*C*). The stimulation of the chitinase activity was higher in roots challenged by extract from an aggressive fungal strain (*Ag*), whereas a nonaggressive strain (*N-Ag*) induced a moderate stimulation. Chitinase activities were observed under UV illumination as dark lytic zones against the fluorescent background of glycol chitin. (After Albrecht et al. 1994a)

B. Elicitors

In order to initiate a response, a host plant needs to perceive the presence of the microorganism. Perception in host cells is mediated by molecules or elicitors released by the microbe. Receptors are believed to be the first step in signal transduction processes, leading to the gene induction and/or protein activation (Dixon and Lamb 1990). Elicitors involved in plant-microbe interactions include a wide range of compounds such as oligosaccharides, fatty acids, proteins and polysaccharides (Aldington et al. 1991; Chap. 5, this Vol.). To date, there is limited information on the induction of plant defense reaction by elicitors from mycorrhizal fungi. Chitinase activities were stimulated by cell-free extracts from various ectomycorrhizal fungi applied either to suspension-cultured cells of *Picea abies*, or onto *Eucalyptus* roots in vitro (Fig. 10; Sauter and Hager 1989; Campbell and Ellis 1992; Schwacke and Hager 1992; Albrecht et al. 1994a). Campbell and Ellis (1992) showed that the content of cell wall phenolics of pine cells is affected by ectomycorrhizal elicitors. Host specificity among *P. tinctorius* eliciting extracts has recently been detected (C. Albrecht and F. Lapeyrie, unpubl.) and ectomycorrhizal fungal elicitors from various strains are presently being purified and characterized.

V. Conclusions

The application of molecular analysis to the development of ectomycorrhizae is all quite recent, but it is beginning to reveal the fine details of fungus-host interactions. During the past few years, emphasis has been placed on analyzing protein patterns and isolating genes regulated by the mycorrhiza development. Already, SR-proteins and SR-genes have been identified during the development and the differentiation of the symbiosis (Fig. 11). Alterations of gene expression is a common feature of both endo- and ectomycorrhizas. Several genes highly expressed in symbiotic tissues (e.g., hydrophobins, *EgPar*) appear to have a morphogenetic and structural function. Other abundant transcripts found in ectomycorrhizas belong to a set of genes induced in plant-microbe interactions. However, much remains to be learned about the mechanisms that are responsible for the symbiotic tissue differentiation and those required for the activation of cell-specific gene sets within specialized fungal and root tissues of mycorrhiza. We hope that the available molecular markers will allow these mechanisms to be deciphered. We can expect that new components of the developmental pathways will soon be identified, facilitating the categorization of genes that may have been identified in other systems (transcriptional factors, homeogenes, signaling pathway kinases), but are not yet suspected of being induced during mycorrhiza development. Thus far, no genes critical to the regulation of the symbiosis development and function have been identified. Genetic screens are just beginning to be exploited in the study of mycorrhiza development (e.g., Myc^--mutants: Duc et al. 1989), and this approach should rapidly provide genes that initiate the differentiation and development of mycorrhiza.

A major theme in future studies in this research area will be molecular signaling involved in host recognition and the coordinated development of the specialized symbiotic structures. The signals identified might be analogous to those that regulate formation of other symbiotic tissues, and may well reveal insights into the processes that regulate organogenesis in fungi and plants.

These new approaches should allow us to make steady progress in answering the questions posed at the beginning of this chapter. If the development of the mycorrhizal symbiosis is to be understood, some of the earliest events in plant and fungus evolution must be contemplated.

Acknowledgments. We are grateful to C. Albrecht, T. Béguiristain, T. Burgess, M.E. Carnero Diaz, D. de Carvalho, M. Hodges, P. Laurent, U. Nehls, P. Murphy, and C. Voiblet (all presently or formerly of the Forest Microbiology team) for communicating data before publication, and to many other colleagues in the field for valuable discussions. We thank L. Peterson for supplying

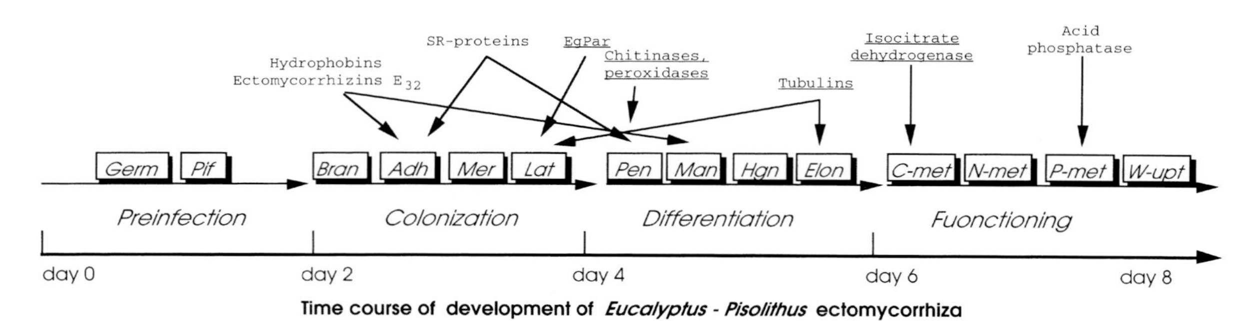

Fig. 11. Changes in gene expression observed during development of *E. globulus-P. tinctorius* ectomycorrhizal. Identified symbiosis-related proteins and genes are indicated *above* the different phnotypic stages observed during the development of the eucalypt ectomycorrhizal. SR-plant proteins are *underlined*. *Germ* Spore germination; *Pif* preinfection: hyphal growth; *Bran* hyphal branching; *Adh* adhesion; *Mer* root meristem stops; *Lat* lateral root formation; *Pen* hyphal penetration; *Man* mantle formation; *Hgn* Hartig net formation; *Elon* root cell elongation; *C-met* carbon metabolism; *N-met* nitrogen metabolism; *P-met* phosphate metabolism; *W-upt* water uptake

Fig. 2. The work referenced from our laboratory was supported by grants from the *Eureka-Eurosilva Cooperation Programme on Tree Physiology*, the Groupement de Recherches et d'Etude des Génomes and the INRA.

References

Altschul SF, Boguski MS, Gish W Wootton JC (1994) Issues in searching molecular sequence databases. Nature Genet 6:119–129

Albrecht C, Asselin A, Piché Y, Lapeyrie F (1994a) Chitinase activities are induced in *Eucalyptus globulus* roots by ectomycorrhizal or pathogenic fungi, during early colonization. Physiol Plant 91:104–110

Albrecht C, Burgess T, Dell B, Lapeyrie F (1994b) *Eucalyptus* root chitinase and peroxidase activities are induced by Australian ectomycorrhizal strains of *Pisolithus* sp. according to aggressiveness. New Phytol 127:217–222

Aldington S, McDougall GJ, Fry SC (1991) Structure-activity relationships of biologically active oligosaccharides. Plant Cell Environ 14:625–636

Allen MF (1991) The ecology of Mycorrhizae. Cambridge University Press, Cambridge

Arines J, Palma JM, Vilariño A (1993) Comparison of protein patterns in non-mycorrhizal and vesicular-arbuscular mycorrhizal roots of red clover. New Phytol 123:763–768

Bécard G, Douds DD, Pfeffer PE (1992) Extensive in vitro hyphal growth of vesicular-arbuscular mycorrhizal fungi in the presence of CO_2 and flavonols. Appl Environ Microbiol 58:821–825

Bécard G, Taylor LP, Douds DD, Pfeffer PE, Doner LW (1995) Flavonoids are not necessary plant signal compounds in arbuscular mycorrhizal symbioses. Mol Plant-Microb Interact 2:252–258

Béguiristain T, Lapeyrie F (1997) Host plant stimulates hypaphorine accumulation in *Pisolithus tinctorius* hyphae during ectomycorrhizal infection while excreted fungal hypaphorine controls root hair development. New Phytol (in press)

Beguiristain T, Côté R, Rubini P, Jay-Allemand C, Lapeyrie F (1995) Hypaphorine accumulation in hyphae of the ectomycorrhizal fungus *Pisolithus tinctorius*. Phytochemistry 40:1089–1091

Beyrle H (1995) The role of phytohormones in the function and biology of mycorrhizas. In: Varma AK, Hock B (eds) Mycorrhiza: structure, molecular biology and function. Springer, Berlin Heidelberg New York, pp 365–390

Bonfante P (1994) Alteration of host cell surfaces by mycorrhizal fungi. In: Petrini O, Ouellette GB (eds) Host wall alteration by parasitic fungi. APS Press, St Paul, Minnesota, pp 103–114

Bonfante P, Perotto S (1992) Plants and endomycorrhizal fungi: the cellular and molecular basis of their interaction. In: Verma DPS (ed) Molecular signals in plant microbe communications. CRC Press, Boca Raton, pp 445–470

Bonfante-Fasolo P (1988) The role of the cell wall as a signal in mycorrhizal associations. In: Scannerini S, Smith D, Bonfante-Fasolo P, Gianinazzi-Pearson V (eds) Cell to cell signals in plant, animal and microbial symbiosis. NATO ISI Series, H17. Springer, Berlin Heidelberg New York, pp 219–235

Botton B, Chalot M (1995) Nitrogen assimilation: enzymology in ectomycorrhizas In: Varma AK, Hock B (eds) Mycorrhiza: Structure, molecular biology and function. Springer, Berlin Heidelberg New York, pp 325–364

Burgess T (1995) Changes in protein biosynthesis associated with the development of *Pisolithus-Eucalyptus grandis* ectomycorrhizas. PhD Thesis, Murdoch University, Western Australia

Burgess T, Laurent P, Dell B, Malajczuk N, Martin F (1995) Effect of the fungal isolate infectivity on the biosynthesis of symbiosis-related polypeptides in differentiating eucalypt ectomycorrhiza. Planta 195:408–417

Campbell MM, Ellis BE (1992) Fungal elicitor-mediated responses in pine cell cultures: cell wall-bound phenolics. Phytochemistry 31:737–742

Carnero Diaz M, Martin F, Tagu D (1996) Eucalypt α-tubulin: cDNA cloning and increased level of transcripts in ectomycorrhizal root system. Plant Mol Biol 31:905–910

Chabot S, Bel-Rhlid R, Chênevert R, Piché Y (1992) Hyphal growth promotion in vitro of the VA mycorrhizal fungus *Gigaspora margarita* by the activity of structurally specific flavonoid compounds under CO_2 enrichment conditions. New Phytol 122:461–467

Chasan R, Walbot V (1993) Mechanisms of plant reproduction: questions and approaches. Plant Cell 5:1139–1146

Codignola A, Verotta L, Spanu P, Maffei M, Scannerini S, Bonfante-Fasolo P (1989) Cell wall-bound phenols in roots of vesicular-arbuscular mycorrhizal plants. New Phytol 112:221–228

De Carvalho D (1994) Contribution à l'étude des protéines régulées par la symbiose ectomycorhizienne. PhD Thesis, Ecole Nationale du Génie Rural et Forestier, Nancy, France

De Vries OMH, Fekkes MP, Wosten HAB, Wessels JGH (1993) Insoluble hydrophobin complexes in the walls of *Schizophyllum commune* and other filamentous fungi. Arch Microbiol 159:330–335

Dexheimer J, Pargney JC (1991) Comparative anatomy of the host-fungus interface in mycorrhizas. Experientia 47:312–320

Dixon RA, Lamb CJ (1990) Molecular communication in interactions between plants and microbial pathogens. Annu Rev Plant Physiol Plant Mol Biol 41:339–367

Duc G, Trouvelot A, Gianinazzi-Pearson V, Gianinazzi S (1989) First report of non-mycorrhizal plant mutants (Myc^-) obtained in pea (*Pisum sativum* L.) and fababean (*Vicia faba* L.). Plant Sci 60:215–222

Duchesne LC (1989) Protein synthesis in *Pinus resinosa* and the ectomycorrhizal fungus *Paxillus involutus* prior to ectomycorrhiza formation. Trees 3:73–77

Dumas E, Tahiri-Alaoui A, Gianinazzi S, Gianinazzi-Pearson V (1989) Observations on modifications in gene expression with VA endomycorrhiza development in tobacco: qualitative and quantitative changes in protein profiles. In: Nardon P, Gianinazzi-Pearson V, Grenier AM, Margulis L, Smith DC (eds) Endocytobiology IV, INRA Press, Paris, pp 153–157

Dumas-Gaudot E, Grenier J, Furlan V, Asselin A (1992) Chitinase, chitosanase and β-1,3-glucanase activities in *Allium* and *Pisum* roots colonized by *Glomus* species. Plant Sci 84:17–24

Fries N, Serck-Hanssen K, Häll- Dimberg L, Theander O

(1987) Abietic acid, an activator of basidiospore germination in ectomycorrhizal species of the genus *Suillus* (Boletaceae). Exp Mycol 11:360–363

Freeman S, Rodriguez RJ (1993) Genetic conversion of a fungal plant pathogen to a nonpathogenic, endophytic mutualist. Science 260:75–78

Gay G, Normand L, Marmeisse R, Sotta B, Debaud JC (1994) Auxin overproducer mutants of *Hebeloma cylindrosporum* Romagnési have increased mycorrhizal activity. New Phytol 128:645–657

Gea L, Normand L, Vian B, Gay G (1994) Structural aspects of ectomycorrhiza of *Pinus pinaster* (Ait.) Sol. formed by an IAA-overproducer mutant of *Hebeloma cylindrosporum* Romagnési. New Phytol 128:659–670

Gianinazzi-Pearson V, Gianinazzi S (1989) Cellular and genetical aspects of interactions between hosts and fungal symbionts in mycorrhizae. Genome 31:336–341

Gianinazzi-Pearson V, Gianinazzi S (1995) Proteins and protein activities in endomycorrhizal symbioses. In: Varma AK, Hock B (eds) Mycorrhiza: structure, molecular biology and function. Springer, Berlin Heidelberg New York, pp 251–266

Gianinazzi-Pearson V, Branzanti B, Gianinazzi S (1989) In vitro enhancement of spore germination and early hyphal growth of a vesicular-arbuscular mycorrhizal fungus by host root exudates and plant flavonoids. Symbiosis 7:243–255

Giollant M, Guillot J, Damez M, Dusser M, Didier P, Didier E (1993) Characterization of a lectin from *Lactarius deterrimus*: research on the possible involvement of the fungal lectin in recognition between mushroom and spruce during the early stages of mycorrhiza formation. Plant Physiol 101:513–522

Giovannetti M, Sbrana C, Avio L, Citernesi AS, Logi C (1993) Differential hyphal morphogenesis in arbuscular mycorrhizal fungi during preinfection stages. New Phytol 125:587–593

Gogala N (1991) Regulation of mycorrhizal infection by hormonal factors produced by hosts and fungi. Experientia 47:331–340

Guttenberger M, Hampp R (1992) Ectomycorrhizins – symbiosis-specific or artifactual polypeptides from ectomycorrhizas? Planta 188:129–136

Hahn K, Strittmatter G (1994) Pathogen-defence gene *prp1-1* from potato encodes an auxin-responsive glutathione S-transferase. Eur J Biochem 226:619–626

Hampp R, Schaeffer C (1995) Mycorrhiza – carbohydrates and energy meyabolism. In: Varma AK, Hock B (eds) Mycorrhiza: structure, molecular biology and function. Springer, Berlin Heidelberg New York, pp 267–296

Harley JL, Smith SE (1983) Mycorrhizal symbiosis. Academic Press, London

Harrison MJ, Dixon RA (1993) Isoflavonoid accumulation and expression of defense gene transcripts during the establishment of vesicular-arbuscular mycorrhizal associations in roots of *Medicago truncatula*. Mol Plant-Microbe Interact 6:643–654

Harisson MJ, Dixon RA (1994) Spatial patterns of expression of flavonoid/isoflavonoid pathway genes during interactions between roots of *Medicago truncatula* and the mycorrhizal fungus *Glomus versiforme*. Plant J 6:9–20

Harrison MJ, Van Buuren ML (1995) A phosphate transporter from the mycorrhizal fungus *Glomus versiforme*. Nature 378:626–629

Hilbert JL, Martin F (1988) Regulation of gene expression in ectomycorrhizas. I. Protein changes and the presence of ectomycorrhiza-specific polypeptides in the *Pisolithus-Eucalyptus* symbiosis. New Phytol 110:339–346

Hilbert JL, Costa G, Martin F (1991) Regulation of gene expression in ectomycorrhizas. Early ectomycorrhizins and polypeptide cleansing in eucalypt ectomycorrhizas. Plant Physiol 97:977–984

Hirsch AM (1992) Developmental biology of legume nodulation. New Phytol 122:211–237

Hofinger M, Coumans M, Ceulemans E, Gaspar T (1976) Assigning a biological role to hypaphorine and lycine (two betaines). Planta Med 30:303–309

Horan DP (1991) The infection process in eucalypt mycorrhizas. PhD Thesis, Australian National University, Canberra Australia

Horan DP, Chilvers GA (1990) Chemotropism; the key to ectomycorrhizal formation? New Phytol 116:297–301

Horan DP, Chilvers GA, Lapeyrie FF (1988) Time sequence of the infection process in eucalypt ectomycorrhizas. New Phytol 109:451–458

Huang J, Béguiristain T, Lapeyrie F (1997) Net K^+ uptake and H^+ extrusion by *Pisolithus tinctorius* and *Paxillus involutus*, two ectomycorrhizal fungi, are stimulated by IAA and hypaphorine, the betaine of tryptophane. Arch Microbiol (in press)

Jacobs PF, Peterson RL, Massicotte HB (1989) Altered fungal morphogenesis during early stages of ectomycorrhiza formation in *Eucalyptus pilularis*. Scanning Microsc 3:249–255

Jakobek JL, Lindgren PB (1993) Generalized induction of defense responses in bean is not correlated with the induction of the hypersensitive reaction. Plant Cell 5:49–56

Jakobsen I (1991) Carbon metabolism in mycorrhiza. Methods Microbiol 23:149–180

Kottke I, Oberwinkler F (1986) Root-fungus interactions observed on initial stages of mantle formation and Hartig net establishment in mycorrhizas of *Amanita muscaria* on *Picea abies* in pure culture. Can J Bot 64:2348–2354

Kottke I, Oberwinkler F (1987) The cellular structure of the Hartig net: coenocytic and transfer cell-like organization. Nord J Bot 7:85–95

Lamb CJ, Lawton MA, Dron M, Dixon RA (1989) Signals and transduction mechanisms for activation of plant defenses against microbial attack. Cell 56:215–224

Lambais MR, Mehdy MC (1993) Suppression of endochitinase, β-1,3-endoglucanase, and chalcone isomerase expression in bean vesicular-arbuscular mycorrhizal roots under different soil phosphate conditions. Mol Plant-Microbe Interact 6:75–83

Lapeyrie F, Mendgen K (1993) Quantitative estimation of surface carbohydrates of ectomycorrhizal fungi in pure culture and during *Eucalyptus* root infection. Mycol Res 97:603–609

Lapeyrie F, Lei J, Malajczuk N, Dexheimer J (1989) Ultrastructural and biochemical changes at the pre-infection stage of mycorrhizal formation by two isolates of *Pisolithus tinctorius*. Ann Sci For (Paris) 46s:754s–757s

Laurent P (1995) Contribution à l'étude des protéines régulées par la symbiose chez l'ectomycorhize d' *Eucalyptus-Pisolithus*. Caractérisation de mannoprotéines pariétales chez le basidiomycète *Pisolithus tinctorius*. PhD Thesis, Université Henri Poincaré, Nancy 1

Lei J, Lapeyrie F, Malajczuk N, Dexheimer J (1990) Infectivity of pine and eucalypt isolates of *Pisolithus tinctorius* (Pers.) Coker & Couch on roots of *Eucalyptus urophylla* S. T. Blake in vitro. II. Ultrastructural and biochemical changes at the early stage of mycorrhiza formation. New Phytol 116:115–122

Lei J, Wong KK, Piché Y (1991) Extracellular concanavalin A-binding sites during early interactions between *Pinus banksiana* and two closely related genotypes of the ectomycorrhizal basidiomycete *Laccaria bicolor*. Mycol Res 95:357–363

Levina NN, Lew RR, Hyde GJ, Heath IB (1995) The roles of Ca^{2+} and plasma membrane ion channels in hyphal tip growth of *Neurospora crassa*. J Cell Sci 108:3405–3417

Lynn DG, Chang M (1990) Phenolic signals in cohabitation: implications for plant development. Annu Rev Plant Physiol Plant Mol Biol 41:497–526

Malajczuk N, Lapeyrie F, Garbaye J (1990) Infectivity of Pine and Eucalypt isolates of *Pisolithus tinctorius* (Pers.) Coker & Couch on roots of *Eucalyptus urophylla* S.T. Blake *in vitro*. 1- Mycorrhizal formation in model systems. New Phytol 114:627–631

Markham P (1992) Stress management: filamentous fungi as exemplary survivors. FEMS Microbiol Lett 100:379–386

Markham P, Robson GD, Bainbridge BW, Trinci APJ (1993) Choline: its role in the growth of filamentous fungi and the regulation of mycelial morphology. FEMS Microbiol Rev 104:287–300

Martin F, Botton B (1993) Nitrogen metabolism of ectomycorrhizal fungi and ectomycorrhiza. Adv Plant Pathol 9:83–102

Martin F, Hilbert JL (1991) Morphological, biochemical and molecular changes during ectomycorrhiza development. Experientia 47:321–331

Martin F, Tagu D (1995) Ectomycorrhiza development: a molecular perspective. In: Varma AK, Hock B (eds) Mycorrhiza: structure, molecular biology and function. Springer, Berlin Heidelberg New York, pp 29–58

Martin F, Ramstedt M, Söderhäll K (1987) Carbon and nitrogen metabolism in ectomycorrhizal fungi and ectomycorrhizas. Biochimie (Paris) 69:569–581

Martin F, Laurent P, De Carvalho D, Burgess T, Murphy P, Nehls U, Tagu D (1995a) Fungal gene expression during ectomycorrhiza formation. Can J Bot 73 Suppl 1:S541–S547

Martin F, Burgess T, Carnero Diaz ME, De Carvalho D, Laurent P, Murphy P, Nehls U, Tagu D (1995b) Ectomycorrhiza morphogenesis: insights from studies of developmentally regulated genes and proteins. In: Stocchi V, Bonfante P, Nuti M (eds) Biotechnology of ectomycorrhizae: molecular approaches. Plenum Press, New York, pp 53–65

Massicotte HB, Peterson RL, Ashford AE (1987) Ontogeny of *Eucalyptus pilularis-Pisolithus tinctorius* ectomycorrhizae. II. Transmission electron microscopy. Can J Bot 65:1940–1947

Massicotte HB, Peterson RL, Melville LH (1989) Ontogeny of *Alnus rubra-Alpova diplophloeus* ectomycorrhizae. I. Light microscopy and scanning electron microscopy. Can J Bot 67:191–200

Mauch F, Hadwiger LA, Boller T (1988) Antifungal hydrolases in pea tissue. I. Purification and characterization of 2 chitinases and 2 β-1,3-glucanases differentially regulated during development and in response to fungal infection. Plant Physiol 87:325–333

Mendgen K, Deising H (1993) Infection structures of fungal plant pathogens – a cytological and physiological evaluation. New Phytol 124:193–213

Moore AEP, Massicotte HB, Peterson RL (1989) Ectomycorrhiza formation between *Eucalyptus pilularis* Sm. and *Hydnangium carneum* Wallr. in Dietr. New Phytol 112:193–204

Morandi D, Bailey JA, Gianinazzi-Pearson V (1984) Isoflavonoid accumulation in soybean roots infected with vesicular-arbuscular fungi. Physiol Plant Pathol 24:357–364

Murphy PJ, Karakousis A, Smith SE, Langridge P (1995) Cloning functional endomycorrhiza genes: potential for use in plant breeding. In: Stocchi V, Bonfante P, Nuti M (eds) Biotechnology of ectomycorrhizae: molecular approaches. Plenum Press, New York, pp 77–84

Nair MG, Safir GR, Siqueira JO (1991) Isolation and identification of vesicular-arbuscular mycorrhiza stimulatory compounds from clover (*Trifolium repens*) roots. Appl Environ Microbiol 57:434–439

Nehls U, Martin F (1995) Changes in root gene expression in ectomycorrhiza. In: Stocchi V, Bonfante P, Nuti M (eds) Biotechnology of ectomycorrhizae: molecular approaches. Plenum Press, New York, pp 125–137

Pacovski RS (1989) Carbohydrate, protein and amino acid status of *Glycine-Glomus-Bradyrhizobium* symbioses. Physiol Plant 75:346–354

Paris F, Dexheimer J, Lapeyrie F (1993) Cytochemical evidence of a fungal cell wall alteration during infection of *Eucalyptus* roots by the ectomycorrhizal fungus *Cenococcum geophilum*. Arch Microbiol 159:526–529

Piché Y, Peterson RL, Massicotte HB (1988) Host-fungus interactions in ectomycorrhizae. In: Scannerini S, Smith D, Bonfante-Fasolo P, Gianinazzi V (eds) Cell to cell signals in plant, animal and microbial symbiosis. NATO ISI Series H17. Springer, Berlin Heidelberg New York, pp 55–71

Read DJ (1991) Mycorrhizas in ecosystems. Experientia 47:376–390

Read DJ (1992) The role of the mycorrhizal symbiosis in the nutrition of plant communities. The Marcus Wallenberg Foundation Symp Proc 7, Stockholm, pp 27–53

Sauter M, Hager A (1989) The mycorrhizal fungus *Amanita muscaria* induces chitinase activity in roots and in suspension-cultured cells of its host *Picea abies*. Planta 179:61–66

Scheidegger C, Brunner I (1995) Electron microscopy of ectomycorrhiza: Methods, applications, and findings. In: Varma AK, Hock B (eds) Mycorrhiza: structure, molecular biology and function. Springer, Berlin Heidelberg New York, pp 205–228

Schellenbaum L, Gianinazzi S, Gianinazzi-Pearson V (1992) Comparison of acid-soluble protein synthesis in roots of endomycorrhizal wild type *Pisum sativum* and corresponding isogenic mutants. J Plant Physiol 141:2–6

Schwacke R, Hager A (1992) Fungal elicitors induce a transient release of active oxygen species from cultured spruce cells that is dependent on Ca^{2+} and protein-kinase activity. Planta 187:136–141

Shinshi H, Mohnen D, Meins F (1987) Regulation of a plant pathogenesis-related enzyme: inhibition of chitinase and chitinase mRNA accumulation in cultured tobacco tissues by auxin and cytokinin. Proc Natl Acad Sci USA 84:89–93

Sijmons PC (1993) Plant-nematode interactions. Plant Mol Biol 23:917–931

Simon L, Bousquet J, Levesque RC, Lalonde M (1993) Origin and diversification of endomycorrhizal fungi and coincidence with vascular land plants. Nature 363:67–69

Simoneau P, Viemont JD, Moreau JC, Strullu DG (1993) Symbiosis-related polypeptides associated with the early stages of ectomycorrhiza organogenesis in birch (*Betula pendula* Roth). New Phytol 124:495–504

Simoneau P, Juge C, Dupuis JY, Viemont JD, Moreau C, Strullu DG (1994a) Protein biosynthesis changes during mycorrhiza formation in roots of micropropagated birch. Acta Bot Gal 141:429–435

Simoneau P, Louisy-Louis N, Plenchette C, Strullu DG (1994b) Accumulation of new polypeptides in Ri T-DNA transformed roots of tomato (*Lycopersicon esculentum*) during the development of vesicular-arbuscular mycorrhizae. Appl Environ Microbiol 60:1810–1813

Smith SE, Gianinazzi-Pearson V (1988) Physiological interactions between symbionts in vesicular-arbuscular mycorrhizal plants. Annu Rev Plant Physiol 39:221–244

Smith SE, Smith FA (1990) Structure and function of the interfaces in biotrophic symbioses as they relate to nutrient transport. New Phytol 114:1–38

Söderström B (1992) The fungal partner in the mycorrhizal symbiosis. Ecophysiology of ectomycorrhizae of forest trees. The Marcus Wallenberg Foundation Symp Proc 7, Stockholm, pp 5–26

Spanu P, Bonfante-Fasolo P (1988) Cell-wall-bound peroxidase activity in roots of mycorrhizal *Allium porrum*. New Phytol 109:119–124

Spanu P, Boller T, Ludwig A, Wiemken A, Faccio A, Bonfante-Fasolo p (1989) Chitinase in roots of mycorrhizal *Allium porrum*: regulation and localization. Planta 177:447–455

St Leger RJ, Staples RC, Roberts DW (1992) Cloning and regulatory analysis of starvation-stress gene, *ssgA*, encoding a hydrophobin-like protein from the entomopathogenic fungus, *Metarhizium anisopliae*. Gene 120:119–124

Stringer MA, Timberlake WE (1993) Cerato-ulmin, a toxin involved in Dutch elm disease, is a fungal hydrophobin. Plant Cell 5:145–146

Stringer MA, Dean RA, Sewall TC, Timberlake WE (1991) *Rodletless*, a new *Aspergillus* developmental mutant induced by directed gene activation. Genes Dev 5:1161–1171

Tagu D, Martin F (1995) Expressed sequence tags of randomly selected cDNA clones from *Eucalyptus globulus-Pisolithus tinctorius* ectomycorrhiza. Mol Plant-Microb Interact 8:781–783

Tagu D, Martin F (1996) Molecular analysis of cell wall proteins expressed during the early steps of ectomycorrhiza development. New Phytol 133:73–85

Tagu D, Python M, Crétin C, Martin F (1993) Cloning symbiosis-related cDNAs from eucalypt ectomycorrhizas by PCR-assisted differential screening. New Phytol 125:339–343

Tagu D, Nasse B, Martin F (1996) Cloning and characterization of hydrophobins encoding cDNAs from the ectomycorrhizal basidiomycete *Pisolithus tinctorius*. Gene 168:93–97

Takahashi Y, Nagata T (1992) *parB*: an auxin-regulated gene encoding glutathione *S*-transferase. Proc Natl Acad Sci USA 89:56–59

Talbot NJ, Ebbole DJ, Hamer JE (1993) Identification and characterization of *MPG1*, a gene involved in pathogenicity from the rice blast fungus *Magnaporthe grisea*. Plant Cell 5:1575–1590

Timberlake WE (1993) Translational triggering and feedback fixation in the control of fungal development. Plant Cell 5:1453–1460

Tsai SM, Phillips DA (1991) Flavonoids released naturally from alfalfa promote development of symbiotic *Glomus* spores in vitro. Appl Environ Microbiol 57:1485–1488

Van Loon LC (1985) Pathogenesis-related proteins. Plant Mol Biol 4:111–116

Vera P, Lamb C, Doerner PW (1994) Cell-cycle regulation of hydroxyproline-rich glycoprotein *HRGPnt3* gene expression during the initiation of lateral root meristems. Plant J 6:717–727

Volpin H, Elkind Y, Okon Y, Kapulnik Y (1994) A vesicular arbuscular mycorrhizal fungus (*Glomus intraradix*) induces a defense response in alfalfa roots. Plant Physiol 104:683–689

Volpin H, Phillips DA, Okon Y, Kapulnik Y (1995) Suppression of an isoflavonoid phytoalexin defense response in mycorrhizal alfalfa roots. Plant Physiol 108:1449–1454

Wessels JGH (1992) Gene expression during fruiting in *Schizophyllum commune*. Mycol Res 96:609–620

Wessels JGH (1993) Wall growth, protein excretion and morphogenesis in fungi. New Phytol 123:397–413

Wessels JGH (1996) Fungal hydrophobins: proteins that function at an interface. Trends Plant Sci 1:9–15

Wessels JGH, De Vries OMH, Asgeirsdottirs A, Schuren FHJ (1991) Hydrophobin genes involved in formation of aerial hyphae and fruit bodies in *Schizophyllum*. Plant Cell 3:793–799

Wong KK, Piché Y, Montpetit D, Kropp BR (1989) Differences in the colonization of *Pinus banksiana* roots by sib-monokaryotic and dikaryotic strains of ectomycorrhizal *Laccaria bicolor*. Can J Bot 67:1717–1726

Wyss P, Mellor RB, Wiemken A (1990) Vesicular-arbuscular mycorrhizas of wild-type soybean and non-nodulating mutants with *Glomus mosseae* contain symbiosis-specific polypeptides (mycorrhizins), immunologically cross-reactive with nodulins. Planta 182:22–26

Yamamoto YT, Cheng CL, Conkling MA (1990) Root-specific genes from tobacco and *Arabidopsis* homologous to an evolutionarily conserved gene family of membrane channel proteins. Nucleic Acids Res 18:7449

Zhang SQ, Sheng JS, Liu YD, Mehdy MC (1993) Fungal elicitor-induced bean proline-rich protein mRNA down-regulation is due to destabilization that is transcription- and translation-dependent. Plant Cell 5:1089–1099

Biosystematic Index

Subject Index